LONG-WAVELENGTH INFRARED SEMICONDUCTOR LASERS

WILEY SERIES IN LASERS AND APPLICATIONS

D. R. VIJ, Editor
Kurukshetra University

OPTICS OF NANOSTRUCTURED MATERIALS • Vadim Markel

LASER REMOTE SENSING OF THE OCEAN: METHODS AND APPLICATIONS • Alexey B. Bunkin

COHERENCE AND STATISTICS OF PHOTONICS AND ATOMS • Jan Perina

METHODS FOR COMPUTER DESIGN OF DIFFRACTIVE OPTICAL ELEMENTS • Victor A. Soifer

PHASE CONJUGATE LASER OPTICS • Arnaud Brignon and Jean-Pierre Huignard (eds.)

LONG-WAVELENGTH INFRARED SEMICONDUCTOR LASERS • Hong K. Choi (ed.)

LONG-WAVELENGTH INFRARED SEMICONDUCTOR LASERS

EDITED BY

Hong K. Choi

WILEY-
INTERSCIENCE

A JOHN WILEY & SONS, INC., PUBLICATION

Copyright © 2004 by John Wiley & Sons, Inc. All rights reserved.

Published by John Wiley & Sons, Inc., Hoboken, New Jersey.
Published simultaneously in Canada.

No part of this publication may be reproduced, stored in a retrieval system, or transmitted in any form or by any means, electronic, mechanical, photocopying, recording, scanning, or otherwise, except as permitted under Section 107 or 108 of the 1976 United States Copyright Act, without either the prior written permission of the Publisher, or authorization through payment of the appropriate per-copy fee to the Copyright Clearance Center, Inc., 222 Rosewood Drive, Danvers, MA 01923, 978-750-8400, fax 978-646-8600, or on the web at www.copyright.com. Requests to the Publisher for permission should be addressed to the Permissions Department, John Wiley & Sons, Inc., 111 River Street, Hoboken, NJ 07030, (201) 748-6011, fax (201) 748-6008.

Limit of Liability/Disclaimer of Warranty: While the publisher and author have used their best efforts in preparing this book, they make no representations or warranties with respect to the accuracy or completeness of the contents of this book and specifically disclaim any implied warranties of merchantability or fitness for a particular purpose. No warranty may be created or extended by sales representatives or written sales materials. The advice and strategies contained herein may not be suitable for your situation. You should consult with a professional where appropriate. Neither the publisher nor author shall be liable for any loss of profit or any other commercial damages, including but not limited to special, incidental, consequential, or other damages.

For general information on our other products and services please contact our Customer Care Department within the U.S. at 877-762-2974, outside the U.S. at 317-572-3993 or fax 317-572-4002.

Wiley also publishes its books in a variety of electronic formats. Some content that appears in print, however, may not be available in electronic format.

Library of Congress Cataloging-in-Publication Data:
Long-wavelength infrared semiconductor lasers / edited by Hong K. Choi.
    p. cm.—(Wiley series in lasers and applications)
  Includes bibliographical references and index.
  ISBN 0-471-39200-6 (cloth)
1. Semiconductor lasers. 2. Far infrared lasers. I. Choi, Hong Kyun.
II. Series.
  TA1700.L66 2004
  621.36′6—dc22
                                                                 2003027635

Printed in the United States of America.

10 9 8 7 6 5 4 3 2 1

CONTENTS

Preface	xi
Acknowledgments	xiii
Contributors	xv

1. Coherent Semiconductor Sources in the Long-Wavelength Infrared Spectrum   1
Hong K. Choi

1.1	Introduction	1
1.2	Synopsis of Long-Wavelength Coherent Semiconductor Sources	3
	1.2.1 Interband Lasers	4
	1.2.2 Intersubband Quantum Cascade Lasers	8
	1.2.3 Hot-Hole Lasers	10
	1.2.4 Photomixers	11
	1.2.5 Plasmon Emitters	12
1.3	Scope of Book	12
References		13

2. 2-μm Wavelength Lasers Employing InP-based Strained-Layer Quantum Wells   19
Manabu Mitsuhara and Mamoru Oishi

2.1	Introduction	19
2.2	Material Properties of InGaAsP	21
	2.2.1 Composition Dependence of Band-Gap Energy and Lattice Constant	21
	2.2.2 Miscibility Gap	22
2.3	Design Consideration of MQW Active Region	25
	2.3.1 Strain and Quantum Size Effects	25
	2.3.2 Critical Layer Thickness for Strained-Layer Heterostructures	31
	2.3.3 Effects of Well Strain and Barrier Height on Lasing Characteristics	34

2.4	Growth and Characterization of Strained-InGaAs Quantum Wells		38
	2.4.1	InGaAs/InGaAs Multiple Quantum Wells	39
	2.4.2	InGaAs/InGaAsP Multiple Quantum Wells	43
2.5	Lasing Characteristics of 2-μm Wavelength InGaAs-MQW Lasers		49
	2.5.1	Fabry-Perot Lasers	49
	2.5.2	Distributed-Feedback Lasers	53
2.6	Conclusions and Future Prospects		57
Acknowledgments			59
References			59

3. Antimonide Mid-IR Lasers — 69
L. J. Olafsen, I. Vurgaftman, and J. R. Meyer

3.1	Introduction		69
3.2	Antimonide III-V Material System		71
3.3	Antimonide Lasers Emitting in the $2\,\mu m < \lambda < 3\,\mu m$ Range		74
	3.3.1	Historical Development	74
	3.3.2	State of the Art	76
3.4	Antimonide Lasers Emitting in the $\lambda \geq 3\,\mu m$ Range		82
	3.4.1	Historical Development	82
	3.4.2	Double-Heterostructure Lasers	82
	3.4.3	Type-I Quantum-Well Lasers	88
	3.4.4	Type-II Quantum-Well Lasers	94
	3.4.5	Interband Cascade Lasers	103
3.5	Challenges and Issues		107
	3.5.1	Antimonide Growth Immaturity	107
	3.5.2	Nonradiative Recombination and Threshold	110
	3.5.3	Linewidth Enhancement Factor (LEF)	115
	3.5.4	Single-Mode Operation and Wavelength Tuning	118
	3.5.5	Beam Quality	121
	3.5.6	Thermal Management and Thermal Conductivity	123
3.6	Conclusions		125
References			126

4. Lead-Chalcogenide-based Mid-Infrared Diode Lasers — 145
Uwe Peter Schießl, Joachim John, and Patrick J. McCann

4.1	Introduction		145
4.2	Homostructure Lasers		146
	4.2.1	Material Properties	146
	4.2.2	Device Fabrication	147
	4.2.3	Device Characterization	150

4.3	Double-Heterostructure Lasers		154
	4.3.1	$Pb_{1-x}Eu_xSe_yTe_{1-y}$ Lasers	155
	4.3.2	$Pb_{1-x}Eu_xSe$ and $Pb_{1-x}Sr_xSe$ Lasers	157
	4.3.3	$Pb_{1-x}Sn_xTe$ and PbSnSeTe/PbSe Lasers	167
	4.3.4	Alternative Cladding Layer Materials	167
	4.3.5	Quality Control Programs at Laser Components	168
	4.3.6	High-Temperature Operation of Double-Heterostructure Lasers	171
	4.3.7	Index-Guided Double-Heterostructure Lasers	175
4.4	Quantum-Well Lasers		177
4.5	DFB and DBR Lasers		184
	4.5.1	Introduction	184
	4.5.2	Experimental Work	188
4.6	IV–VI Epitaxy on BaF_2 and Silicon		197
	4.6.1	Introduction	197
	4.6.2	Growth and Characterization of IV–VI Layers on BaF_2	200
	4.6.3	Growth and Characterization of IV–VI Layers on Silicon	202
4.7	Conclusion		206
	Acknowledgments		207
	References		207

5. InP and GaAs-based Quantum Cascade Lasers — 217
Jérôme Faist and Carlo Sirtori

5.1	Introduction		217
	5.1.1	Quantum Engineering	217
	5.1.2	Organization of the Chapter	217
5.2	Quantum Cascade Laser Fundamentals		218
	5.2.1	History	218
	5.2.2	Unipolarity and Cascading	219
	5.2.3	Intersubband Transitions	219
5.3	Fundamentals of the Three-Quantum-Well Active-Region Device		220
	5.3.1	Active Region	221
	5.3.2	Doping and Injection/Relaxation Region	223
	5.3.3	Threshold Current Density	224
	5.3.4	Effect of Cascading on the Performances of QC Lasers	226
5.4	Waveguide and Technology		227
	5.4.1	Waveguide	227
	5.4.2	Processing	229

5.5	High-Power, Room-Temperature Operation of Three-Quantum-Well Active Region Designs		229
	5.5.1	High Power at Room Temperature	229
	5.5.2	High Room-Temperature Average Power	232
	5.5.3	Continuous-Wave Operation	233
5.6	GaAs-based QC Lasers		233
	5.6.1	Active Region Design	233
	5.6.2	Waveguide Design	234
5.7	Role of the Conduction-Band Discontinuity		237
	5.7.1	Strain-Compensated InGaAs/AlInAs Lasers for 3–5 μm Operation	239
	5.7.2	Role of ΔE_c on the High-Temperature Performances of GaAs QC Lasers	240
5.8	Spectral Characteristics of QC Lasers		245
	5.8.1	Pulsed Operation	245
	5.8.2	Continuous-Wave Operation	248
5.9	Distributed Feedback Quantum Cascade Lasers		251
	5.9.1	Metalized Top Grating	252
	5.9.2	Index-Coupled Lasers	256
	5.9.3	Lateral-Injection Surface-Grating Lasers	259
5.10	Microstructured QC Lasers		263
	5.10.1	Microdisk QC Laser Resonators	263
	5.10.2	High Reflectors	266
	5.10.3	Buried-Heterostructure Lasers	267
5.11	Outlook on Active Region Designs and Conclusions		270
Acknowledgments			272
References			273

6. Widely Tunable Far-Infrared Hot-Hole Semiconductor Lasers 279
Erik Bründermann

6.1	Introduction		279
	6.1.1	Tunable Germanium Lasers	281
	6.1.2	Motivation	281
	6.1.3	Applications	282
6.2	Hot-Hole Laser Model		283
	6.2.1	Semiclassical Model	283
	6.2.2	Magneto-Phonon and Magneto-Impurity Effects	288
	6.2.3	Scattering Mechanisms	296
	6.2.4	Optical Gain	298
	6.2.5	Hall Effect and Device Geometry	301
	6.2.6	Quantum Mechanical Model	306
	6.2.7	Uniaxial Stress	310
	6.2.8	Impurities and Self-absorption Processes	312

	6.2.9	Monte Carlo Simulations	315
	6.2.10	Thermal Effects	316
6.3	Laser Material Fabrication		319
	6.3.1	Growth of Germanium Laser Material	319
	6.3.2	Characterization	320
	6.3.3	Doping by Diffusion	320
	6.3.4	Ohmic Contacts for Germanium and Silicon	322
	6.3.5	Laser Devices with Opposing Contacts	322
	6.3.6	Laser Devices with Coplanar Contacts	323
	6.3.7	Laser Devices with Multiple Contacts	323
6.4	Technology		324
	6.4.1	Electric Field Generation	324
	6.4.2	Magnetic Field Generation	324
	6.4.3	Cooling Systems	326
	6.4.4	Mode-Locking	328
6.5	Laser Emission		330
	6.5.1	Germanium Laser Resonators	330
	6.5.2	Spectra and Mode Structure	335
	6.5.3	Output Power	337
6.6	Future Trends		338
	6.6.1	Continuous-Wave Germanium Lasers	338
	6.6.2	Picosecond Pulsed Germanium Lasers	339
	6.6.3	Hot-Hole Silicon, Diamond, and III–V Lasers	340
	6.6.4	Lasers without Magnetic Field under Uniaxial Stress	341
	6.6.5	Lasers in Parallel Electric and Magnetic Fields	341
6.7	Summary		342
Acknowledgments			342
References			343

7. Continous THz Generation with Optical Heterodyning 351
J. C. Pearson, K. A. McIntosh, and S. Verghese

7.1	Introduction		351
	7.1.1	Scientific Interest in THz Waves	351
	7.1.2	Source Problem	352
	7.1.3	THz Generation with Photomixers	353
7.2	Requirements for Photomixing Systems		353
	7.2.1	Laser Selection	354
	7.2.2	Frequency and Spectral Control	356
	7.2.3	THz-Wave Verification	367
7.3	Design Trade-offs for Photomixers		370
	7.3.1	Basic Operation	372
	7.3.2	Role of Photocarrier Lifetime	373

7.4	Antenna Design		377
	7.4.1	Resonant Designs	377
	7.4.2	Broadband Distributed Designs	379
7.5	Applications		381
	7.5.1	Local Oscillators	381
	7.5.2	Transceiver for Spectroscopy	382
7.6	Summary		383
Acknowledgments			383
References			384

Index   **387**

PREFACE

The long-wavelength infrared spectrum spanning from 2μm to 1000μm is technologically significant because most molecules have strong absorption lines in this spectral range. Compact semiconductor lasers emitting near room temperature are very much needed for many applications ranging from ultra-sensitive detection of molecules, to the study of fine structures of molecules including liquid, gas, dielectric, semiconductor, and DNA, to the study of the origin of the universe. Until recently, the only available semiconductor lasers were lead-salt lasers, which emit between 3 and 30μm, but operate only at low temperatures. No coherent semiconductor source was available beyond 30μm. However, significant progress has been made in the past ten years. In particular, intersubband quantum cascade lasers have exhibited impressive progress. Significant progress has also been made in the shorter wavelength segment (mid-infrared) through the employment of strained quantum-well structures in the III–V compound semiconductors. In addition, several novel approaches have been developed for generating longer wavelength (far-infrared) lasers.

This book is a compilation of the current status of coherent semiconductor sources that emit in this important spectral region. Chapter 1 provides an overview of the different approaches for long-wavelength infrared lasers and coherent sources. Chapters 2–4 cover interband semiconductor lasers using InP-based, antimonide, and lead-salt materials. Chapters 5, 6 and 7 describe intersubband quantum cascade lasers, hot-hole lasers, and photomixers, respectively.

ACKNOWLEDGMENTS

I would like to thank the following contributors for their excellent work: M. Mitsuhara, M. Oishi, L. J. Olafsen, I. Vurgaftman, J. R. Meyer, U. Schießl, J. John, P. J. McCann, J. Faist, C. Sirtori, E. Bründermann, J. C. Pearson, K. A. McIntosh, and S. Verghese. I am deeply indebted to the management and my colleagues at M.I.T. Lincoln Laboratory for their encouragement and collaboration while I was working in the area of long-wavelength semiconductor lasers. I would also like to thank Kopin Corporation for providing the necessary resources to edit this publication.

CONTRIBUTORS

Erik Bründermann, Ruhr-Unversit, D-44780 Bochum, Germany

Hong K. Choi, Kopin Corporation, Taunton, MA 02780, USA

Jérôme Faist, University of Neuchatel, Switzerland

Joachim John, IMEC, B-3001 Leuven, Belgium

Patrick J. McCann, University of Oklahoma, Norman, OK 73019, USA

K. A. McIntosh, Lincoln Laboratory, Massachusetts Institute of Technology, Lexington, MA 02420, USA

J. R. Meyer, Naval Research Laboratory, Washington, DC 20375, USA

Manabu Mitsuhara, NTT Photonics Laboratories, Atsugi, Kanagawa 243-0198, Japan

Mamoru Oishi, NTT Photonics Laboratories, Atsugi, Kanagawa 243-0198, Japan

Linda J. Olafsen, University of Kansas, Lawrence, KS 66045, USA

John C. Pearson, Jet Propulsion Laboratory, Pasadena, CA 91009, USA

Uwe Peter Schießl, Texas Instruments GmbH, Germany

Carlo Sirtori, Central de Recherche, Thales, Domaine de Corberville, France

I. Vurgaftman, Naval Research Laboratory, Washington, DC 20375, USA

Simon Verghese, Lincoln Laboratory, Massachusetts Institute of Technology, Lexington, MA 02420, USA

CHAPTER 1

Coherent Semiconductor Sources in the Long-Wavelength Infrared Spectrum

HONG K. CHOI

1.1 INTRODUCTION

Tremendous advancements have been made in semiconductor lasers since they were first demonstrated in 1962 by four different groups in the United States. The compact size, high efficiency, and mass producibility of semiconductor lasers have all contributed to many far-reaching applications, such as in optical fiber communications, data storage, and materials processing.

The different applications require different laser wavelengths. For example, optical fiber communications utilize semiconductor lasers emitting between 1.3 and 1.6 µm to take advantage of the extremely low transmission loss of silica-based optical fibers. For data storage applications, visible and near-infrared (NIR) lasers emitting between 0.6 and 0.8 µm have been widely used; violet lasers emitting at 0.4 µm are expected to be used in the near future to obtain higher storage density. As is often the case, the availability of inexpensive and reliable sources may open up new applications. It is thus particularly desirable that the efficient and compact semiconductor lasers emit in all wavelength regions, including the long-wavelength infrared spectrum. (See Table 1.1 for the definition of different spectral ranges.)

The long-wavelength infrared spectrum, which includes mid-infrared (MIR) and far-infrared (FIR) spectra, is technologically important because it contains fundamental absorption lines of most molecules associated with their vibrational and rotational motions. The absorption strength of these fundamental lines in the MIR and FIR is two to three orders of magnitude larger than those of overtones in the NIR, enabling much more sensitive detection of trace amounts of molecules. In addition there are atmospheric transmission

Long-Wavelength Infrared Semiconductor Lasers, Edited by Hong K. Choi
ISBN 0-471-39200-6   Copyright © 2004 John Wiley & Sons, Inc.

TABLE 1.1 Definition of spectral ranges. The relationships between wavelength (λ), energy (E), frequency (f), and wavenumbers (ω) are: λ (μm) = 1.24/E (eV), f(THz) = 300/λ (μm), ω (cm^{-1}) = 10,000/λ (μm).

	Visible	Near Infrared (NIR)	Mid Infrared (MIR)	Far Infrared (FIR) or THz	mm Wave
Wavelength (μm)	0.4–0.7	0.7–2.0	2.0–20	20–1000	>1000
Energy (eV)	1.7–3.1	0.6–1.7	0.06–0.6	0.001–0.06	<0.001
Frequency (THz)	400–750	150–400	15–150	0.3–15	<0.3
Wavenumber (cm^{-1})	14,000–25,000	5,000–14,000	500–5,000	10–500	<10

windows in the MIR spectrum (2–5 μm and 8–14 μm) that do not have the appreciable absorption losses typically caused by water vapor.

The primary applications of MIR coherent sources (mostly lasers) are high-resolution, ultra-sensitive trace gas analysis. For such applications it is desirable to have a single-frequency laser that operates continuous wave (cw) at room temperature or at least Peltier-cooled temperatures (>200 K). Practical applications include pollution monitoring, toxic gas sensing, industrial and chemical process monitoring, and biomedical sensing. In addition efficient, high-power MIR semiconductor lasers are useful for secure open-space communications and infrared countermeasures against heat-seeking missiles that target airplanes.

The FIR or terahertz (THz) sources are used in the study of fine structures of molecules, including those of liquids, gas, dielectrics, semiconductors, and DNA. Since the THz signal can penetrate packaging materials such as leather or boxes, it is being used to distinguish different chemical compositions in order to detect hidden materials, such as explosives or bioagents, inside these materials. THz sources are also useful for biomedical imaging. In addition THz sources are essential as a local oscillator for the study of the origin of our universe since interstellar molecular clouds absorb short-wavelength radiation and radiate most of it in the THz region.

MIR semiconductor lasers have been around for a long time, but progress in this area has been rather slow, in sharp contrast to the explosive development of NIR and visible semiconductor lasers. The first MIR semiconductor laser was demonstrated in 1963 from InAs, a III-V compound semiconductor [1], but it operated only under pulsed conditions at cryogenic temperatures. It was soon discovered, however, that IV-VI compound semiconductors such as PbS, PbSn, and PbSe (called lead-salt or chalcogenide) were more promising for the MIR lasers because they could be operated at higher temperatures [2]. Indeed, lead-salt lasers have been used extensively for high-precision spectroscopy because they cover a wide emission wavelength range between 3 and 30 μm, and they can operate cw in a single wavelength, which can be easily tuned by adjusting the temperature [3]. However, their use has been mostly confined to scientific applications because they still operate only at low temperatures, typically around the liquid-nitrogen temperature (77 K). In the 1980s there were renewed efforts to use III-V compound semiconductors, mostly GaSb-based materials, for the MIR lasers [4]. Room-temperature cw operation was demonstrated up to 2.4 µm [5], but their performance degraded rapidly as the wavelength was increased, primarily because of increasingly higher nonradiative recombination rates.

In the very long wavelength region of the spectrum (longer than 1 mm), electronic sources that rely on fast charge oscillations (e.g., Gunn and Impatt diodes) or solid state multiplier circuits (usually based on Schottky varacters) have been available. However, the power of such sources also decreases rapidly as the frequency (wavelength) is increased (decreased). As a result there was no compact coherent semiconductor source in a large region of the intermediate electromagnetic spectrum, the so-called FIR or THz region.

In the past decade, however, there has been rapid progress in the MIR and FIR coherent semiconductor sources. In particular, impressive advances have been made with intersubband quantum cascade lasers, first demonstrated in 1994 [6]. The employment of strained quantum-well structures in the III-V compound semiconductors has produced significant improvements in MIR laser performance [7]. In addition several novel approaches to generate FIR lasers have been developed. In this chapter, I will attempt to give a brief overview of the current status of long-wavelength coherent semiconductor sources. Other recent developments are described in subsequent chapters of this book.

1.2 SYNOPSIS OF LONG-WAVELENGTH COHERENT SEMICONDUCTOR SOURCES

Because the MIR and FIR spectra cover a very wide wavelength range, from 2 to 1000 µm, corresponding to band-gap energies of 0.6 eV to 1.2 meV, it is very difficult for a single semiconductor approach to realize coherent sources covering the entire long-wavelength spectral range. Figure 1.1 shows several approaches used to generate long-wavelength coherent sources: (1) interband lasers utilizing the band-to-band transitions, (2) intersubband lasers utilizing transitions between quantized subband states within quantum wells, (3) hot-hole lasers utilizing transitions between light- and heavy-hole bands, (4) photomixing of two NIR lasers with a small wavelength offset in a very fast photodetector, and (5) plasmon emitters formed by exciting a semiconductor surface by ultrafast femtosecond pulses. The interband lasers have exhibited the best performance at the low end of the MIR spectrum where the band-to-band transition is the dominant recombination mechanism. The intersubband lasers have exhibited the best performance in the mid to high end of the MIR

4 COHERENT SEMICONDUCTOR SOURCES IN THE LONG-WAVELENGTH IR SPECTRUM

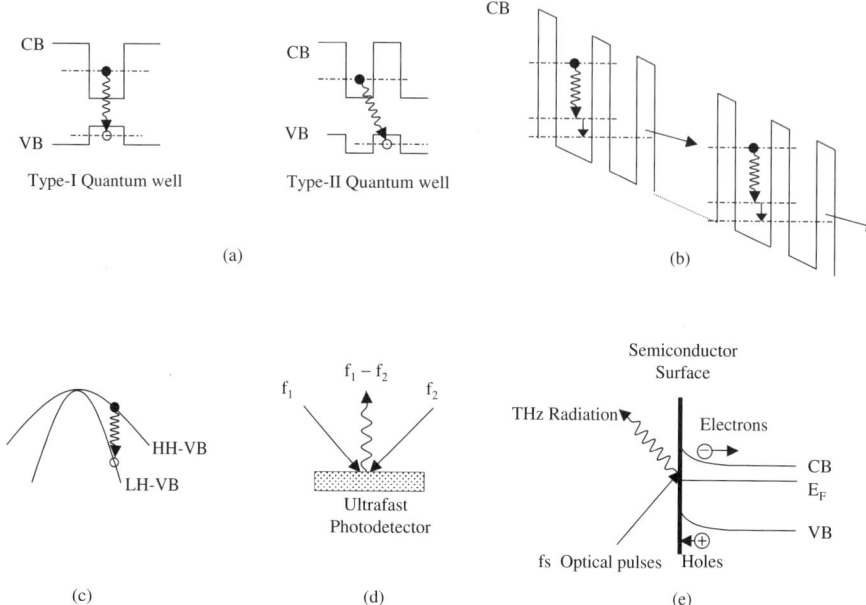

Fig. 1.1 Five approaches to generate coherent infrared sources (*a*) interband type-I and type-II quantum-well lasers. The type-II quantum well can be formed when one material has conduction-band and valence-band positions much lower than those of the other material; (*b*) intersubband quantum cascade laser in which an electron is recycled to generate multiple photons, (*c*) hot-hole laser utilizing transitions between heavy-hole valence band (HH-VB) and light-hole valence band (LH-VB), (*d*) photomixing of near-infrared lasers with a small offset frequency to generate difference-frequency photons in far infrared, and (*e*) plasmon emitter by illuminating a semiconductor surface with femtosecond (fs) near-infrared optical pulses to generate far-infrared radiation.

spectrum, but their emission wavelength now extends to the FIR range. The other three approaches have been used for generating FIR coherent emission.

1.2.1 Interband Lasers

The emission wavelength of the interband laser is primarily determined by the band-gap energy of the material used for the active layer. (One exception is type-II quantum-well lasers, which will be described in more detail later. See Fig. 1.1*a* for the difference between type-I and type-II structures.) The wavelength coverage of the interband lasers is shown in Fig. 1.2. Among III-V compound materials, InP-based alloys ($In_xGa_{1-x}As$) and GaSb-based alloys ($Ga_{1-x}In_xAs_ySb_{1-y}$) have band-gap energies in the MIR range. In addition IV-VI compound semiconductors such as PbS, PbSn, PbSe, PbTe, PbSr, PbEu, and their alloys have been used for MIR emission.

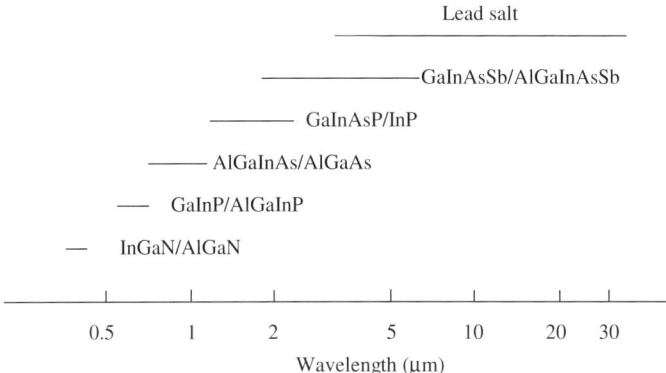

Fig. 1.2 Wavelength ranges of interband lasers fabricated from III-V and IV-VI compound semiconductor materials.

Over the past decade the performance of interband lasers using III-V compound semiconductor materials was substantially improved by the use of strained quantum-well structures. When a thin layer is grown on a substrate with a different lattice constant, the layer is initially strained to match the lattice constant of the substrate instead of forming dislocations, until its thickness reaches a critical value (see Chapter 2 for more details). The benefits of the strained layer are (1) removal of the degeneracy in the valence band, which results in a lower threshold current, (2) extension of wavelength beyond the value determined by the band gap of the lattice-matched active layer material, and (3) reduced Auger recombination rates. Especially for the MIR lasers, the reduction of Auger recombination is very important because it becomes the dominant loss mechanism as the emission wavelength becomes longer.

Diode lasers grown on InP substrates cover the lowest end of the MIR spectrum. Although the longest band-gap wavelength of the lattice-matched alloy ($In_{0.53}Ga_{0.47}As$) is 1.67 μm at room temperature, the emission wavelengths could be extended beyond 2 μm by employing strained InGaAs quantum-well active layers with a higher In composition. The advantages of this material system are (1) the epitaxial growth technique is very well established, and (2) it is relatively easy to realize single-frequency lasers by using distributed feedback (DFB) structures because the device processing techniques have been extensively developed for the telecom lasers. Cw laser emission up to 2.07 μm was demonstrated [8]. With a DFB structure, single-frequency operation was obtained at about 2.05 μm, with cw power up to 10 mW at room temperature [9]. However, it becomes exceedingly difficult to extend the wavelength further because the increased strain with higher In compositions creates defects in the materials, which degrade the laser performance dramatically.

The band gap of the GaInAsSb alloy grown lattice matched on a GaSb substrate allows the emission wavelength between 1.7 and 4.2 μm at room temperature. The best performance is obtained near 2 μm for a strained quantum-well structure with Ga-rich GaInAsSb wells [10]. Room-temperature cw power of more than 1 W has been obtained for a 100-μm aperture. By increasing the In composition in GaInAsSb to nearly 30%, room-temperature cw operation was obtained up to 2.7 μm [11]. However, extending the room-temperature operation to a longer wavelength using a similar structure proved to be very challenging because GaInAsSb with an In content higher than 30% becomes exceedingly difficult to grow, since a miscibility gap occurs in the intermediate In content range. As a result longer wavelength lasers were primarily implemented with InAs-rich alloys, in particular, InAsSb. However, because InAsSb has very high Auger recombination rates, the maximum cw operation temperature is only 175 K at 3.4 μm [12].

To circumvent the problem, type-II quantum-well structures using GaInSb and InAs have been utilized. The type-II quantum-well structure is possible because the conduction-band position of InAs is lower than the valence-band position of GaSb or InSb. Benefits of the type-II structures are (1) an extended wavelength beyond 5 μm because the effective band-gap energy is a strong function of the thickness of InAs, with a smaller band gap for thicker InAs, and (2) reduced Auger recombination rates compared to type I structures [13]. The type-II quantum-well structures consisting of GaInSb and InAs exhibited substantially improved performance by optical pumping, with cw operation at 3 μm up to 290 K [14] and at 6.1 μm up to 210 K [15]. However, electrically pumped type-II quantum-well diode lasers have exhibited somewhat inferior characteristics because the carrier transport (especially the holes) is poor due to a very large valence-band offset between GaInSb and InAs. Nonetheless, the type-II structure yielded the first room-temperature pulsed operation of an electrically pumped III-V interband laser beyond 3 μm [16], and cw operation at 3.25 μm up to 195 K [17].

The staggered band alignment allows cascading the active regions by utilizing tunneling of holes from the valence band to the conduction band of the next stage (Fig. 1.3) [18]. By cascading the active layers, electrons can be recycled for multiple photon generation. This device, called interband cascade lasers (as opposed to the intersubband quantum cascade lasers), has exhibited very low-threshold current density at low temperatures (44 A/cm^2 at 80 K) [19] and high differential quantum efficiency (460% with 23 stages) [20]. Although the cascading scheme trades increased quantum efficiency for increased voltage, it still provides much higher power efficiency, especially at longer wavelengths with smaller photon energies, by minimizing the effect of parasitic resistance. High peak power (4 W/facet at 80 K) and high cw power conversion efficiency (9% at 80 K) were obtained [20]. As with other interband lasers, however, the threshold current increases rapidly with temperature due to Auger recombination. A maximum pulsed operation temperature of 250 K was obtained at 3.9 μm [20].

An aternative approach to obtain high mid-IR power is to optically pump the antimonide laser structure with NIR diode laser arrays. Although it is not the most convenient approach, the optical pumping provides high power

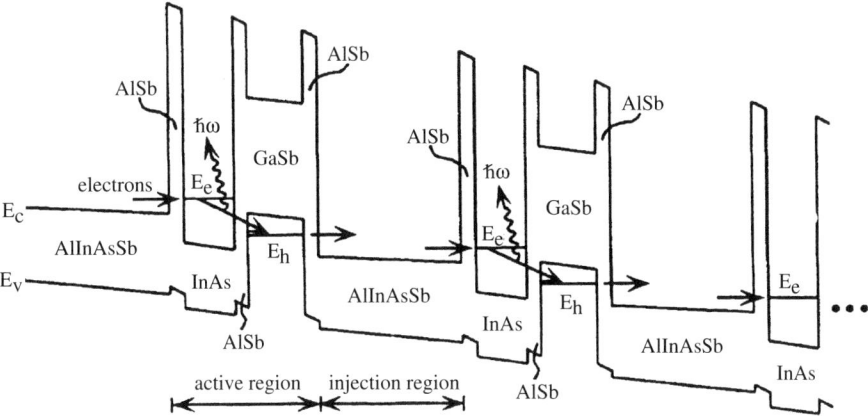

Fig. 1.3 Schematic band structure of an interband cascade laser (from Ref. [18] with permission). Because of the staggered band line up between InAs and GaSb, cascading is obtained by tunneling of electrons through the valence band to the conduction band of the next stage.

efficiency because the optical loss is lower (no doping in the cladding is required) and the equivalent voltage is fixed by the pump photon energy (0.6 V for 2-µm photon) regardless of the pumping level. For efficient use of the pump power, however, the absorption in the active region should be fairly high. This is easily achieved for the double heterostructures with a thick active layer. Quasi-cw power (~100 µs pulses) of several hundred mW was obtained for 4-µm lasers at 77 K [21]. Quantum-well structures optimized to improve the pump beam absorbance, while not increasing the optical loss, have exhibited better power efficiency. Pumped by a 1.85-µm diode array, differential quantum efficiency of 9.8% and peak output power of 2.1 W were obtained at 77 K [22].

Despite substantial advances in III-V interband lasers, however, extending the wavelength beyond 6 µm seems very difficult because the Auger recombination becomes increasingly more dominant. On the other hand, lead-salt lasers have been operating up to 30 µm since early days because they have much lower Auger recombination rates due to a very different band structure [23].

Lead-salt lasers have been commercially available for more than three decades, and used extensively for spectroscopy applications. Homojunction lasers are still useful, especially for wavelengths longer than 10 µm. However, homojunction lasers typically operate at temperatures below 77 K. Higher operating temperatures are obtained for double-heterostructure lasers. Cw operation up to 224 K was demonstrated at nearly 4 µm for a PbEuSeTe/PbTe laser [24]. In general, the maximum operating temperature of lead-salt lasers peaks at between 4 and 5 µm, and decreases at either shorter or longer wavelengths.

Under cw operation many lead-salt lasers with cleaved facets operate in a single frequency. As the temperature is changed, however, the frequency hops to the next longitudinal mode, generating substantial noise. DFB lead-salt lasers were also fabricated to obtain single-frequency operation without mode hopping. Because the lead-salt material is soft, even embossing from a Si master grating can be used to fabricate DFB lasers [25].

Recently lead-salt lasers operated above room temperature under narrow-pulse (10 ns) operation. A maximum operating temperature of 85°C and a maximum output power of 100 mW at room temperature were achieved for a double-heterostructure laser with PbSe active and PbSrSe cladding layers [26]. Under pulsed operation, however, the emission spectrum was fairly broad.

Unlike III-V compound lasers, employment of a quantum-well active region has not shown improved performance for the lead-salt lasers. The primary reason may be due to the fourfold degeneracy in the quantum-confined energy levels when a (100) substrate is used. Although this degeneracy can be removed if a (111) substrate is used [27], forming feedback mirrors is difficult because there is no cleavage plane. No electrically pumped lead-salt lasers grown on (111) substrates have been demonstrated. However, optically pumped vertical cavity lasers grown on a (111) BaF_2 substrate operated pulsed up to 280 K with emission wavelength close to 4.4 µm [28].

1.2.2 Intersubband Quantum Cascade Lasers

The concept of utilizing intersubband transitions for lasers was proposed in 1972 by Kasarinov and Suris [29]. Many attempts were made to realize the intersubband lasers, but little progress was made for 20 years due to several theoretical and technical difficulties. The most prominent problem was figuring out how to achieve population inversion despite the very short electron lifetime (~1 ps), which is caused by phonon scattering. In 1994 Faist et al. [6] demonstrated the first intersubband laser emitting at 4.3 µm, ironically with the help of phonon scattering to achieve population inversion (see below). Since then the intersubband laser has shown impressive progress, and demonstrated room-temperature operation with high power in the MIR [30–38]. The emission wavelength has been recently extended to the FIR [39,40].

The intersubband laser has two important differences from the interband laser. First, the emission wavelength of an intersubband laser is not determined by the band-gap energy of the material used, but by the exact design of the layer structures. As a result it has utilized well-developed materials such as InGaAs/InAlAs grown on InP and GaAs/AlGaAs grown on GaAs substrates. The fact that no materials development is required is a very significant advantage of the intersubband laser.

Second, the intersubband laser is a unipolar device requiring only electrons. As a result the cascading of the active layers to recycle electrons for multiple

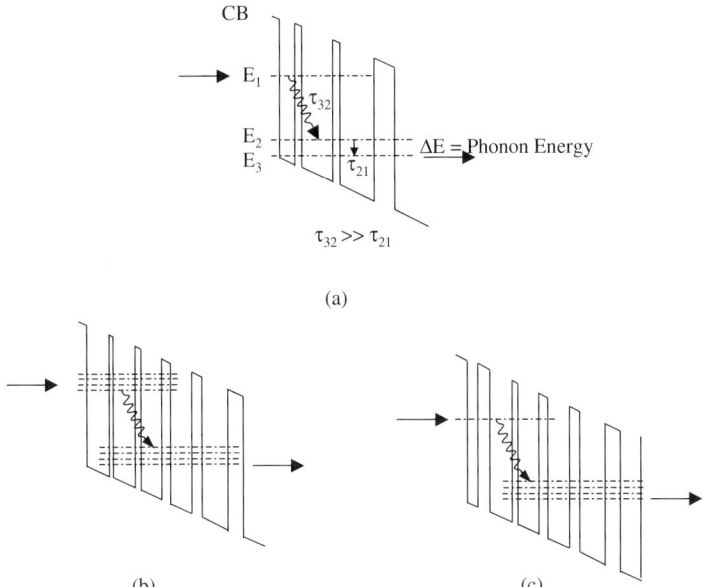

Fig. 1.4 Three successful active region designs for quantum-cascade lasers: (*a*) A three-quantum-well design in which a ground state is introduced below the lower laser level with a separation of one phonon energy (Ref. [6]). The resonant transfer of electrons from the lower laser level reduces the lifetime of the lower laser level sufficiency to obtain population inversion. (*b*) A chirped superlattice design in which the laser transition is between minibands (Ref. [32]). (*c*) A bound-to-continuum design in which the upper laser level is a discrete state and the lower laser level is a miniband (Ref. [33]). There is rapid thermalization in the miniband, lowering the lifetime of the topmost level.

photon generation can be easily implemented. By using 25 stages, external quantum efficiency of 478% has been obtained on QC lasers [30]. Very high pulsed power (>1 W) was obtained at room temperature from a relatively narrow (~40 μm) emitting aperture [31].

Faist et al. [6] achieved population inversion by using a three-level laser concept. In order to overcome the very short lifetime of intersubband transitions, a ground level, separated from the lower laser level by the phonon energy (~35 meV for InGaAs), was introduced with a proper design of the three-quantum-well active region (see Fig 1.4*a*). The resonant transfer of electrons by phonon scattering reduces the lifetime of the lower laser level sufficiently shorter than that of the upper laser level to enable the population inversion. New designs such as chirped (varying period) superlattice [32] and bound-to-continuum transition [33] utilize the top of a miniband as the lower laser level. A short lower-state lifetime is obtained because electrons quickly cool down to the bottom of the miniband (see Fig 1.4*b* and *c*). These new designs are especially advantageous for FIR lasers whose photon energy is smaller than the phonon energy [39,40].

The threshold current density of the QC laser is fairly high (>1 kA/cm^2), even at low temperatures, due to a short upper state lifetime. However, the threshold current is much less sensitive to the temperature change than that of the MIR interband lasers because the rate of phonon scattering does not increase rapidly. In addition the threshold current is not very sensitive to the wavelength. Pulsed operation has been obtained between 4.6 and 16 μm at room temperature and above [30–38], with maximum operation temperature of about 150°C at 9 and 11 μm [33,37]. Pulsed operation was also obtained at 3.4 μm up to 280 K [41], at 24 μm up to 130 K [42], and 67 μm up to 45 K [40].

However, cw operation is much more challenging because of the high power dissipation (typically above 10 kW/cm^2). Nonetheless, cw operation of quantum cascade lasers has been demonstrated at −27°C by using a smaller device with high-reflection facet coating and better heatsinking [43].

Single-frequency operation has been obtained by using DFB lasers [33,36]. Under cw operation a free-running linewidth of nearly 1 MHz over 1 ms integration time was observed [44]. With frequency stabilization using electronic feedback, the linewidth could be reduced to less than 20 kHz [44]. Under pulsed excitation, the single-frequency operation is maintained. However, there is a thermal chirp during the pulse, broadening the average linewidth. By narrowing the pulse width, the linewidth could be reduced. For a 5.3-μm laser, the linewidth of 0.15 cm^{-1} was obtained for 20-ns pulses [33]. While this is not narrow enough for very sensitive detection, it is sufficiently narrow for a large class of applications. The wavelength could be tuned by changing the temperature over a wide range.

1.2.3 Hot-Hole Lasers

The use of hot holes to generate THz emission was proposed in the 1950s [45], but it was not until the 1980s that hot-hole lasers were demonstrated [46]. Hot-hole lasers commonly utilize transitions between light- and heavy-hole bands, although laser emission between Landau levels of the light-hole band has also been obtained [47].

Typically electric and magnetic fields applied perpendicular to each other are used to accelerate the holes. At low temperatures and high electric fields, holes are accelerated without scattering until they reach the optical phonon energy. Because of a higher effective mass, heavy holes reach the optical phonon energy at a smaller electric field/magnetic field (E/B) ratio than light holes. At a certain range of the E/B ratio, typically 0.7 to 2 kV/cm-Telsa, the heavy holes reach the optical phonon energy and get scattered into the light-hole band, creating the population inversion.

Bulk p-type Ge crystals have been the most successful for hot-hole lasers because high-purity Ge crystals are available with very good control of acceptor concentration and donor compensation. Initially Ge crystals with shallow dopants such as Ga were used. High peak power up to 10W was reported for a crystal volume of 1 cm^3, with power conversion efficiency of more than 10^{-4} [47]. The duty cycle was typically very low (~10^{-5}). The operation of these Ge lasers required cryogenic temperatures (~4K) and a superconducting magnet, which was a big barrier for practical applications.

Substantial improvements in performance were obtained by utilizing a Ge crystal doped with Be, which has a higher gain. Be-doped Ge lasers have exhibited operating temperatures up to 40K, and at 4K, pulse widths up to 32 µs, duty cycles up to 5%, and pulse repetition rates up to 45 kHz [48,49]. The emission wavelength of Be-doped Ge lasers could be widely tunable from 1 to 4THz by changing the magnetic field strength [50]. In addition compact Ge lasers have been demonstrated in a closed-cycle cryogenic cooler by using a small permanent magnet [51].

1.2.4 Photomixers

A photomixer comprises an ultrafast photoconductive detector (typically fabricated on low-temperature-grown (LTG) GaAs), which is preferably connected to the input of an antenna (also fabricated on LTG GaAs) to couple the THz power out [52]. The photomixing process occurs by illuminating the photomixer with two frequency-offset (a few nm) diode lasers having photon energies just above the band gap of GaAs (~850nm). The beating of two laser frequencies creates modulation of incident power at the difference frequency, which in turn modulates the current in the photomixer.

The advantages of the photomixing technique are (1) it uses two off-the-shelf single-frequency diode lasers (tens of mW) and a LTG GaAs photoconductive detector fabricated with standard processing steps, (2) it provides useful cw power (~1µW) at room temperature at around 1THz, and (3) the frequency can be easily tunable over a wide range by adjusting the frequency offset. At higher frequencies the power decreases at a rate of 6dB/octave.

To obtain high photomixing efficiency at the THz frequency, the photodetector should have a very short minority carrier lifetime of less than a picosecond. The optimum lifetime for a given operating frequency f is around $1/(2\pi f)$. The output power is typically limited by overheating from the laser pump and dc bias supply [53]. If there is no limit on the available laser power, higher THz power can be obtained by shortening the carrier's lifetime. For a given optical power, the THz power can be increased by up to seven times by fabricating a thin LTG photoconductive detector on a dielectric mirror to enhance the absorption [54].

The linewidth of the THz signal is almost entirely determined by the stability of the pump lasers. Since free-running diode lasers do not have the frequency stability and accuracy required to generate a stable THz signal, special attention should be given to stabilizing the laser frequency by techniques such as operation of the laser in an external cavity, electrical feedback, and precision temperature control. THz emission with linewidth less than 1 MHz has been obtained by utilizing very stable pump lasers [55].

1.2.5 Plasmon Emitters

Generation of THz radiation from a semiconductor by utilizing the built-in electric field at the surface (typically caused by Fermi-level pinning due to surface states) was first reported by Zhang et al. in 1990 [56]. The THz radiation is generated when electrons and holes created by high-power, femtosecond optical pulses are accelerated by the electric field in opposite directions, creating transient current pulses. The resulting THz emission is coherent and diffraction limited, but its spectrum is very broad. Since the semiconductor material requires no chemical processes or microfabrication, it is a viable candidate for a practical broad-band THz source.

The intensity and frequency of the emitted THz radiation depend on basic semiconductor material properties, on the doping level, and on the electric field in the depletion/accumulation region near the surface. The power is proportional to the power of the excitation source. In addition the power depends on the incident angle of the excitation beam, with the maximum power obtained at the Brewster angle [56].

By applying an external magnetic field, substantially higher power can be obtained. Average powers of up to 650 µW were obtained from InAs by exciting with a 1.5-W Ti:Sapphire laser emitting at 800 nm in a 1.7-Telsa magnetic field [57]. The power typically increases proportional to the square of the magnetic field at low magnetic field strengths. The enhancement is explained by the change in the direction of carrier acceleration, which is induced by the Lorenz force in a magnetic field, to increase the coupling through the surface [58]. The enhancement is also inversely proportional to the electron effective mass. Enhancements exceeding 100 have been observed in InSb, which has the smallest effective mass among III-V compound semiconductor materials [59]. It is important that the InSb was excited with 1560-nm pulses from a mode-locked fiber laser, a much more compact source than the Ti:Sapphire laser.

1.3 SCOPE OF BOOK

Chapters 2 to 4 describe interband MIR lasers using three different material systems. In Chapter 2, M. Mitsuhara and M. Oishi review the status of InP-based strained quantum-well lasers emitting above 2 µm, including material properties of the InGaAsP alloys, design considerations of using strained quantum-wells and the limitation of maximum wavelength, growth of laser structures using metalorganic molecular beam epitaxy, and the lasing performance.

Chapter 3, by L. Olafson, I. Vurgaftman, and J. R. Meyer, provides a comprehensive overview of antimonide-based MIR lasers. The authors discuss the status of lasers emitting between 2 and 3 µm, which are dominated by type-I lasers, as well as lasers emitting beyond 3 µm that are either type-I or type-II lasers.

In Chapter 4, U. P. Schießl, J. John, and P. J. McCann review lead-salt laser technology of the homojunction, heterojunction, and quantum-well types. New manufacturing procedures to increase the reliability and the yield of single-mode emission are described. Recent developments including the growth of lead-salt epitaxial layers on barium fluoride and silicon are also discussed.

In Chapter 5, J. Faist and C. Sirtori provide an in-depth discussion of intersubband quantum cascade lasers. They explain the unique features of intersubband lasers, the fundamentals of active region and injector design, and the device processing and performance of both InP-based and GaAs-based quantum cascade lasers, including high-power pulsed and cw operation as well as single-frequency operation with DFB structures.

In Chapter 6, E. Bründermann describes hot-hole lasers. He gives a thorough analysis of hot-hole laser operation using both semiclassical and quantum mechanical approaches. He also gives a detailed discussion on material preparation, resonator designs and magnetic field generation as well as laser characteristics.

In Chapter 7, J. C. Pearson, K. A. McIntosh, and S. Vergese describe the process of photomixing to generate cw coherent THz sources. They explain techniques for stabilizing the NIR lasers to obtain a very narrow linewidth, design trade-offs for photomixers, and the antenna design for efficient radiation of the THz signal. Examples of using THz signal for spectroscopy are also given.

Since these chapters have been written, major progress has been made in the field of cascade lasers. Room-temperature cw operation has been reported at about 9 µm [60], and FIR emission has been obtained including cw operation at 87 µm (3.5 THz) at temperatures up to 55 K [61,62]. Interband cascade lasers have also achieved room-temperature pulsed operation at 3.5 µm, with cw power efficiency of 17% at 77 K [63]. These results are exciting as they will stimulate further development and open up the possibility for new applications in the MIR and FIR spectral range.

REFERENCES

1. I. Melngailis, "Maser action in InAs diodes," *Appl. Phys. Lett.* **2**, 176 (1963).
2. J. O. Dimmock, I. Melngailis, and A. J. Strauss, "Band structure and laser action in $Pb_xSn_{1-x}Te$," *Phys. Rev. Lett.* **16**, 1193 (1966).
3. K. J. Linden and A. W. Mantz, "Tunable diode lasers and laser systems for the 3 to 30 µm infrared spectral region," *SPIE Proc.* **320**, 109 (1982).
4. A. E. Bochkarev, L. M. Dolginov, A. E. Drakin, L. V. Druzhinina, P. G. Eliseev, and B. N. Sverdlov, "Injection InGaSbAs lasers emitting radiation of wavelengths 1.9–2.3 µm at room temperature," *Kvant. Elektron.* **12**, 1309 (1985) [*Sov. J. Quant. Electron.* **15**, 869 (1985)].

5. A. E. Bochkarev, L. M. Dolginov, A. E. Drakin, P. G. Eliseev, and B. N. Sverdlov, "Continuous-wave lasing at room temperature in InGaSbAs/GaAlSbAs injection heterostructures emitting in the spectral range 2.2–2.4 µm," *Kvant. Electron.* **15**, 2171 (1988) [*Sov. J. Quantum Electron.* **18**, 1362 (1988)].
6. J. Faist, F. Capasso, D. L. Sivco, C. Sirtori, A. L. Hutchinson, and A. Y. Cho, "Quantum cascade laser," *Science* **264**, 553 (1994).
7. H. K. Choi and S. J. Eglash, "High-power multiple-quantum-well GaInAsSb/ AlGaAsSb diode lasers emitting at 2.1 µm with low threshold current density," *Appl. Phys. Lett.* **61**, 1154 (1992).
8. M. Mitsuhara, M. Ogasawara, M. Oishi, and H. Sugiura, "Metalorganic molecular-beam-epitaxy-grown $In_{0.77}Ga_{0.23}As$/InGaAs multiple quantum well lasers emitting at 2.07 µm wavelength," *Appl. Phys. Lett.* **72**, 3106 (1998).
9. M. Mitsuhara, M. Ogasawara, M. Oishi, H. Sugiura, and K. Kasaya, "2.05-µm wavelength InGaAs-InGaAs distributed-feedback multiquantum-well lasers with 10-mW output power," *IEEE Photon. Technol. Lett.* **11**, 33 (1999).
10. G. W. Turner, H. K. Choi, and M. J. Manfra, "Ultralow-threshold (50 A/cm^2) strained single-quantum-well GaInAsSb/AlGaAsSb lasers emitting at 2.05 µm," *Appl. Phys. Lett.* **72**, 876 (1998).
11. D. Garbuzov, H. Lee, V. Khalfin, R. Martinelli, J. C. Connolly, and G. L. Belenky, "2.3–2.7-µm room temperature CW operation of InGaAsSb-AlGaAsSb broad waveguide SCH-QW diode lasers," *IEEE Photon. Technol. Lett.* **11**, 794 (1999).
12. H. K. Choi, G. W. Turner, M. J. Manfra, and M. K. Connors, "175 K continuous wave operation of InAsSb/InAlAsSb quantum-well diode lasers emitting at 3.5 µm," *Appl. Phys. Lett.* **68**, 2936 (1996).
13. J. R. Meyer, C. L. Felix, W. W. Bewley, I. Vurgaftman, E. H. Aifer, L. J. Olafsen, J. R. Lindle, C. A. Hoffman, M.-J. Yang, B. R. Bennett, B. V. Shanabrook, H. Lee, C.-H. Lin, S. S. Pei, and R. H. Miles, "Auger coefficients in type-II InAs/$Ga_{1-x}In_xSb$ quantum wells," *Appl. Phys. Lett.* **73**, 2857 (1998).
14. W. W. Bewley, C. L. Felix, I. Vurgaftman, D. W. Stokes, E. H. Aifer, L. J. Olafsen, J. R. Meyer, M. J. Yang, B. V. Shanabrook, H. Lee, R. U. Martinelli, and A. R. Sugg, "High-temperature continuous-wave 3–6.1 µm 'W' lasers with diamond-pressure-bond heat sinking," *Appl. Phys. Lett.* **74**, 1075 (1999).
15. D. W. Stokes, L. J. Olafsen, W. W. Bewley, I. Vurgaftman, C. L. Felix, E. H. Aifer, J. R. Meyer, and M. J. Yang, "Type-II quantum-well 'W' lasers emitting at λ = 5.4–7.3 µm," *J. Appl. Phys.* **86**, 4729 (1999).
16. H. Lee, L. J. Olafsen, R. J. Menna, W. W. Bewley, R. U. Martinelli, I. Vurgaftman, D. Z. Garbuzov, C. L. Felix, M. Maiorov, J. R. Meyer, J. C. Connolly, A. R. Sugg, and G. H. Olsen, "Room-temperature type-II W quantum well diode laser with broadened waveguide emitting at λ = 3.30 µm," *Electron. Lett.* **35**, 1743 (1999).
17. W. W. Bewley, H. Lee, I. Vurgaftman, R. J. Menna, C. L. Felix, R. U. Martinelli, D. W. Stokes, D. Z. Garbuzov, J. R. Meyer, M. Maiorov, J. C. Connolly, A. R. Sugg, and G. H. Olsen, "Continuous-wave operation of λ = 3.25 µm broadened-waveguide W quantum-well diode lasers up to T = 195 K," *Appl. Phys. Lett.* **76**, 256 (2000).
18. R. Q. Yang and S. S. Pei, "Novel type-II quantum cascade lasers," *J. Appl. Phys. Lett.* **79**, 8197 (1996).

19. R. Q. Yang, J. D. Bruno, J. L. Bradshaw, J. T. Pham, and D. E. Wortman, "High-power interband cascade lasers with quantum efficiency >450%," *Electron. Lett.* **35**, 1254 (1999).
20. J. D. Bruno, J. L. Bradshaw, Rui Q. Yang, J. T. Pham, and D. E. Wortman, "Low-threshold interband cascade lasers with power efficiency exceeding 9%," *Appl. Phys. Lett.* **76**, 3167 (2000).
21. H. Q. Le, G. W. Turner, and J. R. Ochoa, "High-power high-efficiency quasi-CW Sb-based mid-IR lasers using 1.9-µm laser diode pumping," *IEEE Photon. Technol. Lett.* **10**, 663 (1998).
22. H. K. Choi, A. K. Goyal, S. C. Buchter, G. W. Turner, M. J. Manfra, and S. D. Calawa, "High-power optically pumped GaInSb/InAs quantum well lasers with GaInAsSb integrated absorber layers emitting at 4µm," Conference on Lasers and Electro-Optics (San Francisco, May 7–12, 2000), Tech. Dig., 63.
23. P. C. Findlay, C. R. Pidgeon, R. Kotitschke, A. Hollingworth, B. N. Murdin, C. J. G. M. Langerak, A. F. G. van der Meer, C. M. Ciesla, J. Oswald, A. Homer, G. Springholz, and G. Bauer, "Auger recombination of lead salts under picosecond free-electron-laser-excitations," *Phys. Rev.* B **58**, 12908 (1998).
24. Z. Feit, M. McDonald, R. J. Woods, V. Archambault, and P. Mak, "Low threshold PbEuSeTe/PbTe separate confinement buried heterostructure diode lasers," *Appl. Phys. Lett.* **68**, 738 (1996).
25. J. John, A. Fach, H. Böttner, and M. Tacke, "Embossed monomode single heterostructure distributed feedback lead chalcogenide diode lasers," *Electron. Lett.* **28**, 2180 (1992).
26. U. P. Schießl, M. Birle, and H.-E. Wagner, "Recent results about room temperature operation of lead salt laser diodes," *Optische Analysentechnik in Industrie und Umwelt—heute und morgen*, Tagung Düsseldorf, March 20–21, 2000 /VDI/ VDE-Gesellschaft Mess- und Automatisierungstechnik. VDI Verlag Düsseldorf, 2000, pp. 113–118.
27. A. Ishida, S. Matsuura, M. Mizuno, and H. Fujiyasu, "Observation of quantum size effects in optical transmission spectra of $PbTe/Pb_{1-x}Eu_xTe$ superlattices," *Appl. Phys. Lett.* **51**, 478 (1987).
28. C. L. Felix, W. W. Bewley, I. Vurgaftman, J. R. Lindle, J. R. Meyer, H. Z. Wu, G. Xu, S. Khosravani, and Z. Shi, "Low-threshold optically pumped (~4.4µm) vertical-cavity surface-emitting laser with a PbSe quantum-well active region," *Appl. Phys. Lett.* **78**, 3770 (2001).
29. R. F. Kazarinov and R. A. Suris, "Possibility of the amplification of electromagnetic waves in a semiconductor with a superlattice," *Sov. Phys. Semicond.* **5**, 707 (1971).
30. J. Faist, A. Trdicucci, F. Capasso, C. Sirtori, D. L. Sivco, J. N. Baillargeon, A. L. Hutchinson, A. Y. Cho, "High-power continuous-wave quantum cascade lasers," *IEEE J. Quantum. Electron.* **34**, 336 (1998).
31. D. Hofstetter, M. Beck, T. Aellen, and J. Faist, "High-temperature operation of distributed feedback quantum-cascade lasers at 5.3µm," *Appl. Phys. Lett.* **78**, 396 (2001).
32. A. Tredicucci, F. Capasso, C. Gmachl, D. L. Sivco, A. L. Huchinson, and A. Y. Cho, "High performance interminiband quantum cascade lasers with graded superlattices," *Appl. Phys. Lett.* **73**, 2101 (1998).

33. J. Faist, M. Beck, T. Aellen, and E. Gini, "Quantum cascade lasers based on a bound-to-continuum transition," *Appl. Phys. Lett.* **78**, 147 (2001).
34. R. Köhler, C. Gmachl, A. Tredicucci, F. Capasso, D. L. Sivco, S. N. G. Chu, and A. Y. Cho, "Single-mode tunable, pulsed, and continuous wave quantum-cascade distributed feedback lasers at $\lambda \sim 4.6$–$4.7\,\mu m$," *Appl. Phys. Lett.* **76**, 1092 (2000).
35. C. Gmachl, A. Tredicucci, F. Capasso, A. L. Hutchinson, D. L. Sivco, J. N. Baillargeon, and A. Y. Cho, "High-power $\lambda \sim 8\,\mu m$ quantum cascade lasers with near optimum performance," *Appl. Phys. Lett.* **72**, 3130 (1998).
36. J. Faist, C. Sirtori, F. Capasso, D. L. Sivco, A. L. Hutchinson, and A. Y. Cho, "High-power long-wavelength ($\lambda = 11.5\,\mu m$) quantum cascade lasers operating above room temperature," *IEEE. Photon. Technol. Lett.* **10**, 1100 (1998).
37. A. Tahraoui, A. Matlis, S. Slivken, J. Diaz, and M. Razeghi, "High-performance quantum cascade lasers ($\lambda \sim 11\,\mu m$) operating at high temperature ($T > 425\,K$)," *Appl. Phys. Lett.* **78**, 416 (2001).
38. M. Rochat, D. Hofstetter, M. Beck, and J. Faist, "Long-wavlength ($\lambda \sim 16\,\mu m$) room-temperature single-frequency quantum-cascade lasers based on a bound-to-continuum transition," *Appl. Phys. Lett.* **79**, 4271 (2001).
39. R. Koller, A. Tredicucci, F. Beltram, H. E. Beere, E. H. Linfield, G. Davies, D. A. Ritchie, R. C. Iotti, and F. Rossi, "Terahertz semiconductor-heterostructure laser," *Nature* **417**, 156 (2002).
40. Michel Rochat, Lassaad Ajili, Harald Willenberg, Jérôme Faist, Harvey Beere, Giles Davies, Edmund Linfield, and David Ritchie, "Low-threshold terahertz quantum cascade lasers," *Appl. Phys. Lett.* **81**, 1381 (2002).
41. J. Faist, F. Capasso, D. L. Sivco, A. L. Hutchinson, S. N. G. Chu, and A. Y. Cho, "Short-wavlength ($\lambda = 3.4\,\mu m$) quantum cascade laser based on strain compensated InGaAs/AlInAs," *Appl. Phys. Lett.* **72**, 680 (1998).
42. R. Colombelli, F. Capasso, C. Gmachl, A. L. Hutchinson, D. L. Sivco, A. Tredicucci, M. C. Wanke, A. M. Sergent, and A. Y. Cho, "Far-infrared surface-plasmon quantum-cascade lasers at $21.5\,\mu m$ and $24\,\mu m$ wavelengths," *Appl. Phys. Lett.* **78**, 2620 (2001).
43. D. Hofstetter, M. Beck, T. Aellen, J. Faist, U. Oesterle, M. Ilegems, E. Gini, and H. Melchior, "Continuous wave operation of a $9.3\,\mu m$ quantum cascade laser on a Peltier cooler," *Appl. Phys. Lett.* **78**, 1964 (2001).
44. R. M. Williams, J. F. Kelly, J. S. Harman, S. W. Sharpe, M. S. Taubman, J. L. Hal, F. Capasso, C. Gmachl, D. L. Sivco, J. N. Baillargeon, and A. Y. Cho, "Kilohertz linewidth from frequency-stabilized mid-infrared quantum cascade lasers," *Opt. Lett.* **24**, 1844 (1999).
45. H. Kroemer, "Proposed negative mass microwave amplifier," *Phys. Rev.* **109**, 1856 (1958).
46. A. Andronov, I. V. Zverev, V. A. Kozlov, Yu. N. Nozdrin, S. A. Pavlov, and V. N. Shastin, "Stimulated emission in the long-wavelength IR region from hot holes in Ge in crossed electric and magnetic field," *Pis'ma Zh. Eksp. Thro. Fiz.* **40**, 69 (1984) [*Sov. Phys. JETP Lett.* **40**, 804 (1984)].
47. Y. L. Ivanov, "Generation of cyclotron radiation by light holes in germanium," *Opt. Quantum Electron.* **23**, S253 (1991).
48. E. Bruendermann, D. R. Chanberlin, and E. E. Haller, "High duty cycle and continuous terahertz emission from germanium," *Appl. Phys. Lett.* **76**, 2991 (2000).

49. E. Bruendermann, D. R. Chanberlin, and E. E. Haller, "Thermal effects in widely tunable germanium terhertz lasers," *Appl. Phys. Lett.* **73**, 2757 (1998).
50. L. A. Reichhertz, O. D. Dubon, G. Sirmain, E. Bruendermann, W. L. Hansen, D. R. Chamberlin, A. M. Linhart, H. P. Rosser, and E. E. Haller, "Stimulated far-infrared emission from combined cyclotron resonances in germanium," *Phys. Rev.* **B56**, 12069 (1997).
51. E. Bruendermann and H. P. Roser, "First operation of a far-infrared p-germanium laser in a standard closed-cycle machine at 15 K," *Infrared Phys. Technol.* **38**, 201 (1997).
52. E. R. Brown, F. W. Smith, and K. A. McIntosh, "Coherent millimeter-wave generation by heterodyne conversion," *J. Appl. Phys.* **1480** (1993).
53. S. Vergese, K. A. McIntosh, and R. R. Brown, "Highly tunable fiber-coupled photomixsers with coherent terahertz with diode lasers in low-temperature-grown GaAs," *IEEE Trans. Microw. Theory Tech.* **45**, 1301 (1997).
54. E. R. Brown, "A photoconductive model for superior GaAs THz photomixers," *Appl. Phys. Lett.* **75**, 769 (1999).
55. S. Vergese, K. A. McIntosh, S. D. Calawa, C.-Y. E. Tong, R. Kimberk, and R. Blundell, "A photomixer local oscillator for a 630-GHz heterodyne receiver," *IEEE Microw. Guided Wave Lett.* **9**, 245 (1999).
56. X.-C. Zhang, B. B. Hu, J. T. Darrow, and D. H. Auston, "Generation of femtosecond electromagnetic pulses from semiconductor surfaces," *Appl. Phys. Lett.* **56**, 1011 (1990).
57. N. Sarukura, H. Ohtake, S. Izumida, and Z. Liu, "High average THz radiation from femtosecond laser-irradiated InAs in a magnetic field and its elliptical polarization characteristics," *J. App. Phys.* **84**, 654 (1998).
58. J. Heyman, P. Niocleous, D. Herbert, P. A. Crowell, T. Muller, and K. Unterrainer, "Terahertz emission from GaAs and InAs in a magnetic field," *Phys. Rev. B* **64**, 085202 (2001).
59. H. Takahashi, Y. Suzuki, M. Sakai, S. Ono, N. Sarukura, T. Sugiura, T. Hirosumi, and M. Yoshida, "Significant enhancement of terahertz radiation from InSb by use of a compact fiber laser and an external magnetic field," *Appl. Phys. Lett.* **82**, 2005 (2003).
60. M. Beck, D. Hofstetter, T. Aellen, J. Faist, U. Oesterle, M. Ilegems, E. Gini, and H. Melchior, "Continuous wave operation of a mid-infrared semiconductor laser at room temperature," *Science* **295**, 301 (2002).
61. L. Ajili, G. Scalari, D. Hofstetter, M. Beck, J. Faist, H. Beere, G. Davies, E. Linfield, and D. Ritchie, "Continuous-wave operation of far-infrared quantum cascade lasers," *Electron. Lett.* **38**, 1675 (2002).
62. R. Koller, A. Tredicucci, F. Beltram, H. E. Beere, E. H. Linfield, G. Davies, D. A. Ritchie, S. Dillion, and C. Sirtori, "High-performance continuous-wave operation of superlattice terahertz quantum-cascade lasers," *Appl. Phys. Lett.* **82**, 1518 (2003).
63. R. Q. Yang, J. L. Bradshaw, J. D. Bruno, J. T. Pham, D. E. Wortman, and R. L. Tober, "Room-temperature type-II interband cascade laser," Appl. Phys. Lett. **81**, 397 (2002).

CHAPTER 2

2-μm Wavelength Lasers Employing InP-based Strained-Layer Quantum Wells

MANABU MITSUHARA and MAMORU OISHI

2.1 INTRODUCTION

Semiconductor laser diodes emitting in the mid-infrared wavelength region (2–20μm) are very useful for sensitive trace-gas monitoring applications in medical diagnostics, environmental and toxic gas monitoring, and factory-process control. Figure 2.1 shows the emission wavelength ranges near room temperature presently attainable with III–V semiconductor lasers, namely GaAs-based, InP-based, Sb-based, and the quantum cascade (QC) lasers [1–8]. Semiconductor lasers fabricated from InP-based and Sb-based materials emit light in the 2-μm wavelength range. Room-temperature and single-mode operation characteristics are favored in laser-based spectroscopy. The mature processing technology of InP-based materials makes it easy to fabricate single-mode lasers. In addition, the high thermal conductivity of InP substrate (Table 2.1) is effective in dissipating the Joule heat generated near the active region of the laser. On the other hand, it is difficult for InP-based lasers to attain emission wavelength longer than 2μm because the band-gap wavelength of InP-based materials is shorter than 1.7μm when the layer is lattice-matched to InP substrate.

For InP-based lasers, emission wavelengths longer than 2μm were first demonstrated in 1989 using an active region of bulk $In_{0.87}Ga_{0.13}As$ on compositionally graded InGaAs and InAsP layers [9]. This laser had an emission wavelength of 2.45μm, but its operating temperature was limited to 210K. The 2-μm wavelength InP-based lasers were quickly improved by using strained-layer quantum wells (QWs) as the active regions. Adams [10] and Yablonovitch

---

Long-Wavelength Infrared Semiconductor Lasers, Edited by Hong K. Choi
ISBN 0-471-39200-6 Copyright © 2004 John Wiley & Sons, Inc.

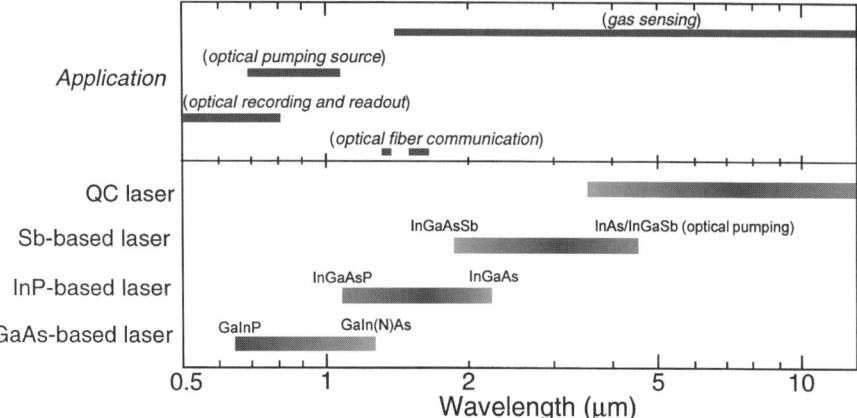

Fig. 2.1 Emission wavelength ranges presently attainable with III–V semiconductor lasers near room temperature. The compounds on each side of the bars represent the materials used in the active regions that give the shortest and longest wavelengths.

TABLE 2.1 Thermal conductivity and coefficients of thermal expansion for InP, GaAs, InAs, and GaP near room temperature.

Compound	Thermal Conductivity ($W\,cm^{-1}\,K^{-1}$)	Coefficient of Thermal Expansion ($10^{-6}/°C$)
InP	0.68	4.75
InAs	0.273	4.52
GaAs	0.46	6.86
GaP	0.77	4.5
InSb	0.166	5.37
GaSb	0.39	7.75

and Kane [11] predicted that the lasing threshold current could be reduced by using compressively strained QW active region in 1986, and Thijs and Dongen succeeded in fabricating 1.5-µm wavelength multiple quantum-well (MQW) lasers in 1989 [12]. Increasing the compressive strain in the well layers has also attracted much interest as a way of increasing the emission wavelength of InP-based laser [13–15]. In 1992 the first InP-based QW laser with emission wavelength longer than 2µm was reported by Forouhar et al. [14]. Significant improvements were obtained in the device characteristics of the 2-µm wavelength lasers by using the mature processing technologies of the telecommunication lasers between 1993 and 1995. During this time, Major et al. [16] fabricated a monolithic laser array with an output power as high as 5.5W at 10°C. Martinelli et al. [17] demonstrated single-mode operation of a ridge-waveguide laser with a distributed-feedback (DFB) structure. Ochiai et al. [18] obtained a threshold current of 17 mA for Fabry-Perot (F-P) lasers by using buried-heterostructure current-blocking layers. More recently the reported device characteristics of 2-μm wavelength InP-based lasers have become comparable to those of telecommunication lasers with the emission wavelengths of 1.3 or 1.55 μm [19–21].

This chapter gives a review of the 2-μm wavelength lasers that use a strained MQW layer for their active region. Section 2.2 describes the material properties of a bulk InGaAsP alloy system. Section 2.3 covers design aspects of the InGaAs QW structure when it is used as the active region of a 2-μm wavelength laser. In this section, we will pay particular attention to well strain, barrier height, and number of wells. Section 2.4 describes the features of epitaxial growth of strained MQW by metalorganic molecular beam epitaxy (MOMBE). We present some experimental results that show how the growth conditions of the barrier layers affect the structural and optical properties of the MQW. Section 2.5 describes the device characteristics of the F-P and the DFB lasers. Section 2.6 draws some conclusions and mentions future prospects.

2.2 MATERIAL PROPERTIES OF InGaAsP

2.2.1 Composition Dependence of Band-Gap Energy and Lattice Constant

As a starting point for any determination of the laser structure, precise expressions of the band-gap energy and the lattice constant for each layer are required. InP-based compounds are some of the most extensively studied alloy compounds and several reviews have summarized the existing knowledge on them [22,23]. A linear interpolation scheme, using the values of the related binary compounds, is generally adopted when deriving many of the physical parameters of the alloy compounds. The parameter of $In_{1-x}Ga_xAs_yP_{1-y}$, $p(x, y)$, can be obtained from the respective values of the four binaries, InP, InAs, GaAs and GaP, in accordance with

$$p(x,y) = (1-x)(1-y)p_{InP} + (1-x)yp_{InAs} + xyp_{GaAs} + x(1-y)p_{GaP}. \quad (2.1)$$

Table 2.2 lists the lattice constant, band-gap energy, spin-orbit splitting energy, elastic stiffness coefficient, pressure coefficient of band-gap energy, and deformation potential of these binaries. From Eq. (2.1), the lattice constant of $In_{1-x}Ga_xAs_yP_{1-y}$, $a(x, y)$, is given by

$$a(x,y) = 5.8688 - 0.4176x + 0.1895y - 0.0126xy. \quad (2.2)$$

The relationship between the band-gap energy $E_g(x, y)$ and the composition can be written

TABLE 2.2 Material parameters for InP, GaAs, InAs, and GaP.

Parameter[a]		InP	GaAs	InAs	GaP
Lattice constant (Å)	a_0	5.8688	5.6533	6.0584	5.4512
Direct energy-gap (eV)	E_g	1.35	1.42	0.36	2.74
Spin-orbit splitting energy (eV)	Δ	0.1	0.34	0.37	0.1
Elastic stiffness coefficient	C_{11}	10.22	11.88	8.33	14.12
(10^{11} dyne·cm^{-2})	C_{12}	5.76	5.38	4.53	6.25
Pressure coefficient of E_g		8.5	11.5	10.0	11.0
(10^{-12} eV/dyne·cm^{-2}) dE_g/dP					
Hydrostatic deformation potential[b] (eV)	a	−6.16	−8.68	−5.79	−9.76
Shear deformation potential (eV)	b	−1.6	−1.7	−1.8	−1.5

[a] The parameters for binary compounds are taken from [24–27].

[b] $a = -\dfrac{(C_{11}+2C_{12})}{3} \dfrac{dE_g}{dP}$.

$$E_g(x,y) = 1.35 + 0.672x - 1.091y + 0.758x^2 + 0.101y^2 \\ + 0.111xy - 0.58x^2y - 0.159xy^2 + 0.268x^2y^2, \quad (2.3)$$

in which the term in $xy(1-x)(1-y)$ is added to the formula of Moon et al. [28]. Figure 2.2 shows the contours of the band-gap wavelength λ_g (μm) [= $1.2407/E_g$ (eV)] and the lattice mismatch to InP, $\Delta a/a$ (%) [= {($a(x, y)$ − a_{InP})/a_{InP}} × 100], for In$_{1-x}$Ga$_x$As$_y$P$_{1-y}$ calculated from Eqs. (2.2) and (2.3). On condition that In$_{1-x}$Ga$_x$As$_y$P$_{1-y}$ layer is lattice-matched to InP, the longest band-gap wavelength is 1.65 μm, as shown in Fig. 2.2. The band-gap wavelength and lattice mismatch of InGaAsP increase as the mole fractions of In and As are increased. An InGaAsP layer with lattice mismatch larger than +0.8% will have a band-gap wavelength longer than 2 μm. However, a +0.8% lattice mismatch is insufficient for a band-gap wavelength of 2 μm owing to the energy shift induced by the strain (see Section 2.3.1). It is difficult to increase the thickness of the epitaxial layer with large lattice mismatch because relaxation occurs when a layer exceeds a certain critical thickness. The critical layer thickness of InGaAs with a lattice mismatch of +0.8% is less than 0.1 μm. For a double-heterostructure laser, the active layer thickness should be larger than 0.1 μm to obtain good laser performance [29]. Therefore strained QW lasers are so far the only ones with emission wavelengths longer than 2 μm.

2.2.2 Miscibility Gap

In a highly strained QW structure composed of InGaAsP alloys, the stability of the growing surface is influenced by the compositional modulation in the well and barrier layers [30–32]. This is because compositional modulations generate distribution of elastic stress in the layers and the surfaces under stress are unstable against morphological changes [33,34]. In this section we discuss the compositional stability of InGaAsP used, and barrier layers of the 2-μm wavelength MQW lasers.

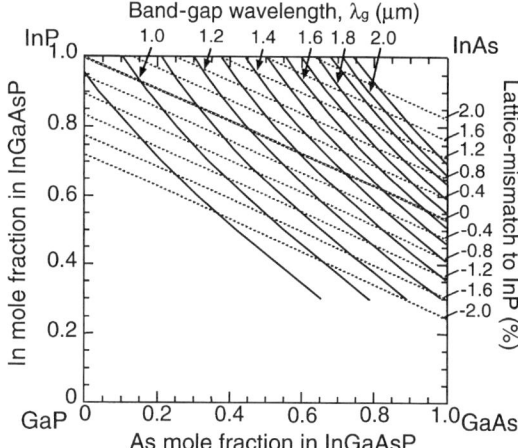

Fig. 2.2 Contour map of band-gap wavelength (solid lines) and lattice mismatch (dashed lines) superimposed on the compositional plane for $In_{1-x}Ga_xAs_yP_{1-y}$ at 300 K.

The unstable region in the phase diagram of InGaAsP alloy is well known as a miscibility gap and has been predicted by thermodynamic calculations [35–37]. Compositional modulations can be observed in InGaAsP layers grown by liquid phase epitaxy (LPE) by using energy dispersive X-ray spectroscopy (EDX) [38,39]. Even in InGaAsP layers grown by gas-source molecular beam epitaxy (GSMBE) and metalorganic vapor phase epitaxy (MOVPE), which are considered to be nonequilibrium growth methods, the existence of the miscibility gap has been confirmed by transmission electron microscopy (TEM), X-ray diffraction, and photoluminescence (PL) measurements [40–43].

On the basis of the thermodynamic model, the compositional fluctuations in the alloys occur when the total curvature of the Gibbs free energy per mole, $G(x, y)$, is negative. The spinodal curve, which limits the immiscible region in the compositional plane at a certain temperature, is given by [44]

$$\frac{\partial^2 G}{\partial x^2} \frac{\partial^2 G}{\partial y^2} - \left(\frac{\partial^2 G}{\partial x \partial y}\right)^2 = 0. \tag{2.4}$$

Using the strictly regular solution approximation based on Ref. [37], $G(x, y)$ of $In_{1-x}Ga_xAs_yP_{1-y}$ can be written as

$$\begin{aligned}G(x,y) = &(1-x)(1-y)\omega_{InP} + (1-x)y\omega_{InAs} + x(1-y)\omega_{GaP} + xy\omega_{GaAs} \\&+ (1-x)x(1-y)\alpha_{InP\text{-}GaP} + (1-x)xy\alpha_{InAs\text{-}GaAs} + (1-x)y(1-y)\alpha_{InP\text{-}InAs} \\&+ xy(1-y)\alpha_{GaP\text{-}GaAs} + RT[(1-x)\ln(1-x) + x\ln x \\&+ (1-y)\ln(1-y) + y\ln y],\end{aligned} \tag{2.5}$$

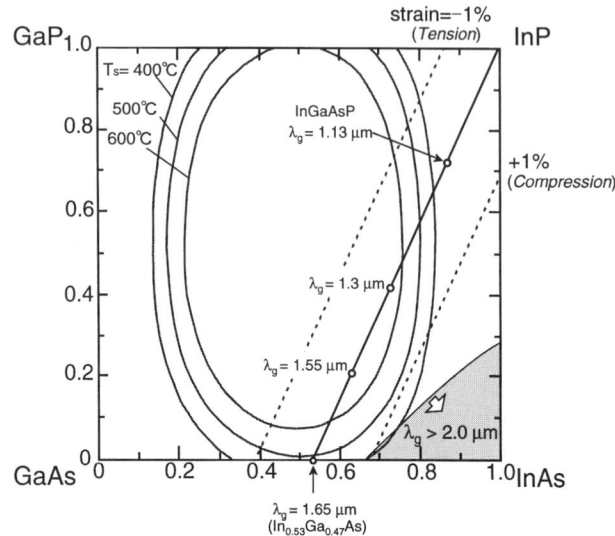

Fig. 2.3 Calculated spinodal isotherms for InGaAsP alloys. The region within an isotherm represents the miscibility gap. The solid line shows the composition of InGaAsP lattice-matched to InP, and the dashed lines show the composition of InGaAsP with a lattice mismatch to InP of +1% and −1%.

where the ω values are the sum of the first and second nearest-neighbor interaction energies of a binary compound, the α values are the ternary interaction parameters, T is the absolute temperature, and R is the gas constant. Substituting Eq. (2.5) into Eq. (2.4) gives the relation

$$\{RT - 2(1-x)x[(1-y)\alpha_{\text{InAs-GaAs}} + y\alpha_{\text{InP-GaP}}]\} \times \{RT - 2(1-y)y[(1-x)\alpha_{\text{InAs-InP}} + x\alpha_{\text{GaAs-GaP}}]\} - (1-x)x(1-y)y(\omega_Q - \alpha_Q)^2 = 0, \qquad (2.6)$$

where

$$\omega_Q \equiv \omega_{\text{InAs}} - \omega_{\text{InP}} - \omega_{\text{GaAs}} + \omega_{\text{GaP}}, \qquad (2.7)$$

$$\alpha_Q \equiv (1-2x)(\alpha_{\text{InP-GaP}} - \alpha_{\text{InAs-GaAs}}) + (1-2y)(\alpha_{\text{GaAs-GaP}} - \alpha_{\text{InAs-InP}}). \qquad (2.8)$$

Figure 2.3 shows the calculated result of the miscibility isotherms of InGaAsP alloys for 400, 500, and 600°C. The shaded area shows the composition of InGaAsP for the band-gap wavelengths longer than 2 μm. The solid line is the composition of InGaAsP lattice-matched to InP. The broken lines show the composition of InGaAsP with a lattice mismatch to InP of +1% and −1%, and InGaAsP layers whose compositions fall inside these lines are usually used as the barrier layers. As can be seen in Fig. 2.3, the region of overlap between the miscibility gap and the shaded area is small. This suggests that the compositional modulations rarely occur in the well layers of 2-µm wavelength MQW lasers. In contrast, a large part of the composition for InGaAsP layer with the band-gap wavelength from 1.13 to 1.65 µm is inside the miscibility gap. This suggests that the compositional modulations frequently occur in the barrier layers. Furthermore the composition of InGaAsP can be put inside the region of the miscibility gap by changing the sign of the lattice mismatch from plus to minus. Therefore the strain-compensated technique [45], in which the barrier layer is strained in the opposite sense to the well strain, is difficult to implement in the 2-µm wavelength MQW laser. The effect of the compositional modulation on the crystalline quality of the MQW will be discussed in Section 2.4.2.

2.3 DESIGN CONSIDERATION OF MQW ACTIVE REGION

2.3.1 Strain and Quantum Size Effects

The design procedure of strained QW laser structures must include calculations of the band-gap energy and critical layer thickness of the QW. We first consider the strain and quantum size effects on the emission wavelength of a strained-InGaAs QW.

In the cubic crystal the stress tensor (σ) is expressed by the elastic constants tensor (C) and the strain tensor (ε) as follows:

$$\begin{bmatrix} \sigma_{xx} \\ \sigma_{yy} \\ \sigma_{zz} \\ \sigma_{yz} \\ \sigma_{zx} \\ \sigma_{xy} \end{bmatrix} = \begin{bmatrix} C_{11} & C_{12} & C_{12} & 0 & 0 & 0 \\ C_{12} & C_{11} & C_{12} & 0 & 0 & 0 \\ C_{12} & C_{12} & C_{11} & 0 & 0 & 0 \\ 0 & 0 & 0 & C_{44} & 0 & 0 \\ 0 & 0 & 0 & 0 & C_{44} & 0 \\ 0 & 0 & 0 & 0 & 0 & C_{44} \end{bmatrix} \begin{bmatrix} \varepsilon_{xx} \\ \varepsilon_{yy} \\ \varepsilon_{zz} \\ \varepsilon_{yz} \\ \varepsilon_{zx} \\ \varepsilon_{xy} \end{bmatrix}. \qquad (2.9)$$

If the mismatch strain between the epitaxial layer and substrate is small, the epitaxial layer is tetragonally distorted to accommodate the mismatch strain. This situation is illustrated in Fig. 2.4. In response to the biaxial stress, the layer relaxes along the growth direction, while the lattice constant in the growth plane $a_{||}$ ($= a_{xx} = a_{yy}$) is the same throughout the structure and is equal to that of substrate. In this case the stress components and the strain components are given by

$$\sigma_{zz} = \sigma_{yz} = \sigma_{zx} = \sigma_{xy} = 0, \qquad (2.10)$$

$$\varepsilon_{xx} = \varepsilon_{yy} = \frac{a_{||} - a_e}{a_e} = \frac{a_s - a_e}{a_e} \approx -\frac{a_e - a_s}{a_s} = -\varepsilon, \qquad (2.11)$$

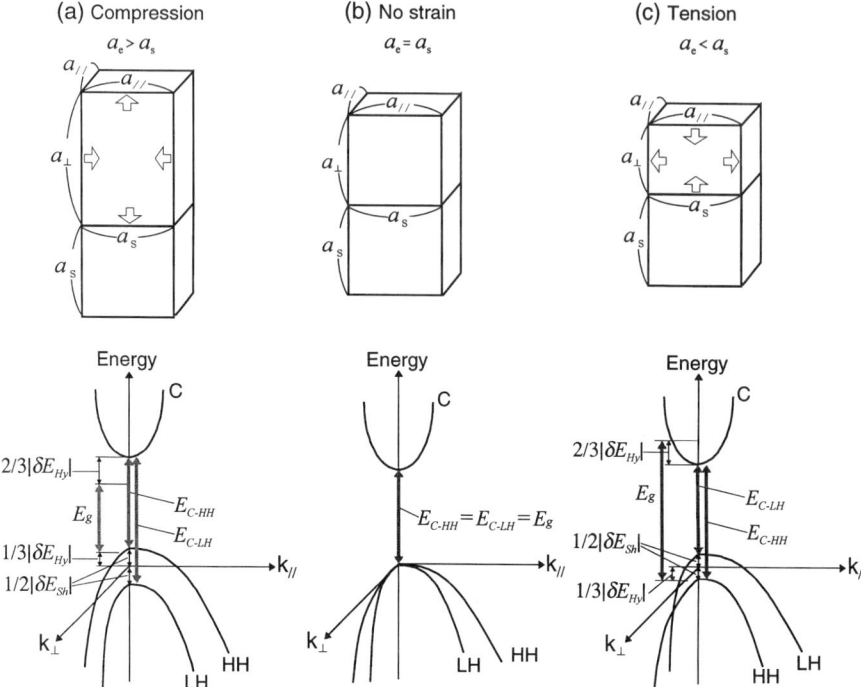

Fig. 2.4 Schematic representations of the lattice (upper part) and the band structure (lower part) for bulk InGaAs layers under (*a*) compressive strain, (*b*) lattice-matching condition, and (*c*) tensile strain. $E_{C\text{-}HH}$ ($E_{C\text{-}LH}$) is the band-gap energy difference between bottom of the conduction band and the top of the heavy-hole (light-hole) band, and E_g represents the unstrained band-gap energy. δE_{Hy} is the total band edge shift arising from the hydrostatic strain for the conduction and valence bands, and one-third of δE_{Hy} is assigned to the valence band [49]. δE_{Sh} is the energy separation of heavy hole and light hole induced by the uniaxial strain.

$$\varepsilon_{zz} = \frac{a_\perp - a_e}{a_e} \approx \frac{a_\perp - a_s}{a_s} - \frac{a_e - a_s}{a_s} = \varepsilon_\perp - \varepsilon, \quad (2.12)$$

$$\varepsilon_{yz} = \varepsilon_{zx} = \varepsilon_{xy} = 0, \quad (2.13)$$

where ε is the mismatch strain defined as the difference between the natural unstrained lattice constant of epitaxial layer (a_e) and substrate (a_s) divided by a_s. Substituting Eq. (2.9) into Eq. (2.10), we have

$$\varepsilon_{zz} = -\frac{2C_{12}}{C_{11}} \varepsilon_{xx}. \quad (2.14)$$

Using Eqs. (2.12) and (2.14), we write the lattice constant along the growth direction as

$$a_\perp = (1+\varepsilon_\perp)a_s = \left(1 + \frac{C_{11}+2C_{12}}{C_{11}}\varepsilon\right)a_s. \tag{2.15}$$

As the value of $(C_{11} + C_{12})/C_{11}$ is about two for InGaAsP alloys, the mismatch strain along the growth direction ($\varepsilon_\perp$) is twice as much as ε.

The elastic strain modifies the energy difference between the conduction and valence bands and the band structures [46]. In the unstrained situation the heavy-hole (HH) and light-hole (LH) bands degenerate at the Γ point, as shown in Fig. 2.4b. When the tetragonal distortion is introduced into the lattice, the symmetry of the crystal is reduced and the degeneracy of the valence bands is lifted. The band structure modifications induced by the strain for bulk InGaAs layers are schematically shown in Fig. 2.4a and c. The energy shift at the Γ point induced by the strain can be calculated using the strain Hamiltonian proposed by Pikus and Bir [47]. The energy shifts can be decomposed into two components, δE_{Hy} and δE_{Sh}, induced by the hydrostatic strain ($\varepsilon_{xx} + \varepsilon_{yy} + \varepsilon_{zz}$) and the uniaxial strain ($\varepsilon_{zz} - \varepsilon_{xx}$), respectively. δE_{Hy} and δE_{Sh} can be written to first order in the strain as [48]

$$\delta E_{Hy} = -2a\frac{C_{11}-C_{12}}{C_{11}}\varepsilon, \tag{2.16}$$

$$\delta E_{Sh} = -2b\frac{C_{11}+2C_{12}}{C_{11}}\varepsilon, \tag{2.17}$$

where a is the hydrostatic deformation potential, and b is the shear deformation potential listed in Table 2.2. By using δE_{Hy} and δE_{Sh}, we can give the energy gaps between the conduction and valence bands by

$$E_{C-HH} = E_g + \delta E_{Hy} - \frac{\delta E_{Sh}}{2}, \tag{2.18}$$

$$E_{C-LH} = E_g + \delta E_{Hy} + \frac{\delta E_{Sh}}{2}, \tag{2.19}$$

where E_{C-HH} (E_{C-LH}) is the energy difference between the conduction and the heavy-hole (light-hole) valence bands, and E_g is the unstrained band-gap energy given by Eq. (2.3). As the shear deformation potential (b) is a negative quantity, the heavy-hole band is above the light-hole band in the compressively strained state, and the light-hole band is above the valence band in the tensilely strained state. Figure 2.5 shows the calculated band-gap wavelength [λ (μm) = 1.2407/E (eV)] for a bulk InGaAs layer using Eqs. (2.18) and (2.19). For InGaAsP with a band-gap wavelength longer than 2 μm, the

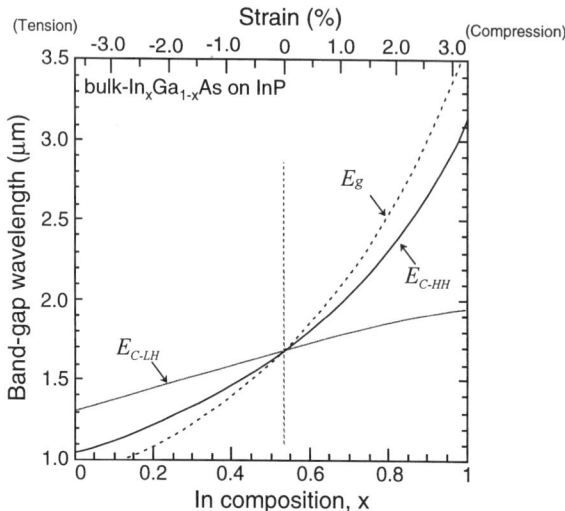

Fig. 2.5 Calculated band-gap wavelength for strained and unstrained $In_xGa_{1-x}As$ on InP as a function of In composition. The upper horizontal axis shows the mismatch strain of $In_xGa_{1-x}As$.

electron-hole recombination is dominated by transition from the conduction band to the heavy-hole band.

To calculate the emission wavelength of the strained QW structure, we need to include the quantum size effect on the band-gap energy in addition to the energy shift induced by the strain. The electron energy levels in a quantum-well structure can be determined to a good approximation by the effective mass equation [50]

$$\left(-\frac{\hbar^2}{2m^*(z)}\frac{\partial^2}{\partial z^2}+V_c\right)\chi_c(z)=E_c\chi_c(z), \quad (2.20)$$

where $m^*(z)$ is the electron effective mass, V_c the potential well depth, $\chi_c(z)$ the envelope function, and E_c the confinement energy of electron. Assuming a symmetric structure around the center of the well, we see that the solution wave function of Eq. (2.20) can only be even or odd. By using the conduction band offset value ΔE_c and the boundary conditions that $\chi_c(z)$ and $[1/m^*(z)][\partial\chi_c(z)/\partial z]$ should be continuous at both of the well-barrier interfaces, we can obtain the characteristic equations for the even and odd functions as follows:

$$\tan\left(k_z\frac{L_w}{2}\right)=\frac{m_w^*}{m_b^*}\frac{\alpha_z}{k_z} \quad \text{(for even functions)}, \quad (2.21)$$

DESIGN CONSIDERATION OF MQW ACTIVE REGION

$$\cot\left(k_z \frac{L_w}{2}\right) = -\frac{m_w^* \alpha_z}{m_b^* k_z} \quad \text{(for odd functions)}, \quad (2.22)$$

where L_w is the well thickness, m_w^* (m_b^*) the electron effective mass in the well (barrier), and

$$k_z = \sqrt{\frac{2m_w^*}{\hbar^2} E_c}, \quad \alpha_z = \sqrt{\frac{2m_b^*}{\hbar^2}(\Delta E_c - E_c)}. \quad (2.23)$$

Equation (2.21) can only be satisfied for a discrete set of energies $E_{c,n}$, where n (= 1, 3, 5, ...) denotes the number of quantized electron levels in the well, and Eq. (2.22) corresponds to the case of even-numbered electron levels.

The hole quantization problem is much more complicated, because the heavy-hole and light-hole bands are close to each other at the Γ point. A precise calculation of the valence band structure requires the Luttinger-Kohn Hamiltonian [51], which includes the interactions of each hole band. However, the effective mass equation also gives a good approximation of the quantized energy levels of the hole bands [52,53] because the effective mass in the growth direction is not affected by the strain [10,54]. The energy levels for heavy hole and light hole can be calculated by substituting the hole effective masses and the valence band offset value into Eqs. (2.21) to (2.23). As a result the band-gap energy of the strained QW structure is given by

$$E_n = E_g + \delta E_{Hy} - \frac{\delta E_{Sh}}{2} + E_{e,n} + E_{HH,n} \quad \text{(for heavy hole band)}, \quad (2.24)$$

$$E_n = E_g + \delta E_{Hy} + \frac{\delta E_{Sh}}{2} + E_{e,n} + E_{LH,n} \quad \text{(for light hole band)}. \quad (2.25)$$

Figure 2.6 shows the calculated band-gap wavelength as a function of well thickness for a compressively strained-InGaAs QW with $In_{0.53}Ga_{0.47}As$ barriers. Here, the band gap in the QW corresponds to the total transition energy for recombination from an $n = 1$ electron state to an $n = 1$ heavy hole state. The conduction offset value ΔE_c is assumed to be 0.4 ΔE_g [55,56], where ΔE_g is the band-gap energy difference between the well and the barrier layers. As shown in Fig. 2.6, a well strain larger than +1.5% is required to obtain a band-gap wavelength longer than 2 μm. As the shift in the band-gap wavelength becomes smaller with increasing well thickness, a large compressive strain in the well is needed to increase the band-gap wavelength. However, the increase of compressive strain in the well induces structural defects such as misfit dislocations and surface roughening [57,58]. Therefore the epitaxial growth must be rigorously controlled when fabricating 2-μm wavelength MQW lasers.

The band-gap wavelength of QW also depends on the barrier composition. Figure 2.7 shows the calculated band-gap wavelength as a function of well

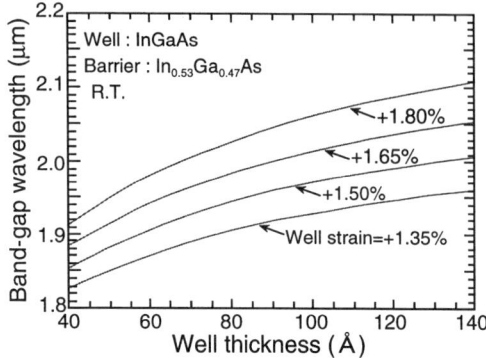

Fig. 2.6 Calculated band-gap wavelength as a function of well thickness for InGaAs QWs sandwiched between $In_{0.53}Ga_{0.47}As$ barriers with well strains of +1.35, +1.5, +1.65, and +1.8%.

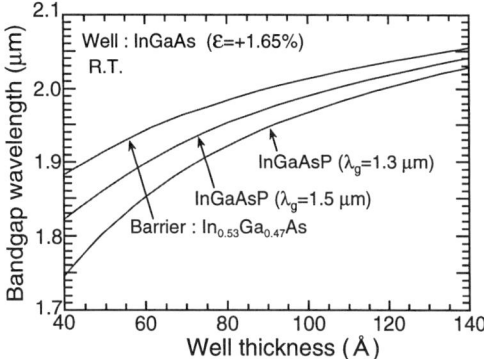

Fig. 2.7 Calculated band-gap wavelength as a function of well thickness for a +1.65%-strained InGaAs QW sandwiched by $In_{0.53}Ga_{0.47}As$ barriers and InGaAsP barriers with band-gap wavelengths of 1.3 and 1.5 μm.

thickness for a +1.65%-strained InGaAs QW sandwiched by $In_{0.53}Ga_{0.47}As$ barriers and InGaAsP barriers with band-gap wavelength of 1.3 μm and 1.5 μm. The $In_{0.53}Ga_{0.47}As$ barrier increases the band-gap wavelength of QW, but the energy barrier height of $In_{0.53}Ga_{0.47}As$ is too low to prevent electron overflow from the well. On the other hand, the InGaAsP barriers, which have a relatively large potential energy, can suppress the electron overflow. However, the large barrier height of InGaAsP induces the nonuniform carrier distribution in the MQW active region. The effect of the barrier height on the laser characteristics will be discussed in Section 2.3.3.

2.3.2 Critical Layer Thickness for Strained-Layer Heterostructures

In strained-layer heterostructures the strain energy associated with the mismatch strain can be accommodated by elastic distortion of the lattice until the epitaxial layer reaches a certain critical thickness. On the other hand, above the critical layer thickness, the epitaxial layer will be relaxed, generating misfit dislocations or changing the growth mode from two-dimensional (2D) to three-dimensional (3D). Various theoretical and experimental investigations have been carried out to determine the critical layer thickness. In this section we consider the relation between the mismatch strain and the critical layer thickness for InGaAs on InP.

Theories of the critical layer thickness are based on either the mechanical equilibrium model or the energy balance model. Matthews and Blakeslee [59,60] calculated the critical layer thicknesses for single-layer and multi-layer structures by considering the mechanical equilibrium between force F_ε exerted by the mismatch strain and tension F_l in a dislocation line. The calculated critical layer thickness for a single-layer structure is about half that for a multi-layer structure, and thus the critical layer thicknesses for both structures should be limited by the thickness of a single-layer structure. So, we explain this model for a single-layer structure. If force F_ε exceeds tension F_l, the migration of a threading dislocation results in the generation of a misfit dislocation. For the misfit 60° type of dislocation [61], critical layer thickness h_c for a single strained layer is given by

$$h_c = \frac{b}{4\pi\varepsilon} \frac{1-v/4}{1+v}\left(\ln\frac{h_c}{b}+1\right), \tag{2.26}$$

where ε is the mismatch strain defined in Eq. (2.11), $b\,(= a_e/\sqrt{2} \sim 0.4\,\text{nm})$ is the Burger's vector, and $v\,(= C_{12}/(C_{11} + C_{12}) \sim 0.33)$ is Poisson's ratio. The energy balance model was initially proposed by van der Merwe [62] and extended by People and Bean [63,64]. The critical layer thickness is obtained by the condition that the sum of the elastic strain areal energy density and the areal energy corresponding to a grid of misfit dislocation is minimized. For the misfit 60° type of dislocation, the critical layer thickness based on the energy balance model is given by [65]

$$h_c = \frac{b^2}{32\sqrt{3}\pi^2 a_e}\frac{2-v}{1+v}\frac{1}{\varepsilon^2}\left(\ln\frac{h_c}{b}+1.67\right). \tag{2.27}$$

Figure 2.8 shows the critical layer thicknesses for InGaAs on InP calculated using Eqs. (2.26) and (2.27). According to the mechanical equilibrium model, Eq. (2.26) gives critical layer thicknesses of about 8 nm for +1.5% strain and 7 nm for +1.7% strain. The critical layer thicknesses calculated using the mechanical equilibrium model and the energy balance model differ by as much as a factor of two. The experimentally determined critical layer thickness for

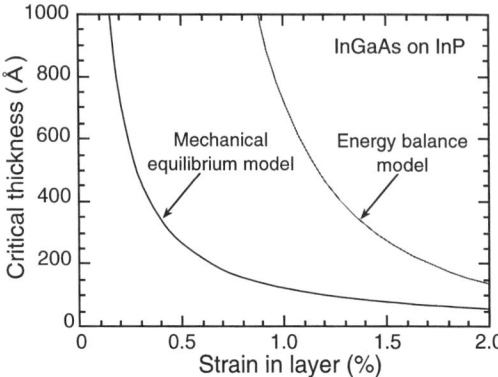

Fig. 2.8 Critical layer thicknesses of InGaAs on InP calculated using mechanical equilibrium model and energy balance model.

InGaAs on InP lies between the two curves [66]. Modified models in which the dependence of the growth temperature is considered have been proposed to explain the experimental results [65,67]. Kim et al. [65] introduced the thermal strain contributions to the energy balance model and obtained reasonable agreement between the experimentally determined critical layer thickness and the calculated one. Their expression for the critical layer thickness is approximately given by

$$h_c = \frac{b^2}{32\sqrt{3}\pi^2 a_e} \frac{2-v}{1+v} \times \frac{\ln(h_c/b)+1.67}{\varepsilon^2 + 2\varepsilon\alpha\Delta T}, \qquad (2.28)$$

where α is the thermal expansion coefficient (Table 2.1) and ΔT is the difference between the growth temperature and room temperature. Figure 2.9 shows the critical layer thickness for InGaAs on InP calculated using Eq. (2.28). The critical layer thickness increases with decreasing growth temperature. For example, the critical layer thickness of InGaAs with +1.7% strain is 12.5 nm at a growth temperature of 600°C and 14 nm at 500°C. This suggests that the epitaxial growth technique with a low growth temperature is effective in increasing the critical layer thickness [68].

The preceding discussion was based on the generation of misfit dislocations in a strained layer grown under the 2D growth mode. However, it has been demonstrated that large elastic stress in a strained layer induces 3D growth in the form of islands [69,70] and wavy structures [71,72]. The 3D growth can provide sources for stress-relieving dislocations. Several experimental studies have been carried out to determine the critical layer thickness in the case of 3D growth [57,73,74]. Gendry et al. [57] evaluated the critical layer thickness for InGaAs on InP dominated by 3D island dislocations using the reflection high-energy electron diffraction (RHEED) technique. Figure 2.10 shows the critical layer thickness for InGaAs on InP grown at 525°C. Curve h_{MD} is the

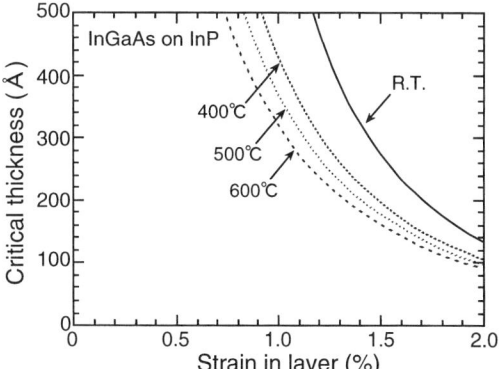

Fig. 2.9 Critical layer thickness of InGaAs on InP against different growth temperature calculated by Eq. (2.28).

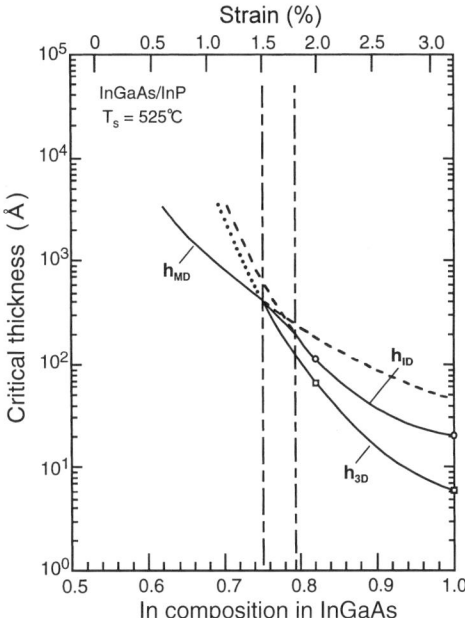

Fig. 2.10 Critical layer thickness of InGaAs on InP measured by RHEED as a function of In mole fraction: h_{MD} is the thickness above which misfit dislocations under the 2D growth mode are generated, h_{ID} is the thickness above which strain relaxation through 3D island dislocations occurs, and h_{MD} corresponds to onset of the 3D growth. The growth temperature is 525°C. (From Gendry et al. [57]. Reproduced by permission of the American Institute of Physics.)

critical layer thickness above which misfit dislocations under the 2D growth mode are generated, and h_{ID} is critical layer thickness above which strain relaxation through 3D island dislocations occurs. As shown in Fig. 2.10, 3D island dislocations are the dominant mechanism of strain relaxation for mismatch strain larger than +1.8%. This suggests that a well strain smaller than +1.8% and a growth temperature lower than 525°C are effective in suppressing 3D island growth of InGaAs.

2.3.3 Effects of Well Strain and Barrier Height on Lasing Characteristics

In this part, we consider the well-strain and barrier-height dependences of device characteristics for InP-based MQW lasers. We first discuss the device characteristics of 2-μm wavelength InP-based lasers derived from fundamental semiconductor laser theories. For semiconductor lasers the rate of increase of output power (P) to injection current (I), defined as the slope efficiency, is given by [75]

$$\frac{dP}{dI} = \eta_D \frac{\hbar\omega}{2e}, \qquad (2.29)$$

where η_D is the differential quantum efficiency of the laser, $\hbar\omega$ the photon energy, and e the electron charge. Assuming that the differential quantum efficiency and the threshold current of the 2-μm wavelength laser ($\hbar\omega = 0.620\,\text{eV}$) are equal to those of the 1.3-μm wavelength laser ($\hbar\omega = 0.954\,\text{eV}$), the output power of the former is 35% smaller than that of the latter under the same injection current. The differential quantum efficiency can be written as [76]

$$\eta_D^{-1} = \eta_i^{-1}\left[1 + \frac{\alpha_{int} L}{\ln(1/R)}\right], \qquad (2.30)$$

where η_i is the internal quantum efficiency, α_{int} the internal loss, L the cavity length, and R the mirror reflectivity. The total internal loss α_{int} can be described in terms of absorption from three separate factors such that [77]

$$\alpha_{int} = \Gamma\alpha_{active} + (1-\Gamma)\alpha_{cladding} + \alpha_{scattering}, \qquad (2.31)$$

where Γ is the optical confinement factor in the active region, α_{active} ($\alpha_{cladding}$) the absorption in the active (cladding) region, and $\alpha_{scattering}$ the scattering loss; α_{active} and $\alpha_{cladding}$ are mainly due to intervalence band absorption. The intervalence band absorption in InGaAsP alloy increases with increasing transmitting light wavelength [78,79]. Then the total internal loss increases with increasing emission wavelength. As a result the differential quantum efficiency of 2-μm wavelength InP-based lasers is smaller than that of lasers with 1.3- or 1.55-μm emission wavelength.

Next we consider the threshold current of the 2-μm wavelength lasers. At the threshold condition for lasing, the modal gain is equal to total internal loss α_{int} and the mirror loss ($= \ln(1/R)/L$). In this case the active region gain, which is a function of carrier density, is given by [75,80]

$$g(n_{th}) = \frac{\alpha_{int} + \ln(1/R)/L}{\Gamma}, \tag{2.32}$$

where n_{th} is the threshold carrier density. The threshold current density is the sum of the current densities related to radiative recombination and nonradiative recombination in the laser at threshold carrier density n_{th}. If an undoped active region and the absence of leakage currents are assumed, the threshold current density J_{th} is given by [80,81]

$$J_{th} = ed(An_{th} + Bn_{th}^2 + Cn_{th}^3), \tag{2.33}$$

where d is the total thickness of the wells, A the defect-mediated nonradiative recombination coefficient, B the radiative recombination coefficient, and C the Auger coefficient. The current density associated with nonradiative Auger recombination can be written as [75]

$$J_{Auger} = edCn^3 = edC'n^3 \exp\left(\frac{-E_a}{k_B T}\right), \tag{2.34}$$

where E_a is the activation energy, C' a constant associated with Auger recombination, and k_B the Boltzmann constant. In Auger recombination, an electron and hole recombine across the band gap and in the process excite a third carrier to higher energy instead of emitting a photon. Various processes can be considered in Auger recombination [82]. Among them, CHSH Auger recombination is a dominant process in InGaAs [83–85], where an electron in the conduction band and a heavy hole in the valence band recombine, exciting an electron in the split-off band into an empty heavy hole, as shown in Fig. 2.11. In the nondegenerate approximation with parabolic conduction and valence bands, the activation energy for the CHSH Auger recombination can be written as [82]

$$E_a = \frac{m_{so}}{2m_{hh} + m_c - m_{so}}(E_g - \Delta), \tag{2.35}$$

where m_{hh}, m_c, and m_{so} are the heavy-hole, conduction band electron, and spin-orbit band electron effective masses, respectively, and Δ is the spin-orbit splitting energy. For InGaAs, the term $(E_g - \Delta)$ in Eq. (2.35) decreases with increasing In mole fraction. The decrease of $(E_g - \Delta)$ results in the increase of the Auger coefficient. Therefore the Auger coefficient of an InGaAs well in

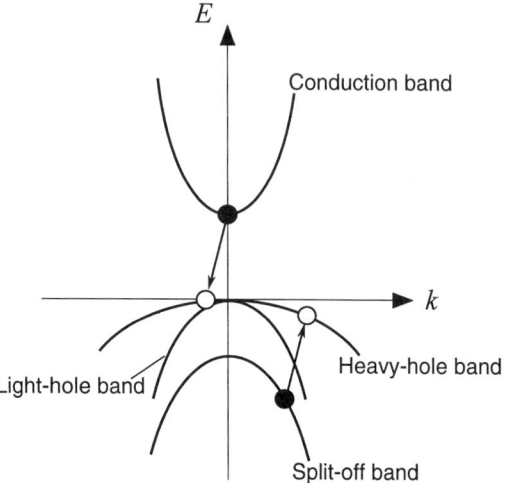

Fig. 2.11 Schematic diagram of the CHSH Auger recombination process.

the strained MQW active region increases with increasing compressive strain. On the other hand, theoretical predictions [10,11] and experimental results [12] have revealed that introducing compressive strain into InGaAs QW laser causes the threshold current density to decrease. Loehr et al. [86] calculated the current densities associated with radiative recombination and nonradiative Auger recombination for strained-InGaAs MQW lasers using an eight-band tight binding model. Figure 2.12 shows the radiative current density (J_{Rad}), Auger current density (J_{Auger}), and the threshold current density ($J_{th} = J_{Rad} + J_{Auger}$) calculated using that model. The threshold current density decreases as compressive strain in the well is increased up to about +1%. After the well strain reaches +1.5%, however, the threshold current density rapidly increases with increasing well strain. This trend has been experimentally confirmed in strained-InGaAs MQW lasers [20,87,88]. These results indicate that the optimization of strain is required for the InGaAs well in a strained-MQW laser in order to reduce the threshold current.

As described in Section 2.3.1, an InGaAs-QW laser requires well strain larger than +1.5% to obtain emission wavelength longer than 2 μm. However, introducing large compressive strain into the well layer results in an increase in threshold current density. In addition the 3D growth easily occurs in InGaAs layer with strain larger than +1.8%, as described in Section 2.3.2. Therefore an InGaAs well layer with compressive strain from 1.5% to 1.8% is considered to be suitable for the 2-μm wavelength MQW laser. In this study an InGaAs layer with compressive strain of 1.65% is chosen for the well layer.

The barrier height of the MQW laser has strong influence on emission wavelength as well as other device characteristics. A small barrier height is

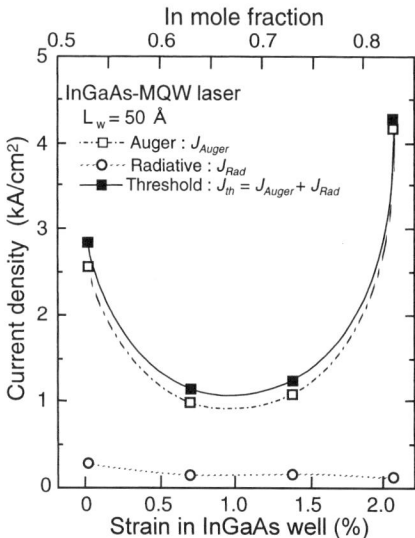

Fig. 2.12 Calculated radiative, Auger, and threshold current densities for strained-InGaAs MQW lasers as a function of strain in InGaAs well. (From Loehr et al. [86]. Reproduced by permission, copyright 1993, IEEE.)

effective in increasing the emission wavelength of a QW laser, but the potential energy is insufficient to prevent electron overflow from the wells. Figure 2.13a schematically shows the energy diagram of a single QW laser with small barrier height. The electron overflow results in decreases in the operating temperature and output power [14]. Increasing the number of wells can suppress electron overflow, since the subband against the injection current fills more slowly in MQWs than in SQWs [89]. However, doing so enhances the optical loss in the active region, which then results in high threshold current. Therefore the number of wells should be optimized to obtain a low threshold current.

In contrast, a large barrier height suppresses electron overflow from wells, but it decreases the emission wavelength of MQW laser. In this case, increasing the number of wells is effective in increasing the emission wavelength due to the suppression of subband filling against the injection current. However, an MQW laser with both a large barrier height and a large number of wells exhibits nonuniform carrier distribution between QWs. That is, low-mobility holes are captured in the wells closest to the p-cladding layer due to the large valence band discontinuity, and the electrons tend to follow the hole concentration profile. Figure 2.13b is a schematic energy diagram of an MQW laser with a large barrier height. Various theoretical and experimental investigations have been carried out to clarify the effects of the nonuniform carrier distribution on the performance of the 1.3- and 1.55-µm wavelength MQW lasers

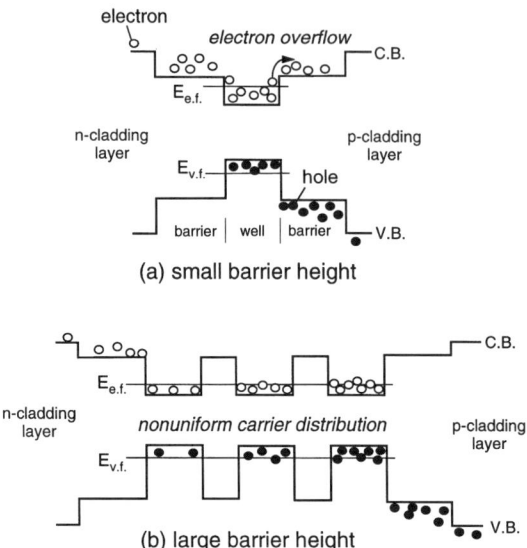

Fig. 2.13 Schematic diagrams of the energy bands and carrier distributions for laser structures with (*a*) small barrier height and small number of wells and (*b*) large barrier height and large number of wells.

[90–95]. For example, it has been reported for 1.3-μm wavelength MQW lasers that the threshold current increases [93] and maximum output power decreases [96] when the energy difference between the well and barrier layers is larger than 250 meV. From this result we can say that the InGaAsP barrier with band-gap wavelength shorter than 1.4 μm induces the nonuniform carrier distribution in 2-μm wavelength MQW lasers. A comparison with the experimental results will be presented in Section 2.5.1.

2.4 GROWTH AND CHARACTERIZATION OF STRAINED-InGaAs QUANTUM WELLS

At present, the methods commonly used to produce InP-based optoelectronic devices containing QW structures are metalorganic vapor phase epitaxy (MOVPE), gas source molecular beam epitaxy (GSMBE), and metalorganic molecular beam epitaxy (MOMBE) which is also known as chemical beam epitaxy (CBE) [97]. Among them, MOVPE is the most popular due to its adaptability to large-scale production systems. In MOVPE growth, gas source materials are thermally cracked on the heated substrate. The MOVPE growth temperature for InP-based material is higher than 600°C, since group-V hydrides such as arsine (AsH_3) and phosphine (PH_3) will not decompose at lower temperatures [36]. In contrast, hydrides are pre-cracked in GSMBE and

MOMBE systems before they reach the substrate, so the growth temperature of GSMBE and MOMBE for the growth of InP-based material is about 100°C lower than that of MOVPE [98,99]. As discussed in Section 2.3.2, epitaxial growth methods with low growth temperatures are favorable for preparing strained-layer heterostructures. MOMBE is an attractive growth method for strained QW structures, not only for its low growth temperature but also because the InGaAsP composition can be easily controlled [100–103]. In this section we describe the MOMBE growth of strained-InGaAs MQW structures for use as the active regions of 2-μm wavelength InP-based lasers.

MOMBE growth of strained-InGaAs QW was performed at substrate temperature of 510°C using In-free holders on InP (100) substrates [104]. The group-III sources were trimethyl-indium (TMIn) and triethyl-gallium (TEGa). Molecular beams of As_2 and P_2 were produced by thermal decomposition of AsH_3 and PH_3 in a low-pressure cell heated to 900°C. The sample structure used in this study was a separate-confinement-heterostructure (SCH) MQW, which consisted of strained MQW embedded in InGaAsP waveguide layers with band-gap wavelength of 1.3 μm. The misfit strains for the well and barrier layers were determined by fitting the double-crystal X-ray diffraction patterns of MQW with a kinematical step model [105].

2.4.1 InGaAs/InGaAs Multiple Quantum Wells

In an InGaAs/InGaAs MQW laser, laser performance can be improved by increasing the number of wells, as described in Section 2.3.3. Figure 2.14 shows the room-temperature photoluminescence (PL) spectra of +1.65%-strained InGaAs/InGaAs QWs with well number (N_w) of one, two, and four. The barrier composition is $In_{0.53}Ga_{0.47}As$ lattice-matched to InP. The well and barrier thicknesses are 11.5 and 18.5 nm, respectively. The single QW structure has a strong PL emission with peak wavelength of 2.04 μm. The PL peak wavelength agrees well with the calculation in Fig. 2.6. However, the PL peak intensity decreases with increasing well number. No PL emission is detectable for the four wells. This indicates that the elastic stress that accumulates as well number is increased generates structural defects that act as nonradiative recombination centers. Figure 2.15a shows the cross-sectional transmission electron microscope (TEM) photograph of the MQW with four wells. A threading dislocation parallel to the growth direction, which is the cause of the PL degradation, is introduced into the upper part of the MQW region and the upper InGaAsP waveguide layer. At the bottom of the threading dislocation, thickness undulations can be seen. Such a dislocation is related to the propagation of sessile-type edge dislocation due to island coalescence [106].

The thickness undulation can be suppressed by increasing the group-V elements on the growing surface [107,108], since doing so reduces the migration lengths of the group-III elements [109]. Figure 2.15b is a cross-sectional TEM photograph of the MQW with four wells, taken after the AsH_3 beam pressure for the barrier layer was changed from 1.2×10^{-4} to 3.8×10^{-4} Torr. The inter-

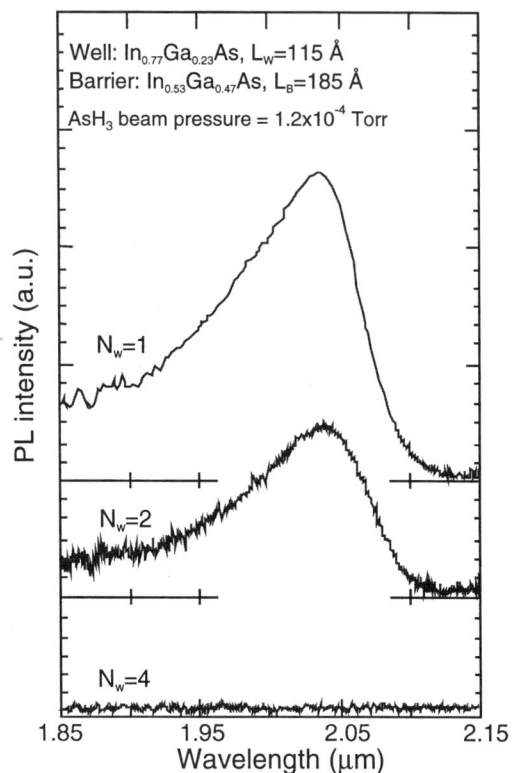

Fig. 2.14 Room-temperature PL spectra of $In_{0.77}Ga_{0.23}As/In_{0.53}Ga_{0.47}As$ QW structures with different numbers of wells. Both well and barrier layers were grown under AsH_3 beam pressure of 1.2×10^{-4} Torr.

faces between the well and barrier layers are flat and no dislocations are seen in the MQW region. Figure 2.16 plots the PL peak intensity as a function of the number of wells for the InGaAs/InGaAs MQW in the case of using AsH_3 beam pressures of 1.2, 2.6, and 3.8×10^{-4} Torr for barrier layer growth. The PL peak intensity increases with increasing well number and drastically drops for a certain number of wells. The number of wells to obtain the largest PL intensity increases with increasing AsH_3 beam pressure. Therefore, increasing the group-V elements on the growing surface is effective in increasing the number of wells [110]. A similar effect can be obtained by decreasing the growth temperature [107].

The strain-compensation technique [45] has been widely used to increase the number of wells in strained MQW lasers [68,88]. In a strain-compensated MQW structure, the barrier layers are strained in the opposite sense to the well layers to reduce the net strain. The net stain ε^* is given by

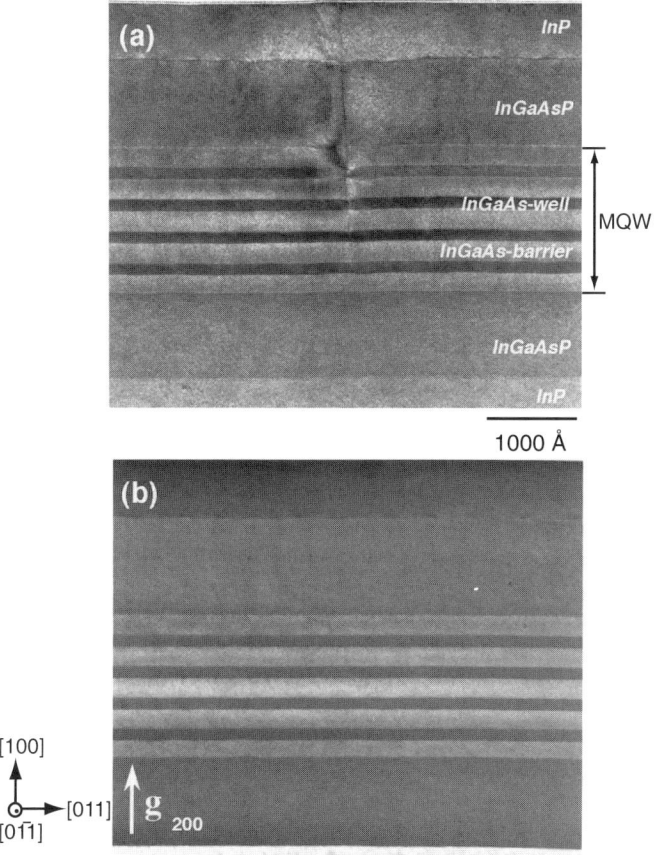

Fig. 2.15 Cross-sectional TEM photographs for $In_{0.77}Ga_{0.23}As/In_{0.53}Ga_{0.47}As$ MQWs with four wells. The barrier layers were grown under AsH_3 beam pressures of (*a*) 1.2×10^{-4} and (*b*) 3.8×10^{-4} Torr.

$$\varepsilon^* = \frac{\varepsilon_w L_w + \varepsilon_b L_b}{L_w + L_b}, \qquad (2.36)$$

where $\varepsilon_w, L_w, \varepsilon_b,$ and L_b are well strain, well thickness, barrier strain, and barrier thickness, respectively. On the other hand, the strain-compensation technique is not always effective in an MQW with well strain larger than +1% [111,112]. As discussed in Section 2.3.1, mismatch strain has to be larger than +1.5% in the well layer of an InP-based MQW laser to obtain emission wavelength longer than 2 μm. The net strain in MQW is altered by strain in the barrier layer. Therefore the barrier strain is an important factor to improve the crystalline quality of the 2-μm wavelength MQW lasers. Figure 2.17 shows the plot

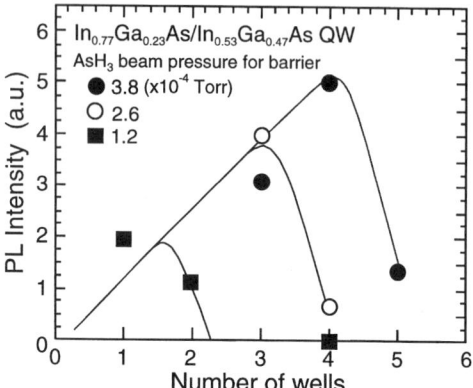

Fig. 2.16 Comparison of PL peak intensity and well number for $In_{0.77}Ga_{0.23}As/In_{0.53}Ga_{0.47}As$ QW structures. The barrier layers were grown under AsH_3 beam pressures of 1.2×10^{-4} (*closed squares*), 2.6×10^{-4} (*open circles*), and 3.8×10^{-4} Torr (*closed circles*).

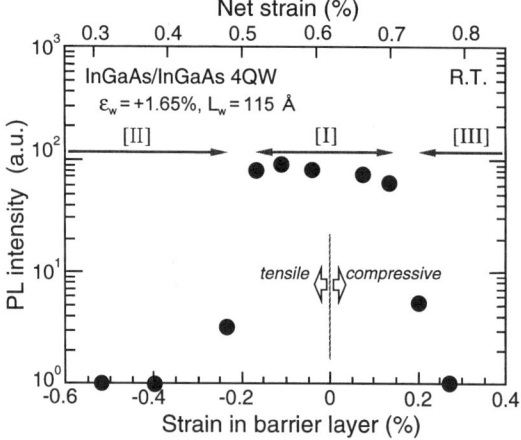

Fig. 2.17 Plot of the PL peak intensity of strained-InGaAs/InGaAs MQW versus the strain in the barrier layer. (From Mitsuhara et al. [114]. Reproduced by permission of Elsevier Science)

of PL peak intensity versus the barrier strain for a +1.65%-strained InGaAs/InGaAs MQW with four wells. The upper horizontal axis shows the net strain of the MQW. As shown in the figure, the PL peak intensity dependence on the barrier strain is divided into three regions: the region of barrier strain from −0.17% to +0.14%, where PL peak intensity is large (denoted as [I]); strain below −0.23%, where the intensity is small ([II]); and strain above +0.20%, where again the intensity is small ([III]). In region [I], which is close to the lattice-matching condition, all samples have PL peaks at 2.06 ± 0.01 μm and PL-FWHM values of only 35 meV. In regions [II] and [III], the PL peak intensity is one to two orders of magnitude smaller than that in region [I]. The decreases of the PL peak intensity in regions [II] and [III] indicate the generation of structural defects that act as nonradiative recombination centers.

Figure 2.18a to c shows cross-sectional TEM photographs of MQWs with barrier strains of +0.27%, −0.04%, and −0.52%, respectively. The sample with compressively +0.27%-strained barriers exhibits flat interfaces between the well and barrier layers (Fig. 2.18a). However, a pair of defects originating from the misfit dislocation [59,66], which are arranged along the (111) glide plane, can be seen in the top and bottom interfaces between the MQW and the InGaAsP waveguide layers. For the MQW with −0.04%-strained barriers in Fig. 2.18b, flat interfaces and no defects are seen. For the tensilely 0.52%-strained barriers in Fig. 2.18c, thickness undulations in the well and barrier layers are observed. In this sample the amplitude of the undulation is considerably enhanced as the number of wells is increased. As can be seen in Fig. 2.18c, the defects are generated at the points where the thickness undulations occur. This is because the thickness undulations increase the number of step edges on the growing surface and strain relief readily occurs at the step edge [113]. Consequently it is difficult to achieve strain-compensation structures for strained-InGaAs MQW containing highly compressive strained wells [114].

For InP-based lasers a buried heterostructure (BH) is generally used to obtain a low threshold current and lateral mode control. The thermal stability of strained MQW is an important issue from the viewpoint of high-temperature regrowth for BH laser fabrication [115,116]. We examined the thermal stability of strained MQWs grown by MOMBE at 510°C. Thermal annealing was carried out at 620°C for 2.5h under PH_3/H_2 atmosphere. Figure 2.19 shows the double-crystal X-ray diffraction patterns before and after annealing an $In_{0.77}Ga_{0.23}As/In_{0.54}Ga_{0.46}As$ MQW with four wells. The well and barrier strains are +1.65% and +0.04%, respectively. The X-ray diffraction patterns before and after annealing are the same. In addition it was confirmed that PL intensity was not deteriorated by the thermal annealing [4]. Such good thermal stability guarantees that no degradations occur in the structural and optical properties of the strained MQW even after the BH processes.

2.4.2 InGaAs/InGaAsP Multiple Quantum Wells

The critical layer thicknesses of strained-layer heterostructures can be increased by decreasing growth temperature as described in Section 2.3.2. However, the compositional range of the miscibility gap is enlarged with decreasing growth temperature as shown in Fig. 2.3. In addition it has been confirmed that thickness undulations accompanied by compositional modula-

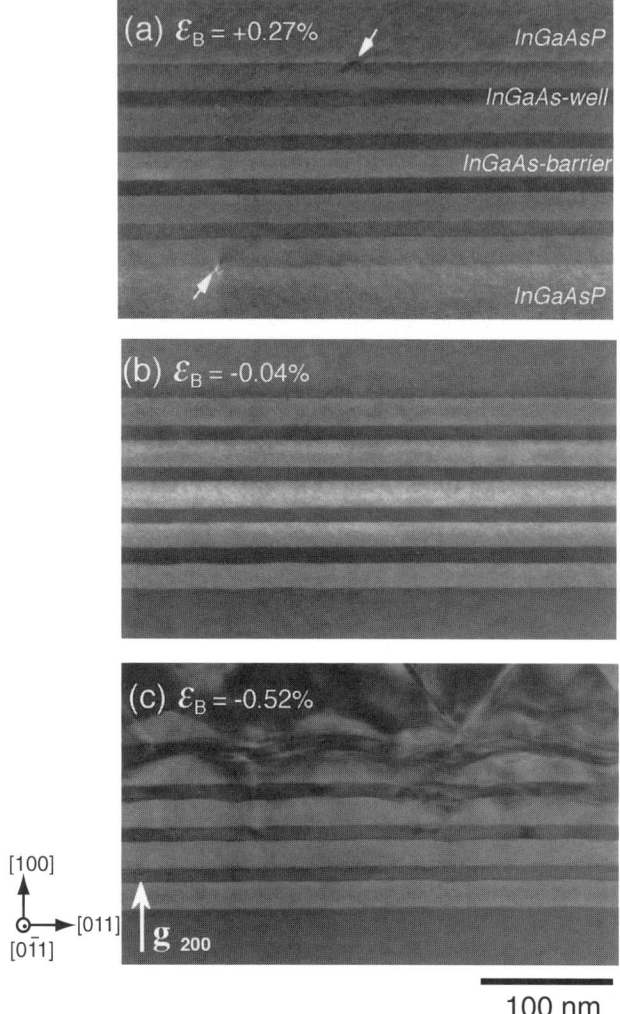

Fig. 2.18 Cross-sectional TEM photographs of the MQWs observed with diffraction vector $g = (200)$ for the strain in barrier layers of (a) +0.27%, (b) −0.04%, and (c) −0.52%. The arrows indicate a pair of dislocations at the interfaces between the MQW and the InGaAsP waveguide layers.

tions occur in strain-compensated MQW [31,32,43]. In this section we show how the crystalline qualities of strained-InGaAs MQW are affected by the compositional stability of the InGaAsP barrier.

To investigate the effect of the barrier composition on optical properties, strained-InGaAs MQWs with different InGaAsP barrier compositions were grown. The compositions of the barrier layers were $In_{0.53}Ga_{0.47}As$ and InGaAsP

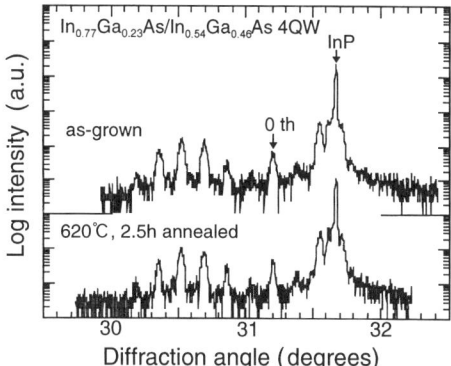

Fig. 2.19 Double-crystal X-ray diffraction patterns for the MQW (*a*) before and (*b*) after annealing. (From Mitsuhara et al. [4]. Reproduced by permission of the American Institute of Physics)

lattice-matched to InP with the band-gap wavelengths of 1.13 and 1.55 μm (denoted as 1.13Q and 1.55Q). All samples had four InGaAs wells with compressive strain of 1.65% and thickness of 11.5 nm. The barrier layer thickness was fixed at 19 nm. Figure 2.20 shows the room-temperature PL spectra of these samples. For the sample with 1.55Q barriers, the PL peak intensity is much smaller than that for the sample with 1.13Q barriers and $In_{0.53}Ga_{0.47}As$ barriers. As shown in Fig. 2.3, the 1.55Q barrier has composition in the miscibility gap at 500°C. This suggests that the decrease of PL intensity is related to the compositional modulations in the barrier layers.

As described in Section 2.2.2, the composition of the InGaAsP layer enters the miscibility gap by changing the sign of the strain from compression to tension. To investigate the relation between the PL intensity of the MQW and the compositional modulations in the barrier layers, we grew strained-InGaAs MQWs having InGaAsP barriers with different mismatch strains. The mismatch strain in the barrier layer was altered by changing the group-III compositions. The band-gap wavelength of the barrier layer was 1.5 μm on condition that the layer was lattice-matched to InP. The strain and thickness of the well layers were fixed at +1.65% and 10.2 nm, and the barrier thickness was 20.0 nm. Figure 2.21 plots the PL peak intensity versus the barrier strain for the MQW with four wells. Barrier strain from −0.02% to +0.21% is needed in order to obtain large PL intensity. The optimal range of the strain for the InGaAsP barrier shifts to the compressive side for InGaAs barrier shown in Fig. 2.17. This indicates that strong compositional modulations in InGaAsP barriers are the reason for the decrease in the PL intensity. The compositional modulations in an MQW can be confirmed by cross-sectional TEM observations with diffraction vector $g = (022)$, since the observation using $g = (022)$ is sensitive to the variation of the composition and the lattice spacing along the

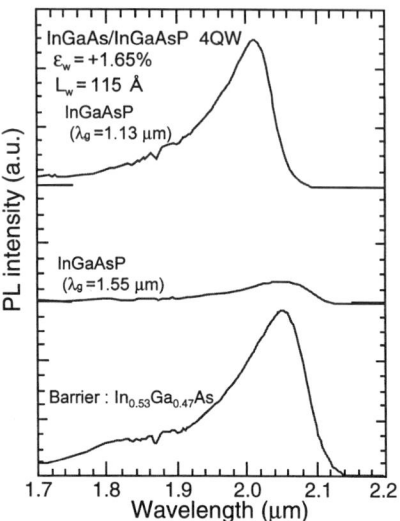

Fig. 2.20 Room-temperature PL spectra of strained-InGaAs MQWs with different compositions of barrier layers: (*a*) InGaAsP lattice-matched to InP with band-gap wavelength of 1.13 μm (1.13 Q), (*b*) InGaAsP lattice-matched to InP with band-gap wavelength of 1.55 μm (1.55 Q), and (*c*) $In_{0.53}Ga_{0.47}As$. The growth rates of 1.13 Q, 1.55 Q, and the $In_{0.53}Ga_{0.47}As$ layer were 0.43, 0.44, and 0.58 nm/s, respectively.

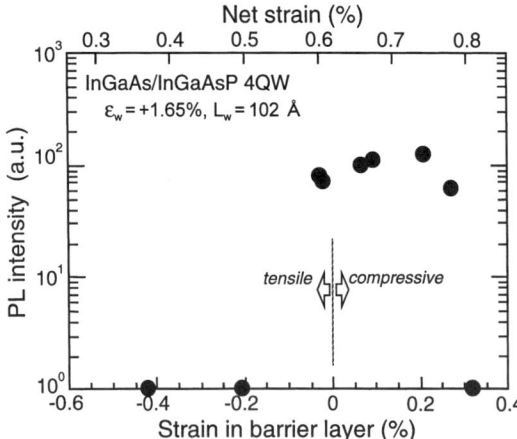

Fig. 2.21 Plot of the PL peak intensity of strained-InGaAs/InGaAsP MQW versus the strain in the barrier layer. The band-gap wavelength of the barrier layer was 1.5 μm on the condition that the layer was lattice-matched to InP.

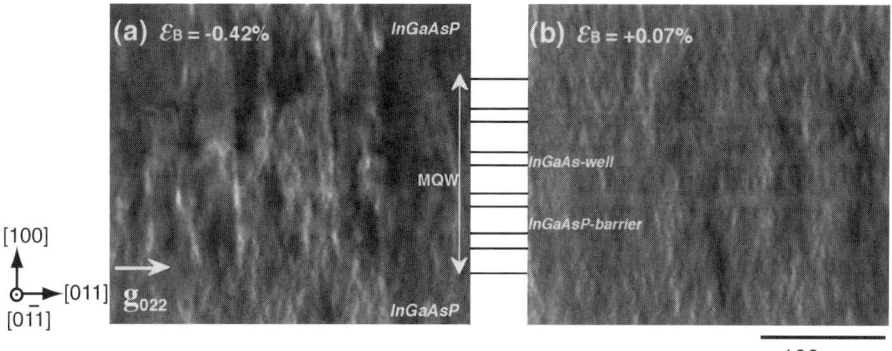

Fig. 2.22 Cross-sectional TEM photographs of strained–InGaAs/InGaAsP MQWs observed with diffraction vector of $g = (022)$ for the strain in the barrier layers of (a) –0.42% and (b) +0.07%.

(011) direction [30,40]. Figure 2.22a and b are cross-sectional TEM photographs of MQWs for barrier strains of –0.42% and +0.07%. In the sample with tensilely 0.42%-strained barriers in Fig. 2.22a, sharp variations in the contrast can be seen. This indicates that strong compositional modulations exist in the MQW region. In contrast, the variation in the contrast is weak for the sample with compressively 0.04% strained barriers in Fig. 2.22b. These results indicate that increasing tensile strain in the barrier layers enhances the compositional modulation in the MQW region. It has been reported that the compositional modulations induce the thickness undulations of the well and barrier layers [32,114,117]. Figures 2.23a and b are cross-sectional TEM photographs of the MQWs for barrier strains of –0.42% and +0.07% observed with diffraction vector $g = (200)$. In the case of the –0.42%-strained barrier in Fig. 2.23a, threading dislocations originating from the thickness undulations are observed. In contrast, flat interfaces and no defects are observed in the MQW with +0.07%-strained barriers in Fig. 2.23b. Consequently strain compensation is more difficult in strained-InGaAs MQW with InGaAsP barriers owing to the compositional modulations in the barrier layers.

The compositional modulations can be suppressed by reducing the migration length of group-III elements on the growing surface. The migration length of group-III element is reduced by increasing the growth rate [118]. Increasing the growth rate is also effective in suppressing 3D growth in highly strained MQW [119,120]. To investigate the effect of the growth rate on the optical property, we grew strained MQWs using InGaAsP barriers at different growth rates. The barrier layer for each sample has the same composition, namely an InGaAsP lattice-matched to InP with a band-gap wavelength of 1.55 μm. The barrier thickness is 19.5 nm. The strain and thickness for the well layers are fixed at +1.65% and 11.5 nm. Figure 2.24 plots the PL peak intensity versus the growth rate of the barrier layer for the InGaAs/InGaAsP MQW with four

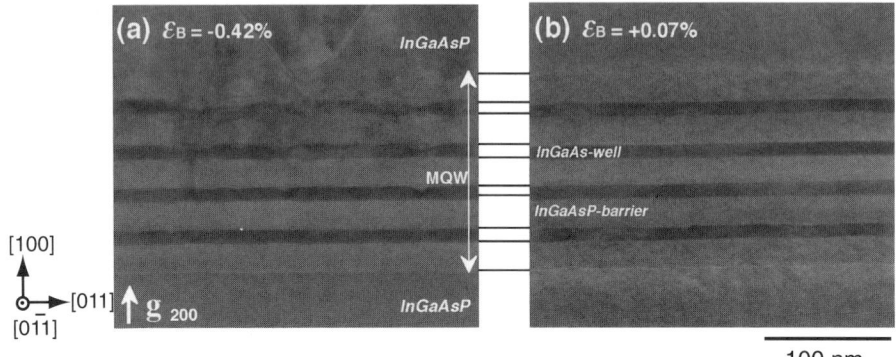

Fig. 2.23 Cross-sectional TEM photographs of strained-InGaAs/InGaAsP MQWs observed with diffraction vector of $g = (200)$ for the strain in the barrier layers of (*a*) −0.42% and (*b*) +0.07%.

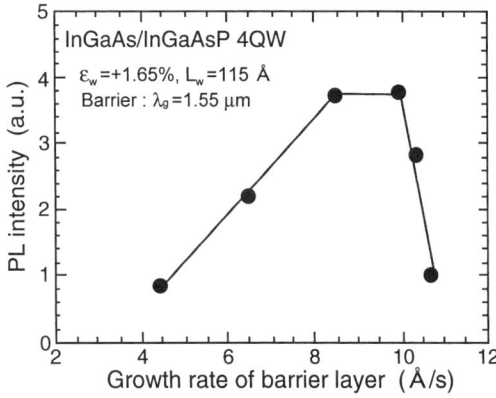

Fig. 2.24 Plot of the PL peak intensity of strained-InGaAs/InGaAsP MQW versus the growth rate of the barrier layer.

wells. The PL peak intensity increases with increasing the growth rate in the range from 0.45 to 0.85 nm/s, peaks in the range from 0.85 to 1.0 nm/s, and then sharply decreases at 1.05 nm/s. The cross-sectional TEM observation revealed that the thickness undulation in the MQW region was suppressed by increasing the growth rate of the barrier layer [121]. On the other hand, surface defects caused by the lack of group-V elements were observed on samples at the growth rate higher than 1.05 nm/s. Consequently the optimization of the growth rate is important in improving the crystalline quality of strained MQWs with InGaAsP barriers.

2.5 LASING CHARACTERISTICS OF 2-μm WAVELENGTH InGaAs-MQW LASERS

For MQW lasers there are a variety of the optimum device structures. Which device is chosen depends on the purpose of the laser, such as a low threshold current or a high maximum output power. For example, the threshold currents of laser having an MQW active region with a small number of wells [88,122] and high-reflectivity facet coatings [123,124] are low, but their output power is difficult to increase since the electron and hole subbands fill up rapidly against the injection currents. Coating facets with high-reflectivity films also suppresses output power [123]. Thus the device structure should be modified in response to the purpose of the laser.

In this section we start off by describing the characteristics of Fabry-Perot (F-P) lasers with emission wavelengths near 2 μm. We consider the effects of barrier composition and well number on a threshold current and emission wavelength. We also describe a high-power laser array using broad-area lasers. Then we present the characteristics of 2-μm wavelength distributed-feedback (DFB) lasers.

2.5.1 Fabry-Perot Lasers

As described in Section 2.3.3, the energy difference between the well and barrier layers influences the threshold current density of the 2-μm wavelength MQW laser. Figure 2.25 shows the threshold current densities of the ridge-waveguide lasers fabricated from MOMBE-grown MQWs with different barrier layer compositions. The barrier layers were $In_{0.53}Ga_{0.47}As$ and InGaAsP with band-gap wavelengths of 1.3, 1.2, and 1.13 μm (denoted as 1.3, 1.2, and 1.13 Q, respectively). The well layer was InGaAs with strain of $+1.68 \pm 0.05\%$ and thickness of 10.5 ± 1.0 nm, and the number of wells is four. The sample structure was an SCH-MQW consisting of a MQW embedded by InGaAsP waveguide layers. As shown in Fig. 2.25, for the MQW lasers using InGaAsP barriers with band-gap wavelength shorter than 1.3 μm, the threshold current densities increase significantly with increasing band-gap energy of the barrier layer. This indicates that nonuniform carrier distributions between QWs occur, as predicted in Section 2.3.3. On the other hand, the threshold current density of the laser with $In_{0.53}Ga_{0.47}As$ barriers is about 1.5 times higher than that of the laser with 1.3 Q barriers. This is because the optical loss in the active region, which originates from the intervalence band absorption, increases with decreasing band-gap energy of barrier layer [93].

From the results mentioned above, a lower threshold current is expected if the band-gap energy of the barrier layer is chosen so as to avoid both nonuniform carrier distributions and the increase of internal loss. Oishi et al. [19] reported threshold current densities as low as $0.6 kA/cm^2$ for buried-heterostructure InGaAs/InGaAsP MQW lasers with four wells where the band-gap wavelength of the barrier layer was 1.5 μm. In addition, reducing

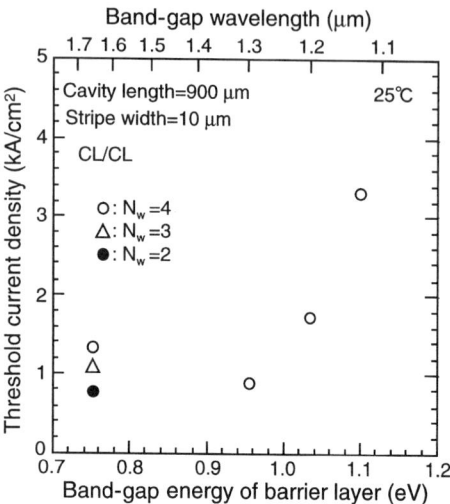

Fig. 2.25 Variation of threshold current density of the 2-μm wavelength MQW laser as a function of band-gap energy of barrier layer. The strain for InGaAs well is +1.68 ± 0.05%, and the thickness is 10.5 ± 1.0 nm.

the number of wells results in low-threshold current density, since the internal loss in the active regions decreases with decreasing well number. As can be seen in Fig. 2.25, the threshold current density for MQW lasers with $In_{0.53}Ga_{0.47}As$ barriers is reduced as the number of wells decreases from four to two. Using double quantum well (DQW) structures as active regions, Dong et al. [20] obtained threshold currents as low as 10 mA for buried-heterostructure lasers.

The emission wavelength is the most important issue for 2-μm wavelength InP-based lasers. InGaAs barriers with small potential energy to the wells can extend the wavelength range of laser. Figure 2.26 shows the peak emission wavelength as a function of cavity length for InGaAs/InGaAs DQW lasers. The active region consisted of DQW (two 11.5-nm-thick InGaAs wells with 1.65% compressive strain and three 20-nm-thick $In_{0.53}Ga_{0.47}As$ barriers) sandwiched by InGaAsP ($\lambda_g = 1.3$ μm) waveguide layers. The PL peak wavelength of this wafer was 2.04 μm. The emission wavelength shifts from 1.97 to 2 μm when the cavity length is increased from 600 to 900 μm, but remains constant at 2 μm for cavity length between 900 and 1500 μm. Because the subband filling against the injected currents rapidly occurs in an MQW laser with a small number of wells [89], the emission wavelength strongly depends on the cavity length of the laser and is shorter than the PL peak wavelength by 40 nm. Increasing the number of wells effectively increases the emission wavelength. Figure 2.27 shows the emission spectra of InGaAs/InGaAs MQW lasers with different numbers of wells [21]. Ridge-waveguide lasers with stripe width of 6 μm were used for two and three wells, and a buried heterostructure laser with stripe width of 1.5 μm was used for four wells. The peak emission wave-

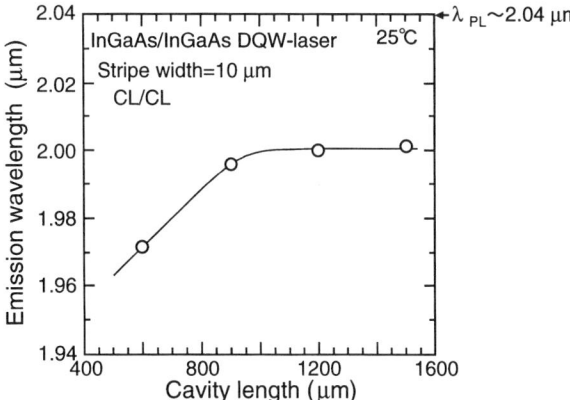

Fig. 2.26 Variation of emission peak wavelength of an InGaAs/InGaAs DQW laser as a function of cavity length. The strain and thickness of the InGaAs well are +1.65% and 11.5 nm, respectively.

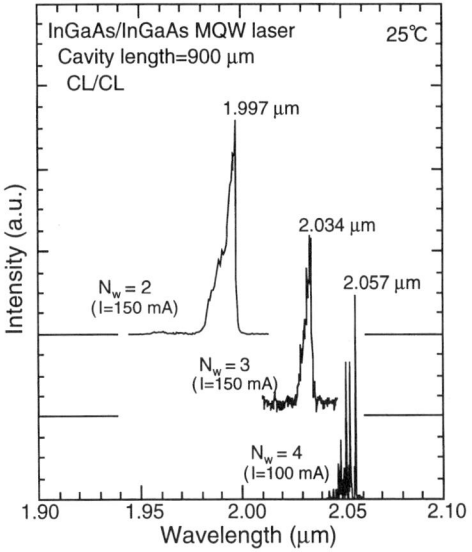

Fig. 2.27 Emission spectra of InGaAs/InGaAs MQW lasers with 2, 3, and 4 wells under continuous operation at 25°C. (From Mitsuhara et al. [21]. Reproduced by permission, copyright 1999, IEEE)

lengths of the lasers with two, three, and four wells are 1.997, 2.034, and 2.057 μm, respectively. This indicates that increasing the number of wells suppresses the subband filling effect. Increasing the injection currents and operating temperature can further increase the emission wavelength [125]. Figure 2.28 shows changes in the peak wavelength of the laser with four wells against

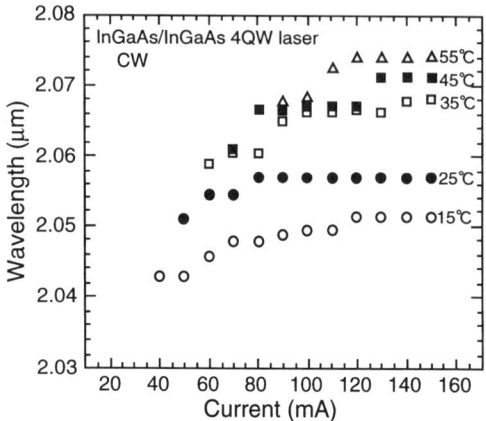

Fig. 2.28 Peak emission wavelength of the MQW laser with four wells as a function of injection current for operating temperatures from 15 to 55°C. (From Mitsuhara et al. [4]. Reproduced by permission of the American Institute of Physics)

the injection current at heat sink temperatures of 15 to 55°C [4]. The peak wavelength shifts to a longer wavelength with increasing injection current and heat sink temperature, and reaches 2.074 μm at 120 to 140 mA of injection current at 55°C.

High-power lasers are desirable as the light sources for laser-based spectroscopy and the optical pumping source of solid-state lasers. Major et al. [16] have demonstrated high-power operation of a monolithic laser array with the emission wavelength of about 1.94 μm. The laser structure was grown by MOVPE and consisted of a 3-μm-thick n-InP cladding layer, a 7.5-nm-thick $In_{0.75}Ga_{0.25}As$ DQW active region sandwiched by InGaAsP ($\lambda_g = 1.3$ μm) waveguide layers, a 1.8-μm-thick p-InP cladding layer, and a 0.2-μm-thick p-InGaAsP ($\lambda_g = 1.55$ μm) contact layer [126]. The laser bar was composed of 12 broad-area lasers with stripe width of 200 μm spaced on 800-μm-wide centers. The laser array had a cavity length of 1 mm and was fabricated with low- and high-reflective coatings on the front and back facet, respectively. For a constituent broad-area laser, the threshold current was 0.5 A with a differential quantum efficiency of 43% at 10°C. Figure 2.29 shows the cw light output power versus injection current (L-I) characteristics for a monolithic array at heatsink temperatures of −10 and 10°C. The maximum output power is 8.5 and 5.5 W at −10 and 10°C, respectively. Figure 2.30 shows the results of reliability testing for the constituent broad-area lasers [16]. The lasers were aged at constant output powers of 0.5 or 0.8 W at 20°C. The lasers operating at 0.5 W did not fail. Even at 0.8 W, three of the four lasers did not fail. The reliability of 2-μm wavelength InP-based lasers is expected to be good because InP-based materials do not suffer catastrophic optical damage [127].

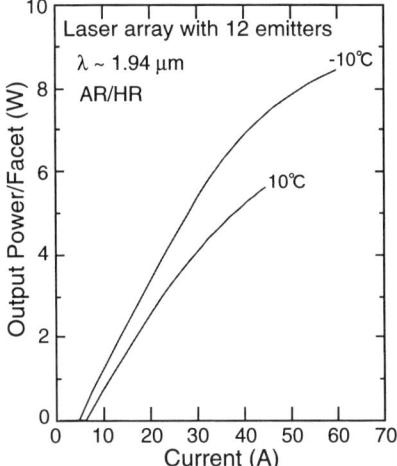

Fig. 2.29 Continuous-wave output power versus injection current characteristics of monolithic array of 12 InGaAs/InGaAsP DQW lasers. The stripe width and cavity length of constituent broad-area laser are 200 μm and 1 mm, respectively. (From Major et al. [16]. Reproduced by permission of IEE Publishing)

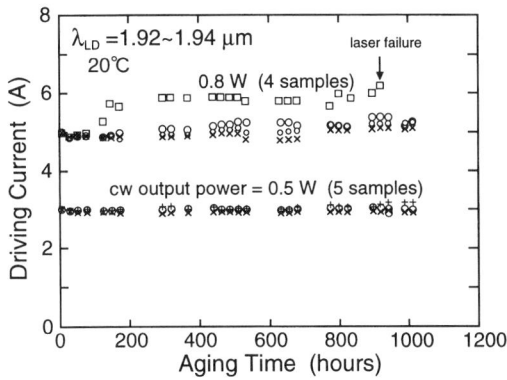

Fig. 2.30 Aging results of broad-area lasers with stripe width of 200 μm and cavity length of 1 mm at operating temperature of 20°C. (From Major et al. [16]. Reproduced by permission of IEE Publishing)

2.5.2 Distributed-Feedback Lasers

InGaAsP/InP distributed-feedback (DFB) lasers have been extensively developed with wavelengths of 1.3 and 1.55 μm for optic fiber communications over the last two decades. Similar processing techniques can be used for fabrication of single-mode lasers operating in the 2-μm wavelength range. In 1994 Martinelli et al. [17] demonstrated the first successful fabrication of a 2-μm

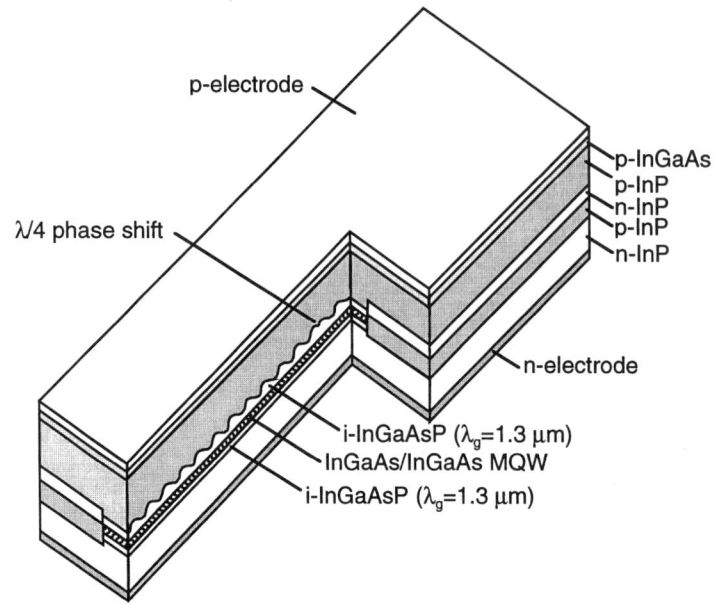

Fig. 2.31 Schematic structure of the 2-μm wavelength DFB laser. The quarter-wavelength phase shift is placed at the center of the cavity.

wavelength DFB laser. The active region of that laser consisted of two 12-nm-thick $In_{0.75}Ga_{0.25}As$ QWs separated by a 20-nm-thick InGaAsP (λ_g = 1.3μm) barrier. The emission wavelength of the DFB laser was 1.95μm at 273 K, and the maximum output power was about 0.1 mW. Recent progress in epitaxial growth techniques has made it possible to increase the emission wavelength and the output power of 2-μm wavelength DFB lasers [19,20,21,128,129].

Mitsuhara et al. [21] demonstrated DFB lasers with emission wavelength longer than 2.05μm and output power higher than 10 mW. The laser structure was a SCH-MQW grown by MOMBE at 510°C, and consisted of a MQW sandwiched between 50-nm-thick (lower) and 100-nm-thick (upper) InGaAsP (λ_g = 1.3μm) waveguide layers. The MQW active region consisted of four 11.5-nm-thick $In_{0.77}Ga_{0.23}As$ wells and five 18.5-nm-thick $In_{0.53}Ga_{0.47}As$ barriers, and the PL peak wavelength was 2.05μm at 25°C. The schematic structure of the DFB laser is shown in Fig. 2.31. To ensure stable single-longitudinal-mode operation, the quarter-wavelength (λ/4) phase shift was placed at the center of the cavity [130,131]. First-order corrugation patterns were formed on the upper InGaAsP waveguide layer with electron beam lithography and wet etching. This was followed by a three-step regrowth process using MOVPE at 600°C. First, a p-InP cladding layer was grown on the corrugated surface. Then, to stabilize the lateral mode and to reduce the threshold current, a 1.5μm-

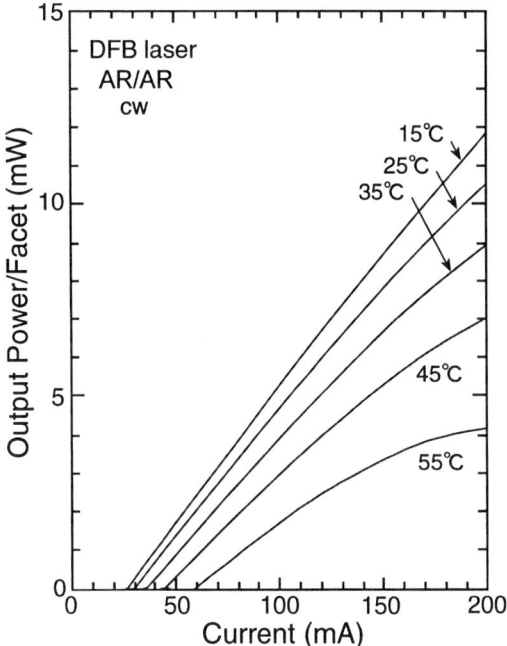

Fig. 2.32 Continuous-wave output power versus current characteristics of the DFB laser taken for heat sink temperatures between 15 to 55°C.

wide active region was buried by InP current blocking layers. After removing the mesa-defining SiO_2 mask, a p-InP cladding layer and a p-InGaAs contact layer were grown. After contact metallization the wafer was cleaved into bars with a 900-μm-long cavity. Both facets of the laser bar were coated with SiN antireflection (AR) films and separated into individual devices.

Figure 2.32 shows the cw light output power versus current characteristics for the DFB laser at heatsink temperatures of 15 to 55°C. The threshold current density is 26 mA at 25°C. The threshold characteristic temperature (T_0) is measured as 50 K, which is comparable to that of the 1.55-μm wavelength DFB laser [132]. The cw output power at an injection current of 200 mA is 10.5 mW at 25°C and 4.1 mW at 55°C. The output power of this laser makes it suitable for the light source in trace-gas monitoring [133].

Figure 2.33 shows the variation of the slope efficiency with heatsink temperature for the DFB laser at an output power of 3 mW/facet. The slope efficiency for both facets slightly decreases from 0.14 to 0.11 mW/mA as the temperature increases from 15 to 45°C, while it drastically decreases to 0.08 mW/mA at 55°C. For the 1.55-μm wavelength DFB lasers, more stable slope efficiencies were obtained even at 60°C [134]. The decrease of the slope efficiency at high temperature is significant for the 2-μm wavelength laser,

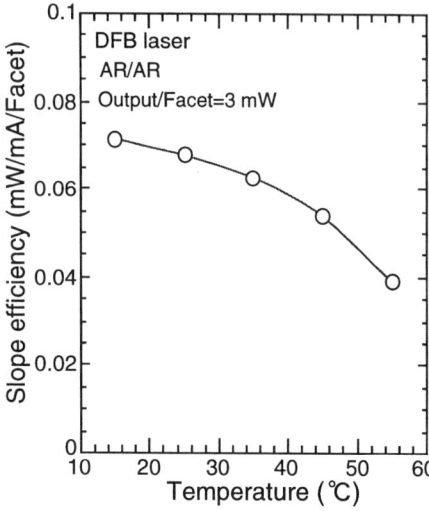

Fig. 2.33 Temperature dependence of slope efficiency for the DFB laser.

since the Auger recombination rates increase with increasing band-gap wavelength of the active region and operating temperature [75].

Figure 2.34 shows the emission spectrum of the DFB laser obtained under an injection current of 50 mA at 25°C. The emission wavelength of the DFB mode is 2.052 μm. The side-mode suppression ratio (SMSR) is as high as 32 dB, which is limited by the dark current of the InGaAs photodetector with cutoff wavelength of 2.55 μm (Hamamatsu Photonics, G5853-21). Figure 2.35 shows the emission wavelengths of DFB and Fabry-Perot lasers as a function of heatsink temperature for a constant injection current of 100 mA. The emission wavelength of the DFB laser linearly changes from 2.051 to 2.056 μm with increasing heatsink temperature. The temperature dependence of emission wavelength is +0.125 nm/K for the DFB laser and +0.48 nm/K for the F-P laser. This indicates that the emission wavelength of the DFB laser is mainly determined by the refractive index, not the gain peak shift due to heating [135]. The emission wavelength of a DFB laser can be expressed as $\lambda_B = 2n_{eff} \Lambda$, where λ_B the Bragg wavelength, n_{eff} the equivalent refractive index, and Λ the corrugation periodicity of the grating. A longer emission wavelength is expected by detuning the Bragg wavelength to the long-wavelength side of the material gain peak (~2.05 μm at 25°C).

Figure 2.36 shows the relationship between the emission wavelength of the DFB laser and the injection current at heat sink temperatures of 15 to 55°C. The current-tuning rate of the wavelength is constant at +0.0025 nm/mA in the temperature range from 15 to 45°C, and the rate at 55°C is also constant at −0.002 nm/mA. The emission wavelength of the DFB laser is well controlled by varying the injection currents.

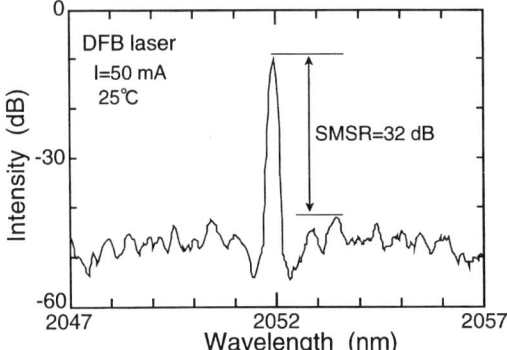

Fig. 2.34 Longitudinal-mode spectrum of the DFB laser at an injection current of 50 mA and 25°C. (From Mitsuhara et al. [21]. Reproduced by permission, copyright 1999, IEEE)

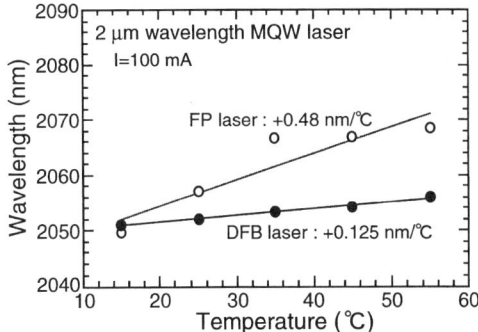

Fig. 2.35 Temperature-tuning characteristics of wavelength for the DFB and the Fabry-Perot lasers under an injection current of 100 mA. (From Mitsuhara et al. [21]. Reproduced by permission, copyright 1999, IEEE)

Single-mode emission with a narrow linewidth (<100 MHz) and continuous tuning of the emission wavelength are required for the light source in trace-gas monitoring. The 2-μm wavelength DFB laser presented in this section satisfies these requirements. Consequently the present data confirm that the 2-μm wavelength DFB laser employing strained-InGaAs MQW is promising candidate for the light source for trace-gas monitoring.

2.6 CONCLUSIONS AND FUTURE PROSPECTS

For InP-based MQW lasers, a well strain larger than +1.5% is indispensable in achieving an emission wavelength longer than 2 μm. Strain-compensation

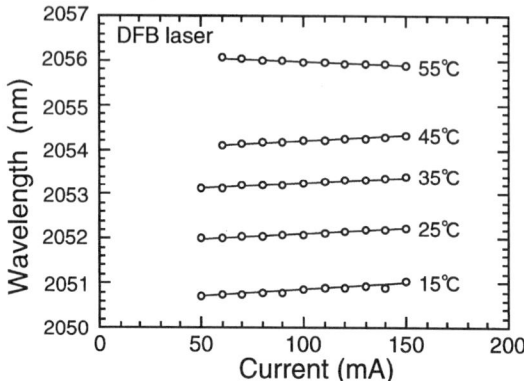

Fig. 2.36 Dependence of a DFB mode wavelength on injection current under heat sink temperature between 15 to 55°C. (From Mitsuhara et al. [21]. Reproduced by permission, copyright 1999, IEEE)

techniques have been widely used for 1.3- and 1.55-μm wavelength MQW lasers to suppress the generations of defects originating from the strain relaxation. However, strain compensation is more difficult in the 2-μm wavelength MQW lasers, since the dominant mechanism of strain relaxation is related to the three-dimensional growth. Not only the well strain but also the number of wells and the composition of the barrier affect the generation of structural defects in the MQW. The structural and optical properties of the MQW are greatly improved by optimizing the barrier layer growth conditions, and the number of wells can be increased without any crystalline degradation.

The fabrication processes for 2-μm wavelength InP-based laser are similar to those for the telecommunication lasers. For low-threshold operation of 2-μm wavelength MQW lasers, the band-gap energy of the barrier layer should be chosen so that nonuniform carrier distribution in the quantum wells will not occur. Decreasing the number of wells is also effective in reducing the threshold current density, but doing so decreases the emission wavelength of laser due to the large subband filling against the injection current. A low threshold current, high output power, and emission wavelength longer than 2.07 μm have been achieved for the 2-μm wavelength InP-based lasers by optimizing the device structures in response to each intended purpose of the laser. In addition the DFB lasers have demonstrated single-mode operation with emission wavelength longer than 2.05 μm and output power higher than 10 mW. These laser performances confirm that InP-based MQW lasers are practical light sources in the wavelength range near 2 μm.

The mismatch strain in the InGaAs layer larger than +1.8% easily induces three-dimensional growth. Therefore emission wavelength longer than 2.2 μm will be impossible in strained-InGaAs MQW lasers. The emission wavelength of strained-InGaAs MQW lasers could be increased by adding nitrogen (N)

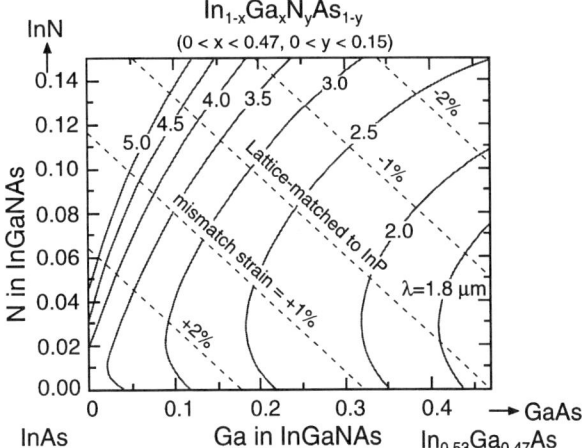

Fig. 2.37 Calculated band-gap wavelength for $In_{1-x}Ga_xN_yAs_{1-y}$ on InP as a function of Ga and N mole fractions at room temperature. (From Gokhale et al. [136]. Reproduced by permission of the author and the American Institute of Physics)

to the InGaAs wells, since large band-gap bowing occurs in the N-containing III–V materials. Figure 2.37 shows the calculated band-gap wavelength of InGaNAs as a function of Ga and N mole fractions [136]. On condition that the mismatch strain of the InGaNAs layer is constant, incorporating a small amount of nitrogen results in a large increase of the band-gap wavelength. Several experimental results have demonstrated that introducing nitrogen into InP-based MQWs results in a longer band-gap wavelength [136,137]. By using the InGaAsN wells, the InP-based MQW lasers are expected to become promising light sources in the wavelength range up to 3 µm.

ACKNOWLEDGMENTS

We are indebted to H. Sugiura, M. Ogasawara, K. Kasaya, and M. Nakao for technical discussions and cooperation, and to Y. Hirano, J. Satoi, J. Asaoka, and K. Yamaguchi for their assistance in device processing. We gratefully acknowledge Y. Kadota, M. Yamamoto, and H. Iwamura for their continual encouragement. We also thank the referenced authors and journals for the permission to include their work and figures in this chapter.

REFERENCES

1. A. Valster, C. J. van der Poel, M. N. Finke, and M. J. B. Boermans, "Effect of strain on the threshold current of GaInP/AlGaInP quantum well lasers emitting at

633 nm," in *13th Int. Semiconductor Laser Conference Digest*, Takamatsu, Japan, 1992, p. 152.
2. T. Kitatani, K. Nakahara, M. Kondow, K. Uomi, and T. Tanaka, "1.3-µm GaInNAs/GaAs single-quantum-well laser diode with a high characteristic temperature over 200 K," *Jpn. J. Appl. Phys.* **39**, L86 (2000).
3. J. J. Hsieh, "Room-temperature operation of GaInAsP/InP double-heterostructure diode lasers emitting at 1.1 µm," *Appl. Phys. Lett.* **28**, 283 (1978).
4. M. Mitsuhara, M. Ogasawara, M. Oishi, and H. Sugiura, "Metalorganic molecular-beam-epitaxy-grown $In_{0.77}Ga_{0.23}As$/InGaAs multiple quantum well lasers emitting at 2.07 µm wavelength," *Appl. Phys. Lett.* **72**, 3106 (1998).
5. H. K. Choi, G. W. Turner, and S. J. Eglash, "High-power GaInAsSb-AlGaAsSb multiple-quantum-well diode lasers emitting at 1.9 µm," *IEEE Photon. Technol. Lett.* **6**, 7 (1994).
6. C. L. Felix, J. R. Meyer, Vurgaftman, C. H. Lin, S. J. Murry, D. Zhang, and S. S. Pei, "High-temperature 4.5-µm type-II quantum-well laser with Auger suppression," *IEEE Photon. Technol. Lett.* **6**, 734 (1997).
7. J. Faist, F. Capasso, D. L. Sivco, A. L. Hutchinson, S. G. Chu, and A. Y. Cho, "Short wavelength ($\lambda = 3.4$ µm) quantum cascade laser based on strained compensated InGaAs/AlInAs," *Appl. Phys. Lett.* **72**, 680 (1998).
8. J. Faist, C. Sirtoni, F. Capasso, D. L. Sivco, J. N. Baillargeon, A. L. Hutchinson, and A. Y. Cho, "High-power long-wavelength ($\lambda = 11.5$ µm) quantum cascade lasers operating above room temperature," *IEEE Photon. Technol. Lett.* **10**, 1100 (1998).
9. R. U. Martinelli, T. J. Zamerowski, and P. A. Longeway, "InGaAs/InAsP lasers with output wavelengths of 1.58–2.45 µm," *Appl. Phys. Lett.* **54**, 277 (1989).
10. A. R. Adams, "Band-structure engineering for low-threshold high-efficiency semiconductor lasers," *Electron. Lett.* **27**, 249 (1986).
11. E. Yablonovitch and E. O. Kane, "Reduction of lasing threshold current density by the lowering of valence band effective mass," *IEEE J. Lightwave Technol.* **LT-4**, 504 (1986).
12. P. J. A. Thijs and T. V. Dongen, "High quantum efficiency, high power, modulation doped GaInAs strained-layer quantum well laser diodes emitting at 1.5 µm," *Electron. Lett.* **25**, 1735 (1989).
13. D. P. Bour, R. U. Martinelli, R. E. Enstrom, T. R. Stewart, N. G. DiGiuseppe, F. Z. Hawrylo, and D. B. Cooper, "$1.5 < \lambda < 1.7$ µm strained multiquantum well InGaAs/InGaAsP diode lasers," *Electron. Lett.* **28**, 37 (1992).
14. S. Forouhar, A. Ksendzov, A. Larsson, and H. Temkin, "InGaAs/InGaAsP/InP strained-layer quantum well lasers at approximately 2 µm," *Electron. Lett.* **28**, 1431 (1992).
15. M. Davies, M. Dion, D. C. Houghton, J. Z. Sedivy, and C. M. Vigneron, "Long-wavelength high-efficiency low-threshold InGaAsP/InP MQW lasers with compressive strain," *Electron. Lett.* **28**, 2004 (1992).
16. J. S. Major Jr. J. S. Osinski, and D. F. Welch, "8.5 W CW 2.0 µm InGaAsP laser diodes," *Electron. Lett.* **29**, 2112 (1993).
17. R. U. Martinelli, R. J. Menna, G. H. Olsen, and J. S. Vermaak, "1.95-µm strained InGaAs-InGaAsP-InP distributed-feedback quantum-well lasers," *IEEE Photon. Technol. Lett.* **6**, 1415 (1994).

18. M. Ochiai, H. Temkin, S. Forouhar, and R. A. Logan, "InGaAs-InGaAsP buried heterostructure lasers operating at 2.0 µm," *IEEE Photon. Technol. Lett.* **8**, 825 (1995).
19. M. Oishi, M. Yamamoto, and K. Kasaya, "2.0-µm single-mode operation of InGaAs-InGaAsP distributed-feedback buried-heterostructure quantum-well lasers," *IEEE Photon. Technol. Lett.* **9**, 431 (1997).
20. J. Dong, A. Ubukata, and K. Matsumoto, "Characteristics dependence on confinement structure and single-mode operation in 2-µm compressively strained InGaAs-lnGaAsP quantum-well lasers," *IEEE Photon. Technol. Lett.* **10**, 513 (1998).
21. M. Mitsuhara, M. Ogasawara, M. Oishi, H. Sugiura, and K. Kasaya, "2.05-µm wavelength InGaAs-InGaAs distributed-feedback multiquantum-well lasers with 10-mW output power," *IEEE Photon. Technol. Lett.* **11**, 33 (1999).
22. T. P. Pearsall, ed., *GaInAsP Alloy Semiconductors*, Wiley, New York, 1982.
23. A. Katz, ed., *Indium Phosphide and Related Materials: Processing, Technology, and Devices*, Artech House, Boston, 1992.
24. S. M. Sze, *Physics of Semiconductor Devices*, 2nd ed., Wiley, New York, 1981.
25. S. Adachi, "Material parameters of $In_{1-x}Ga_xAs_yP_{1-y}$ and related binaries," *J. Appl. Phys.* **53**, 8775 (1982).
26. S. Seki, T. Yamanaka, W. Lui, Y. Yoshikuni, and K. Yokoyama, "Theoretical analysis of pure effects of strain and quantum confinement on differential gain in InGaAsP/lnP strained-layer quantum-well lasers," *IEEE J. Quantum Electron.* **30**, 500 (1992).
27. M. G. Holland, in R. K. Willardson and C. A. Beer, eds., *Semiconductor and Semimetals*, Vol. 2, Academic Press, New York, 1966.
28. R. L. Moon, G. A. Antypas, and L. W. James, "Bandgap and lattice constant of GaInAsP as a function of alloy composition," *J. Electron. Mater.* **3**, 635 (1974).
29. Y. Itaya, Y. Suematsu, S. Katayama, K. Kishino, and S. Arai, "Low threshold current density (100) GaInAsP/InP double-heterostructure lasers for wavelength 1.3 µm," *Jpn. J. Appl. Phys.* **18**, 1795 (1979).
30. A. Ponchet, A. Rocher, J. Y. Emery, C. Starck, and L. Goldstein, "Lateral modulations in zero-net-strained GaInAsP multilayers grown by gas source molecular-beam epitaxy," *J. Appl. Phys.* **74**, 3778 (1993).
31. U. Bangert, A. J. Harvey, V. A. Wilkinson, C. Dieker, J. M. Jowett, A. D. Smith, S. D. Perrin, and C. J. Gibbins, "Evidence for strain relaxation via composition fluctuations in strained quaternary/quaternary and quaternary/ternary multiple quantum well structures," *J. Cryst. Growth* **132**, 231 (1993).
32. H. Sugiura, M. Mitsuhara, M. Ogasawara, M. Itoh, and H. Kamada, "Structural and optical properties of 1.3 mu m wavelength tensile-strained InGaAsP multi-quantum wells grown by metalorganic molecular beam epitaxy," *J. Appl. Phys.* **81**, 1427 (1997).
33. F. Glas, "Elastic state and thermodynamical properties of inhomogeneous epitaxial layers: Application to immiscible III–V alloys," *J. Appl. Phys.* **62**, 3201 (1987).
34. J. E. Guyer and P. W. Voorhee, "Morphological stability of alloy thin films," *Phys. Rev. Lett.* **74**, 4031 (1995).

35. G. B. Stringfellow, *Organometallic Vapor Phase Epitaxy: Theory and Practice*, Academic Press, San Diego, CA, 1989.
36. B. de Cremoux, P. Hirtz, and J. Ricciardi, "On the presence of a solid immiscibility domain in the GaInAsP phase diagram," *Inst. Phys. Conf. Ser.* **56**, 115 (1981).
37. K. Onabe, "Calculation of miscibility gap in quaternary InGaPAs with strictly regular solution approximation," *Jpn. J. Appl. Phys.* **21**, 797 (1982).
38. P. Henoc, A. Izrael, M. Quillec, and H. Launois, "Composition modulation in liquid phase epitaxial $In_{1-x}Ga_xAs_yP_{1-y}$ layers lattice matched to InP substrates," *Appl. Phys. Lett.* **40**, 963 (1982).
39. O. Ueda, S. Isozumi, and S. Komiya, "Composition-modulated structures in InGaAsP and InGaP liquid phase epitaxial layers grown on (001) GaAs substrates," *Jpn. J. Appl. Phys.* **23**, L241 (1984).
40. R. R. LaPierre, T. Okada, B. J. Robinson, D. A. Thompson, and G. C. Weatherly, "Spinodal-like decomposition of InGaAsP/(100) InP grown by gas source molecular beam epitaxy," *J. Cryst. Growth* **155**, 1 (1995).
41. K. Tappura and J. Laurila, "Unstable regions in the growth of GaInAsP by gas-source molecular beam epitaxy," *J. Cryst. Growth* **131**, 309 (1993).
42. T. Okada and G. C. Weatherly, "The role of strain and composition on the morphology of InGaAsP layers grown on (001) InP substrates," *J. Cryst. Growth* **179**, 339 (1997).
43. R. W. Glew, K. Scarrott, A. T. R. Briggs, A. D. Smith, V. A. Wilkinson, X. Zhou, and M. Silver, "Elimination of wavy layer growth phenomena in strain-compensated GaInAsP/GaInAsP multiple quantum well stacks," *J. Cryst. Growth* **145**, 764 (1994).
44. G. B. Stringfellow, "Spinodal decomposition and clustering in III/V alloys," *J. Electron. Mater.* **11**, 903 (1982).
45. B. I. Miller, U. Koren, M. G. Young, and M. D. Chien, "Strain-compensated strained-layer superlattices for 1.5 μm wavelength lasers," *Appl. Phys. Lett.* **58**, 1952 (1991).
46. S. L. Chuang, "Efficient band structure calculations of strained quantum wells," *Phys. Rev.* **B43**, 9649 (1991).
47. G. E. Pikus and G. L. Bir, "Effect of deformation on the hole energy spectrum of germanium and silicon," *Sov. Phys. Solid State* **1**, 1502 (1960).
48. H. Asai and K. Oe, "Energy band-gap shift with elastic strain in $Ga_xIn_{1-x}P$ epitaxial layers on (001) GaAs substrates," *J. Appl. Phys.* **54**, 2052 (1983).
49. K. Nishi, K. Hirose, and T. Mizutani, "Optical characterization of InGaAs-InAlAs strained-layer superlattices grown by molecular beam epitaxy," *Appl. Phys. Lett.* **49**, 794 (1986).
50. G. Bastard, "Superlattice band structure in the envelope-function approximation," *Phys. Rev.* **B24**, 5693 (1981).
51. J. M. Luttinger and W. Kohn, "Motion of electrons and holes in perturbed periodic fields," *Phys. Rev.* **97**, 869 (1955).
52. M. Irikawa, I. J. Murgatroyd, T. Ijichi, N. Matsumoto, A. Nakai, and S. Kashiwa, "Sharp interfaces in GaInAsP/InP single quantum wells grown by MOVPE," *J. Cryst. Growth* **93**, 370 (1988).
53. W. Xiaoliang, S. Dianzhao, K. Meiying, H. Xun, and Z. Yiping, "GSMBE growth and PL investigation of lattice-matched InGaAs/InP quantum wells," *J. Cryst. Growth* **164**, 281 (1996).

54. M. Houng and Y. C. Chang, "Electronic structures of $In_{1-x}Ga_xAs/InP$ strained layer quantum wells," *J. Appl. Phys.* **65**, 3096 (1989).

55. R. E. Cavicchi, D. V. Lang, D. Gershoni, A. M. Sergent, J. M. Vandenberg, S. N. G. Chu, and M. B. Panish, "Admittance spectroscopy measurement of band offsets in strained layers of $In_xGa_{1-x}As$ grown on InP," *Appl. Phys. Lett.* **54**, 739 (1989).

56. B. R. Nag and S. Mukhopadhyay, "Band offset in $InP/Ga_{0.47}In_{0.53}As$ heterostructures," *Appl. Phys. Lett.* **58**, 1056 (1991).

57. M. Gendry, V. Drouot, C. Santinelli, and G. Hollinger, "Critical thicknesses of highly strained InGaAs layers grown on InP by molecular beam epitaxy," *Appl. Phys. Lett.* **60**, 2249 (1992).

58. J. Tersoff and F. K. LeGoues, "Competing relaxation mechanisms in strained layers," *Phys. Rev. Lett.* **72**, 3570 (1994).

59. J. W. Matthews and A. E. Blakeslee, "Defects in epitaxial multilayers: I. Misfit dislocations," *J. Cryst. Growth* **27**, 118 (1974).

60. J. W. Matthews, S. Mader, and T. B. Light, "Accommodation of misfit across the interface between crystals of semiconducting elements or compounds," *J. Appl. Phys.* **41**, 3800 (1970).

61. M. J. Maree, J. C. Barbour, J. F. van der Veen, K. L. Kavanagh, C. W. T. Bulle-Lieuwma, and M. P. A. Viegers, "Generation of misfit dislocations in semiconductors," *J. Appl. Phys.* **62**, 4413 (1987).

62. J. H. van der Merwe, "Crystal interfaces. Part II. Finite overgrowths," *J. Appl. Phys.* **34**, 123 (1962).

63. R. People and J. C. Bean, "Calculation of critical layer thickness versus lattice mismatch for Ge_xSi_{1-x}/Si strained-layer heterostructures," *Appl. Phys. Lett.* **47**, 322 (1985).

64. R. People and J. C. Bean, "Erratum: Calculation of critical layer thickness versus lattice mismatch for Ge_xSi_{1-x}/Si strained layer heterostructures," *Appl. Phys. Lett.* **49**, 229 (1986).

65. K. Kim and Y. H. Lee, "Temperature-dependent critical layer thickness for strained-layer heterostructures," *Appl. Phys. Lett.* **67**, 2212 (1995).

66. H. Temkin, D. G. Gershoni, S. N. Chu, J. M. Vandenberg, R. A. Hamm, and M. B. Panish, "Critical layer thickness in strained $Ga_{1-x}In_xAs/InP$ quantum wells," *Appl. Phys. Lett.* **55**, 1668 (1989).

67. G. L. Price, "Critical-thickness and growth-mode transitions in highly strained $In_xGa_{1-x}As$ films," *Phys. Rev. Lett.* **66**, 469 (1991).

68. H. Sugiura, M. Ogasawara, M. Mitsuhara, H. Oohashi, and T. Amano, "Metalorganic molecular beam epitaxy of strain-compensated InAsP/InGaAsP multi-quantum-well lasers," *J. Appl. Phys.* **79**, 1233 (1996).

69. P. Krapf, Y. Robach, M. Gendry, and L. Porte, "Role of the step curvature in the stabilizationof coherently strained epitaxial structures," *Phys. Rev.* **B55**, 10229 (1997).

70. A. Ponchet, A. Le Corre, H. L'Haridon, B. Lambert, S. Salaün, D. Alquier, D. Lacombe, and L. Durand, "Effect of the growth procedure and the InAs amount on the formation of strain-induced islands in the InAs/InP(001) system," *Appl. Surf. Sci.* **123/124**, 751 (1998).

71. D. E. Jesson, S. J. Pennycock, J. M. Baribeau, and D. C. Houghton, "Direct imaging of surface cusp evolution during strained-layer epitaxy and implications for strain relaxation," *Phys. Rev. Lett.* **71**, 1744 (1993).
72. W. H. Yang and D. J. Srolovitz, "Cracklike surface instabilities in stressed solids," *Phys. Rev. Lett.* **71**, 1593 (1993).
73. S. M. Wang, T. G. Andersson, and M. J. Ekenstedt, "Temperature-dependent transition from two-dimensional to three-dimensional growth in highly strained $In_xGa_{1-x}As/GaAs$ (0.36 < x < 1) single quantum wells," *Appl. Phys. Lett.* **61**, (1992).
74. T. Fujii and S. Yamazaki, "A new lattice relaxation mode in InGaAs on GaAs," *J. Cryst. Growth* **146**, 489 (1995).
75. G. P. Agrawal and N. K. Dutta, *Semiconductor Lasers*, 2nd ed., Van Nostrand Reinhold, New York, 1993.
76. H. C. Casey Jr. and M. B. Panish, *Heterostructure Lasers, Parts A and B*, Academic Press, New York, 1978.
77. U. Koren, B. I. Miller, Y. K. Su, T. L. Koch, and J. E. Bower, "Low internal loss separate confinement heterostructure InGaAs/InGaAsP quantum well laser," *Appl. Phys. Lett.* **51**, 1744 (1987).
78. C. H. Henry, R. A. Logan, F. R. Merritt, and J. P. Luongo, "The effect of intervalence band absorption on the thermal behavior of InGaAsP lasers," *IEEE J. Quantum Electron.* **19**, 947 (1983).
79. M. Asada, A. R. Adams, K. E. Stubkjaer, Y. Itaya, and S. Arai, "The temperature dependence of the threshold current of GaInAsP/InP DH lasers," *IEEE J. Quantum Electron.* **17**, 611 (1981).
80. R. W. H. Engelmann, C. Shieh, and C. Shu, in P. S. Zory Jr., ed., *Quantum Well Lasers*, Academic Press, San Diego, CA, 1993, ch. 3.
81. A. Haug, "Theory of the temperature dependence of the threshold current of an InGaAsP laser." *IEEE J. Quantum Electron.* **21**, 716 (1985).
82. R. J. Nelson and N. K. Dutta, "Calculated Auger rates and temperature dependence of threshold for semiconductor lasers emitting at 1.3 and 1.55 µm," *J. Appl. Phys.* **54**, 2923 (1983).
83. S. Hausser, G. Fucks, A. Hangleiter, K. Streubel, and W. T. Tsang, "Auger recombination in bulk and quantum well InGaAs," *Appl. Phys. Lett.* **56**, 913 (1990).
84. G. Fucks, C. Schiedel, A. Hangleiter, V. Härle, and F. Scholz, "Auger recombination in strained and unstrained InGaAs/InGaAsP multiple quantum-well lasers," *Appl. Phys. Lett.* **62**, 396 (1993).
85. P. Rees, P. Blood, M. J. H. Vanhommerig, G. J. Davies, and P. J. Skevington, "The temperature dependence of threshold current of chemical beam epitaxy grown InGaAs-InP lasers," *J. Appl. Phys.* **78**, 1804 (1995).
86. J. P. Loehr and J. Singh, "Effect of strain on CHSH Auger recombination in strained $In_{0.53+x}Ga_{0.47-x}As$ on InP," *IEEE J. Quantum Electron.* **29**, 2583 (1993).
87. D. Nichols, M. Sherwin, G. Munns, J. Pamulapati, J. Loehr, J. Singh, P. Bhattacharya, and M. Ludowise, "Theoretical and experimental studies of the effects of compressive and tensile strain on the performance of InP-InGaAs multiquantum-well lasers," *IEEE J. Quantum Electron.* **28**, 1239 (1992).

88. P. J. A. Thijs, L. F. Tiemeijer, J. J. M. Binsma, and T. van Dongen, "Progress in long-wavelength strained-layer quantum-well lasers and amplifiers," *IEEE J. Quantum Electron.* **30**, 477 (1994).

89. J. Z. Wilcox, S. Ou, J. J. Yang, M. Jansen, and G. L. Peterson, "Dependence of emission wavelength on cavity length and facet refectivities in multiple quantum well semiconductor lasers," *Appl. Phys. Lett.* **54**, 2174 (1989).

90. N. Tessler and G. Eisenstein, "Carrier injection and gain dynamics in quantum well lasers," *IEEE J. Quantum Electron.* **29**, 1586 (1993).

91. K. Fröjdh, S. Marcinkevičius, U. Olin, C. Silfvenius, B. Stålnacke, and G. Landgren, "Interwell carrier transport in InGaAsP multiple quantum well laser structures," *Appl. Phys. Lett.* **69**, 3695 (1996).

92. H. Yamazaki, A. Tomita, M. Yamaguchi, and Y. Sasaki, "Evidence of nonuniform carrier distribution in multiple quantum well lasers," *Appl. Phys. Lett.* **71**, 767 (1997).

93. J. F. Hazell, J. G. Simmons, J. D. Evans, and C. Blaauw, "The effect of varying barrier height on the operational characteristics of 1.3-μm strained-layer MQW lasers," *IEEE J. Quantum Electron.* **34**, 2358 (1998).

94. J. Piprek, P. Abraham, and J. E. Bowers, "Carrier nonuniformity effects on the internal efficiency of multiquantum-well lasers," *Appl. Phys. Lett.* **74**, 489 (1999).

95. C. Silfvenius, G. Landgren, and Marcinkevicius, "Hole distribution in InGaAsP 1.3-μm multiple-quantum-well laser structures with different hole confinement energies," *IEEE J. Quantum Electron.* **35**, 603 (1999).

96. B. B. Elenkrig, S. Smetona, J. G. Simmons, T. Makino, and J. D. Evans, "Maximum operating power of 1.3μm strained layer multiple quantum well InGaAsP lasers, *J. Appl. Phys.* **85**, 2367 (1999).

97. W. T. Tsang, "Advances in MOVPE, MBE, and CBE," *J. Cryst. Growth* **120**, 1 (1992).

98. D. C. Houghton, M. Davies, T. S. Rao, and M. Dion, "Comparison of chemical beam epitaxy and metalorganic chemical vapour deposition for highly strained multiple quantum well InGaAsP/InP μm lasers," *J. Cryst. Growth* **136**, 56 (1994).

99. J. Dong, A. Ubukata, and K. Matsumoto, "1.95-μm-wavelength InGaAs/InGaAsP laser with compressively strained quantum well active layer," *Jpn. J. Appl. Phys.* **36**, 5468 (1997).

100. J. F. Carlin, A. V. Syrbu, C. A. Berseth, J. Behrend, A. Rudra, and E. Kapon, "Low threshold 1.55μm wavelength InAsP/InGaAs strained multiquantum well laser diode grown by chemical beam epitaxy," *Appl. Phys. Lett.* **71**, 13 (1997).

101. W. T. Tsang and T. H. Chiu, in J. S. Foord, G. J. Davies, and W. T. Tsang, eds., *Chemical Beam Epitaxy and Related Techniques*, Wiley, New York, 1997, ch. 6.

102. H. Heinecke, B. Baur, E. Emeis, and M. Scier, "Growth of highly uniform InP/GaInAs/GaInAsP heterostructures by MOMBE for device integration," *J. Cryst. Growth* **120**, 140 (1992).

103. A. Rudra, J. F. Carlin, P. Ruterana, M. Gailhanou, J. L. Staehli, and M. Ilegems, "Growth of GaInAs(P) and GaInAsP/GaInAs MQW structures by CBE," *J. Cryst. Growth* **120**, 338 (1992).

104. H. Sugiura, M. Mitsuhara, R. Iga, and N. Yamamoto, "Metalorganic molecular beam epitaxial growth of highly uniform InGaAsP quantum well structures using an indium-free holder," *J. Cryst. Growth* **141**, 299 (1994).

105. J. M. Vandenberg, R. A. Hamm, M. B. Panish, and H. Temkin, "High-resolution x-ray diffraction studies of InGaAs(P)/InP Superlattices grown by gas-source molecular-beam epitaxy," *J. Appl. Phys.* **62**, 1278 (1987).

106. K. H. Chang, R. Gibala, D. Srolovitz, P. K. Bhattacharya, and J. F. Mansfield, "Crosshatched surface morphology in strained III–V semiconductor films," *J. Appl. Phys.* **67**, 4093 (1990).

107. A. D. Smith, A. T. R. Briggs, K. Scarrott, X. Zhou, and U. Bangert, "Optimization of growth conditions for strain compensated $Ga_{0.32}In_{0.68}As/Ga_{0.61}In_{0.39}As$ multiple quantum wells," *Appl. Phys. Lett.* **65**, 2311 (1994).

108. U. Bangert, A. Harvey, C. Dieker, and H. Hardtdegen, "Suppression of wavy growth in metalorganic vapor phase epitaxy grown GaInAs/InP superlattices," *Appl. Phys. Lett.* **69**, 2101 (1996).

109. T. Isu, M. Hata, Y. Morishita, Y. Nomura, and Y. Katayama, "Surface diffusion length during MBE and MOMBE measured from distribution of growth rates," *J. Cryst. Growth* **115**, 423 (1991).

110. M. Mitsuhara, M. Ogasawara, M. Oishi, H. Sugiura, and K. Kasaya, "2.05-µm wavelength InGaAs-InGaAs distributed-feedback multiquantum-well lasers with 10-mW output power," *3rd Int. Conf. Mid-infrared Optoelectronics Materials and Devices*, Aachen, Germany, 1999, paper I6.

111. A. Ponchet, A. Rocher, J. Y. Emery, C. Starck, and L. Goldstein, "Direct measurement of lateral elastic modulations in a zero-net strained GaInAsP/InP multilayer," *J. Appl. Phys.* **77**, 1977 (1995).

112. T. Tsuchiya, M. Komori, R. Tsuneta, and H. Kakibayashi, "Investigation of effect of strain-compensated structure and compensation limit in strained-layer multiple quantum wells," *J. Cryst. Growth* **145**, 371 (1994).

113. J. Y. Yao, T. G. Andersson, and G. L. Dunlop, "The interfacial morphology of strained epitaxial $In_xGa_{1-x}As/GaAs$," *J. Appl. Phys.* **69**, 2224 (1991).

114. M. Mitsuhara, M. Ogasawara, and H. Sugiura, "Effect of strain in the barrier layer on structural and optical properties of highly strained $In_{0.77}Ga_{0.23}As/InGaAs$ multiple quantum wells," *J. Cryst. Growth* **210**, 463 (2000).

115. A. V. Syrbu, J. Behrend, J. Fernandez, J. F. Carlin, C. A. Berseth, V. P. Iakovlev, A. Rudra, and E. Kapon, "Thermal stability of InP-based structures for wafer fused laser diodes," *J. Cryst. Growth* **188**, 338 (1998).

116. K. Naniwae, S. Sugou, and T. Anan, "Effect of strain compensation on crystalline quality for InGaAs/InAlP strained multiple quantum well structures on InP grown by gas-source molecular beam epitaxy," *Jpn. J. Appl. Phys.* **33**, L156 (1994).

117. A. Ponchet, A. Rocher, A. Qugazzaden, and A. Mircea, "Self-induced laterally modulated GaInP/InAsP structure grown by metal-organic vapor-phase epitaxy," *J. Appl. Phys.* **75**, 7881 (1994).

118. J. H. Neave, P. J. Dobson, B. A. Joyce, and J. Zhang, "Reflection high-energy electron diffraction oscillations from vicinal surfaces—a new approach to surface diffusion measurements," *Appl. Phys. Lett.* **47**, 100 (1985).

119. N. Grandjean, J. Massies, M. Leroux, J. Leymarie, A. Vasson, and A. M. Vasson, "Improved GaInAs/GaAs heterostructures by high growth rate molecular beam epitaxy," *Appl. Phys. Lett.* **64**, 2664 (1994).
120. M. Ogasawara, M. Mitsuhara, M. Itoh, and T. Amano, "Effects of increased growth rate of well layer on 2.1% compressive strained InAsP-MQWs grown by metalorganic molecular beam epitaxy (MOMBE)," *J. Cryst. Growth* **205**, 489 (1999).
121. M. Mitsuhara, M. Ogasawara, and H. Sugiura, unpublished results.
122. J. S. Osinski, P. G. rodzinski, Y. Zou, and P. D. Dapkus, "Threshold current analysis of compressive strain (0–1.8%) in low-threshold, long-wavelength quantum well lasers," *IEEE J. Quantum Electron.* **29**, 1576 (1993).
123. H. Temkin, N. K. Dutta, T. Tanbun-Ek, R. A. Logan, and A. M. Sergent, "InGaAs/InP quantum well lasers with sub-mA threshold current," *Appl. Phys. Lett.* **57**, 1610 (1990).
124. C. E. Zah, F. J. Favire, R. Bhat, S. G. Menocal, N. C. Andreadakis, D. M. Hwang, M. Koza, and T. P. Lee, "Submilliampere-threshold 1.5-μm strained-layer multiple quantum well lasers," *IEEE Photon. Technol. Lett.* **2**, 852 (1990).
125. S. Arai, Y. Suematsu, and Y. Itaya, "1.11–1.67 μm (100) GaInAsP/InP injection lasers prepared by liquid phase epitaxy," *IEEE J. Quantum Electron.* **16**, 197 (1980).
126. J. S. Major, Jr., D. W. Nam, J. S. Osinski, and D. F. Welch, "High-power 2.0μm InGaAsP laser diodes," *IEEE Photon. Technol. Lett.* **5**, 594 (1993).
127. M. Fukuda, *Reliability and Degradation of Semiconductor Lasers and LEDs*, Artech House, Boston, Massachusetts, 1991, ch. 7.
128. J. Dong, A. Ubukata, and K. Matsumoto, "Low threshold compressively strained InGaAs/lnGaAsP quantum well distributed feedback laser at 1.95μm," *Electron. Lett.* **33**, 1090 (1997).
129. M. G. Young, S. A. Keo, S. Forouhar, T. Turner, L. Davis, R. Mueller, and P. D. Maker, "Room temperature InGaAs-InGaAsP distributed feedback ridge lasers operating beyond 2μm," *Annual Mtig IEEE/LEOS*, San Francisco, CA, 1997, paper MI5.
130. H. A. Haus and C. V. Shank, "Antisymmetric taper of distributed feedback lasers," *IEEE J. Quantum Electron.* **12**, 532 (1976).
131. K. Utaka, S. Akiba, K. Sakai, and Y. Matsushima, "l/4-shifted InGaAsP/InP DFB lasers by simultaneous holographic exposure of positive and negative photoresists," *Electron. Lett.* **20**, 1008 (1984).
132. T. Tanbun-Ek, R. A. Logan, S. N. G. Chu, A. M. Sergent, and K. W. Wecht, "Effects of strain in multiple quantum well distributed feedback lasers," *Appl. Phys. Lett.* **57**, 2184 (1990).
133. R. U. Martinelli, R. J. Menna, and D. E. Cooper, "InGaAs/InP distributed-feedback lasers for spectroscopic applications," *Annual Meetings IEEE/LEOS*, Boston, Massachusetts, 1994, p. 95.
134. W. T. Tsang, F. S. Choa, M. C. Wu, Y. K. Chen, R. A. Logan, T. Tanbun-Ek, S. N. G. Chu, K. Tai, A. M. Sergent, and K. W. Wecht, "1.5μm wavelength InGaAs/InGaAsP distributed feedback multi-quantum-well lasers grown by chemical beam epitaxy," *Appl. Phys. Lett.* **59**, 2375 (1991).

135. F. Koyama and S. Arai, in Y. Suematsu and A. R. Adams, eds., *Handbook of Semiconductor Lasers and Photonic Integrated Circuits*, Chapman & Hill, London, UK, 1994, p. 422.
136. M. R. Gokhale, J. Wei, H. Wang, and S. R. Forrest, "Growth and characterization of small band gap (~0.6 eV) InGaAsN layers on InP," *Appl. Phys. Lett.* **74**, 1287 (1999).
137. C. W. Tu, W. G. Bi, Y. Ma, J. P. Zhang, L. W. Wang, and S. T. Ho, "A novel material for long-wavelength lasers: InNAsP," *IEEE J. Select. Topics Quantum Electron.* **4**, 510 (1998).

CHAPTER 3
Antimonide Mid-IR Lasers

L. J. OLAFSEN, I. VURGAFTMAN, and J. R. MEYER

3.1 INTRODUCTION

This chapter will review the dramatic recent technical progress of midwave infrared (mid-IR) semiconductor lasers emitting at wavelengths beyond 2 μm, specifically those employing the antimonide family of narrow-gap III-V semiconductors with lattice constants near 6.1 Å. It should be emphasized at the outset that the challenges associated with producing a high-performance diode laser multiply at longer wavelengths. Nature has made the mid-IR problem inherently more difficult in at least three important ways: (1) narrow-gap semiconductors are energetically predisposed to non-radiative Auger decay rather than radiative recombination, (2) internal losses associated with free carrier absorption also tend to increase rapidly with increasing wavelength, and (3) growth on a GaSb or InAs substrate is dictated by lattice-matching the narrow-gap active region to the substrate, with the consequence that all aspects of the laser fabrication are less mature than for GaAs- and InP-based devices. Although much of what follows may seem primitive when compared directly to near-IR diode laser standards, the mid-IR technology must be judged in the context of these significant challenges.

Even within the mid-IR region, the level of difficulty in achieving high performance tends to increase monotonically with increasing wavelength. Toward the short end of the mid-IR spectrum (λ = 2.0–2.4 μm), the challenges mentioned above have been overcome to a large extent, and uncooled high-power cw devices are already available. It now appears likely that diode lasers emitting to at least 4 μm will also eventually operate cw at room temperature, a prospect that seemed remote only 10 years ago. However, the difficulties listed above become increasingly serious at wavelengths longer than 4 to 5 μm, where perhaps the best one can hope for is to minimize the consequences of the

Long-Wavelength Infrared Semiconductor Lasers, Edited by Hong K. Choi
ISBN 0-471-39200-6 Copyright © 2004 John Wiley & Sons, Inc.

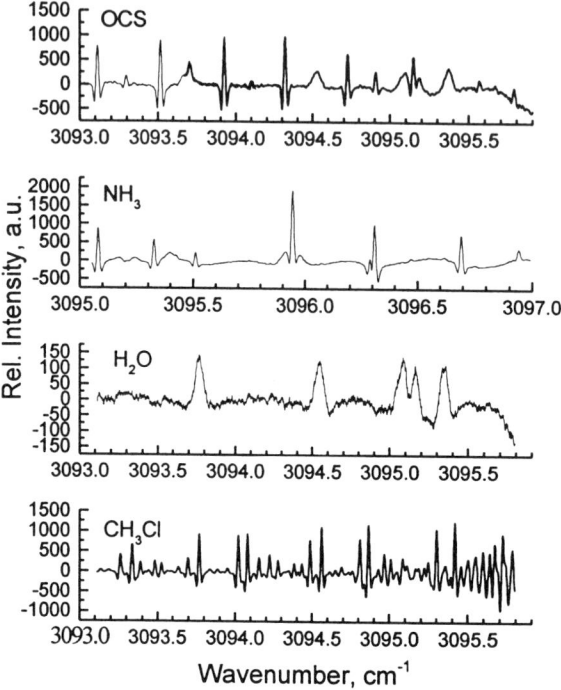

Fig. 3.1 Transition spectra measured for OCS, NH_3, H_2O, and CH_3Cl gases by scanning with a single-mode InAsSb/InAsSbP laser. (Reprinted from [1], with permission)

limiting factors as much as possible. For example, optical pumping is being considered as one possible approach to boosting performance levels, at least in the near term.

Perhaps the broadest application for antimonide mid-IR lasers will be the detection of trace chemicals. Inexpensive, compact, low-maintenance sources could be used in portable detection systems to monitor emissions, pollution, leaks, and industrial processes. Figure 3.1 shows a current-tuned scan of the absorption spectra for four different chemicals by Danilova et al. [1]. Continuous-wave operation in at least the thermoelectric cooler range ($T \geq 240\,K$) will be required for some environments. Potential military uses include IR countermeasures, IR scene projection, and chemical weapons monitoring. There are also medical applications, such as laser surgery and breath analysis.

In the following sections the historical development, present status, and future prospects for antimonide lasers emitting in the mid-IR spectral range will be reviewed. The relation between the optical properties and band structure will be emphasized, since it is especially critical for this material family. For example, whereas virtually all near-IR lasers have had type-I active regions, in which the conduction band minimum and valence band maximum are in the same layer, one of the important antimonide mid-IR laser classes has a type-II active region, in which the conduction band minimum and valence band maximum are in adjacent layers.

The chapter is organized as follows: Section 3.2 overviews the highly versatile band structure and other properties of the antimonide heterostructure family. Section 3.3 reviews the status of lasers emitting in the 2 to 3 µm range, which have already achieved relatively high performance. The more challenging wavelength range beyond 3 µm is covered in Section 3.4. Relevant device classes include double heterostructures, type-I quantum wells, and type-II quantum wells, with both optically pumped and diode devices representing subcategories of each. The novel type-II interband cascade laser is a further permutation of earlier device configurations. Finally Section 3.5 reviews the most critical issues limiting the performance of mid-IR semiconductor lasers, which include nonradiative recombination, internal losses, the linewidth enhancement factor, and thermal management. Single-mode operation and beam quality are also discussed.

3.2 ANTIMONIDE III-V MATERIAL SYSTEM

One of the most versatile semiconductor material families is the antimonide system, which is usually taken to encompass all III-V zinc-blend compounds and alloys with lattice constants of approximately 6.1 Å (including those that contain no antimony). Liquid phase epitaxy (LPE) [2–4], metal-organic chemical vapor deposition (MOCVD) [5–8], and molecular beam epitaxy (MBE) have all been used successfully to grow antimonide mid-IR lasers. More advanced structures are usually grown by MBE, a technique that allows better control of the layer thicknesses and interface compositions. However, antimonide MBE technology is much less mature than for GaAs-based and InP-based optoelectronic devices. Some of the special issues associated with antimonide MBE growth will be discussed in Section 3.5.1.

In designing the active, carrier-confinement, and optical-confinement regions of mid-IR lasers, the antimonide band structure and band alignments introduce both additional challenges and unique opportunities. This is illustrated by Fig. 3.2, which plots on an absolute scale the conduction band minima (CBM, solid curves) and valence band maxima (VBM, dashed curves) for the InAs, GaSb, and AlSb binaries, along with many of the relevant ternaries. Further flexibility is introduced by inclusion of the most important antimonide quaternaries: $(AlAs_{0.08}Sb_{0.92})_z(GaSb)_{1-z}$ and $(GaSb)_z(InAs_{0.91}Sb_{0.09})_{1-z}$ lattice-matched to GaSb, as well as $(GaAs_{0.08}Sb_{0.92})_z(InAs)_{1-z}$ lattice-matched to InAs. The vertical lines in Fig. 3.2 represent the range of band extrema for those quaternaries, while Fig. 3.3 plots energy gaps as a function of composition, z. Dotted portions of the curves in Fig. 3.3 correspond to composition ranges that may be difficult to access due to miscibility gaps, although some of the nominally prohibited materials may be fabricated in practice by using a

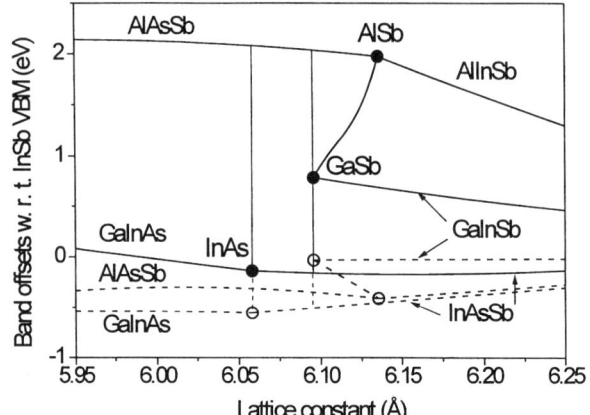

Fig. 3.2 Conduction band minima (CBM, *solid curves*) and valence band maxima (VBM, *dashed curves*) for InAs, GaSb, and AlSb, along with many of the relevant ternaries. The vertical lines represent the range of band extrema for the leading lattice-matched antimonide quaternaries. (Parameters are from [9])

nonequilibrium growth process such as MBE. The energy gaps and valence band offsets employed in generating Figs. 3.2 and 3.3 are from a recent comprehensive review of the literature [9].

In considering the band alignments plotted in Fig. 3.2, note first that the energy gaps for InAsSb alloys can be very narrow, for example, corresponding to wavelengths beyond 5 µm. While this would seem to make InAsSb the most obvious choice for the active region in a mid-IR laser, it is unfortunate that As-rich InAsSb also happens to have one of the lowest VBMs in the entire antimonide material system. It consequently becomes difficult to find a barrier material with a VBM that is low enough to provide adequate electrical confinement for the holes, a factor that has limited the high-temperature performance of most InAsSb-based DH and type-I quantum well (QW) lasers studied to date. In principle, strained InAsSb/InAs QWs can yield marginal electrical confinement of the holes, with the growth conditions sometimes favoring a type-I band alignment over type-II [10–13]. It is promising that somewhat larger band offsets may be realized by alloying InAs(Sb) with InP to form InAsSbP barriers [14,15], although control over the growth of that quaternary remains difficult. A further unfortunate characteristic of the InAsSb band structure is the occurrence of a near resonance between the spin-orbit splitting and the energy gap, which can favor strong non-radiative Auger recombination as will be discussed in Section 3.5.2.

Materials with a substantial AlSb component (e.g., AlAsSb and AlGaAsSb) form a natural barrier to electron flow in antimonide heterostructures and also provide a good hole barrier when coupled with any layer containing a substantial fraction of GaSb (e.g., GaInSb, GaAlSb, and InGaAsSb). For example,

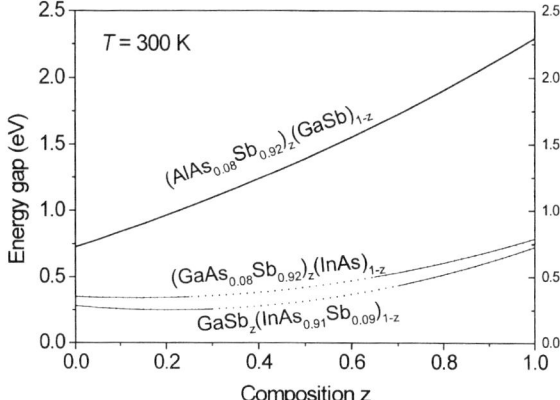

Fig. 3.3 Energy gap as a function of composition, z. Dotted portions of the curves correspond to composition ranges that may be difficult to achieve due to miscibility gaps. (Parameters are from [9])

the combination InGaAsSb/AlGaAsSb, lattice-matched to either GaSb or InAs, has a type-I alignment and in principle can cover a significant segment of the mid-IR spectral range (1.7–4.2 μm). While the valence band offset (VBO) remains rather small, and the miscibility gap for GaInAsSb limits the practical wavelength range, lasers containing metastable InGaAsSb active regions have emitted at wavelengths as long as about 3 μm [16].

Ghiti and O'Reilly studied theoretically the beneficial influence of strain on the operating characteristics of InGaAsSb/AlGaAsSb lasers [17]. They showed that either compressive or tensile strain of the well region could reduce the valence-band density of states, thereby lowering the threshold current density. Strain can also substantially reduce the Auger coefficients in InGaAsSb/AlGaAsSb QWs [18,19]. Compressive strain of the InGaAsSb layers can be accomplished in two different ways: (1) by increasing the In composition for constant As fraction [20], or (2) by decreasing the As composition for constant In fraction [21]. The relative merits and shortcomings of the two approaches have yet to be fully explored. The effect of compressive strain on the position of the photoluminescence (PL) peak is illustrated in Fig. 3.4 [21]. The filled points correspond to strained wells and lattice-matched barriers, whereas the open points are for strain-balanced structures with compressive strain in the wells and tensile strain in the barriers. As the strain is increased by reducing the As fraction in the wells and increasing the As fraction in the barriers, the PL peak is seen to shift from about 2.5 μm to 2.05 μm. Note that a decrease in the In fraction also leads to a blue shift of the laser emission line [20]. The use of strain engineering to tune the emission wavelength and reduce the threshold is ultimately limited by defect formation when the well width exceeds the critical thickness for the formation of dislocations.

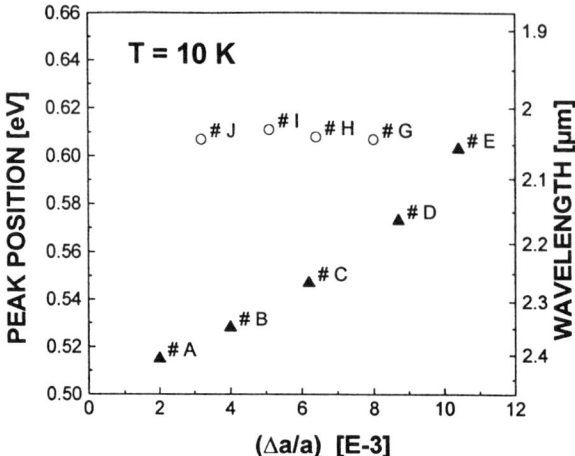

Fig. 3.4 Photoluminescence peak energies as a function of strain at $T = 10\,\text{K}$. Filled triangles (▲) represent quantum wells under compressive strain with lattice-matched barriers, while open circles (○) correspond to quantum wells under constant compressive strain, but with barriers under increasing tensile strain. (Reprinted from [21], with permission from Elsevier Science)

The most striking feature of Fig. 3.2 is the broken-gap alignment, with a semimetallic overlap of about 110 meV, between the InAs conduction band and the GaSb valence band [22]. Since GaSb forms a barrier to electron states in InAs and InAs forms a barrier to hole states in GaSb, decreasing either layer thickness induces quantum confinement (although the effect is much larger for electrons). Only a modest confinement energy is required to convert the semimetal into a type-II semiconductor, which has arbitrarily long emission wavelength when electrons in the InAs recombine with holes in the GaSb. In the opposite limit of very strong quantum confinement, wavelengths shorter than 2 µm can be reached (as long as AlSb-based barriers are introduced to provide enough conduction-band offset). Thus all far-IR and mid-IR wavelengths can be covered in principle by tailoring the layer thicknesses in type-II antimonide quantum wells. It will be seen in Section 3.4.3 that type-II mid-IR lasers become increasingly advantageous over type-I structures for achieving longer emission wavelengths and higher temperatures of operation.

3.3 ANTIMONIDE LASERS EMITTING IN THE 2µm < λ < 3µm RANGE

3.3.1 Historical Development

In 1977 a GaInAsSb DH laser emitting at 2 µm was demonstrated using electron-beam pumping [23]. The first mid-IR diode lasers (λ = 2.2–2.4 µm) operated under pulsed excitation at room temperature in 1985 [24–26], although the threshold current densities of ≥7 kA/cm^2 were too high to permit cw operation at $T > 80$ K. However, reductions in the threshold through increasing the confinement-layer barrier heights [27–29] soon led to cw operation at 220 K [30] and 235 K [31]. Room-temperature cw operation of a GaInAsSb/AlGaAsSb DH laser emitting at $\lambda = 2.4$–2.5 μm was reported in 1988 [32].

In parallel, a group at the Ioffe Institute investigated type-II GaInAsSb/GaSb heterojunctions for emission at a wide range of mid-IR wavelengths (the $\lambda > 3$ μm devices will be reviewed in the following section). The electron and hole confinement resulting from band bending at the heterojunction interface improved the emission characteristics. This work led to the development of room-temperature pulsed diode lasers operating at $\lambda \approx 2.0$ μm [33].

Following these initial demonstrations of the potential for long-wavelength III-V semiconductor lasers in the 1980s, the drive to develop truly practical laser diodes intensified in the early 1990s. Workers at MIT Lincoln Laboratory (LL) reported a GaInAsSb/AlGaAsSb DH device that had a pulsed threshold current density of 1.5 kA/cm^2 and a differential quantum efficiency of 25% per facet at 2.3 μm and 300 K [34,35]. These results were soon followed by the demonstration of room-temperature cw operation with a threshold of 940 A/cm^2 at $\lambda = 2.2$ μm [36]. Pulsed operation up to nearly 400 K was reported by the University of Montpellier group [37].

However, the largest leap in performance occurred when QW active regions were employed to realize a much higher gain per injected carrier than in the bulk-like DH devices. The first (lattice-matched) GaInAsSb/AlGaAsSb QW laser grown by MBE had a cw threshold current density of only 260 A/cm^2, and emitted up to 190 mW/facet cw at room temperature [38]. MOCVD-grown devices were also reported [5,39].

In the middle 1990s the emphasis shifted toward strain-compensated GaInAsSb QWs [40], which were predicted [17] to have superior Auger lifetimes (see Section 3.5.2). LL researchers demonstrated the first strained GaInAsSb/AlGaAsSb multiple-QW laser emitting at 1.9 μm, which had a pulsed threshold current density as low as 143 A/cm^2 and a cw output power of 1.3 W/facet [41]. However, the compressive strain in the wells was not compensated, since the barriers were lattice-matched to the GaSb substrate. The Sarnoff Corporation demonstrated GaInAsSb/AlGaAsSb QW lasers emitting at 2.7 to 2.8 μm, which displayed pulsed operation up to 333 K and a maximum output power of 30 mW [42]. The same device operated cw up to 234 K, which was limited by a relatively high threshold current density (e.g., 10 kA/cm^2 pulsed at 288 K).

Room-temperature pulsed emission at $\lambda = 2.36$ μm, with a threshold current density of 305 A/cm^2, was reported by the University of Montpellier group for a type-II GaInAsSb/GaSb QW structure [43]. GaInSb/GaSb strained QW lasers have also been investigated for emission at 2.0 μm [44].

3.3.2 State of the Art

$\lambda \approx 2.0$–$2.2\,\mu m$

Presently nearly all high-performance antimonide 2-μm diode lasers are grown by MBE on n-GaSb(100) substrates. The active region of a recent Sarnoff Corporation structure [45] contained a single 1% compressively strained $In_{0.19}Ga_{0.81}As_{0.02}Sb_{0.98}$ QW (100 Å) surrounded by two 0.4-μm-thick $Al_{0.25}Ga_{0.75}As_{0.02}Sb_{0.98}$ undoped (low-loss) separate-confinement regions (SCR) that were designed to minimize the overlap of the mode with the doped (high-loss) cladding layers. In this "broadened waveguide" structure, the SCR was much thicker than necessary to support a single confined optical mode, but not so thick as to allow lasing in the second-order mode. Graded transition regions were inserted between the $Al_{0.9}Ga_{0.1}As_{0.07}Sb_{0.93}$ cladding layers and the n-GaSb buffer and p-GaSb cap layers to facilitate electrical transport. A broadened-waveguide single QW (SQW) laser emitting at $\lambda = 2.0\,\mu m$ had a room-temperature threshold current density of $115\,A/cm^2$ for a 3-mm-long cavity [45]. On the other hand, the threshold for similar devices with five QWs was 70% higher, and the internal loss was nearly three times greater than the record low of $2\,cm^{-1}$ for the SQW laser. The observation of an external quantum efficiency of 53% implied that the internal efficiency, governed by the electrical injection of carriers into the well active region, must also have been high. As can be seen from the L-I (light-current) characteristic in Fig. 3.5, a maximum cw output power of 1.9 W was obtained from a device with high reflectivity (HR) and antireflective (AR) facet coatings operating at 288 K. For quasi-cw operation with 10 μs pulses at a repetition rate of 100 Hz, the maximum power was 4 W.

Similar lasers containing two or four $In_{0.15}Ga_{0.85}As_{0.06}Sb_{0.94}$ QWs (105 Å) and $Al_{0.4}Ga_{0.6}As_{0.03}Sb_{0.97}$ barriers were also grown at Sarnoff and tested at the University of New Mexico [46]. The larger Al fraction in the barrier was designed to increase the valence band offset by about 70 meV compared to the SQW structure, to a total of nearly 150 meV. The two-well device achieved a differential quantum efficiency of 74% and internal loss of $2.5\,cm^{-1}$, and the four-well laser had a high characteristic temperature T_0 of 140 K (up to 323 K) [46].

Lincoln Laboratory devices were grown in a similar configuration, except that the well composition was $In_{0.22}Ga_{0.78}As_{0.01}Sb_{0.99}$. Graded transition layers were also inserted between the claddings and the SCR, and the doping of the p-type cladding was lighter in the first 0.2 μm adjacent to the waveguide core in order to further reduce the free-carrier absorption losses. The HR and AR coating reflectivities were 96% and 1%, respectively. A broad-stripe (100 μm) SQW laser displayed a very low room-temperature threshold current density of $50\,A/cm^2$, which is comparable to the best results reported for any QW laser diode [18]. The internal quantum efficiency was remarkably high at 95%, and the internal loss coefficient was $7\,cm^{-1}$. A cw power of 1 W was generated at 283 K. The improved performance characteristics were attrib-

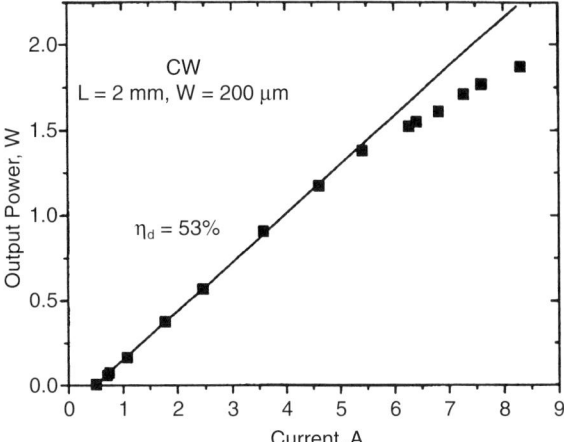

Fig. 3.5 Cw output power versus injection current at $T = 288\,\mathrm{K}$ for a $2\,\mu\mathrm{m}$ AlGaAsSb/InGaAsSb SCR SQW broadened-waveguide laser with 2 mm cavity, 200 μm aperture, and coated facets. (Reprinted from [45], with permission from the American Institute of Physics.)

uted to an even higher compressive strain of 1.4% in the well, as compared to 1% for the Sarnoff structure. Section 3.5.5 will discuss the incorporation of a similar active region into tapered-laser and tapered-laser-array configurations to achieve both near-diffraction-limited beam quality and high output power.

The group at Fraunhofer Institute fabricated GaInAsSb/AlGaAsSb SQW and triple QW diodes emitting at 2.26 μm [47]. The internal quantum efficiencies were 65% and 69%, respectively, and the triple-well structure with a 1-mm long cavity and a 64-μm stripe width emitted a cw output power of 240 mW per facet at 280 K. Figure 3.6 shows the threshold current densities as a function of inverse cavity length for both series of lasers. The extrapolation to infinite cavity length yields thresholds of 55 A/cm^2 for single-well devices and 150 A/cm^2 for triple-well lasers. Similar measurements of the quantum efficiencies *vs.* inverse cavity length yielded internal losses of 5 and 7.7 cm^{-1}, respectively. Because the loss is low, the optical gain necessary to achieve lasing is easily obtained from a single well, which can reach threshold at a lower injection current density.

Antimonide cw diode lasers have also demonstrated excellent spectral selectivity. Single-mode emission at $\lambda \approx 2\,\mu\mathrm{m}$ was first achieved by a LL 1-mm-long ridge-waveguide QW laser [48], which displayed a relatively large side-mode suppression ratio of 20 dB. Thermoelectrically cooled DH Fabry-Perot

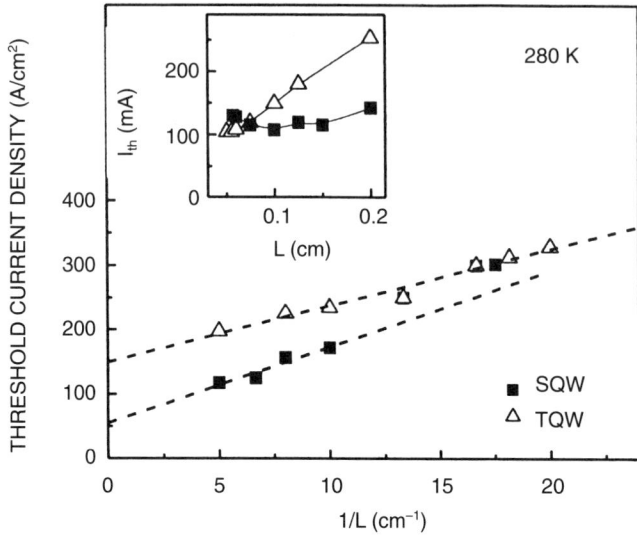

Fig. 3.6 Threshold current densities versus inverse cavity length for both single (■) and triple (△) QW lasers at 280 K. The inset shows threshold versus cavity length for the same lasers. (Reprinted from [47], with permission from the American Institute of Physics)

lasers also exhibited single-mode operation for mesa widths of 6 μm and short cavity lengths between 150 and 300 μm [49]. More recently single-mode operation was reported for GaInAsSb/AlGaAsSb QW lasers at temperatures as high as 403 K. The ridge width in those lasers was 5 μm, and the cavity length was 0.82 mm [50]. The emission spectra for a number of heat-sink temperatures and injection currents are shown in Fig. 3.7. Up to 45 mW/facet were emitted in a single mode at 300 K. Some general considerations concerning single-mode operation will be discussed in Section 3.5.4.

$\lambda \approx 2.3$–$2.7\,\mu m$

Emission from GaInAsSb/AlGaAsSb QWs at wavelengths somewhat longer than 2 μm can be realized by increasing the amount of compressive strain [20,21,51]. The most impressive performance to date was exhibited by Sarnoff broadened-waveguide devices with compressive strains of up to 2.3% [20]. Those diodes had two InGaAsSb QWs to increase the modal gain. With In fractions varying from 0.25 to 0.4 and As fractions held fixed at 2% [20], the compositions were nominally outside the quaternary miscibility gap. The wells were separated by approximately 1000 Å of AlGaAsSb with the same composition as the SCR, and the rest of the growth sequence and processing was quite similar to that of the 2-μm devices discussed above.

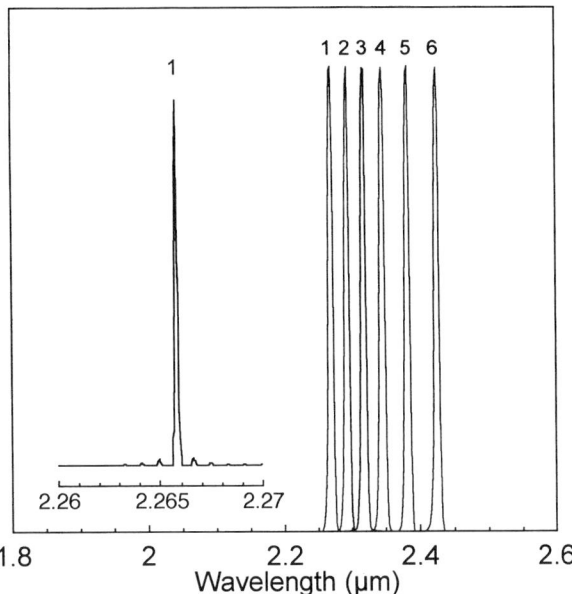

Fig. 3.7 Single-mode emission spectra for a GaInSbAs/GaAlSbAs QW ridge-waveguide laser operating cw at (1) $T = 295$ K, $I = 35$ mA, (2) $T = 313$ K, $I = 50$ mA, (3) $T = 333$ K, $I = 55$ mA, (4) $T = 355$ K, $I = 65$ mA, (5) $T = 378$ K, $I = 105$ mA, and (6) $T = 397$ K, $I = 180$ mA. (Reprinted from [50], with permission from the Institution of Electronics Engineers)

The cw threshold current densities at 290 K varied from 230 A/cm² at 2.3 µm to 400 A/cm² at 2.6 µm, and then to 1.1 kA/cm² at 2.7 µm. The differential quantum efficiency of about 30% from 2.3 µm to 2.6 µm dropped to a little over 20% at 2.7 µm. The L-I curves for 100-µm-wide devices operating in the 2.3 to 2.6 µm range are shown in Fig. 3.8. Maximum cw output powers were 500, 250, and 160 mW at room temperature for lasers emitting at 2.3, 2.5, and 2.6 µm. Thermal resistances in the 3 to 5 K/W range were estimated for those devices.

A high characteristic temperature of $T_0 = 110$ K for the 2.3 µm lasers operating at temperatures below 338 K was attributed to domination of the lasing threshold by Shockley-Read processes [19]. On the other hand, $T_0 \approx 40$–50 K was observed for the same device at higher temperatures and for the 2.7 µm laser at all temperatures. The temperature dependence of the spontaneous emission intensity implied that this much lower T_0 could be associated with Auger recombination. Hole leakage into the n-cladding layer was found not to be a factor limiting the high-temperature operation [52]. A Hakki-Paoli study of the gain spectrum indicated a more pronounced broadening than in near-IR devices.

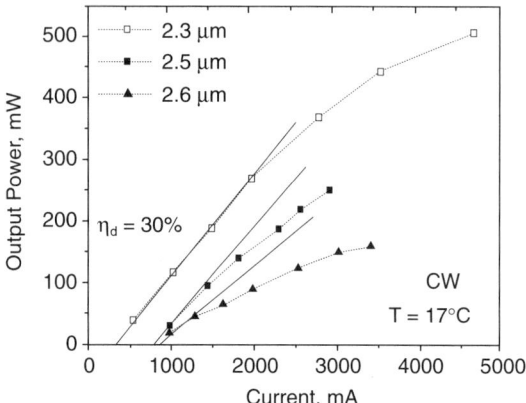

Fig. 3.8 Cw output power versus injection current for Sarnoff diodes emitting at 2.3 μm, 2.5 μm, and 2.6 μm at $T = 290$ K. (Reprinted from [20], with permission from the Institute of Electrical and Electronics Engineers, ©1999.)

Room-temperature pulsed operation at $\lambda = 2.63$ μm was reported by workers at the University of Montpellier for compressively strained GaInAsSb/GaSb active QWs with a marginally type-II band alignment [53]. The slope efficiency was 35 mW/A per facet. A shorter-wavelength (2.38 μm) type-II diode emitted more than 1 mW/facet cw at room temperature [54]. Single-mode operation with a side-mode suppression ratio of 30 dB was achieved using a narrow-ridge Fabry-Perot geometry [55], with a cw output power of 20 mW/facet at room temperature.

VCSELs

Vertical-cavity surface-emitting lasers (VCSELs) are typically formed by growing or depositing distributed Bragg reflector (DBR) mirrors below and above the active region, to form a short vertical cavity that favors emission in a single longitudinal mode. Single-mode operation can then be achieved by restricting the lateral area of the device. Since the active volume can be quite small, very low lasing thresholds have been observed. The circular, low-divergence output beam is quite convenient for many applications, and two-dimensional VCSEL arrays have been demonstrated in the near-IR. It should be noted that whereas the TE polarization of most interband semiconductor lasers is suitable for either edge or surface emission, the TM polarization of the intersubband transitions in a quantum cascade laser is incompatible with the vertical-cavity geometry.

In view of the challenges inherent to long-wavelength devices, it is perhaps surprising that the very first surface-emitting semiconductor laser operated at a mid-IR wavelength of $\lambda = 5.2\,\mu m$ [56]. The multi-mode cavity for that InSb injection device was formed by the polished surfaces of a 220-μm-thick wafer rather than DBR mirrors, a magnetic field was required, and operation was reported only to 10 K.

The achievement of single-longitudinal-mode mid-IR VCSELs with short cavities formed by DBRs is a much more recent occurrence. The first was by the Grenoble group, who fabricated an optically pumped HgCdTe VCSEL emitting at 3.06 μm [57]. The device operated only to $T = 30$ K, and at 10 K it required a high threshold pump intensity of 45 kW/cm^2. A subsequent device emitting at 2.63 μm operated in pulsed mode up to 190 K, and had a much lower threshold of 1.7 kW/cm^2 at 80 K [58]. Optically pumped lead-salt VCSELs have also been demonstrated recently. A University of Oklahoma/Naval Research Laboratory (NRL) collaboration achieved $\lambda = 4.5$–$4.6\,\mu m$ lasing at temperatures up to 290 K [59], while the Linz/Bayreuth team obtained VCSEL operation at lower temperatures, to 25 K at $\lambda = 6.1\,\mu m$ [60] and 85 K at $\lambda = 4.8\,\mu m$ [61], but with narrower spectral and angular emission.

Only two short-cavity III-V VCSELs emitting in the mid-IR have been reported to date, both of which employed antimonide QW active regions. Using the type-II W structure (discussed in Section 3.4.4.3), Felix et al. demonstrated an optically pumped VCSEL whose emission wavelength of $\lambda = 2.9\,\mu m$ was nearly independent of temperature [62]. Pulsed operation was achieved up to 280 K, and >2 W of peak power was obtained from a 600 μm spot at 260 K. In cw mode the device operated up to 160 K and for close to a 5-μm-diameter pump spot had a lasing threshold of only 4 mW at 78 K [63]. The power conversion efficiency at 78 K was 5.6% [64]. The temperature coefficient of the wavelength was quite small (0.09 nm/K), and the spectral linewidth of 2.9 nm was much narrower than for analogous edge-emitting devices. Vurgaftman et al. carried out detailed modeling of electrically and optically pumped type-II VCSELs [65]. Especially attractive was an interband cascade (see Section 3.4.5) VCSEL, emitting at $\lambda = 3.0\,\mu m$, that was projected to operate cw at 300 K with a threshold of 1.1 mA and output power of 1.2 mW.

The Montpellier group demonstrated an electrically pumped VCSEL that emitted near 2.2 μm [66]. The MBE-grown top and bottom DBR mirrors consisted of GaSb/AlAs$_{0.08}$Sb$_{0.92}$ quarter-wavelength stacks, and the active region occupying the one-wavelength-long optical cavity contained six GaInAsSb/GaSb strained QWs. Photoresponse spectra at 80 K and room temperature are shown in Fig. 3.9. At 296 K the cavity mode (1) is clearly observed, and the short-wavelength edge of the stop-band (2) can be seen at both temperatures. The band gap of the active region (3) is also apparent, owing to some absorption of the incident light by the edges of the mesa. For pulsed operation at 296 K, the 200-μm-diameter device had a threshold current density of 2 kA/cm^2 and peak output power of 20 mW.

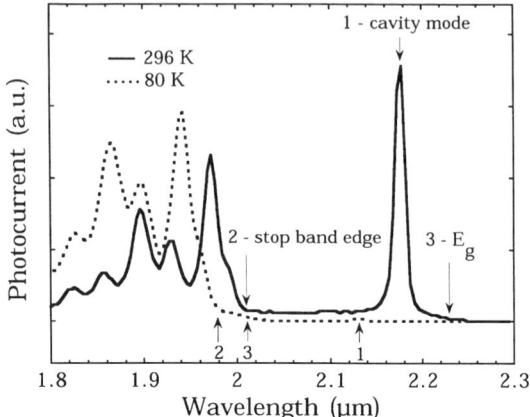

Fig. 3.9 Photoresponse spectra for an electrically pumped VCSEL with six strained GaInAsSb/GaSb QWs at $T = 80$ K (*dotted*) and 296 K (*solid*). (Reprinted from [66], with permission from the Institution of Electronics Engineers)

3.4 ANTIMONIDE LASERS EMITTING IN THE $\lambda \geq 3\,\mu m$ RANGE

3.4.1 Historical Development

Less than a year after the first semiconductor lasers were reported to operate in the near-IR, Melngailis demonstrated an InAs homojunction diode emitting at $\lambda = 3.1\,\mu m$ [67]. Magnetically tunable cw operation was also reported [68], as was the operation of a similar device employing InAsSb [69]. Later a bulk InSb vertical emitter operated at $\lambda \approx 5.2\,\mu m$ [70]. A step toward practical long-wavelength semiconductor lasers was the development of double-heterostructure devices in the 1980s, in particular, those based on InAsSbP/InAsSbP [71], InAsSb/InAsSbP [72–74], InAsSb/AlGaSb [75], and GaInAsSb/InAsSbP [76]. None of these early mid-IR lasers attempted to take advantage of the quantum confinement of carriers in the active region.

3.4.2 Double-Heterostructure Lasers

Double-heterostructure (DH) lasers have a simple geometry that is relatively easy to fabricate. A thick active layer (typically $\approx 1\,\mu m$ for mid-IR devices) consisting of a bulk binary, ternary, or quaternary is surrounded on both sides by higher gap layers to block carrier flow and hence reduce the threshold current density. Since larger band-gap materials tend to have lower refractive indices, the same "barriers" usually provide the optical waveguide as well. The most common active regions are As-rich InAsSb and InGaAsSb, while the barrier layers are usually AlGaAsSb or InAsSbP quaternaries. Mid-IR

DH lasers may be grown by LPE, MBE, or MOCVD on InAs or GaSb substrates.

Diodes

In 1989, researchers at the Ioffe Institute extended the lasing wavelength to 3.55 µm by using an LPE-grown $InAs_{0.93}Sb_{0.07}$ active layer combined with InAsSbP barriers on an n-InAs substrate [76]. The threshold current density was 86 A/cm^2 in pulsed mode, and the maximum operating temperature was 140 K. Cw lasing was achieved at 77 K with a threshold of 130 A/cm^2. Optimal parameters were determined through a systematic investigation of the threshold as a function of active region thickness, dopant concentration, and resonator length [77]. For cw operation, an InAsSb/InAsSbP DH laser with a 20 µm stripe operated to T_{max} = 82 K and produced 10 mW/facet of output power at 3.6 µm [78]. A careful study indicated that individual modes lased at a maximum power of 2 mW each.

The Ioffe group also grew DH lasers containing type-I InAs/InAsSb/InAs separate confinement regions clad by InAsSbP [3]. With pulsed injection, the maximum operating temperature was 203 K for emission at $\lambda \approx 3.4$ µm. Danilova et al. suggested that T_{max} could be increased further by adding more phosphorus to the confining layers or by creating several QWs.

In the mid-1990s, researchers at LL made significant progress with DH lasers grown by MBE on GaSb. For example, a $\lambda = 4$ µm device with an $InAs_{0.91}Sb_{0.09}$ active layer and $AlAs_{0.08}Sb_{0.92}$ claddings operated pulsed to 155 K and cw to 80 K [79]. While the threshold at 77 K was relatively low (158 A/cm^2), T_0 was only 17 K. The subsequent growth of a similar structure with improved lattice match (better than 2.5×10^{-3}) yielded a 77 K threshold of 80 A/cm^2 and $T_0 = 20$ K [80]. Figure 3.10 plots the threshold current density versus temperature for that device. The threshold at 170 K was 8.5 kA/cm^2, which is significantly lower than 24.4 kA/cm^2 for the earlier $\lambda \approx 4$ µm DH laser. The maximum pulsed and cw operating temperatures were 170 and 105 K, respectively [80], and with coated facets the cw output power was 24 mW at 80 K for a 100-µm stripe. Choi et al. pointed out that the structure contained a large potential barrier for electrons between the active and n-cladding layers, which increased the turn-on voltage and series resistance but had the advantage of confining holes to the active region.

In 1994, the LL group reported a $\lambda = 3.0$ µm DH laser with a metastable $Ga_{0.46}In_{0.54}As_{0.48}Sb_{0.52}$ active layer and $Al_{0.9}Ga_{0.1}As_{0.08}Sb_{0.92}$ cladding layers. That device yielded $T_{max} = 255$ K for pulsed operation and $T_{max} = 170$ K for cw operation, which exceeded all earlier results for $\lambda \geq 3$ µm up to that time [16]. The cw output power was 45 mW/facet at 100 K for a 100-µm aperture. Choi et al. suggested that the maximum temperatures and T_0 could be increased further by employing compressively strained QW structures with lower Auger recombination and free-carrier absorption losses.

Northwestern University has also made significant progress with DH lasers using MOCVD growth on InAs substrates. For example, InAsSb/InAsSbP

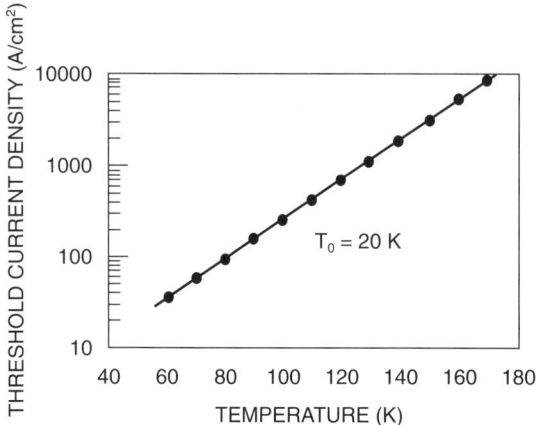

Fig. 3.10 Pulsed threshold current density versus temperature for a 4 µm DH laser with an $InAs_{0.91}Sb_{0.09}$ active layer and $AlAs_{0.08}Sb_{0.92}$ cladding layers. This LL device had a 100 µm × 500 µm cavity. (Reprinted from [80], with permission from the American Institute of Physics.)

laser diodes emitting at $\lambda = 3.2$ µm had a threshold of 40 A/cm^2 at 77 K, and an internal loss of only 3.0 cm^{-1}. They operated to 220 K in pulsed mode [8,81]. The cw output power was 150 mW/facet at 78 K for a 100 µm × 400 µm cavity. Based in part on a photoluminescence study of the minority carrier leakage, the rapid threshold increase around 150 K was attributed to both leakage and Auger recombination [82].

In order to reduce the leakage currents for both electrons and holes (see the discussion in Section 3.2 of the low valence band maximum for InAsSb relative to most of the possible barrier materials), Northwestern proposed an InAsSbP/InAsSb/AlAsSb DH that appears to provide adequate barriers. Figure 3.11 schematically illustrates the energy band diagram for this bandgap-engineered device, which emitted at $\lambda = 3.4$ µm [15]. The relatively smaller gap of the InPAsSb n-type cladding layer results in a low turn-on voltage of 0.36 V. The L-I curves in Fig. 3.12 indicate that at $T = 80$ K the peak pulsed output power was 1.88 W from two facets (Fig. 3.12a), while the cw maximum output power was 350 mW from two facets (Fig. 3.12b). An array of these lasers yielded 3.35 W peak power with a differential efficiency of 34%. However, the highest peak output from a single mid-IR diode was achieved using symmetric InAsSb/InAsSbP double heterostructures [83]. For $\lambda = 3.2$ µm, those lasers emitted approximately 3 W in pulsed mode from two facets at 90 K. The doping profile in the symmetric DH was further optimized to extend the maximum cw output power to 450 mW for two facets [84]. The achievement of higher powers from the symmetric structures was attributed to better heat dissipation for InAsSbP rather than AlAsSb cladding layers.

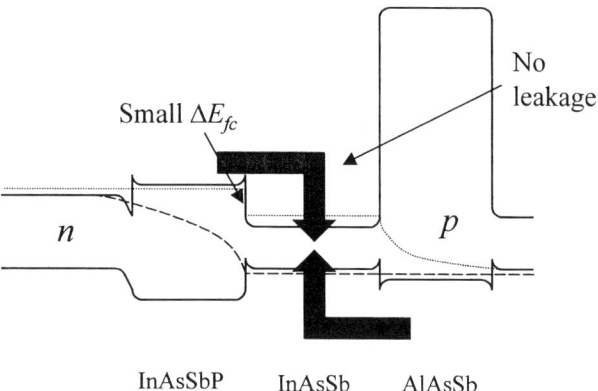

Fig. 3.11 Schematic energy bands for an InAsSbP/InAsSb/AlAsSb double heterostructure laser. Note that while AlAsSb is shown here to have a slightly lower VBM than InAsSb, a thorough review of the literature indicates that the InAsSb VBM should be lower [9]. (Reprinted from [15], with permission from the American Institute of Physics)

Apart from the early InSb homojunctions devices that lased only at very low temperatures [56], the longest wavelength electrically injected III-V interband lasers have been InSb/InAlSb DHs grown by MBE onto InSb or GaInSb substrates at DERA Malvern (strictly speaking, the lattice constants are too large for these structures to fit neatly into the "6.1 Å family" of antimonide lasers that is the focus of the present review). In pulsed mode, these diodes emitting at $\lambda = 5.1\,\mu m$ had a peak output of about 28 mW/facet at 77 K [85]. From the plot of threshold current versus temperature in Fig. 3.13, T_0 was 45 K up to 85 K, but then dropped to nearly 22 K at higher temperatures, possibly indicating the onset of a different loss mechanism [86]. The threshold current density was 1.48 kA/cm^2 at 77 K, and the maximum operating temperature was 110 K. With the application of a 5 T magnetic field (which effectively creates a quasi-quantum wire laser with 1D densities of states), the threshold current density at $T = 4.2\,K$ dropped from 93 to 58 A/cm^2.

Optical Pumping

Although far less convenient in terms of expense, compactness, and portability, optical pumping can be a valuable laboratory tool, since it requires far less processing and avoids the need to dope the cladding layers. Furthermore the superior mid-IR laser performance that is attainable using optical pumping with a near-IR diode array can sometimes outweigh the disadvantages in the most challenging applications, such as those requiring very high cw output powers. Le et al. pointed out that optical pumping with 2-µm radiation is equivalent to a 0.6 V voltage bias for the same quantum efficiency [87]. Insofar as most diodes require higher voltage biases (at least 2–3 V) to produce mid-IR

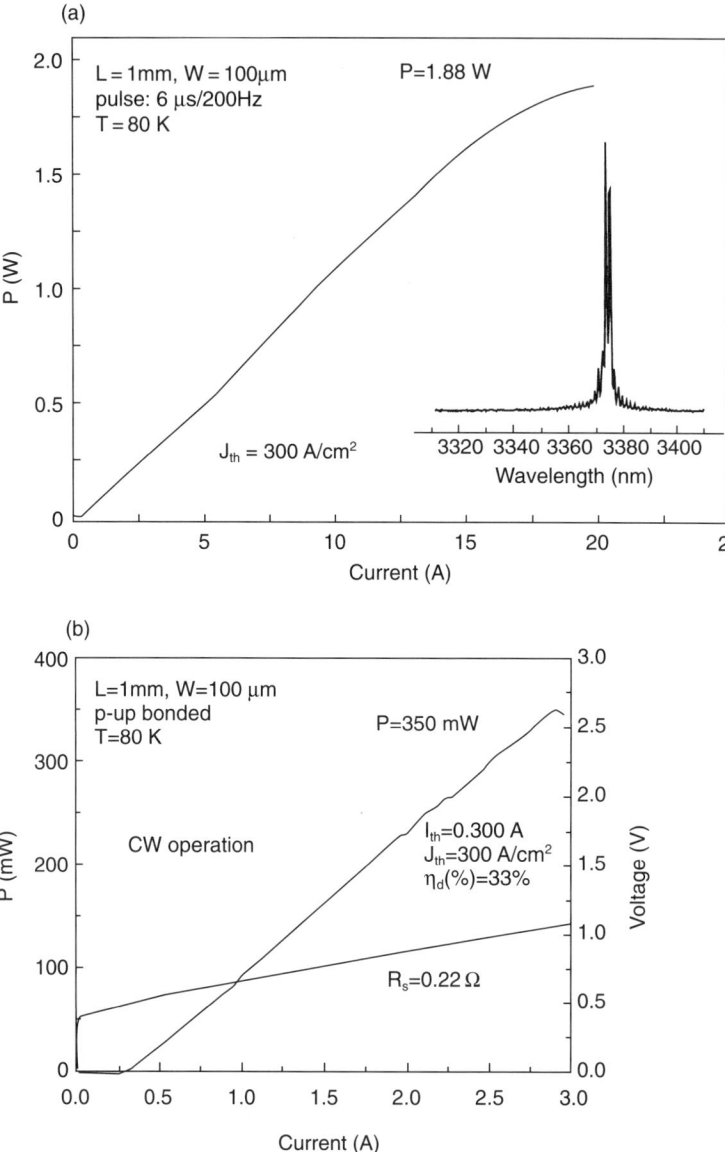

Fig. 3.12 (*a*) Pulsed & (*b*) cw output power per two facets *vs.* injection current for an InAsSb/InPAsSb/AlAsSb laser with the structure shown in Fig. 3.11. (Reprinted from [15], with permission from the American Institute of Physics.)

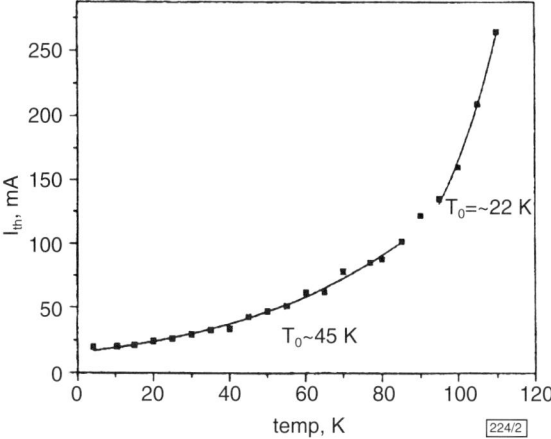

Fig. 3.13 Threshold current versus temperature for an InAlSb on InSb DH laser. (Reprinted from [86], with permission from the Institution of Electronics Engineers)

photons, pumping at 2μm can in principle have a higher power efficiency if most of the pump photons can be absorbed in the active region. One of the earliest optically pumped mid-IR DH lasers had an InAsSb active region and AlGaAsSb cladding [75]. That device emitted at λ = 3.9μm and operated to 125 K with T_0 = 17 K.

Most of the work on optically pumped DH lasers for the mid-IR has been conducted by H. Q. Le and co-workers at MIT Lincoln Laboratory. In one study, a DH with an InAsSb active region and AlGaAsSb cladding produced λ = 3μm emission, while a GaInAsSb active region with the same cladding was used for 4-μm emission. At T = 85 K, diode-laser pumping (0.8 or 0.94μm) yielded 1.5 W peak power and 95 mW average power per facet at 3μm, and 0.8 W peak power and 50 mW average power per facet at 4μm [88]. For very low duty cycles, the 3-μm laser operated to 210 K while the 4μm laser operated to 150 K. In order to increase the efficiency, a second study used longer-wavelength pumping with a 2.1μm Ho:YAG laser [87]. The 4-μm DH laser operated cw to 86.5 K with 11% absorbed power conversion efficiency per facet. The same laser operated in pulsed mode to 211 K.

Le et al. packaged a DH laser with a 2-μm $InAs_{0.91}Sb_{0.09}$ active region sandwiched by 4-μm $AlAs_{0.07}Sb_{0.93}$ cladding layers in a turn-key liquid-nitrogen-cooled system for which pumping was provided by a cw 0.98-μm diode array [89]. The cw output power was 270 mW/facet at 84 K and 360 mW/facet at 68 K. Further enhancement was achieved by using a 1.9-μm diode pump array to decrease the quantum defect ratio [90]. The heat sinking was also improved substantially, by mounting the InAsSb/AlAsSb DH device epitaxial-side-down and pumping through the GaSb substrate (which is strongly absorbing at 0.98μm). Figure 3.14 shows that 1.25 W peak output power and 7% optical-

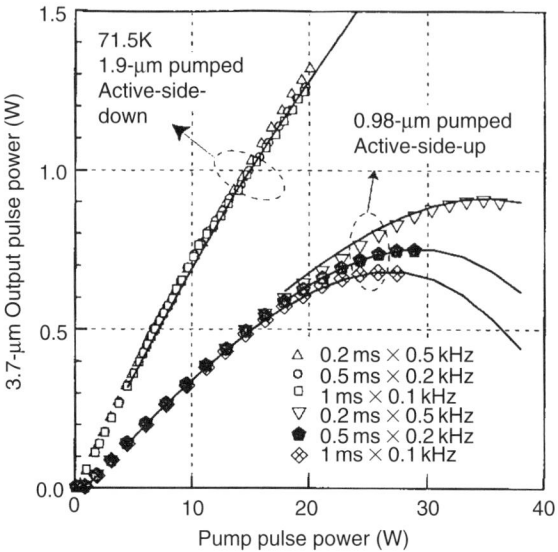

Fig. 3.14 Comparison of power performance for 0.98-μm epitaxial-side-up and 1.9-μm epitaxial-side-down optical pumping of a 3.7-μm InAsSb-AlAsSb DH laser. (Reprinted from [90], with permission from the Institute of Electrical and Electronics Engineers, ©1998.)

to-optical power conversion efficiency were obtained for 1-ms pulses and a 10% duty cycle. The maximum output is seen to be limited primarily by the pump power, since little saturation is evident. The figure also shows that 0.98-μm pumping of an epitaxial-side-up mounted sample produced a maximum power of only 0.67 W (2.7% efficiency). In that case the output was clearly limited by sample heating.

3.4.3 Type-I Quantum-Well Lasers

As described in the introduction, the band alignment is type-I when the conduction band minimum and valence band maximum reside in the same epitaxial layer of the structure. Type-I QW lasers have several advantages over DH lasers, including a two-dimensional (2D) density of states, improved carrier confinement, and the capability to tune the energy levels and hence the emission wavelength with the width and composition of the QWs. Due to the large miscibility gap in GaInAsSb, it is difficult to grow GaInAsSb alloy compositions emitting at wavelengths beyond 3 μm. Auger recombination severely limits the operation of DH lasers, and at least at wavelengths shorter than 3 μm, substantial suppression of the Auger rate has been demonstrated in strained QW structures (see Section 3.5.2). It was also predicted that compressive strain coupled with quantum size effects would lower the in-plane effective mass of the heavy holes, resulting in lower threshold current densities [91,92].

Diodes

The LL group has been responsible for the most significant advances of MBE-grown type-I QW injection lasers emitting beyond 3 μm. In late 1994, InAsSb/InAlAs QW diodes emitting at 4.5 μm operated in pulsed mode up to 85 K, with threshold current densities of 350 A/cm^2 at 50 K and 1.95 kA/cm^2 at 85 K [93]. The performance was thought to be limited by a nonuniform distribution of the injected holes. More uniform injection via optical pumping allowed the same structures to operate to higher temperatures, as will be discussed in the next subsection.

In order to increase the conduction-band offset, Choi et al. replaced the InAlAs barriers with InAlAsSb [94]. The tensile-strained barriers were combined with compressively strained InAsSb QWs to fabricate diodes emitting at $\lambda = 3.9$ μm. Maximum lasing temperatures were 165 K for pulsed operation and 128 K for cw ridge waveguide devices. The cw output power for a broad stripe laser was 30 mW/facet at 80 K, and the differential quantum efficiency was about 16% at 100 K. Lower J_{th} (78 A/cm^2 at 80 K) and higher T_0 (30 K) values relative to analogous DH lasers were attributed to improved carrier confinement and lower Auger rates in the strained QW structures.

LL further improved the temperature, threshold, and efficiency characteristics by growing compressively strained InAs$_{0.935}$Sb$_{0.065}$ MQW lasers with tensile-strained In$_{0.85}$Al$_{0.15}$As$_{0.9}$Sb$_{0.1}$ barriers on InAs substrates [95]. While the earlier devices had been grown on GaSb, it was found that the stable growth region (avoiding the InAlAsSb miscibility gap) could be extended by growing on InAs. Broad-area lasers emitting at 3.2–3.55 μm operated to 225 K in pulsed mode and ridge-waveguide lasers operated cw to 175 K. The pulsed threshold current density at 80 K was as low as 30 A/cm^2, and characteristic temperatures were 30 to 40 K. The observation of high turn-on voltages was attributed to large barriers between the substrate and the n cladding layer, and also between the n cladding and the active region.

A similar $\lambda = 3.4$ μm InAsSb/InAlAsSb strained QW laser, grown on n-InAs and mounted epitaxial-side-down, produced a record cw output power of 215 mW/facet at 80 K [96]. At 150 K, 35 mW/facet was obtained. The plot of pulsed threshold current density versus temperature in Fig. 3.15 indicates that $J_{th} = 44$ A/cm^2 at 80 K. The maximum operating temperature was 220 K, and the characteristic temperature decreased from 50 K at low T to 25 K near T_{max}. From a cavity length study, it was determined that the internal quantum efficiency was 63% and the internal loss was 9 cm^{-1}.

Significant progress has also been demonstrated using MOCVD to grow type-I InAsSb QW lasers. In 1995, Kurtz and co-workers at Sandia grew a pseudomorphic active region consisting of compressively strained InAs$_{0.94}$Sb$_{0.06}$ QWs for enhanced hole confinement and carrier diffusion, surrounded by InAs barriers [97]. The result was a type-I band alignment with a

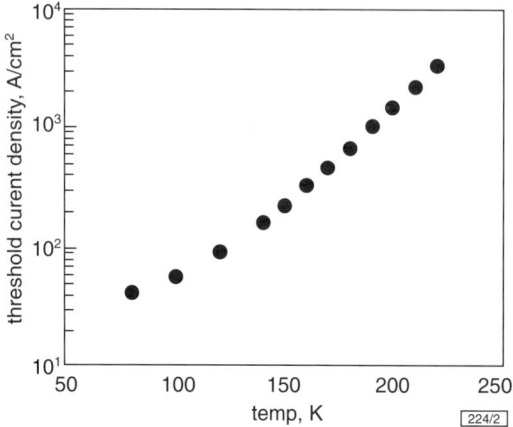

Fig. 3.15 Pulsed threshold current density versus temperature for a 250 μm × 1000 μm InAsSb/InAlAsSb strained QW diode laser emitting at 3.4 μm. (Reprinted from [96], with permission from the Institution of Electronics Engineers.)

valence-band offset of 50 to 100 meV. InPSb cladding layers then provided the optical confinement for this gain-guided injection laser emitting at 3.5 μm. For 100-ns pulses at 10^{-4} duty cycle, the maximum operating temperature was 135 K and the characteristic temperature was 33 K. For a 40-μm stripe, the threshold current at 77 K was 250 A/cm^2, with a turn-on voltage of 2–3 V at low temperature. The authors speculated that the high series resistance (10^5 Ω/cm^2) and turn-on voltage arose from cladding either due to carrier freeze-out or contact resistance. It was also proposed that the characteristic temperature could be increased by substituting InAlSb barriers with larger band offsets.

In 1996, the Sandia group demonstrated the first MOCVD-grown mid-IR diodes with AlAsSb optical cladding layers [6,7,98]. Another novel feature was that a semimetallic p-GaAsSb/n-InAs heterojunction was employed to inject electrons into the strained InAsSb/InAs multiple QW active region. Gain-guided lasers emitted at 3.8 to 3.9 μm and operated to 210 K in pulsed mode, with a characteristic temperature of 30 to 40 K. However, peak powers were only about 1 mW/facet for 0.1% duty cycle, and the longest pulses to produce lasing were 10 μs. For longer pulses and high forward bias, extreme band bending may have depleted the semimetallic injection region.

The Northwestern group used low-pressure MOCVD to grow compressively strained MQW laser diodes [99]. The active regions consisted of 10 InAsSb QWs, embedded in InAs and surrounded by n- and p-InAsSbP cladding layers. At $T = 90$ K and $\lambda = 3.65$ μm, the output was up to 500 mW/facet in pulsed mode (6 μs, 200 Hz). A 100 μm × 700 μm device exhibited a differential quantum efficiency of 70%, which remained >65% for temperatures up to 155 K. The lasers operated up to 200 K in pulsed mode, and thresholds were approximately half of those for DH lasers grown under the same conditions.

In 1999, the Northwestern group reported MOCVD-grown InAsSb/InAsP strained-layer superlattice injection lasers emitting at 4.0 μm [100]. While the active region was not strain-compensated, the thin InAsP barrier layers were designed to allow for a more uniform carrier distribution. The L-I curves in Figs. 3.16a and b illustrate that at 100 K the pulsed and cw peak output powers were 546 and 94 mW (two facets), respectively. Threshold current densities were as low as 100 A/cm^2, and the laser operated to T_{max} = 125 K with T_0 = 27 K. The differential quantum efficiency was >30% in both pulsed and cw modes.

Optical Pumping

For optical pumping, type-I QW mid-IR lasers have in general not demonstrated any dramatic improvements over the performance of analogous DH devices emitting at the same wavelength.

The first report of a type-I optically pumped laser with a superlattice or QW active region was by the Sandia group in 1994 [10]. The MOCVD-grown strain-compensated structures were superlattices with biaxially compressed $InAs_{0.9}Sb_{0.1}$ wells and biaxially tensile-strained $In_{0.93}Ga_{0.07}As$ barriers, surrounded by InPSb optical cladding layers. The lasers emitted at 3.9 μm and operated to 100 K under pumping by 30-ns pulses from a Ti-sapphire laser (825 nm).

Sandia also demonstrated the first mid-IR lasers with MOCVD-grown InAsSb/InAsP strained-layer superlattice (SLS) active regions [14]. For devices pumped by 10-ns pulses from a 1.06-μm Nd:YAG laser, Fig. 3.17a shows the output power per facet *vs.* pump intensity corresponding to low duty cycle and a pump-stripe width of 200 μm. At the maximum operating temperature of 240 K (the highest for any type-I interband III-V laser at this wavelength), the device emitted at λ = 3.86 μm. The characteristic temperature, as determined from the plot of threshold intensity versus T in Fig. 3.17b, was 33 K.

Zhang et al. at Hughes Research Laboratory (HRL) studied optically pumped λ = 3.4 μm lasers with $InAs_{1-x}Sb_x$/InAs superlattice active regions that were grown by MBE [12]. (Even though the active regions of these structures were type-II, these lasers will be discussed here because of their closer kinship to type-I InAsSb superlattice lasers than to the InAs/GaSb-based type-II QW structures discussed below. Actually the distinction between type-I and type-II becomes relatively unimportant when the offsets are small and the electron wavefunctions are not appreciably localized in any one layer.) The substrate was InAs, and the optical cladding layers were AlAsSb. A threshold pump intensity of ≈400 W/cm^2 was measured at the maximum cw operating

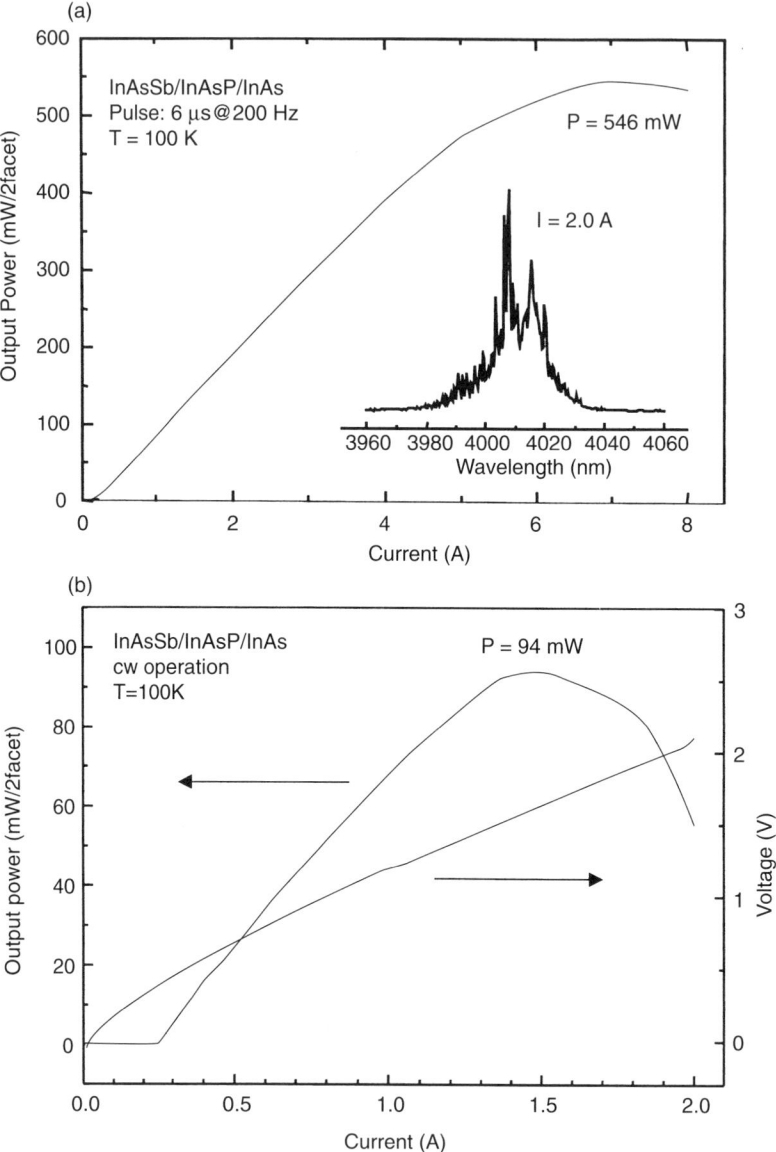

Figs. 3.16 Light output versus injection current in (*a*) pulsed and (*b*) cw modes for an InAsSb/InAsP strained-layer superlattice laser at $T = 100$ K. The inset in part (*a*) shows the emission spectrum at 80 K for $I = 2.0$ A. The cw current voltage characteristic is shown in (*b*), right scale. (Reprinted from [100], with permission from the American Institute of Physics)

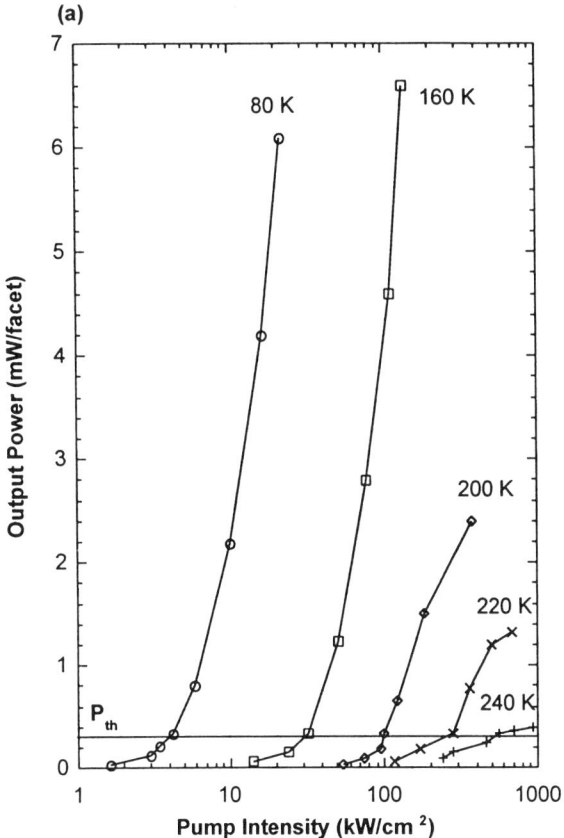

Fig. 3.17 (*a*) Pulsed output power plotted against optical pump intensity at temperatures from 80 to 240 K for an MOCVD-grown InAsSb/InAsP strained-layer superlattice laser. (*b*) Threshold pump intensity against temperature for the same laser. (Reprinted from [14], with permission from the American Institute of Physics.) *Continued on next page.*

temperature of 95 K, which was limited by the available pump intensity rather than by saturation.

At LL, lasers with compressively strained InAsSb QWs and tensile-strained InAlAs barriers were designed to provide higher conduction band offsets than are attainable using InGaAs barriers [93]. For optical pumping by 8-µs pulses from a 0.94-µm diode laser array (300-µm stripes on 2.5-mm cavities), the devices operated to $T_{max} = 144$ K, where the emission wavelength of 4.5 µm was the longest of any InAs-based laser up to that time (1995). The threshold pump intensity at 95 K was only 47 W/cm^2. The peak output power at the same temperature was 0.54 W, which required a pump power of 58 W.

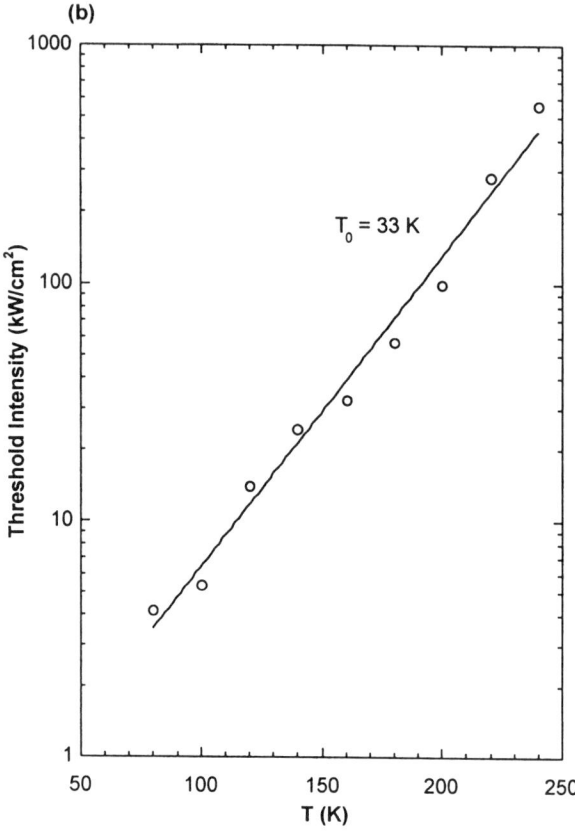

Fig. 3.17 *Continued*

LL researchers also studied strained InAsSb QW lasers on GaSb substrates [101], in which InAlAsSb was used as the barrier material in order to improve the confinement of both electrons and holes. The $\lambda = 3.9\,\mu m$ devices operated to 162 K for pulsed optical pumping. The peak output power was 0.6 W at 110 K, and the maximum cw power at 80 K was 30 mW/facet. Electrical pumping data for the same structure were discussed in the preceding subsection.

3.4.4 Type-II Quantum-Well Lasers

In the preceding discussions of antimonide DH and type-I QW lasers emitting at $\lambda > 3\,\mu m$, we saw that primary factors limiting the performance include (1) inadequate band offsets, particularly due to the low valence-band maximum of InAsSb, (2) rapid Auger nonradiative decay, and (3) restricted spectral range, limited at a given temperature to wavelengths shorter than the cutoff in bulk InAsSb. All of these limitations can, in principle, be removed by employing a type-II quantum heterostructure such as InAs/GaSb. For example, the electrical confinement is far more effective because GaSb has one of the highest valence-band maxima in the antimonide family rather than one of the lowest (see Fig. 3.2). Hole states in the GaSb are bound by the InAs VBO, just as electron states in the InAs are confined by the GaSb CBO. Auger recombination tends to be significantly suppressed in type-II QWs, as will be discussed in Section 3.5.2. Moreover, because the InAs conduction band minimum lies more than 100 meV *below* the GaSb valence band maximum, both arbitrarily long ($\gg 20\,\mu m$) and relatively short ($\leq 2\,\mu m$) wavelengths are attainable by using QW width variations to tune the quantum confinement energies. In this subsection, we will discuss how these potential advantages have been realized in practice.

Type-II Single-Heterojunction Lasers

Kroemer and Griffiths pointed out in 1983 that optical transitions at a type-II interface can produce lasing at a photon energy smaller than the energy gap of either constituent [102]. That possibility was first demonstrated experimentally in 1986, when Baranov et al. at the Ioffe Institute observed lasing at $\lambda = 1.86\,\mu m$ from a single LPE-grown type-II InGaAsSb/GaSb heterojunction [103].

Moiseev et al. from Ioffe used p-GaInAsSb/p-InAs type-II heterojunctions to obtain lasing at $\lambda \approx 3.3\,\mu m$ [2,104]. The temperature characteristics were further improved by inducing a p-n rather than p-p heterojunction to reduce hole leakage while maintaining a large band offset for electron confinement [105]. The p-n heterojunction laser operated in pulsed mode to $T_{max} = 195\,K$, and had a threshold current density of $400\,A/cm^2$ at 77 K with single-mode lasing at $\lambda = 3.2\,\mu m$. A later device of this type, with the threshold current density vs. temperature illustrated in Fig. 3.18, lased to 203 K [106]. In 1998, the Ioffe group investigated MBE- (rather than LPE-) grown type-II QW heterojunctions, and demonstrated electroluminescence in the 3 to 4 µm range at $T = 77\,K$ [107].

Type-II Superlattice Lasers

The most obvious disadvantage of the type-II approach is that the optical matrix elements and gain are inevitably reduced by the weaker overlap of the electron and hole wave functions, since the two carrier types reside primarily in different layers. This is why all semiconductor lasers had a type-I band alignment prior to the antimonide type-II devices. In order to increase the gain to a usable level, the QWs should be as thin as possible, which forces the wave functions to overflow significantly into the neighboring "barrier" layers. This strategy was followed by Smith and Mailhiot in designing type-II superlattice infrared detector active regions with interband absorption coefficients roughly as large as those in bulk HgCdTe [108]. Those authors proposed that one can

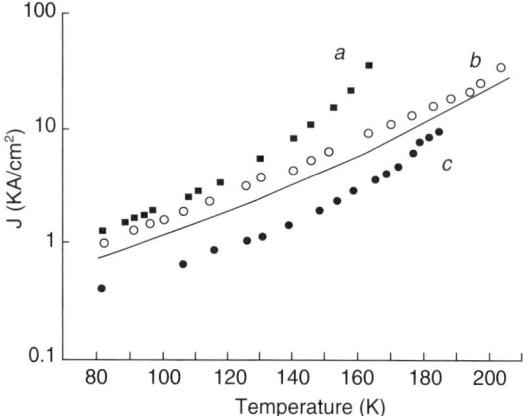

Fig. 3.18 Threshold current density versus temperature for (a) a type-I InAsSb/InAsSbP laser (■), (b) a type-II staggered-alignment InAsSb/InAsSbP laser (○), for which $T_0 = 40$ K and $T_{max} = 203$ K, and (c) a type-II broken gap p-n GaInAsSb/InGaAsSb laser (●). (Reprinted from [106], with permission from the Institution of Electronics Engineers (IEE)).

reduce the QW thicknesses (and enhance the wave function overlap) while maintaining the same type-II energy gap if strain is introduced by incorporating indium into the GaSb. Although the early development of type-II InAs/GaInSb strained-layer superlattices was motivated primarily by IR detector applications [109,110], Grein et al. pointed out in 1993 that type-II superlattice lasers may also have attractive properties [111]. They noted that the advantages would include large conduction and valence band offsets as well as the potential for substantially suppressing Auger recombination.

Within a year of that proposal, the HRL group achieved low-temperature stimulated emission at $\lambda = 3.2\,\mu m$ from a type-II InAs/Ga$_{0.75}$In$_{0.25}$Sb superlattice structure grown by MBE [112], and then lasing up to a maximum operating temperature of 160 K in pulsed mode at $\lambda = 3.5\,\mu m$ [113]. Emission wavelengths were easily tuned by varying the thickness of the superlattice wells. Subsequent HRL work rapidly extended the temperature range to 255 K in pulsed mode and 180 K for cw operation, and the emission wavelength to 4.3 µm [114]. The optical cladding layers were n-doped and p-doped InAs/AlSb superlattices [115], which were strain balanced to achieve an overall lattice match to the GaSb substrate. The active regions consisted of a number of InAs/Ga$_{0.75}$In$_{0.25}$Sb superlattice segments (typically $\approx 4\frac{1}{2}$ periods in each segment) sandwiched between Ga$_{0.75}$In$_{0.25}$As$_{0.22}$Sb$_{0.78}$ barriers, for a total of around 5 to 8 repeats of the segment-plus-barrier substructure. While by some figures of merit the laser performance was at or near the state of the art for its time, later calculations showed that the $4\frac{1}{2}$-period superlattice segments yield a poor wave function overlap between the lowest conduction subband and the highest valence subband [116]. It is actually the third valence subband, lying a few meV below the valence band maximum, that participates in the lasing transitions.

The simplest multiple-period type-II active region is a two-constituent superlattice. Type-II diode lasers based on InAs/GaSb superlattices were investigated at the University of Montpellier [117]. In that study, thin InAs layers were inserted into a GaSb matrix, and the hole confinement was attributed to band bending in the vicinity of the InAs layer. Lasing at 80 K was reported at 2.85 µm, and room-temperature operation was achieved at wavelengths shorter than 2.5 µm.

Type-II superlattice diodes were also studied by a collaboration of NRL and the University of Houston (UH) [118]. Lasers emitting at $\lambda = 2.92$ µm operated to $T = 260$ K in pulsed mode. At 200 K, they had a threshold of 1.1 kA/cm^2 and generated >100 mW per facet of peak power.

Wilk et al. quite recently reported a study of $\lambda = 3.5$ µm diode lasers containing InAs$_{0.92}$Sb$_{0.08}$/InAs multiple QWs with a relatively large type-II band misalignment (115 meV for electrons and 70 meV for holes) [119]. For pulsed operation, the threshold current was 150 A/cm^2 and the maximum operating temperature was 220 K. Cw operation was also achieved, with $T_0 = 40$ K and $T_{max} = 130$ K.

Type-II Quantum-Well Lasers and the "W" Laser

Design Considerations. While the optical matrix elements for a two-constituent type-II superlattice active region can be quite favorable, penetration of the electron wave functions into the thin GaInSb layers is in fact so great that a wide miniband forms along the growth axis. The electrons then have strong energy dispersion in all three dimensions, whereas it is well known that QW lasers with quasi-2D electron and hole populations generally display superior performance to DH lasers with three-dimensional (3D) carriers [120]. This is primarily because the stepped 2D density of states yields a much higher gain per injected carrier at threshold. The most straightforward approach to blocking the formation of a miniband in the superlattice is to add a high electron barrier between the periods, such as InAs/GaInSb/AlSb. However, the drawback of that approach is that the wave function overlap becomes much smaller than in the simple superlattice, resulting in significantly lower gain [121].

On the other hand, if a second electron well is added on the other side of the hole well, the resulting structure with four constituents per period (e.g., InAs/GaInSb/InAs/AlSb) combines the large matrix element of a two-constituent superlattice with the 2D electron and hole confinement of a three-constituent multiple QW [122]. Band profiles, energy levels, and wave functions for the W laser, which takes its name from the shape of the conduction-band profile, are illustrated in Fig. 3.19. Calculations show that the electron wave function ψ_n, which has its maximum in the InAs layers, and the hole wave function ψ_p, which is centered on the GaInSb, overlap sufficiently to yield interband optical matrix elements that are almost 70% as large as those in a typical type-I laser structure. The W laser also maintains the effective electrical confinement and potential for strong Auger suppression of the other type-II approaches. A similar structure with quinternary AlGaInAsSb barriers was later discussed by workers at the University of Iowa [116].

Optical Pumping. As a means of achieving high output powers at $\lambda \geq 3\,\mu m$ and $T > 77\,K$, optically pumped type-II W lasers have surpassed all other semiconductor approaches. Pulsed lasing at $\lambda = 3.9$–$4.1\,\mu m$ was first reported in 1996 by NRL and UH [123]. Within a few months the first room-temperature

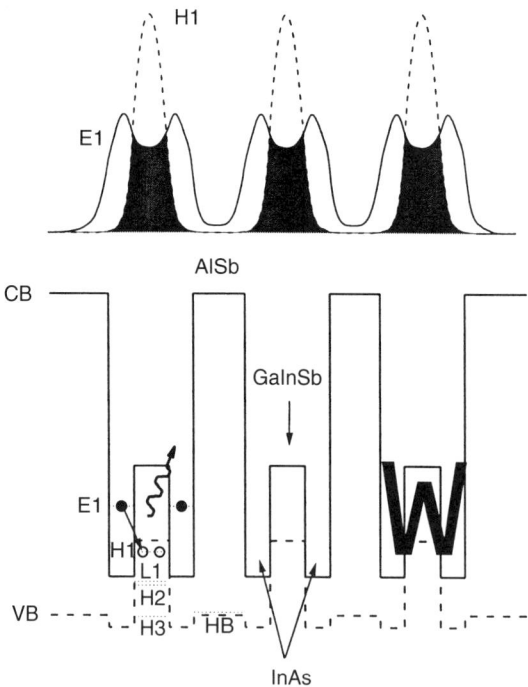

Fig. 3.19 Energy levels and conduction, valence, and split-off band profiles for the four-constituent, type-II W well configuration. At the top are shown electron (*solid*) and hole (*dashed*) wave functions, with their overlap shaded in gray. (Reprinted from [122], with permission from the American Institute of Physics)

interband semiconductor lasers emitting beyond 3μm were demonstrated [124]. VCSEL operation at λ = 2.9μm was also reported by NRL/HRL [62] (see Section 3.3.2 above). Pulsed operating temperatures well above 300 K are now routinely attainable out to at least λ = 4.5μm [125–128]. The maximum wavelength for interband III-V lasers was also extended using the W approach, to 5.2μm by the University of Iowa with HRL and Air Force Research Laboratory [129], and then to 7.3μm by NRL [130]. Even the longest wavelength device operated to 220 K in pulsed mode and to 130 K cw. Flatté et al. [131] and Vurgaftman et al. [132] project high performance at much longer wavelengths, possibly even to λ = 100μm for $T = 4.2$ K operation.

While peak output powers significantly exceeding 1 W are easily attainable using short-pulse excitation (even at 350 K [128]), most applications will require either cw or high-duty-cycle quasi-cw operation. MIT Lincoln Labs with UH pumped a 4-μm GaInSb/InAs type-II W laser at 0.98μm and obtained 1.5 W peak output power at 71 K for 100-μs pulses with a 10% duty cycle [133]. Figure 3.20 shows that the average output power at 82 K was up to 360 mW for 20-μs pulses and a 50% duty cycle. The net power conversion efficiency was between 3.5% and 4%.

NRL/Sarnoff demonstrated cw operation of a 3.4-μm W laser to $T = 220$ K [134], using a 1.06-μm Nd:YAG pump beam and epitaxial-side-up mounting. When a second device from the same wafer was mounted epitaxial-side down, the cw T_{max} increased to 275 K [135]. This was accomplished using the diamond-pressure-bonding (DPB) technique, which is discussed in Section 3.5.6. Using the DPB approach, cw operation at λ = 3.0μm was observed nearly to room temperature (290 K), and even a laser emitting at 6.1μm emitter operated cw to 210 K [135,136]. At that time no other III-V semiconductor laser emitting beyond 3μm had achieved cw lasing above T_{max} = 180 K, though recently cw operation of quantum cascade lasers reached 312 K [137]. The cw output power of 540 mW/facet from a λ = 3.2μm W laser with 80 QW periods was the highest reported for 78 K operation in its wavelength range [138]. The peak output for quasi-cw operation with a 25% duty cycle was 0.76 W, which fell to 0.27 W at 140 K. Despite the rather high output power, the differential power conversion efficiency was only 2.5% at 78 K and 1% at 140 K.

Figure 3.21a schematically illustrates the optical pumping injection cavity (OPIC) approach in which the active region is surrounded by semiconductor Bragg mirrors that form an etalon cavity whose resonance is tuned to the pump wavelength [139]. Increasing the number of passes that the pump beam makes through the active region results in three significant advantages: (1) the pump absorbance can be high for a long pump wavelength (e.g., $\lambda_{pump} \approx 2\mu m$), (2) the lasing threshold is reduced, and (3) the internal loss can be suppressed because fewer wells are needed to obtain a high pump absorbance. Both (1) and (3) contribute to a substantial enhancement of efficiency.

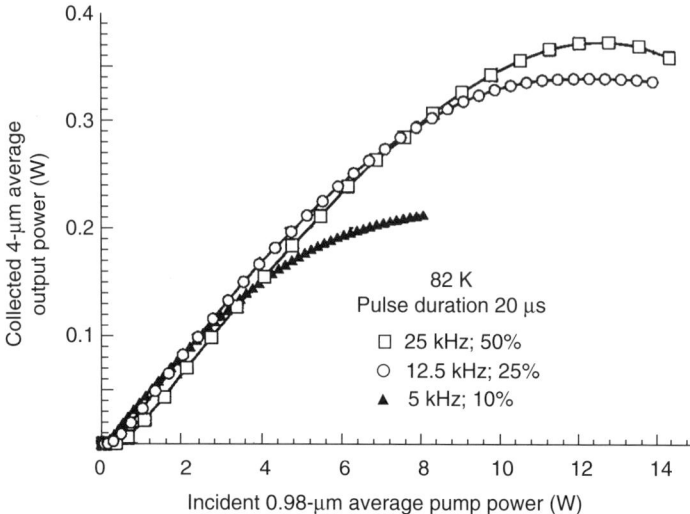

Fig. 3.20 Average output power *vs.* incident pump power (at 0.98 µm) for a 4 µm GaInSb/InAs/AlSb type-II QW laser. (Reprinted from [133], with permission from the American Institute of Physics.)

The filled circles and boxes in Fig. 3.21*b* illustrate pulsed differential power conversion efficiencies for two OPIC lasers emitting in the 3.1–3.4 µm and ~3.7 µm ranges, respectively [140]. Both contain only 10 active W QWs compared to 80 QWs in the conventional W laser, which produced the high cw and quasi-cw output powers discussed above (open circles). The OPIC efficiencies of 9 to 12% per uncoated facet at 78 K degrade only slowly with increasing temperature, for example, to 7.1% at 220 K for one device and 4% at 275 K for the other. The OPIC threshold pump intensities of nearly 8 kW/cm^2 at 300 K were also significantly lower than for the conventional device (≈40 kW/cm^2). Internal losses were significantly suppressed in the OPIC lasers, as will be discussed in Section 3.5.2.

The group at LL has pursued an alternative approach to obtaining high pump-beam absorbance despite a small number of QWs [141]. In the integrated absorber (IA), the active W QWs are separated by intermediate-gap GaInAsSb absorbing layers, which donate optically generated electrons and holes to the active wells. An IA laser with only five QWs separated by around 1600-Å-thick absorber layers was pumped by a 1.85 µm diode array. The differential slope efficiency was 9.8%, and the peak output power was 2.1 W for 35-µs pulses at 2.5% duty cycle. Besides displaying much slower efficiency degradation with increasing temperature, the IA beam quality was much better than for analogous InAsSb/AlAsSb DH lasers with similar geometry, presumably owing to reduction of the threshold gain. Further improvements

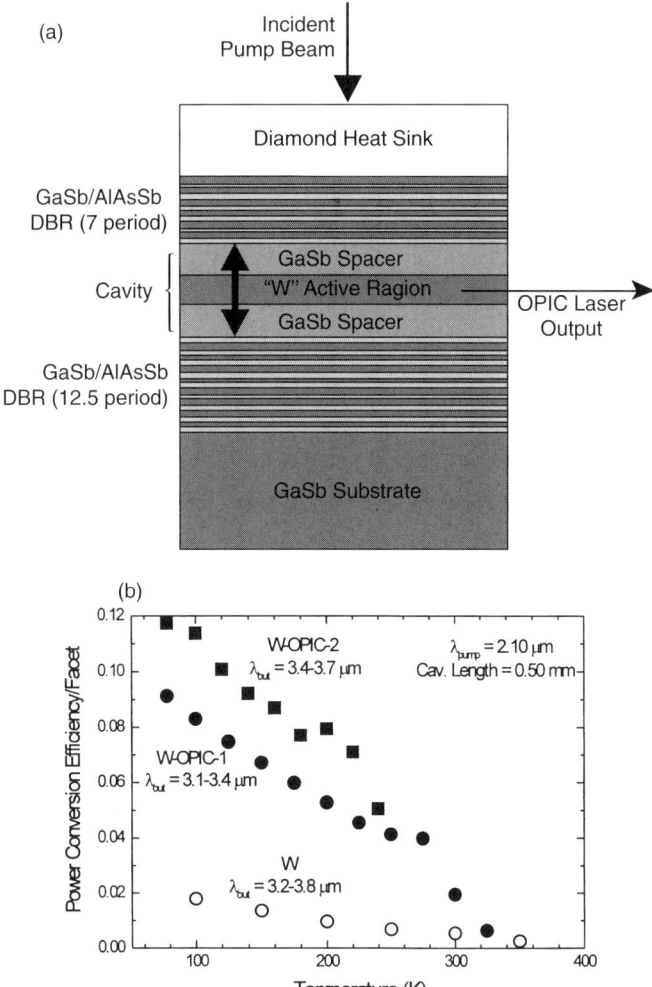

Fig. 3.21 (*a*) Schematic of the W-OPIC laser, in which the W active region is surrounded by GaSb/AlAsSb distributed Bragg reflectors (reprinted from Ref. [138], with permission from the American Institute of Physics). (*b*) Power conversion efficiencies per facet against temperature for a 3.1 to 3.4 μm W-OPIC laser (●), a 3.4 to 3.7 μm W-OPIC laser (■), and an optically pumped 3.2–3.8 μm type-II W laser without a pump cavity (○) (reprinted from [139], with permission from the Institute of Electrical and Electronics Engineers).

to the beam quality for optically pumped W lasers will be discussed in Section 3.5.5.

W Diodes. The main advantage of the four-constituent W structure (e.g., InAs/GaInSb/InAs/AlSb) over a simple two-constituent type-II superlattice (e.g., InAs/GaInSb) is that the high barrier layers (e.g., AlSb) induce a 2D density of states for both electrons and holes. Generalization of the W configuration to electrical pumping is not entirely straightforward, however. This is because holes injected from the p-type cladding layer may not be transported well along the growth direction, and hence hole populations in the various GaInSb QWs could be highly nonuniform. Whereas the holes in an InAs/GaInSb superlattice are required to tunnel through only one InAs electron well at a time, the barrier for the structure in Fig. 3.19 consists of two InAs layers in addition to the AlSb layer. Since the growth-axis hole mass is heavy, the net thickness is far too great to allow any appreciable interwell transfer.

A solution pursued by the NRL/Sarnoff team was to lower the barrier separating the periods of the W structure, so that interwell tunneling can occur via thermally excited light-hole states [142]. Quaternary $Al_{0.15}Ga_{0.85}As_{0.05}Sb_{0.95}$ barriers are employed rather than AlSb or AlAsSb, which brings the first light-hole subband to within 100 meV of the valence-band maximum. Another key design element is the insertion of relatively thick (0.6 μm) undoped $Al_{0.35}Ga_{0.65}As_{0.03}Sb_{0.97}$ "broadened-waveguide" layers on each side of the active region (see Section 3.3.2), in order to minimize free-carrier absorption in the doped $Al_{0.9}Ga_{0.1}As_{0.07}Sb_{0.93}$ cladding layers. Some of the devices also contain InAs/AlSb "hole-blocking" layers, which prevent hole leakage into the n-type broadened waveguide layer.

The resulting W diode configuration yielded the first room-temperature operation of an electrically pumped III-V interband laser beyond 3 μm [143]. For a device with uncoated facets and 10 QWs in the active region, a pulsed-mode T_{max} of 300 K was obtained at λ = 3.3 μm and with a threshold current density of about 20 kA/cm². With the application of facet coatings, the maximum operating temperature was extended to 310 K, with J_{th} = 25 kA/cm² [142]. A 5-QW device with dimensions 100 μm × 2 mm operated in cw mode to 195 K, with a threshold current density of 1.4 kA/cm² and a T_0 of 38 K from 80 K to 195 K [142]. Figure 3.22 shows L-I curves for cw operation at 180 K and 195 K. The turn-on voltages and series resistances of those devices are not yet optimized.

The W configuration has also been applied to MOVPE-grown InAsSb/InAsP/InAsSb/InAsPSb QWs with InAsPSb broadened waveguides. Those λ = 3.3 μm laser diodes with *n*-InAsPSb and *p*-InPSb cladding layers operated in pulsed mode up to 135 K [144]. At 90 K the threshold current density was 120 A/cm² and the efficiency was 31 mW/A per facet.

To determine the optimal active region for an antimonide mid-IR laser, Olesberg et al. applied the figure of merit: $[(\gamma - \alpha_a)/J]_{max}$, where γ is the mate-

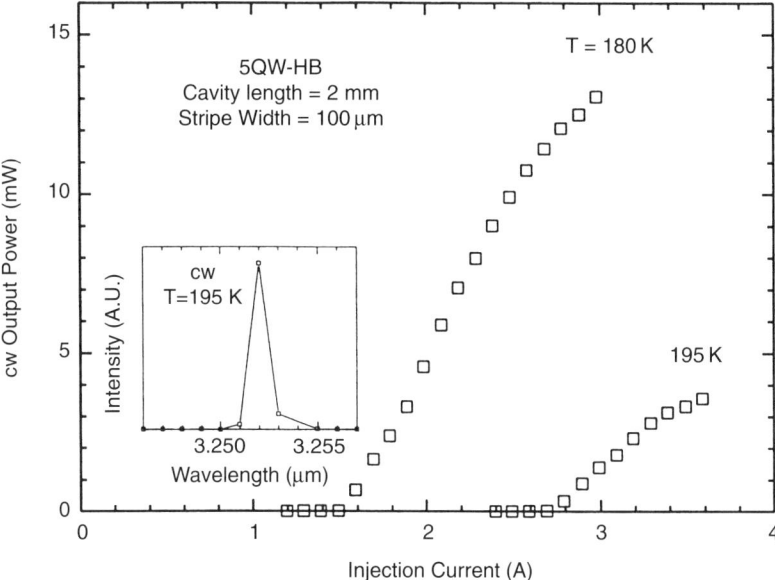

Fig. 3.22 Cw output power versus injection current at 180 and 195 K for a 5 QW W diode laser with broadened waveguide, hole-blocking layers, and facet coatings. The inset shows a spectrum at 195 K and 3.5 A. (Reprinted from [141], with permission from the American Institute of Physics)

rial gain, α_a is the absorption, and J is the volumetric current density [145]. A comparison of seven different InAs/GaInSb-based type-II, type-I QW, and bulk structures led to the conclusion that optimized "four-layer superlattice" (W) type-II configurations should have 2 to 30 times lower room-temperature threshold current densities than the various alternatives. They found that the optimal structure would contain only 3 QW periods, which is considerably fewer than any of the $\lambda \geq 3\,\mu m$ optically pumped and diode antimonide QW lasers have employed to date. A later analysis concluded that out to at least $\lambda = 11\,\mu m$, optimized type-II lasers should outperform intersubband quantum cascade lasers [131]. The same group recently projected even better performance for a new structure with a single-period W active region [146].

3.4.5 Interband Cascade Lasers

Structure

The interband cascade laser (ICL) is a counterpart to the intersubband quantum cascade laser [147,148], and is described more fully in Chapter 5 of this book. It generates multiple photons per injected electron by making an optical transition at each step of a staircase-like QW structure. Rui Yang first proposed the ICL in 1994 [149], shortly after the initial demonstrations of the

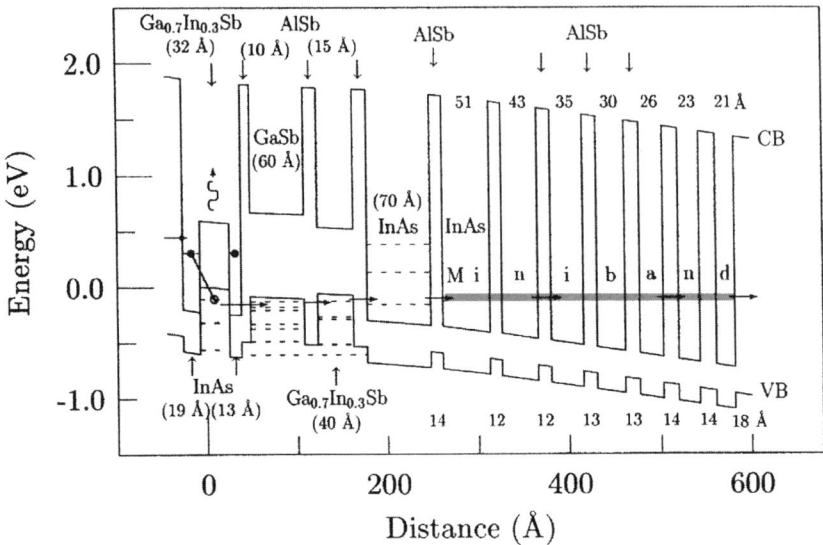

Fig. 3.23 Conduction and valence band profiles for one period of a three-hole-well W-ICL design. Calculated zone-center subband energy levels are indicated. (Reprinted from [154], with permission from the American Institute of Physics)

QCL [147]. While the QCL requires relatively high threshold current densities because of the inherently short lifetimes associated with intersubband phonon scattering, the interband configuration of the ICL eliminates that nonradiative relaxation path. Detailed modeling predicts that ICLs should display low thresholds, and thus high cw operating temperatures throughout the 2.5 to 7 μm range [150,151]. One also expects high output powers per facet as well as differential quantum efficiencies well above the conventional limit of one photon/electron, as has been confirmed by experimental observations [152–154].

Figure 3.23 illustrates conduction- and valence-band profiles for a typical single stage of the ICL staircase. This example [155] employs a W configuration in which two InAs electron QWs (19 and 13 Å) surround the active GaInSb hole QW (32 Å). Electrons tunneling into the InAs QWs from the preceding injection region at left emit a photon by making spatially indirect radiative transitions to the GaInSb valence band. They next tunnel into the valence states of the adjacent GaSb and GaInSb hole QWs, whose function is to provide a thick barrier to prevent electron tunneling from the active InAs QWs directly to the 70-Å InAs well that begins the next injection region [150]. From the final GaInSb hole QW (40 Å), the carriers undergo near-elastic interband scattering into conduction states of the 70-Å InAs QW. Electrons traverse the superlattice miniband of the injection region, which in this case consists of eight digitally graded InAs wells, and finally tunnel into the active electron QWs of the next period. This series of events is repeated at each step of the ICL staircase.

The most critical step in this process is the type-II scattering event (in the example, from hole states in the 40-Å GaInSb layer to electron states in the 70-Å InAs layer), which is key to the recycling of electrons for reuse in the subsequent stages. However, the active lasing transition may be either type-I or type-II [150]. Both have been demonstrated experimentally and will be discussed in what follows. "Cascaded" lasers based on wide-gap tunnel junctions with only two or a few periods were demonstrated some time ago [156].

Type-II ICLs

After a few observations of electroluminescence [157,158], the first interband cascade laser ($\lambda = 3.8\,\mu m$) was demonstrated by researchers at UH and Sandia early in 1997 [159]. That device with 20 periods utilized a type-II active region containing a single electron QW. It lased in pulsed mode up to 170 K. The maximum operating temperature was soon extended to 225 K when NRL/UH employed a W design with double-electron wells as in Fig. 3.23 [152]. The peak output power was 170 mW/facet at 180 K [152]. UH reported 480 mW/facet from a non-W ICL at 80 K [153], and later an efficiency exceeding 200% [160].

Since the threshold current densities of 100 to 200 A/cm^2 at 80 K and about 2.7 kA/cm^2 at 220 K were always considerably higher than the theoretical projections [151], a third hole QW was added to the structure illustrated in Fig. 3.23 in order to suppress tunneling leakage currents. While the low-temperature results were slightly worse than the best earlier findings, at somewhat higher temperature the thresholds for this NRL/UH device were the lowest ever reported to that time for a diode laser emitting beyond 3 μm (e.g., 1.8 kA/cm^2 at 220 K) [155]. Consequently laser emission at $\lambda \approx 3.6\,\mu m$ was observed up to 286 K in pulsed mode.

The group at the Army Research Laboratory (ARL) has recently reported substantial further advances in ICL performance [154,161,162]. Devices emitting at 3.6 to 3.9 μm have displayed slope efficiencies in excess of 750 mW/A per facet, corresponding to 4.6 photons per injected electron (or a 460% quantum efficiency) [154]. The peak output power of >4 W/facet was far greater than that for any other mid-IR semiconductor diode emitting beyond 3 μm.

ARL has also reported cw ICLs with low threshold current densities, such as 56 A/cm^2 at 80 K [162]. Figure 3.24 shows that the external wall-plug efficiency at 60 K exceeded 9% for a type-II ICL emitting at 3.6 to 3.8 μm. The maximum lasing temperature for epitaxial-side-up cw operation with uncoated facets was 127 K, and the peak output power was approximately 100 mW/facet at 80 K. In pulsed mode, the devices lased to 250 K, and exhibited a peak power efficiency of >11% at 80 K.

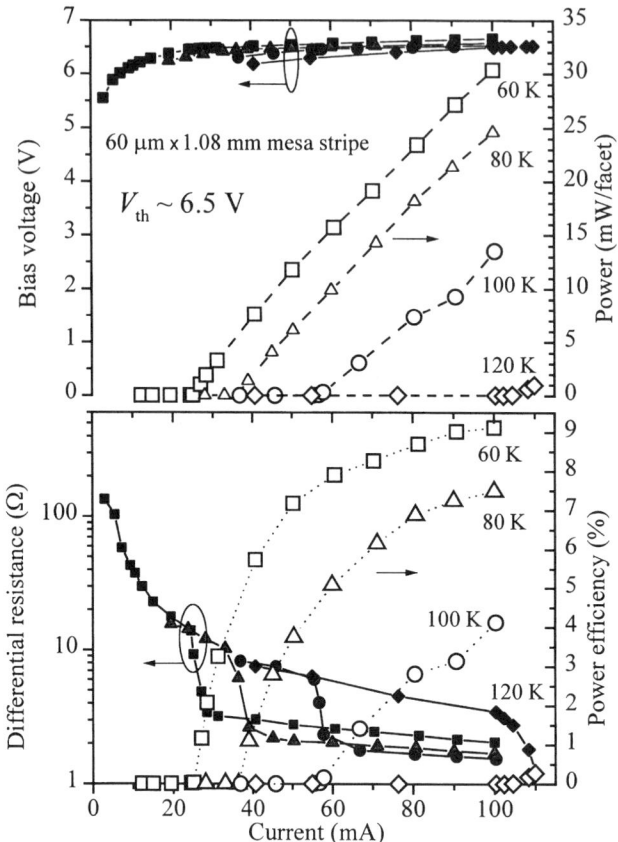

Fig. 3.24 Bias voltage and output power versus injection current (*top*) and differential resistance and power efficiency against injection current (*bottom*) at several temperatures for a mesa-stripe InAs/GaInSb type-II ICL. (Reprinted from [161], with permission from the American Institute of Physics)

There have been several reports of interband electroluminescence from cascaded structures at longer wavelengths in the 5 to 15 μm range [158,163,164]. Theoretical studies of antimonide *intersubband* lasers [149,165] have projected lower thresholds than for the analogous GaAs-based and InP-based QCLs. Ohtani and Ohno observed intersubband electroluminescence at $\lambda \approx 5$ μm from InAs/AlSb [166] and InAs/GaSb/AlSb [167] cascaded structures.

Type-I ICLs

It was pointed out above that the ICL active transitions may be either type-I or type-II [150]. While most of the research has focused on type-II devices, in 1998 Allerman et al. demonstrated the first type-I ICLs with InAsSb active

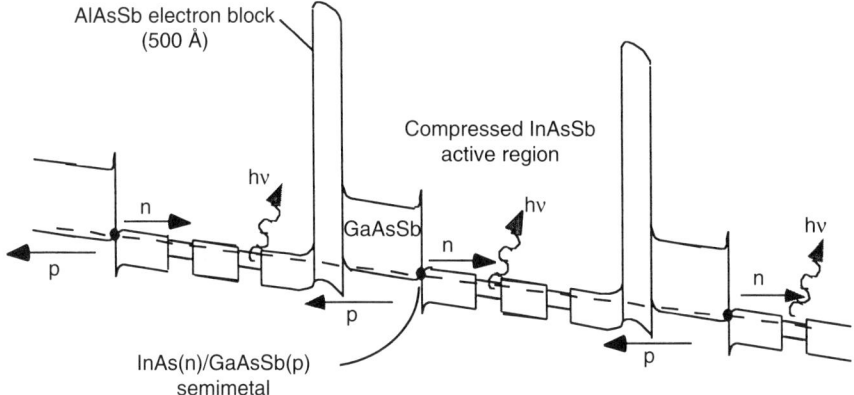

Fig. 3.25 Band profile for a type-I ICL with compressed InAsSb QWs separated by InAsP barriers. Electron-hole generation occurs at the InAs/GaInSb semimetal heterojunction, and an AlAsSb layer blocks electron leakage. (Reprinted from [168], with permission from the American Institute of Physics)

QWs [168]. Another distinction is that in contrast to virtually all earlier cascaded laser structures that were grown by MBE, these devices were grown by MOCVD. Conduction- and valence-band profiles for two periods of the active region are illustrated in Fig. 3.25 [169]. Photons are created via type-I optical transitions in the InAsSb QWs, while a type-II GaAsSb/InAs interface recycles electrons from the valence band of one stage to the conduction band of the next. A 10-stage structure with three 85-Å $InAs_{0.88}Sb_{0.12}$ QWs separated by 85-Å $InAs_{0.76}P_{0.24}$ barriers yielded stimulated emission up to 170 K under pulsed conditions. A slightly modified 10-stage device with five 94-Å n-$InAs_{0.85}Sb_{0.15}$ wells separated by six 95-Å n-$InAs_{0.67}P_{0.33}$ barriers operated to 180 K. The peak output power was >100 mW, and the slope efficiencies were as high as 48% (4.8% per stage) for these gain-guided devices.

3.5 CHALLENGES AND ISSUES

3.5.1 Antimonide Growth Immaturity

Unlike the GaAs- and InP-based systems, antimonide MBE has not yet reached the stage where the routine and reproducible growth of high-quality heterostructures may be taken for granted. One issue is strain-induced defects. Whereas the entire GaAs/AlGaAs/AlAs system is closely lattice-matched, and each layer in an $In_{0.53}Ga_{0.47}As/In_{0.52}Al_{0.48}As$/InP structure can in principle have exactly the same lattice constant, the differences between the lattice constants of InAs, GaSb, and AlSb are nonnegligible. Moreover layers such as $Ga_{0.7}In_{0.3}Sb$ with somewhat higher strains are often incorporated, so that the compressive and tensile strains must be closely compensated if useful device quality is to be obtained [101]. While some of the early optically pumped type-II lasers were strain-compensated to an AlSb buffer layer that was deposited onto a GaSb substrate [170], that approach is probably not suitable for diode lasers. One recent mid-IR laser was grown on a compliant universal substrate [171], which has the added advantage that the substrate is transparent to an optical pump beam.

LPE has the advantage that plastic deformation of the substrate during growth relieves a large part of the misfit stress [172,173]. However, that method is not well suited to fabricating the more advanced structures with thin QW layers. While MOCVD offers greater control than LPE, and very low dislocation density in the active region of laser structures has been achieved [97], the use of MOCVD is also limited by the difficulty of growing layers as thin as those attainable with MBE. This is especially important in the case of type-II lasers. It is also challenging to grow MOCVD structures containing AlSb.

Another issue that complicates the growth of antimonide heterostructures is the large miscibility gap for quaternary alloys such as $Ga_{1-x}In_xAs_ySb_{1-y}$ [16]. This gap tends to become increasingly less stable as x approaches 0.5, limiting the variety of quaternary compositions that can be realized. Turner et al. avoided the InAlAsSb miscibility gap by growing InAsSb/InAlAsSb QW active regions on InAs substrates with oppositely strained wells and barriers [174].

The evaluation of type-II laser heterostructures by PL and X-ray diffraction has shown that the MBE-grown material quality is very sensitive to both growth temperature and interfacial bond type [175]. There is some debate as to the nature of the InAs/GaSb interface, which has no common cation nor anion and may therefore have GaAs-like, InSb-like, or mixed interfacial bonds [176]. In addition to changing the balance of the epilayer strain, the bond type appears to influence the growth quality [175] and possibly the valence band offset [9]. Since the extent to which the atomic constituents diffuse across interfaces is also problematic and of concern [93,177,178], several microscopic techniques have been employed to study the interfaces in antimonide-based heterostructures. The composition and interdiffusion properties of the GaSb-on-InAs interface have been studied with cross-sectional scanning tunneling microscopy (STM) [179,180]. The STM image in Fig. 3.26 illustrates that the interfaces are rather rough due to a high density of defects. The defects are associated with both As/Sb cross-incorporation and interface-specific random substitutional defects. Gray et al. investigated a GaInAsSb/AlGaAsSb multiple QW laser structure using high-resolution X-ray diffractometry (HRXRD) and transmission electron microscopy (TEM) [181]. Those methods determined the chemical composition of the quaternary QWs. Interface composition and growth dynamics have also been studied and are becoming better understood through the use of in situ STM [182,183].

There is also concern about defects originating on the GaSb substrates [184], which are related in part to the surface preparation before growth. The

CHALLENGES AND ISSUES 109

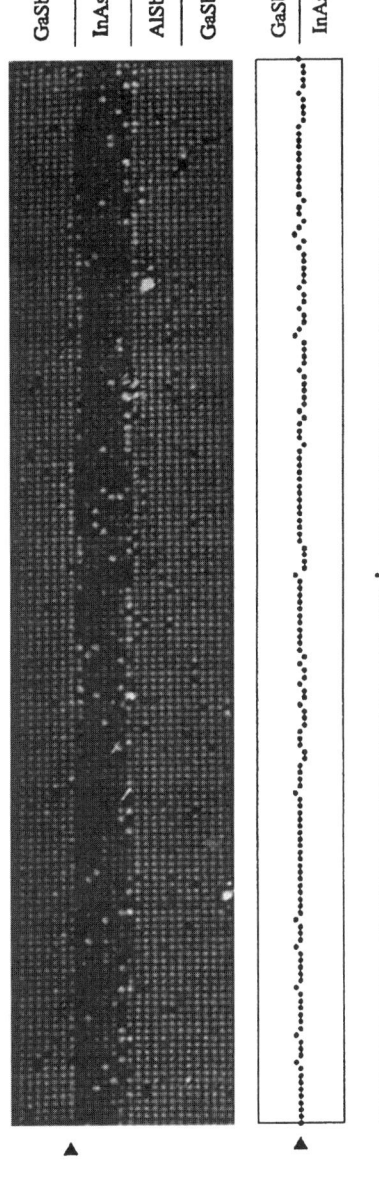

Fig. 3.26 Cross-sectional STM image of GaSb-on-InAs, InAs-on-AlSb, and AlSb-on-GaSb interfaces. The growth direction is toward the top. The bottom portion shows the profile at the GaSb-on-InAs interface, discretized in units of the $(1\bar{1}0)$ surface mesh. (Reprinted from [179], with permission from the American Institute of Physics)

defect density in antimonide-based heterostructures has been observed to track the defects on the substrate before growth.

3.5.2 Nonradiative Recombination and Threshold

While the low thresholds for near-IR diode lasers tend to be limited primarily by radiative recombination lifetimes, at longer wavelengths (small energy gaps) the nonradiative Shockley-Read (S-R) and Auger mechanisms become increasingly important. Typically the single-carrier S-R process governs the low-temperature thresholds (e.g., <150 K), while the multi-carrier Auger process dominates at higher temperature. Suppressing the Auger decay rate to a point where reasonable thresholds can be maintained at temperatures approaching ambient represents one of the primary challenges of antimonide mid-IR laser development.

Shockley-Read Recombination
The Shockley-Read lifetime is quite sensitive to material quality, since S-R recombination occurs when electrons and holes are captured by defect levels deep in the energy gap. Although the S-R recombination rate is often assumed to be independent of carrier density and temperature, in general, it can vary strongly with both [185]. S-R lifetimes in antimonide mid-IR laser materials typically fall in the 1 to 200 ns range [126,186–189].

Auger Recombination
In an Auger event the energy and momentum of the recombining electron-hole pair is transferred to a third carrier, which scatters to a higher lying state in either the same or a different band. Because three different carriers must simultaneously interact, the recombination rate scales as n^3 whenever $n \approx p$ as is usually a good approximation under lasing conditions. The lifetime can therefore be written in the form $\tau_A = 1/\gamma_3 n^2$, where γ_3 is the Auger coefficient. For nondegenerate statistics, γ_3 tends to be nearly independent of density, although it saturates somewhat when both electrons and holes become degenerate [190].

The Auger rate can be suppressed substantially if the material is band-structure engineered so as to minimize the density of final states available to the third (nonrecombining) carrier [111]. This is more easily accomplished if that carrier must transfer to a discrete ladder of *confined* subband states in a QW, rather than to a *delocalized* quasi-continuum above or close to the barrier energy. Because most type-II antimonide QWs have larger conduction- and valence-band offsets than type-I structures with the same energy gap, they tend to have confined final Auger states in contrast to the delocalized type-I final states.

It is also important to distinguish whether the third carrier is an electron or a hole. In the case of the so-called CCCH process, which involves three conduction-band electrons and one heavy hole, the recombination is activated,

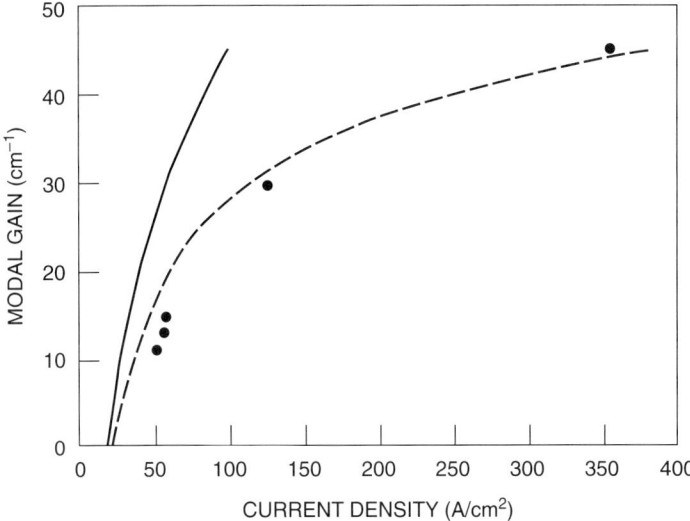

Fig. 3.27 Modal gain versus current density for a SQW broadened-waveguide GaInAsSb/AlGaAsSb diode laser emitting at 2.05 µm. The filled circles denote experimental points (sum of mirror loss and internal loss), while the solid line is the intrinsic theoretical gain using a 6 × 6 **k·p** model. The dotted line includes an Auger current component, assuming $\gamma_3 \approx 5 \times 10^{-29}$ cm^6/s. (Reprinted from [18], with permission from the American Institute of Physics.)

$\gamma_3(T) \propto \exp(-E_A/k_B T)$, because one of the initial carriers (usually the hole) must have a relatively high energy in order to conserve momentum. For parabolic bulk bands [191], $E_A \approx E_g m_c/(m_c + m_h)$, where m_c and m_h are the electron and hole masses and E_g is the energy gap. Thus bulk narrow-gap III-V materials with $m_h/m_c \gg 1$ invariably have small activation energies and hence large γ_3. While QWs tend to have a much smaller in-plane mass ratios near the band edge (typically $m_h \approx m_c$), γ_3 is reduced only if the hole mass remains small over a substantial range of energies [192]. In near-IR QW lasers, such as $\lambda = $ 1.3–1.55 µm InGaAsP devices, this has been accomplished by using strain to increase the splitting between the heavy- and light-hole subbands [193,194]. A similar reduction has been achieved in strained InGaAsSb/AlGaAsSb QW lasers emitting near 2 µm, with the recent analysis illustrated in Fig. 3.27 yielding very small γ_3 ($\approx 5 \times 10^{-29}$ cm^6/s) [18]. Although no quantitative γ_3 results have been reported, low-threshold InGaAsSb/AlGaAsSb strained QW lasers emitting at $\lambda = $ 2.3–2.6 µm probably take advantage of a substantial Auger suppression as well [19]. For type-I QW lasers emitting beyond 3 µm, however, it may be difficult to apply enough strain to induce the required band structure modification. In such structures only the initial states, and not the final states, can be engineered because the band offsets are small. The heavy/light splitting is usually far greater in type-II QWs, due to thinner wells and also to a much larger valence-band offset. Consequently the CCCH Auger coefficient can be significantly suppressed [111,195]. Moreover the lighter hole mass and lower band-edge density of states decrease the threshold carrier density, which further reduces the Auger rate due to the scaling of the Auger lifetime with $1/n^2$.

The analogous process involving multiple H1 heavy holes (CHHH) tends to be quite weak, since the activation energy becomes prohibitive when m_h is in the numerator of the above expression rather than the denominator. However, the Auger hole can also be excited to the split-off hole band (CHHS), or to a confined heavy or light subband (H2, H3, L1, L2, etc.) that is separated from the valence-band maximum by an energy E_Δ. The activation energy for this process is proportional to $|E_g - E_\Delta|$ [196,197], which vanishes at resonance because then both energy and momentum are easily conserved. In bulk materials such as GaSb, InAs, and InAs-rich InAsSb alloys, the CHHS process sometimes dominates [173,198–200], although the topic remains controversial [201,202].

While the resonance with the split-off gap is completely removed in InAs/Ga(In)Sb-based type-II QWs, it is replaced by the potential for resonances with a variety of lower-lying heavy and light hole subbands. Grein et al. pointed out the possibility of designing the QWs such that all intervalence resonances are avoided [203], an approach to Auger suppression which has been analyzed comprehensively by Flatté and collaborators at the University of Iowa and elsewhere [145,204–207]. The suppression of Auger recombination in type-II heterostructures has also been treated by Zegrya and Andreev [208–210].

Figure 3.28 summarizes experimental room-temperature Auger coefficients as a function of band gap wavelength (λ_g) for a variety of III-V type-I materials (both bulk and QWs, open points) as well as type-II QWs (filled points) [195]. The figure also gives 300 K Auger coefficients for bulk $Hg_{1-x}Cd_xTe$ (upside-down triangles) with a range of alloy compositions [211–215]. Most of the data for bulk materials [18,216–218] and nonoptimized type-I QWs are seen to lay relatively near the dashed curve in Fig. 3.28, which is a guide to the eye. However, modification of the band structure (see above) led to significant suppression of the CCCH-dominated Auger rate in some of the shorter wavelength type-I strained systems such as InGaAs [218] and InGaAsSb [18] QWs.

Figure 3.28 also confirms a significant suppression for the type-II QWs, since γ_3 is as much as an order of magnitude lower than for type-I structures with the same λ_g. The data are from photoconductive response [186,219] (filled circles) and lasing-threshold [124,125,128,130] (filled boxes) measurements performed at NRL, and from pump-probe and photoluminescence upconversion experiments [129,187,190,220] (filled triangles) performed at the University of Iowa. The solid line is a guide to the eye passing through the NRL data, whose consistency for samples obtained from four different sources and measured by two different techniques is perhaps surprising in view of the valence-intersubband-resonance considerations discussed above. While, in principle,

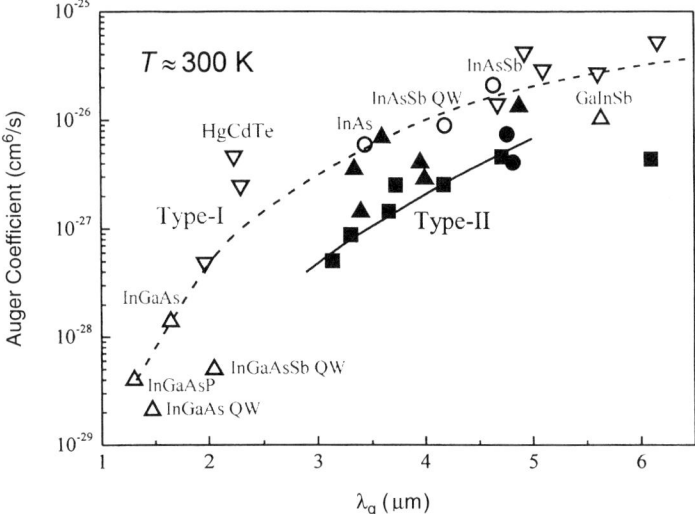

Fig. 3.28 Auger coefficient versus emission wavelength at 300 K for type-I (*open points*) and type-II (*filled points*) materials. A few recent data points have been added to the original figure, which is reprinted from [194], with permission from the American Institute of Physics.

the recombination rate should be sensitive to the details of the band structure, recent Auger calculations have found that much of that sensitivity may be lost at higher temperatures where several different classes of transitions contribute simultaneously [206].

Internal Losses

In addition to Auger recombination, a second fundamental issue limiting the mid-IR laser performance is a rapid increase of the internal absorption loss (α_{int}) with increasing wavelength. At present this probably represents in fact the greatest impediment to practical, higher temperature lasers emitting beyond 3 μm [221]. High loss leads to low external quantum efficiency and hence low output power, and in cases where the differential loss exceeds the differential gain it can prevent a device from lasing at all at high T [130,188].

Most losses in mid-IR lasers may be associated with free-carrier absorption. While other processes such as scattering at rough interfaces can also occur, at present there is no evidence for those ever becoming dominant in this spectral region. Contributions to the loss come from both injected carriers in the active region and extrinsic electrons and holes in the doped cladding layers of diodes. The Drude contribution to α_{int}, whose cross section scales as λ^2 or λ^3 [222], is generally not excessive at $\lambda \leq 5$ μm. However, there are many opportunities for resonant interband and intersubband transitions, especially involving holes. While conduction intersubband absorption is generally negligible due to the edge-emitting laser's TE-polarization selection rule, heavy-hole intersubband transitions are allowed because strong interband mixing relaxes that rule.

By analogy with Auger recombination, intervalence absorption can be enhanced if the split-off gap is nearly equal to the energy gap. Gun'ko et al. carried out a detailed theoretical analysis of intervalence absorption between the heavy-hole and split-off bands in type-I InAs-based structures, and its effect on the projected mid-IR laser performance [223]. They found that the absorption feature is quite strong (≈ 100–$300\,\text{cm}^{-1}$) and broad (≈ 0–$100\,\text{meV}$), and that there is a large effect on the external quantum efficiency if the photon energy is too close to this resonance.

Although this absorption process is favored in bulk GaSb and InAs, it can be suppressed in InGaAsSb or strained GaInAsSb-based QWs when the energy gap becomes smaller than the spin-orbit splitting. In that regime, heavy-hole to light-hole transitions can dominate [3]. While at $T \approx 80\,\text{K}$ the α_{int} measured for bulk DH and type-I QW lasers emitting at $\lambda \geq 3\,\mu\text{m}$ have sometimes been as high as 25 to $130\,\text{cm}^{-1}$ [8,78,224], values in the 3 to $10\,\text{cm}^{-1}$ range are attainable in optimized structures [81,96,174,225]. To our knowledge, the only published temperature dependence was reported by Le et al., who found that α_{int} increased from $7\,\text{cm}^{-1}$ at $T = 87\,\text{K}$ to $18\,\text{cm}^{-1}$ at $162\,\text{K}$ [87].

In optimized diode lasers for the 2 to $2.7\,\mu\text{m}$ range, interband absorption resonances appear to have been eliminated as a major issue, and the losses tend to be dominated by Drude or intervalence band absorption in the doped cladding layers. That loss may be minimized using the broadened-waveguide configuration, in which the active region is surrounded by relatively thick *undoped* separate confinement regions (SCR) with a high refractive index [18,45]. For example, GaSb or a low-Al-fraction AlGaAsSb quaternary alloy may be employed as the SCR, in combination with a high-Al-content quaternary as the cladding layer [45,55]. The dependence of the internal loss on SCR layer thickness was investigated in detail by researchers at Sarnoff [226]. In that study the internal loss decreased from $32\,\text{cm}^{-1}$ for a 0.12-µm-thick waveguide to $2\,\text{cm}^{-1}$ for a 0.88-µm-thick waveguide core. Another study reported an internal loss of $2.5\,\text{cm}^{-1}$ at room temperature [46].

For the case of type-II mid-IR lasers, there have been several detailed theoretical and experimental investigations of the internal losses, which are believed to be dominated by valence intersubband resonances [116,188]. According to the TE-polarization selection rules derived from a four-band $\mathbf{k} \cdot \mathbf{p}$ Hamiltonian [227,228], transitions between H1 and any lower-lying heavy-hole subband vanish at zero in-plane wave vector. However, interband mixing relaxes this selection rule at finite $k_\parallel$, so that the matrix elements for such processes as H1 $\rightarrow$ H3 are substantial at wave vectors on the order of $0.03\,\text{Å}^{-1}$. One consequence is that the cross sections for intra-heavy-hole transitions tend to increase rapidly with temperature. Transitions from H1 to L1, L3, L5, ..., which require a change of parity, similarly become allowed only at finite $k_\parallel$. On the other hand, transitions from H1 to L2, L4, L6, ..., are allowed even at $k_\parallel = 0$ because they *do not* involve a change of parity, and hence they remain strong at *all* temperatures (they are analogous to the large interband cross sections for bulk semiconductors). The latter class of transitions is the most important to avoid.

Simulations of the wavelength-dependent free-hole absorption cross sections by several groups [116,129,188,205,207,221] predict that it should be possible to design type-II laser structures that avoid most intervalence resonances, and therefore have relatively low α_{int} up to ambient temperature (although Lorentzian broadening may contribute a nonnegligible background even for the best designs). Theoretical absorption spectra associated with various intervalence transitions in an InAs/GaInSb superlattice are shown in Fig. 3.29 [205]. The calculations imply the appearance of a low-absorption window, which in principle can be made to coincide with the lasing photon energy at the design operating temperature.

To date, we are aware of only one explicit experimental characterization of multiple valence subbands in an antimonide laser material [229]. An empirical-pseudopotential-method calculation gave good agreement with the data for an InAs/GaSb superlattice, whereas the $\mathbf{k} \cdot \mathbf{p}$ approach yielded less accurate results. However, further study is needed before the input parameters become sufficiently trustworthy to rely fully on theoretical predictions. At present, one cannot yet confidently design type-II laser structures that are optimized for low internal loss.

Optically pumped type-II structures with 50 to 80 QWs (chosen for strong absorptance of the pump beam) often display internal losses as low as nearly 10 cm^{-1} at 80 K [128,133,160,187,221]. However, in contrast to the theoretical projections, a strong monotonic increase of α_{int} with increasing temperature has always been reported for these devices. This is illustrated by the circles of Fig. 3.30, which indicate a loss of 90 cm^{-1} at 300 K [188]. Since the primary mechanism is free-carrier absorption in the active region, the high-temperature α_{int} can be suppressed substantially if the number of QWs is reduced so as to minimize the number of carriers lying along the path of the optical mode. The squares of Fig. 3.30 illustrate a reduction of the loss to 20 cm^{-1} at 240 K as a result of using only 10 active type-II W QW periods [140] in an OPIC laser (see Section 3.4.4). A similar reduction should be possible up to high temperatures using the IA approach (also discussed in Section 3.4.4) with only a few active QW periods [141].

3.5.3 Linewidth Enhancement Factor (LEF)

The linewidth enhancement factor quantifies the focusing (LEF < 0) or defocusing (LEF > 0) of the optical wave induced by small fluctuations in the carrier density. A large positive value causes small intensity perturbations to grow exponentially, leading to filamentation. For this reason the LEF plays a crucial role in governing a semiconductor laser's beam quality. Typical values for optimized near-IR devices are around 2 for QW structures and around 5

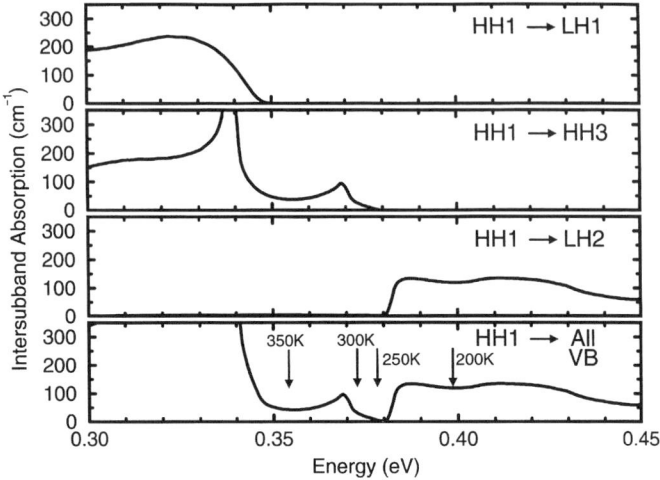

Fig. 3.29 Valence intersubband absorption versus photon energy for a 12.5 Å InAs/39 Å $In_{0.25}Ga_{0.75}Sb$ superlattice at 300 K and the threshold carrier density. The top three plots show individual contributions from the relevant final subbands, while the bottom plot shows the total intervalence absorption. The arrows in the bottom plot indicate lasing wavelengths at the denoted temperatures. (Reprinted from [205], with permission from the American Institute of Physics.)

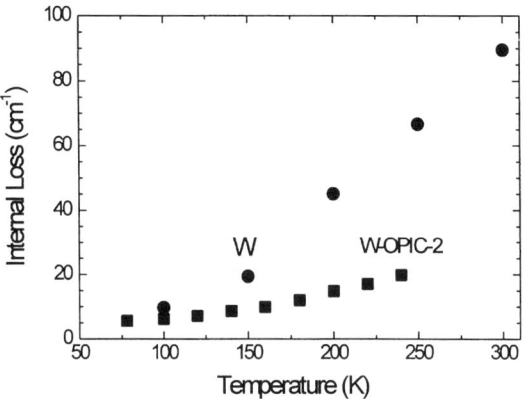

Fig. 3.30 Internal loss versus temperature for an optically pumped type-II W laser emitting at 3.2 to 3.8 μm (●) and a W-OPIC emitting at 3.4 to 3.7 μm (■). Note that at 240 K, the internal loss in the W-OPIC is reduced by a factor of three. (Reprinted from [139], with permission from the Institute of Electrical and Electronics Engineers)

for double heterostructures. Furthermore the LEF always enters the formalism in terms of its product with the threshold modal gain, implying that this product rather than the LEF alone is the proper figure of merit. Whereas in near-IR lasers the threshold gain is governed by the reflection loss, the mid-IR threshold gain is typically dominated by the internal loss. Therefore lossier devices are expected to exhibit poorer beam quality.

The LEF is defined as the ratio between the real and imaginary components of the carrier-induced variation of the linear optical susceptibility. In principle, the LEF depends on the carrier density (N), although at low temperatures and densities it is roughly constant. It may be written in the form [230]

$$LEF(\lambda) = -\frac{4\pi}{\lambda} \frac{dn/dN}{dg/dN},$$

where g is the gain and n is the refractive index. The differential index in the numerator is sometimes obtained from the optical gain spectrum, via the Kramers-Kronig relation. However, that procedure ignores the plasma contribution to n, which is nonnegligible at mid-IR wavelengths. It is more rigorous to measure the differential modal index directly by determining the spectral shift of the Fabry-Perot modes as a function of pump level.

While the LEF vanishes at the peak of the *differential* gain (apart from a small plasma shift), the more relevant wavelength is the gain peak where lasing usually occurs (barring strong spectral selectivity of the cavity loss). Most semiconductor lasers have an appreciable LEF because these two peak wavelengths are mismatched as a consequence of the asymmetric conduction- and valence-band densities of states in the active region.

Baranov obtained LEF = 3.1 for $\lambda \approx 2\,\mu m$ lasers with DH (bulk-like) active regions at 82 K, from measurements of the spontaneous emission spectra below threshold [231]. The LEF was tentatively extrapolated to 6 at room temperature. A smaller LEF was observed for a laser based on the InGaAsSb/GaSb type-II junction, although the exact numerical value was not given.

Olesberg et al. recently calculated LEFs (ignoring the plasma contribution to the refractive index) for a variety of mid-IR lasers [232]. The results for bulk InAsSb, InAsSb/AlInAsSb, and InAsSb/InAsP type-I QWs, and a type-II W structure with a quinternary barrier are shown in Fig. 3.31, along with gains and differential gains. Whereas LEF = 6.5 is obtained for bulk InAsSb, the value for the optimal type-I InAsSb/AlInAsSb QW is 2.5. The best result (LEF = 0.9) is obtained for the InAs/GaInSb/InAs/AlGaInAsSb W structure. That agrees well with an experimental value obtained from differential gain spectra in conjunction with *dn/dN* derived from a Kramers-Kronig transformation of the absorption data [207]. The improvement for QWs over bulk is due in part to the higher differential gain associated with the 2D density of states. Type-II QWs with comparable electron and hole masses should be particularly favorable because the conduction and valence densities of states are nearly

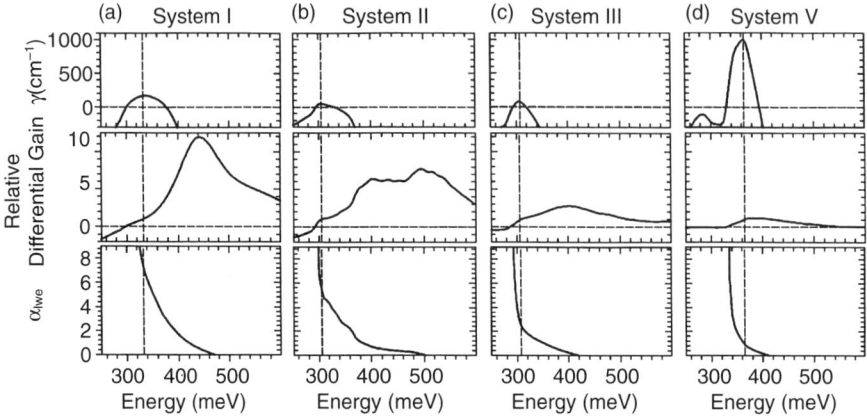

Fig. 3.31 Gain, relative differential gain, and linewidth enhancement factor versus transition energy for (a) bulk $InAs_{0.91}Sb_{0.09}$, (b) an InAsSb/AlInAsSb QW, (c) an InAsSb/InAsP QW, and (d) a type-II W structure with a quinternary barrier. Curves are calculated at the carrier density for which the LEF at the peak of the gain spectrum is minimized, and the vertical dashed lines denote energy peaks of the gain spectra. (Reprinted from [232], with permission from the American Institute of Physics.)

equal. Olesberg et al. pointed out that even smaller LEFs may be achievable by detuning the emission wavelength to the blue (toward the peak of the differential gain spectrum), such as by a distributed feedback (DFB) design [232]. An NRL calculation of the LEF for a W structure [122] yielded a somewhat larger value, due in part to inclusion of the plasma contribution.

Recent indirect evidence based on the beam quality of mid-IR α-DFB lasers with W active regions suggested consistency with a somewhat larger LEF (≥4) at $T = 150\,K$ [233]. The earlier theoretical projections of small LEFs in such structures assumed homogeneous broadening of the gain spectra and favorable alignments of the intervalence resonances. However, if layer-thickness fluctuations contribute additional broadening of the gain spectrum, the differential gain decreases faster than the differential index and a larger LEF results. Alternatively, if a strong intervalence resonance such as the H1–L2 transition has a smaller energy than the band gap, the observed LEF will exceed projections based on the interband gain spectrum alone. A preliminary analysis of Hakki-Paoli measurements [234] at $T = 77\,K$ yielded LEFs of 2.5 and 5.2 for two optically-pumped type-II W lasers emitting at 3.9 μm and 4.6 μm [235].

3.5.4 Single-Mode Operation and Wavelength Tuning

For spectroscopic applications such as chemical detection, single-mode operation with a narrow spectral linewidth is indispensable. The most straightforward approach is to shrink the device along all three spatial dimensions so that only a single optical mode falls within the rather broad gain spectrum. Although this approach is compatible with the VCSEL geometry, none of the mid-IR VCSELs discussed in Section 3.3.2 employed an active diameter small enough to ensure a single lateral mode. Stable single-mode operation also generally requires index confinement rather than the gain guiding used in some of the studies.

Mid-IR semiconductor lasers usually lase in a single mode along the growth axis, and one can pattern a mesa no more than a few wavelengths wide that permits only a single lateral mode. While it is not practical to fabricate a cavity short enough to exclude all but a single longitudinal mode, the same objective can be accomplished by introducing a wavelength dependence to the cavity loss. At shorter wavelengths, the most common approach is the DFB configuration, which employs a diffraction grating etched into either a high-index separate confinement region or the top cladding layer, followed by epitaxial regrowth. It is perhaps surprising, however, that no antimonide DFB lasers emitting beyond 2μm have been reported to date (the α-DFB devices discussed below have somewhat different properties). Vicet et al. achieved longitudinal wavelength selectivity by using an external cavity with a grating. A GaInAsSb/GaSb QW laser with a 7-μm-wide mesa produced single-mode output at λ = 2.42μm, with 20dB side-mode suppression [236].

While it is difficult to eliminate mode hopping completely without the spectral selectivity of a DFB or external grating, single-mode operation has also been reported for both the 2–3μm and 3–4μm wavelength regions using short cavities, in which the mode separation is given by

$$\Delta\lambda = \frac{\lambda^2}{2nL}.$$

Since this splitting is larger than in the near-IR, the prospects for single-mode output are good given a narrow (homogeneously broadened) gain spectrum, a short Fabry-Perot cavity, and a narrow-ridge waveguide. For the wavelength range λ = 2.1 − 2.4μm, ridge waveguide devices (6–10μm wide) with short cavities (150–1070μm) have operated in a single mode at current densities of up to three times threshold [48,50,54,237]. The single-mode output power was up to 100mW cw at room temperature [50].

Despite a greater longitudinal mode splitting, single-mode operation at $\lambda \geq 3$μm tends to be more challenging. This is due in part to the frequent occurrence of larger linewidths associated with increased (inhomogeneous) broadening of the gain spectrum, an obstacle that can be overcome in principle by employing a wavelength-selective optical cavity. Nonetheless, the Ioffe group has reported single-mode operation in ridge-waveguide DH Fabry-Perot lasers with short cavities [1,78,238–242]. The devices operating near 77K typically had stripe widths of 10 to 20μm, cavity lengths of 75 to 600μm, and emission wavelengths of 3.0 to 3.6μm. The maximum reported single-mode cw

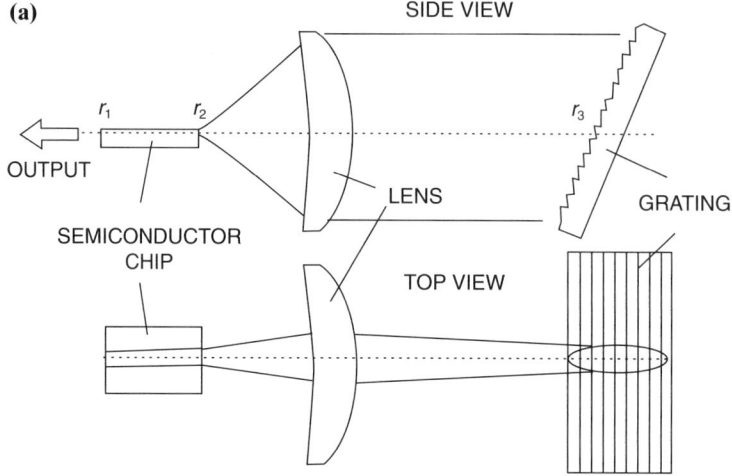

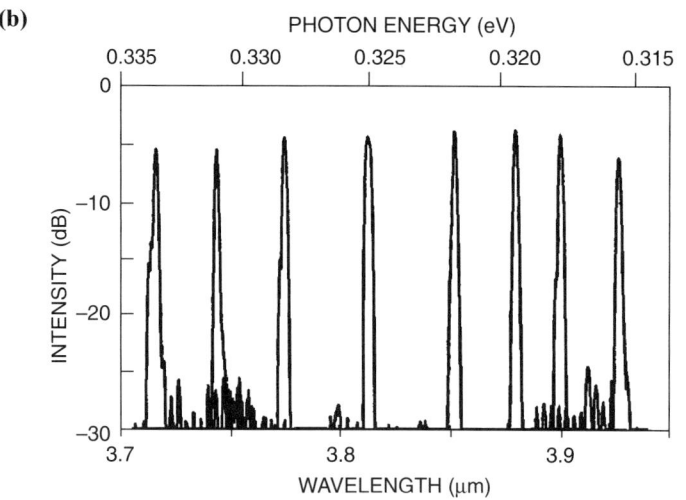

Fig. 3.32 (a) External cavity laser configuration. In the side view, which shows divergence along the fast axis, r_1 and r_2 are the facet reflectance while r_3 is the effective external cavity reflectance. The top view shows divergence along the slow axis. (b) A grating-tuned output spectrum from the GaSb/InAsSb-based external cavity laser. (Reprinted from [243], with permission from the American Institute of Physics.)

output power was 2 mW [78,241]. Danilova et al. demonstrated 10 nm tuning of the wavelength by varying the current [1].

Le et al. used the grating-coupled external-cavity configuration shown in Fig. 3.32a to achieve much greater wavelength tunability for optical pumping [243]. The InAs/InAsSb superlattice and InAsSb DH lasers, which were mounted in a liquid nitrogen dewar with AR-coated windows, were pumped either cw or quasi-cw by a 0.97-μm diode array. A lens with 0.55 numerical aperture and focal length of 14 mm coupled the beam to an external grating that could be rotated to achieve the wavelength tuning. The grating could also be translated for cavity-length adjustment to accommodate wavelengths from 3.35 μm to 3.91 μm. Figure 3.32b illustrates the tuning by about 250 nm of a laser with center wavelength $\lambda \approx 3.82$ μm. Although the output was multi-mode, the linewidth was only 1 to 2 nm, and up to 20 mW average power was produced at 80 K.

3.5.5 Beam Quality

For practical applications, mid-IR lasers must have a stable beam profile that does not vary rapidly with temperature or drive current. In comparing InAsSb/InPAsSb DH and MQW lasers, Northwestern found that the MQW devices had much more stable far-field patterns [244]. While the DH far field broadened with increasing temperature, the MQW profile remained nearly constant under a variety of operating conditions. The DH data were attributed to fluctuations in the refractive index due to InAsSb alloy compositional variations resulting from a phase separation on the 100-nm scale. The Ioffe group also studied variations of the far-field profile with current [245]. They proposed to narrow the lateral far-field profile by coupling two modes [246].

For applications requiring enhanced brightness, high output power must be combined with near-diffraction-limited beam quality. While a single lateral mode is emitted when the gain stripe is sufficiently narrow, the available output power is limited. On the other hand, the multiple lateral modes excited by a wide stripe produce a broad and double-lobed far-field pattern. That is because self-modulation of the refractive index, as quantified by the linewidth-enhancement-factor/threshold-gain product discussed above, leads to a loss of phase coherence between the various modes.

One approach to overcoming the poor beam quality of broad-area semiconductor lasers is the tapered-waveguide configuration [247,248], in which the light from a single-mode ridge waveguide is allowed to diffract into a tapered amplifier section. Figure 3.33a schematically illustrates a tapered-waveguide design used by LL in conjunction with a GaInAsSb/AlGaAsSb QW diode active region emitting at $\lambda = 2.05$ μm. Cavity-spoiling grooves and an antireflection coating at the output facet adjacent to the amplifying section suppress parasitic Fabry-Perot oscillations. The cw output power of 0.6 W was emitted from the 100 μm aperture into a nearly-diffraction-limited beam whose far-field characteristic is shown in Fig. 3.33b. In order to boost the power further, light from a nine-element array was collimated by anamorphic GaP microlenses fabricated by mass transport [249]. The array, which is shown schematically in Fig. 3.34, emitted 1.7 W into a full-cone angle of 65 mrad, corresponding to twice the diffraction limit.

As an alternative approach to enhancing the beam quality for broad-stripe visible and near-IR diodes, SDL developed the angled-grating distributed

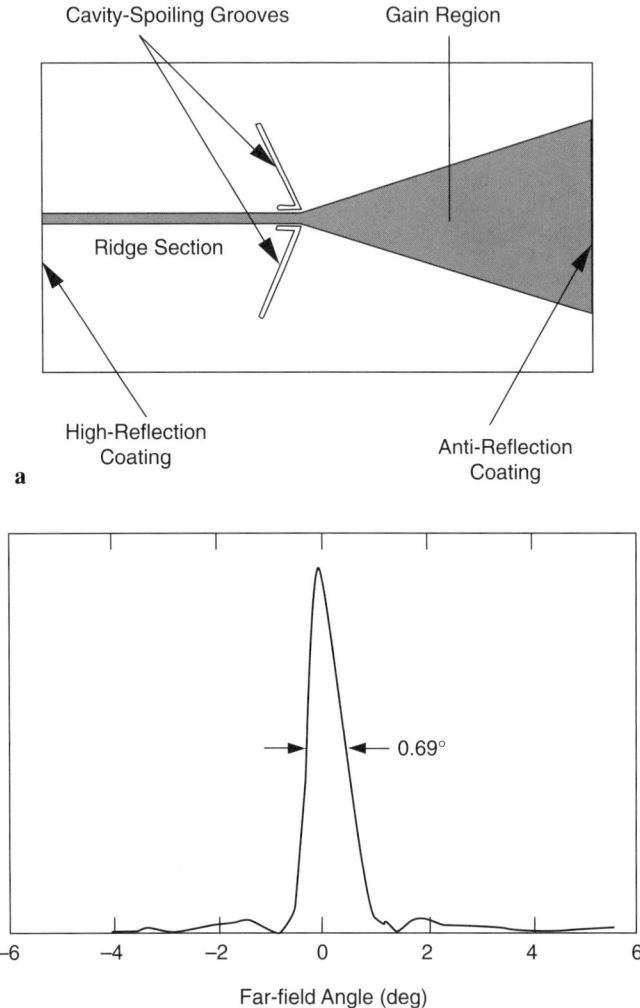

Fig. 3.33 (*a*) Schematic of a tapered laser fabricated from a GaInAsSb-AlGaAsSb SQW structure. (*b*) Far-field profile of the tapered laser output at 3 A cw and $T = 289.4$ K. (Reprinted from [248], with permission from the Institute of Electrical and Electronics Engineers, ©1998.)

feedback (α-DFB) laser [250,251]. As shown schematically in Fig. 3.35*a*, diffraction from the grating causes the lasing mode to propagate in a zigzag pattern along the angled stripe. There is very low angular divergence of the output beam at the facet because only those components of the optical wave that impinge nearly at normal incidence have appreciable feedback. A type-II W α-DFB laser emitting at $\lambda \approx 3.4\,\mu$m was recently demonstrated [252].

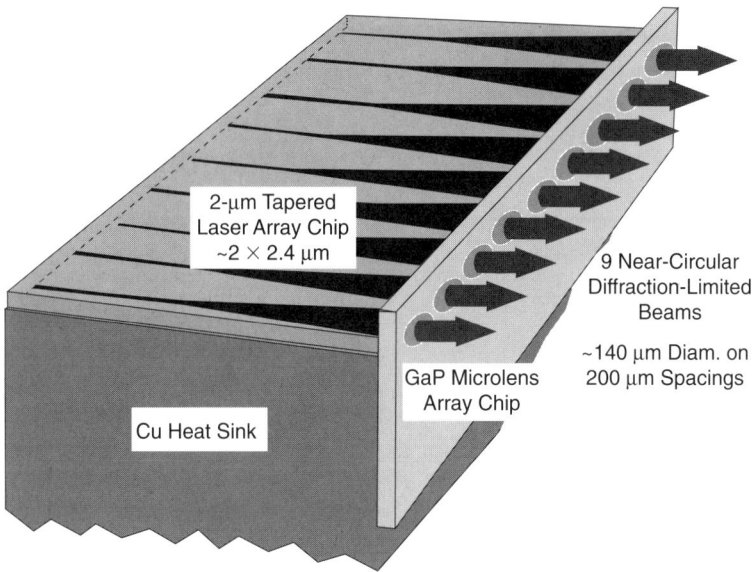

Fig. 3.34 Schematic of a nine-element tapered laser array mounted on copper heat sinks. A GaP microlens array collimates the light output. (Reprinted from [249], with permission from the Institute of Electrical and Electronics Engineers, ©1999.)

Figure 3.35b illustrates that under pulsed optical excitation with a 50 µm stripe width (filled points), the far-field divergence angle was nearly diffraction limited (1.4° FWHM). The far-field profile for a Fabry-Perot laser with the same stripe width was double-lobed and more than an order of magnitude wider. At the expense of a modest reduction in efficiency ($\approx 64\%$ of that for the Fabry-Perot laser), the beam quality was improved for stripes as wide as 800 µm.

3.5.6 Thermal Management and Thermal Conductivity

Since the antimonide mid-IR laser characteristics tend to depend rather strongly on temperature, efficient thermal management is crucial. Sample-mounting and heat-sinking methods are in most ways similar to those used for shorter wavelength semiconductor lasers, although the details vary and antimonide processing is generally much less mature. Epitaxial-side-down mounting is strongly preferred in order to place the active region as close as possible to the heat sink rather than requiring heat extraction through a thick substrate. Usually the heat sink (e.g., copper) and epitaxial surface are both metallized, and a solder such as indium is used to attach the two. Care must be taken to prevent voids in the solder, as well as contamination of the laser facets by solder or flux.

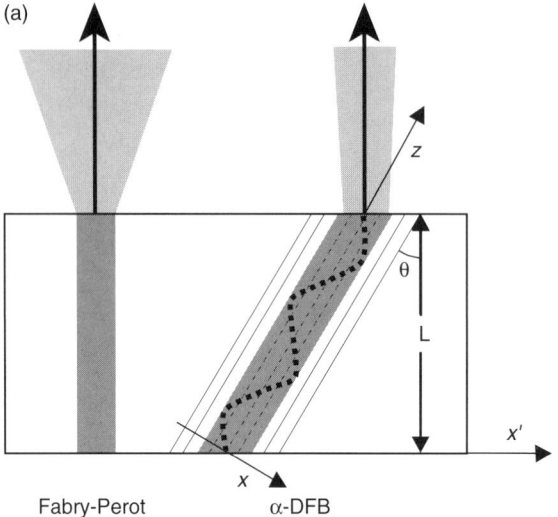

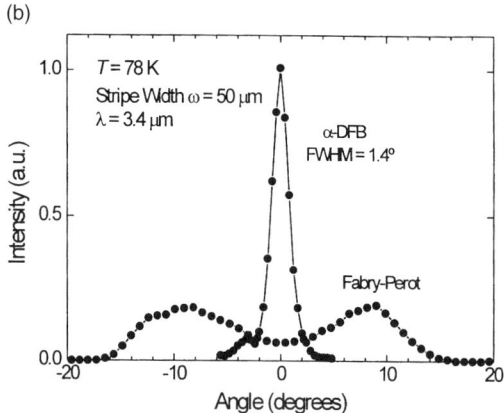

Figs. 3.35 (*a*) Schematic of optically pumped Fabry-Perot and α-DFB lasers (top view). (*b*) Lateral far field patterns for the Fabry-Perot and α-DFB lasers, acquired at eight times the threshold pump intensities. Areas under the curves reflect the relative efficiencies of the two devices. (Reprinted from [252], with permission from the American Institute of Physics.)

Even a device mounted epitaxial-side-down requires heat flow through the active and top optical cladding layers. As long as those layers are binary compounds or ternary alloys, their net thermal resistance is probably not a primary factor limiting the heat flow. However, many of the more complex structures employ lattice-matched quaternary alloys [19] or short-period superlattices [115], for which few thermal-conductivity measurements have been performed to date [253]. In particular, the thermal conductivities for short-period superlattices such as InAs/AlSb can be more than an order of magnitude lower than those of the binary bulk constituents at 300 K [254], and the difference becomes even larger when the device is operated at lower temperatures. At that point such layers are likely to dominate the thermal resistance, and maximizing the thermal transport in the optical cladding becomes a primary design consideration.

One approach to determining the specific thermal resistance is to correlate the cw-pump-intensity variation of the emission wavelength with its dependence on heat sink temperature at low duty cycle [255]. Using this method, the active regions of W lasers with cw optical pumping were found to operate at 30 to 70 K above the heat sink temperature [138]. A wavelength-chirp method has also been successful [90]. Gmachl et al. found that at the maximum cw operating temperature, the active region heated to approximately the same temperature as the maximum for low-duty-cycle operation [256].

NRL demonstrated an alternative approach to heat sinking called the diamond pressure bond (DPB) [135], in which the epitaxial side of the cleaved laser is placed in mechanical contact with a flat industrial diamond and the thermal bond is accomplished through pressure alone. Specific thermal resistances were measured to be as low as $1.3 \text{ K} \cdot \text{cm}^2/\text{kW}$ [138], which compares favorably with earlier results for solder mounting [89,255]. Besides greatly simplifying the processing, an additional advantage of the DPB technique is that it uniquely allows top optical access to the sample by pumping through the diamond. However, it is not yet clear whether the method will prove practical for commercial implementation, since periodic readjustment of the applied pressure would need to be accomplished automatically.

3.6 CONCLUSIONS

In this chapter we surveyed the current status of mid-IR lasers based on the antimonide material system, and summarized the main issues that limit the performance of practical devices.

The most mature antimonide lasers are strained type-I GaInAsSb/AlGaAsSb QW structures that lase cw at room temperature in the vicinity of 2 µm. Very low thresholds have been achieved by using compressive strain and quantum confinement to suppress Auger recombination. Internal losses as low as 2 cm^{-1} were demonstrated by reducing the overlap of the optical mode with the doped cladding layers in a broadened-waveguide geometry. Room-temperature cw operation was achieved at wavelengths as long as 2.7 µm by increasing the strain, although it remains unclear how much farther the wavelength can be extended using that approach.

In the 3 to 5 µm spectral range, InAsSb-based double heterostructure lasers served for a number of years as the performance benchmarks at 77 K. High powers were generated using optical pumping, although the inefficient 3D density of states has restricted the maximum operating temperature. Workers at the Ioffe Institute have demonstrated a number of single-mode DH devices, which have been applied to chemical spectroscopy.

Type-I InAsSb QW diodes yielded lower thresholds, higher characteristic temperatures, and higher cw output powers than the analogous DH devices. Differential quantum efficiencies of up to 70% have been reported. Many of the type-I QW laser designs have suffered from inadequate band offsets, and T_{max} will probably be limited by Auger recombination unless some means can be found to significantly suppress it.

Type-II InAs/GaInSb-based structures have exhibited more promising results at higher temperatures, due in part to larger band offsets and substantial Auger suppression. Whereas early devices employed single heterojunctions and two-constituent superlattices, major advances have been achieved using multiple QWs based on the four-constituent W structure. With optical pumping, cw operation has been obtained to 290 K at $\lambda = 3.0\,\mu m$ and to 210 K at 6.1 µm. The OPIC scheme has yielded high-power conversion efficiencies and low losses up to relatively high temperatures, and the integrated absorber configuration also has exhibited low loss and enhanced efficiencies. W diodes have operated to room temperature in pulsed mode and to 195 K cw.

The interband cascade laser provides a promising avenue to increased output power, since each injected electron is recycled to produce multiple photons. While the epitaxial growth is inherently challenging due to a large number of active layers, differential quantum efficiencies as high as 460% have been demonstrated. ARL reported a phenomenal pulsed output power of >4 W/facet at 80 K. The maximum operating temperature for pulsed operation has reached 286 K.

The ideal mid-IR semiconductor laser technology would provide compact and inexpensive diodes that operate at ambient temperature and emit high cw powers into a single mode. While this goal is already attainable at $\lambda \approx 2\,\mu m$, for the foreseeable future it is probably not realistic to project complete success over the entire 2 to 5 µm spectral range. The greatest challenges include improvement of the mode stability in addition to the suppression of Auger recombination, internal losses, and the linewidth enhancement factor.

REFERENCES

1. A. P. Danilova, A. N. Imenkov, N. M. Kolchanova, S. Civis, V. V. Sherstnev, and Yu. P. Yakovlev, "Single-mode InAsSb/InAsSbP laser ($\lambda \approx 3.2\,\mu m$) tunable over 100 Å," *Fiz. Tekh. Poluprovodn.* **34**, 243 (2000) [*Semiconductors* **34**, 237 (2000)].
2. K. D. Moiseev, M. P. Mikhaĭlova, O. G. Ershov, and Yu. P. Yakovlev, "Tunnel-injection laser based on a single p-GaInAsSb/p-InAs type-II broken-gap heterojunction," *Fiz. Tekh. Poluprovodn.* **30**, 399 (1996) [*Semiconductors* **30**, 223 (1996)].

3. T. N. Danilova, O. G. Ershov, A. N. Imenkov, M. V. Stepanov, V. V. Sherstnev, and Yu. P. Yakovlev, "Maximum working temperature of InAsSb/InAsSbP diode lasers," *Fiz. Tekh. Poluprovodn.* **30**, 1265 (1996) [*Semiconductors* **30**, 667 (1996)].

4. T. N. Danilova, O. I. Evseenko, A. N. Imenkov, N. M. Kolchanova, M. V. Stepanov, V. V. Sherstnev, and Yu. P. Yakovlev, "Influence of charge carriers on tuning in InAsSb lasers," *Fiz. Tekh. Poluprovodn.* **31**, 662 (1997) [*Semiconductors* **31**, 563 (1997)].

5. C. A. Wang, K. F. Jensen, A. C. Jones, and H. K. Choi, "n-AlGaSb and GaSb/AlGaSb double-heterostructure lasers grown by organometallic vapor phase epitaxy," *Appl. Phys. Lett.* **68**, 400 (1996).

6. R. M. Biefeld, A. A. Allerman, and S. R. Kurtz, "The growth of mid-infrared lasers and $AlAs_xSb_{1-x}$ by MOCVD," *J. Cryst. Growth* **174**, 593 (1997).

7. R. M. Biefeld, S. R. Kurtz, and A. A. Allerman, "The metalorganic chemical vapor deposition growth of AlAsSb and InAsSb/InAs using novel source materials for infrared emitters," *J. Electron. Mater.* **26**, 903 (1997).

8. J. Diaz, H. Yi, A. Rybaltowski, B. Lane, G. Lukas, D. Wu, S. Lim, M. Erdtmann, E. Kaas, and M. Razeghi, "InAsSbP/InAsSb/InAs laser diodes ($\lambda = 3.2\,\mu m$) grown by low-pressure metal-organic chemical-vapor deposition," *Appl. Phys. Lett.* **70**, 40 (1997).

9. I. Vurgaftman, J. R. Meyer, and L. R. Ram-Mohan, "Band parameters for III-V compound semiconductors and their alloys," *J. Appl. Phys.* **89**, 5815 (2001).

10. S. R. Kurtz, R. M. Biefeld, L. R. Dawson, K. C. Baucom, and A. J. Howard, "Midwave ($4\,\mu m$) infrared lasers and light-emitting diodes with biaxially compressed InAsSb active regions," *Appl. Phys. Lett.* **64**, 812 (1994).

11. S. R. Kurtz and R. M. Biefeld, "Magnetophotoluminescence of biaxially compressed InAsSb quantum wells," *Appl. Phys. Lett.* **66**, 364 (1995).

12. Y.-H. Zhang, R. H. Miles, and D. H. Chow, "InAs-InAsSb type-II superlattice midwave infrared lasers grown on InAs substrates," *IEEE J. Sel. Topics Quant. Electron.* **1**, 749 (1995).

13. S.-H. Wei and A. Zunger, "InAsSb/InAs: A type-I or a type-II band alignment," *Phys. Rev. B* **52**, 12039 (1995).

14. S. R. Kurtz, A. A. Allerman, and R. M. Biefeld, "Midinfrared lasers and light-emitting diodes with InAsSb/InAsP strained-layer superlattice active regions," *Appl. Phys. Lett.* **70**, 3188 (1997).

15. D. Wu, B. Lane, H. Mohseni, J. Diaz, and M. Razeghi, "High power asymmetrical InAsSb/InAsSbP/AlAsSb double heterostructure lasers emitting at $3.4\,\mu m$," *Appl. Phys. Lett.* **74**, 1194 (1999).

16. H. K. Choi, S. J. Eglash, and G. W. Turner, "Double-heterostructure diode lasers emitting at $3\,\mu m$ with a metastable GaInAsSb active layer and AlGaAsSb cladding layers," *Appl. Phys. Lett.* **64**, 2474 (1994).

17. A. Ghiti and E. P. O'Reilly, "Antimony-based strained-layer 2–2.5 μm quantum well lasers," *Semicond. Sci. Technol.* **8**, 1655 (1993).

18. G. W. Turner, H. K. Choi, and M. J. Manfra, "Ultralow-threshold ($50\,A/cm^2$) strained single-quantum-well GaInAsSb/AlGaAsSb lasers emitting at $2.05\,\mu m$," *Appl. Phys. Lett.* **72**, 876 (1998).

19. D. Garbuzov, M. Maiorov, H. Lee, V. Khalfin, R. Martinelli, and J. Connolly, "Temperature dependence of continuous wave threshold current for 2.3–2.6 µm InGaAsSb/AlGaAsSb separate confinement heterostructure quantum well semiconductor diode lasers," *Appl. Phys. Lett.* **74**, 2990 (1999).
20. D. Garbuzov, H. Lee, V. Khalfin, R. Martinelli, J. C. Connolly, and G. L. Belenky, "2.3–2.7-µm room temperature CW operation of InGaAsSb-AlGaAsSb broad waveguide SCH-QW diode lasers," *IEEE Photon. Technol. Lett.* **11**, 794 (1999).
21. S. Simanowski, N. Herres, C. Mermelstein, R. Kiefer, J. Schmitz, M. Walther, J. Wagner, and G. Weimann, "Strain adjustment in (GaIn)(AsSb)/(AlGa)(AsSb) QWs for 2.3–2.7 µm laser structures," *J. Cryst. Growth* **209**, 15 (2000).
22. G. A. Sai-Halasz, L. Esaki, and W. A. Harrison, "InAs-GaSb superlattice energy structure and its semiconductor-semimetal transition," *Phys. Rev. B* **18**, 2812 (1978).
23. L. M. Dolginov, L. V. Druzhinina, P. G. Eliseev, I. V. Kryukova, V. I. Leskovich, M. G. Milvidskii, B. N. Sverdlov, and E. G. Shevchenko, "Multicomponent solid-solution semiconductor lasers," *IEEE J. Quantum Electron.* **13**, 609 (1977).
24. A. E. Bochkarev, L. M. Dolginov, A. E. Drakin, L. V. Druzhinina, P. G. Eliseev, and B. N. Sverdlov, "Injection InGaSbAs lasers emitting radiation of wavelengths 1.9–2.3 µm at room temperature," *Kvant. Elektron.* **12**, 1309 (1985) [*Sov. J. Quantum Electron.* **15**, 869 (1985)].
25. C. Caneau, A. K. Srivastava, A. G. Dentai, J. L. Zyskind, and M. A. Pollack, "Room-temperature GaInAsSb/AlGaAsSb DH injection lasers at 2.2 µm," *Electron. Lett.* **21**, 815 (1985).
26. A. E. Bochkarev, L. M. Dolginov, A. E. Drakin, L. V. Druzhinina, P. G. Eliseev, B. N. Sverdlov, and V. A. Skripkin, "Injection InGaSbAs laser emitting at 2.4 µm (300 K)," *Kvant. Electron.* **13**, 2119 (1986) [*Sov. J. Quantum Electron.* **16**, 1397 (1986)].
27. T. H. Chiu, W. T. Tsang, J. A. Ditzenberger, and J. P. van der Ziel, "Room-temperature operation of InGaAsSb/AlGaSb double heterostructure lasers near 2.2 µm prepared by molecular beam epitaxy," *Appl. Phys. Lett.* **49**, 1052 (1986).
28. I. V. Akimova, A. E. Bochkarev, L. M. Dolginov, A. E. Drakin, L. V. Druzhinina, P. G. Eliseev, B. N. Sverdlov, and V. A. Skripkin, "Room temperature 2.0–2.4 µm injection lasers," *Zh. Tekh. Fiz.* **58**, 701 (1988) [*Sov. Phys. Tech. Phys.* **33**, 429 (1988)].
29. J. L. Zyskind, J. C. Dewinter, C. A. Burrus, J. C. Centanni, A. G. Dentai, and M. A. Pollack, "Highly uniform, high quantum efficiency GaInAsSb/AlGaAsSb double heterostructure lasers emitting at 2.2 µm," *Electron. Lett.* **25**, 568 (1989).
30. A. N. Baranov, E. A. Grebenshchikova, B. E. Dzhurtanov, T. N. Danilova, A. N. Imenkov, and Yu. P. Yakovlev, "Long-wavelength lasers based on GaInAsSb solid solutions near the immiscibility boundary," *Pis'ma Zh. Tekh. Fiz.* **14**, 1839 (1988) [*Sov. Tech. Phys. Lett.* **14**, 798 (1988)].
31. C. Caneau, A. K. Srivastava, J. L. Zyskind, J. W. Sulhoff, A. G. Dentai, and M. A. Pollack, "Reduction of threshold current density of 2.2 µm GaInAsSb/AlGaAsSb injection lasers," *Electron. Lett.* **22**, 992 (1986).
32. A. E. Bochkarev, L. M. Dolginov, A. E. Drakin, P. G. Eliseev, and B. N. Sverdlov, "Continuous-wave lasing at room temperature in InGaSbAs/GaAlSbAs injection heterostructures emitting in the spectral range 2.2–2.4 µm," *Kvant. Electron.* **15**, 2171 (1988) [*Sov. J. Quantum Electron.* **18**, 1362 (1988)].

33. A. N. Baranov, T. N. Danilova, B. E. Dzhurtanov, A. N. Imenkov, S. G. Konnikov, A. M. Litvak, V. E. Usmanskii, and Yu. P. Yakovlev, "Cw lasing in GaInAsSb/GaSb buried channel laser ($T = 20°C, \lambda = 2.0\,\mu m$)," *Pis'ma Zh. Tekh. Fiz.* **14**, 1671 (1988) [*Sov. Tech. Phys. Lett.* **14**, 727 (1988)].

34. S. J. Eglash and H. K. Choi, "Efficient GaInAsSb/AlGaAsSb diode lasers emitting at 2.29 µm," *Appl. Phys. Lett.* **57**, 1292 (1990).

35. H. K. Choi and S. J. Eglash, "High-efficiency high-power GaInAsSb-AlGaAsSb double-heterostructure lasers emitting at 2.3 µm," *IEEE J. Quantum Electron.* **27**, 1555 (1991).

36. H. K. Choi and S. J. Eglash, "Room-temperature cw operation at 2.2 µm of GaInAsSb/AlGaAsSb diode lasers grown by MBE," *Appl. Phys. Lett.* **59**, 1165 (1991).

37. A. N. Baranov, C. Fouillant, P. Grunberg, J. L. Lazzari, S. Gaillard, and A. Joullié, "High temperature operation of GaInAsSb/AlGaAsSb double-heterostructure lasers emitting near 2.1 µm," *Appl. Phys. Lett.* **65**, 616 (1994).

38. H. K. Choi and S. J. Eglash, "High-power multiple-quantum-well GaInAsSb/AlGaAsSb diode lasers emitting at 2.1 µm with low threshold current density," *Appl. Phys. Lett.* **61**, 1154 (1992).

39. C. A. Wang and H. K. Choi, "GaInAsSb/AlGaAsSb multiple-quantum-well diode lasers grown by organometallic vapor phase epitaxy," *Appl. Phys. Lett.* **70**, 802 (1997).

40. W. Z. Shen, S. C. Shen, W. G. Tang, Y. Chang, Y. Zhao, and A. Z. Li, "Demonstration of light-hole behavior in quaternary GaInAsSb/AlGaAsSb quantum wells using infrared photoluminescence spectroscopy," *Appl. Phys. Lett.* **69**, 952 (1996).

41. H. K. Choi, G. W. Turner, and S. J. Eglash, "High-power GaInAsSb-AlGaAsSb multiple quantum well lasers emitting at 1.9 µm," *IEEE Photon. Technol. Lett.* **6**, 7 (1994).

42. H. Lee, P. K. York, R. J. Menna, R. U. Martinelli, D. Z. Garbuzov, S. Y. Narayan, and J. C. Connolly, "Room-temperature 2.78 µm AlGaAsSb/InGaAsSb quantum-well lasers," *Appl. Phys. Lett.* **66**, 1942 (1995).

43. A. N. Baranov, Y. Cuminal, G. Boissier, C. Alibert, and A. Joullié, "Low-threshold laser diodes based on type-II GaInAsSb/GaSb quantum-wells operating at 2.36 µm at room temperature," *Electron. Lett.* **32**, 2279 (1996).

44. N. Bertru, A. Baranov, Y. Cuminal, G. Almuneau, F. Genty, A. Joullie, O. Brandt, A. Mazuelas, and K. H. Ploog, "Long-wavelength (Ga,In)Sb/GaSb strained quantum well lasers grown by molecular beam epitaxy," *Semicond. Sci. Technol.* **13**, 936 (1998).

45. D. Z. Garbuzov, R. U. Martinelli, H. Lee, R. J. Menna, P. K. York, L. A. DiMarco, M. G. Harvey, R. J. Matarese, S. Y. Narayan, and J. C. Connolly, "4 W quasi-continuous-wave output power from 2 µm AlGaAsSb/InGaAsSb single-quantum-well broadened waveguide laser diodes," *Appl. Phys. Lett.* **70**, 2931 (1997).

46. T. Newell, X. Wu, A. L. Gray, S. Dorato, H. Lee, and L. F. Lester, "The effect of increased valence band offset on the operation of 2 µm GaInAsSb-AlGaAsSb lasers," *IEEE Photon. Technol. Lett.* **11**, 30 (1999).

47. C. Mermelstein, S. Simanowski, M. Mayer, R. Kiefer, J. Schmitz, M. Walther, and J. Wagner, "Room-temperature low-threshold low-loss continuous-wave opera-

tion of 2.26 μm GaInAsSb/AlGaAsSb quantum-well laser diodes," *Appl. Phys. Lett.* **77**, 1581 (2000).
48. H. K. Choi, S. J. Eglash, and M. K. Connors, "Single-frequency GaInAsSb/AlGaAsSb quantum-well ridge-waveguide lasers emitting at 2.1 μm," *Appl. Phys. Lett.* **63**, 3271 (1993).
49. A. A. Popov, V. V. Sherstnev, A. N. Baranov, C. Alibert, and Yu. P. Yakovlev, "Continuous-wave operation of singlemode GaInAsSb lasers emitting near 2.2 μm at Peltier temperatures," *Electron. Lett.* **34**, 1398 (1998).
50. D. A. Yarekha, G. Glastre, A. Perona, Y. Rouillard, F. Genty, E. M. Skouri, G. Boissier, P. Grech, A. Joullié, C. Alibert, and A. N. Baranov, "High temperature GaInSbAs/GaAlSbAs quantum well single mode continuous wave lasers emitting near 2.3 μm," *Electron. Lett.* **36**, 537 (2000).
51. S. Simanowski, M. Walther, J. Schmitz, R. Kiefer, N. Herres, F. Fuchs, M. Maier, C. Mermelstein, J. Wagner, and G. Weimann, "Arsenic incorporation in molecular beam epitaxy (MBE) grown (AlGaIn)(AsSb) layers for 2.0–2.5 μm laser structures on GaSb substrates," *J. Cryst. Growth* **201/202**, 849 (1999).
52. D. V. Donetsky, G. L. Belenky, D. Z. Garbuzov, H. Lee, R. U. Martinelli, G. Taylor, S. Luryi, and J. C. Connolly, "Direct measurements of heterobarrier leakage current and modal gain in 2.3 μm double QW *p*-substrate InGaAsSb/AlGaAsSb broad area lasers," *Electron. Lett.* **35**, 298 (1999).
53. Y. Cuminal, A. N. Baranov, D. Bec, P. Grech, M. Garcia, G. Boissier, A. Joullié, G. Glastre, and R. Blondeau, "Room-temperature 2.63 μm GaInAsSb/GaSb strained quantum-well laser diodes," *Semicond. Sci. Technol.* **14**, 283 (1999).
54. A. Joullié, G. Glastre, R. Blondeau, J. C. Nicolas, Y. Cuminal, A. N. Baranov, A. Wilk, M. Garcia, P. Grech, and C. Alibert, "Continuous-wave operation of GaInAsSb-GaSb type-II quantum-well ridge-lasers," *IEEE J. Sel. Top. Quantum. Electron.* **5**, 711 (1999).
55. D. A. Yarekha, A. Vicet, A. Perona, G. Glastre, B. Fraisse, Y. Rouillard, E. M. Skouri, G. Boissier, P. Grech, and A. Joullié, "High efficiency GaInSbAs/GaSb type-II quantum well continuous wave lasers," *Semicond. Sci. Technol.* **15**, 390 (2000).
56. I. Melngailis, "Longitudinal Injection-Plasma Laser of InSb," *Appl. Phys. Lett.* **6**, 59 (1965).
57. E. Hadji, J. Bleuse, N. Magnea, and J. L. Pautrat, "Photopumped infrared vertical-cavity surface-emitting laser," *Appl. Phys. Lett.* **68**, 2480 (1996).
58. C. Roux, E. Hadji, and J.-L. Pautrat, "2.6 μm optically pumped vertical-cavity surface-emitting laser in the CdHgTe system," *Appl. Phys. Lett.* **75**, 3763 (1999).
59. W. W. Bewley, C. L. Felix, I. Vurgaftman, J. R. Meyer, G. Xu, and Z. Shi, "Lead-salt vertical-cavity surface-emitting lasers operating at λ = 4.5–4.6 μm," *Electron. Lett.* **36**, 539 (2000).
60. T. Schwarzl, W. Heiss, G. Springholz, M. Aigle, and H. Pascher, "6 μm vertical cavity surface emitting laser based on IV–VI semiconductor compounds," *Electron. Lett.* **36**, 322 (2000).
61. G. Springholz, T. Schwarzl, M. Aigle, H. Pascher, and W. Heiss, "4.8 μm vertical emitting PbTe quantum-well lasers based on high-finesse EuTe/Pb$_{1-x}$Eu$_x$Te microcavities," *Appl. Phys. Lett.* **76**, 1807 (2000).

62. C. L. Felix, W. W. Bewley, I. Vurgaftman, J. R. Meyer, L. Goldberg, D. H. Chow, and E. Selvig, "Midinfrared vertical-cavity surface-emitting laser," *Appl. Phys. Lett.* **71**, 3483 (1997).
63. W. W. Bewley, C. L. Felix, I. Vurgaftman, E. H. Aifer, J. R. Meyer, L. Goldberg, J. R. Lindle, D. H. Chow, and E. Selvig, "Continuous-wave mid-infrared VCSELs," *IEEE Photon. Technol. Lett.* **10**, 660 (1998).
64. W. W. Bewley, C. L. Felix, I. Vurgaftman, E. H. Aifer, L. J. Olafsen, J. R. Meyer, L. Goldberg, and D. H. Chow, "Mid-IR vertical-cavity surface-emitting lasers for chemical sensing," *Appl. Optics* **38**, 1502 (1999).
65. I. Vurgaftman, J. R. Meyer, and L. R. Ram-Mohan, "Mid-IR vertical-cavity surface-emitting lasers," *IEEE J. Quantum Electron.* **34**, 147 (1998).
66. A. N. Baranov, Y. Rouillard, G. Boissier, P. Grech, S. Gaillard, and C. Alibert, "Sb-based monolithic VCSEL operating near 2.2 μm at room temperature," *Electron. Lett.* **34**, 281 (1998).
67. I. Melngailis, "Maser action in InAs diodes," *Appl. Phys. Lett.* **2**, 176 (1963).
68. I. Melngailis, "Magnetically tunable cw InAs diode maser," *Appl. Phys. Lett.* **2**, 202 (1963).
69. N. G. Basov, A. V. Dudenkova, A. I. Krasil'nikov, V. V. Nikitin, and K. P. Fedoseev, "Semiconductor p-n junction lasers in the InAsSb system," *Fiz. Tverd. Tela* **8**, 1060 (1966) [*Sov. Phys. Solid State* **8**, 847 (1966)].
70. N. Menyuk, A. S. Pine, and A. Mooradian, "Efficient InSb laser with resonant longitudinal optical pumping," *IEEE J. Quantum Electron.* **11**, 477 (1975).
71. N. Kobayashi and Y. Horikoshi, "DH lasers fabricated by new III–V semiconductor material InAsPSb," *Jpn. J. Appl. Phys.* **19**, L641 (1980).
72. J. P. van der Ziel, R. A. Logan, R. M. Mikulyak, and A. A. Ballman, "Laser oscillation at 3–4 μm from optically pumped InAsSbP," *IEEE J. Quantum Electron.* **21**, 1827 (1985).
73. N. V. Zotova, S. A. Karandashev, B. A. Matveev, N. M. Stus', and G. N. Talalakin, "Coherent emission at 3.9 μm in InAsSbP p-n structures," *Pis'ma Zh. Tekh. Fiz.* **12**, 1444 (1986) [*Sov. Tech. Phys.* **12**, 599 (1986)].
74. H. Mani, A. Joullié, G. Boissier, E. Tournie, F. Pitard, A.-M. Joullie, and C. Alibert, "New III–V double-heterojunction laser emitting near 3.2 μm," *Electron. Lett.* **24**, 1542 (1988).
75. J. P. van der Ziel, T. H. Chiu, and W. T. Tsang, "Optically pumped laser oscillation at 3.9 μm from $Al_{0.5}Ga_{0.5}Sb/InAs_{0.91}Sb_{0.09}/Al_{0.5}Ga_{0.5}Sb$ double heterostructures grown by MBE on GaSb," *Appl. Phys. Lett.* **48**, 315 (1986).
76. M. Aĭdaraliev, N. V. Zotova, S. A. Karandashev, B. A. Matveev, N. M. Stus', and G. N. Talalakin, "Low-threshold lasers for the interval 3–3.5 μm based on $InAsSbP/In_{1-x}Ga_xAs_{1-y}Sb_y$ double heterostructures," *Pis'ma Zh. Tekh. Fiz.* **15**, 49 (1989) [*Sov. Tech. Phys. Lett.* **15**, 600 (1989)].
77. M. Aĭdaraliev, N. V. Zotova, S. A. Karandashev, B. A. Matveev, N. M. Stus', and G. N. Talalakin, "Long-wavelength low-threshold lasers based on III–V compounds," *Fiz. Tekh. Poluprovodn.* **27**, 21 (1993) [*Semiconductors* **27**, 10 (1993)].
78. A. Popov, V. Sherstnev, Y. Yakovlev, R. Mücke, and P. Werle, "High power InAsSb/InAsSbP double heterostructure laser for continuous wave operation at 3.6 μm," *Appl. Phys. Lett.* **68**, 2790 (1996).

79. S. J. Eglash and H. K. Choi, "InAsSb/AlAsSb double-heterostructure diode lasers emitting at 4 μm," *Appl. Phys. Lett.* **64**, 833 (1994).
80. H. K. Choi, G. W. Turner, and Z. L. Liau, "3.9-μm InAsSb/AlAsSb double-heterostructure lasers with high output power and improved temperature characteristics," *Appl. Phys. Lett.* **65**, 2251 (1994).
81. D. Wu, E. Kaas, J. Diaz, B. Lane, A. Rybaltowski, H. J. Yi, and M. Razeghi, "InAsSbP-InAsSb-InAs diode lasers emitting at 3.2 μm grown by metal-organic chemical vapor deposition," *IEEE Photon. Technol. Lett.* **9**, 173 (1997).
82. B. Lane, D. Wu, H. J. Yi, J. Diaz, A. Rybaltowski, S. Kim, M. Erdtmann, H. Jeon, and M. Razeghi, "Study on the effect of minority carrier leakage in InAsSb/InPAsSb double heterostructure," *Appl. Phys. Lett.* **70**, 1447 (1997).
83. A. Rybaltowski, Y. Xiao, D. Wu, B. Lane, H. Yi, H. Feng, J. Diaz, and M. Razeghi, "High power InAsSb/InPAsSb/InAs mid-infrared lasers," *Appl. Phys. Lett.* **71**, 2430 (1997).
84. B. Lane, S. Tong, J. Diaz, Z. Wu, and M. Razeghi, "High power InAsSb/InAsSbP electrical injection laser diodes emitting between 3 and 5 μm," *Mater. Sci. Eng.* B**74**, 52 (2000).
85. T. Ashley, C. T. Elliott, R. Jefferies, A. D. Johnson, G. J. Pryce, A. M. White, and M. Carroll, "Mid-infrared $In_{1-x}Al_xSb$/InSb heterostructure diode lasers," *Appl. Phys. Lett.* **70**, 931 (1997).
86. R. T. Kotitschke, A. R. Hollingworth, E. P. O'Reilly, A. R. Adams, B. N. Murdin, C. T. Elliott, C. J. G. M. Langerak, P. Findlay, C. R. Pidgeon, T. Ashley, and G. Pryce, "Influence of Auger and LO-phonon scattering on bulk and 'quasi'-quantum wire mid-IR laser diodes," *IEE Proc.-Optoelectron.* **145**, 281 (1998).
87. H. Q. Le, G. W. Turner, J. R. Ochoa, and A. Sanchez, "High-efficiency, high-temperature mid-infrared ($\lambda > 4$ μm) InAsSb/GaSb lasers," *Electron. Lett.* **30**, 1944 (1994).
88. H. Q. Le, G. W. Turner, S. J. Eglash, H. K. Choi, and D. A. Coppeta, "High-power diode-laser-pumped InAsSb/GaSb and GaInAsSb/GaSb lasers emitting from 3 to 4 μm," *Appl. Phys. Lett.* **64**, 152 (1994).
89. H. Q. Le, G. W. Turner, and J. R. Ochoa, "25 Turn-key, liquid-nitrogen-cooled 3.9 μm semiconductor laser package with 0.2 W CW output," *Electron. Lett.* **32**, 2359 (1996).
90. H. Q. Le, G. W. Turner, and J. R. Ochoa, "High-power high-efficiency quasi-cw Sb-based mid-IR lasers using 1.9-μm laser diode pumping," *IEEE Photon. Technol. Lett.* **10**, 663 (1998).
91. S. Colak, R. Eppenga, and M. F. H. Schuurmans, "Band mixing effects on quantum well gain," *IEEE J. Quantum Electron.* **23**, 960 (1987).
92. P. A. Chen and C. Y. Chang, "Analysis of differential gain in GaAs/AlGaAs quantum-well lasers," *J. Appl. Phys.* **76**, 85 (1994).
93. H. K. Choi, G. W. Turner, and H. Q. Le, "InAsSb/InAlAs strained quantum-well lasers emitting at 4.5 μm," *Appl. Phys. Lett.* **66**, 3543 (1995).
94. H. K. Choi and G. W. Turner, "InAsSb/InAlAsSb strained-quantum-well diode lasers emitting at 3.9 μm," *Appl. Phys. Lett.* **67**, 332 (1995).
95. H. K. Choi, G. W. Turner, M. J. Manfra, and M. K. Connors, "175 K continuous wave operation of InAsSb/InAlAsSb quantum-well diode lasers emitting at 3.5 μm," *Appl. Phys. Lett.* **68**, 2936 (1996).

96. H. K. Choi, G. W. Turner, and M. J. Manfra, "High cw power (>200 mW/facet) at 3.4 μm from InAsSb/InAlAsSb strained quantum well diode lasers," *Electron. Lett.* **32**, 1296 (1996).
97. S. R. Kurtz, R. M. Biefeld, A. A. Allerman, A. J. Howard, M. H. Crawford, and M. W. Pelczynski, "Pseudomorphic InAsSb multiple quantum well injection laser emitting at 3.5 μm," *Appl. Phys. Lett.* **68**, 1332 (1996).
98. A. A. Allerman, R. M. Biefeld, and S. R. Kurtz, "InAsSb-based mid-infrared lasers (3.8–3.9 μm) and light-emitting diodes with AlAsSb claddings and semimetal electron injection, grown by metalorganic chemical vapor deposition," *Appl. Phys. Lett.* **69**, 465 (1996).
99. B. Lane, D. Wu, A. Rybaltowski, H. Yi, J. Diaz, and M. Razeghi, "Compressively strained multiple quantum well InAsSb lasers emitting at 3.6 μm grown by metal-organic chemical vapor deposition," *Appl. Phys. Lett.* **70**, 443 (1997).
100. B. Lane, Z. Wu, A. Stein, J. Diaz, and M. Razeghi, "InAsSb/InAsP strained-layer superlattice injection lasers operating at 4.0 μm grown by metal-organic chemical vapor deposition," *Appl. Phys. Lett.* **74**, 3438 (1999).
101. G. W. Turner, H. K. Choi, and H. Q. Le, "Growth of InAsSb quantum wells for long-wavelength (~4 μm) lasers," *J. Vac. Sci. Technol. B* **13**, 699 (1995).
102. H. Kroemer and G. Griffiths, "Staggered-lineup heterojunctions as sources of tunable below-gap radiation: Operating principle and semiconductor selection," *IEEE Electron Dev. Lett.* **4**, 20 (1983).
103. A. N. Baranov, B. E. Dzhurtanov, A. N. Imenkov, A. A. Rogachev, Yu. M. Shernyakov, and Yu. P. Yakovlev, "Generation of coherent radiation in a quantum-well structure with one heterojunction," *Fiz. Tekh. Poluprovodn.* **20**, 2217 (1986) [*Semiconductors* **20**, 1385 (1986)].
104. K. D. Moiseev, M. P. Mikhailova, O. G. Ershov, and Yu. P. Yakovlev, "Long-wavelength laser ($\lambda = 3.26$ μm) with a single isolated *p*-GaInAsSb/*p*-InAs type-II heterojunction in the active region," *Pis'ma Zh. Tekh. Fiz.* **21**, 83 (1995) [*Sov. Tech. Phys. Lett.* **21**, 482 (1995)].
105. K. D. Moiseev, M. P. Mikhaĭlova, O. G. Ershov, and Yu. P. Yakovlev, "Infrared laser ($\lambda = 3.2$ μm) based on broken-gap type II heterojunctions with improved temperature characteristics," *Pis'ma Zh. Tekh. Fiz.* **23**, 55 (1997) [*Sov. Tech. Phys. Lett.* **23**, 151 (1997)].
106. M. P. Mikhailova, K. D. Moiseev, Y. A. Berezovets, R. V. Parfeniev, N. L. Bazhenov, V. A. Smirnov, and Yu. P. Yakovlev, "Interface-induced phenomena in type II antimonide-arsenide heterostructures," *IEE Proc. Optoelectron.* **145**, 268 (1998).
107. K. D. Moiseev, B. Ya. Mel'tser, V. A. Solov'ev, S. V. Ivanov, M. P. Mikhaĭlova, Yu. P. Yakovlev, and P. S. Kop'ev, "Electroluminescence of quantum-well structures on type-II InAs/GaSb heterojunctions," *Pis'ma Zh. Tekh. Fiz.* **24**, 50 (1998) [*Sov. Tech. Phys. Lett.* **24**, 477 (1998)].
108. D. L. Smith and C. Mailhiot, "Proposal for strained type II superlattice infrared detectors," *J. Appl. Phys.* **62**, 2545 (1987).
109. D. H. Chow, R. H. Miles, J. R. Soderstrom, and T. C. McGill, "Growth and characterization of InAs/Ga$_{1-x}$In$_x$Sb strained-layer superlattices," *Appl. Phys. Lett.* **56**, 1418 (1990).
110. R. H. Miles, D. H. Chow, and W. J. Hamilton, "High structural quality Ga$_{1-x}$In$_x$Sb/InAs strained-layer superlattices grown on GaSb substrates," *J. Appl. Phys.* **71**, 211 (1992).

111. C. H. Grein, P. M. Young, and H. Ehrenreich, "Theoretical performance of InAs/In$_x$Ga$_{1-x}$Sb superlattice-based midwave infrared lasers," *J. Appl. Phys.* **76**, 1940 (1994).

112. H. Miles, D. H. Chow, Y.-H. Zhang, P. D. Brewer, and R. G. Wilson, "Midwave infrared stimulated emission from a GaInSb/InAs superlattice," *Appl. Phys. Lett.* **66**, 1921 (1995).

113. T. C. Hasenberg, D. H. Chow, A. R. Kost, R. H. Miles, and L. West, "Demonstration of 3.5 μm Ga$_{1-x}$In$_x$Sb/InAs superlattice diode laser," *Electron. Lett.* **31**, 275 (1995).

114. T. C. Hasenberg, R. H. Miles, A. R. Kost, and L. West, "Recent advances in Sb-based midwave-infrared lasers," *IEEE J. Quantum Electron.* **33**, 1403 (1997).

115. D. H. Chow, R. H. Miles, T. C. Hasenberg, A. R. Kost, Y. H. Zhang, H. L. Dunlap, and L. West, "Mid-wave infrared diode lasers based on GaInSb/InAs and InAs/AlSb superlattices," *Appl. Phys. Lett.* **67**, 3700 (1995).

116. M. E. Flatté, J. T. Olesberg, S. A. Anson, T. F. Boggess, T. C. Hasenberg, R. H. Miles, and C. H. Grein, "Theoretical performance of mid-infrared broken-gap multiplayer superlattice lasers," *Appl. Phys. Lett.* **70**, 3212 (1997).

117. A. N. Baranov, N. Bertru, Y. Cuminal, G. Boissier, C. Alibert, and A. Joullié, "Observation of room-temperature laser emission from type III InAs/GaSb multiple quantum well structures," *Appl. Phys. Lett.* **71**, 735 (1997).

118. W. W. Bewley, E. H. Aifer, C. L. Felix, I. Vurgaftman, J. R. Meyer, C.-H. Lin, S. J. Murry, D. Zhang, and S. S. Pei, "High-temperature type-II superlattice diode laser at λ = 2.9 μm," *Appl. Phys. Lett.* **71**, 3607 (1997).

119. A. Wilk, M. El Gazouli, M. El Skouri, P. Christol, P. Grech, A. N. Baranov, and A. Joullié, "Type-II InAsSb/InAs strained quantum-well laser diodes emitting at 3.5 μm," *Appl. Phys. Lett.* **77**, 2298 (2000).

120. W. T. Tsang, "Extremely low threshold AlGaAs modified multiquantum well heterostructure lasers grown by MBE," *Appl. Phys Lett.* **39**, 786 (1981).

121. L. R. Ram-Mohan and J. R. Meyer, "Multiband finite element modeling of wavefunction-engineered electro-optical devices," *J. Nonlin. Opt. Phys. Mat.* **4**, 191 (1995).

122. J. R. Meyer, C. A. Hoffman, F. J. Bartoli, and L. R. Ram-Mohan, "Type-II quantum-well lasers for the mid-wavelength infrared," *Appl. Phys. Lett.* **67**, 757 (1995).

123. J. I. Malin, J. R. Meyer, C. L. Felix, J. R. Lindle, L. Goldberg, C. A. Hoffman, and F. J. Bartoli, "Type II mid-infrared quantum well lasers," *Appl. Phys. Lett.* **68**, 2976 (1996).

124. J. I. Malin, C. L. Felix, J. R. Meyer, C. A. Hoffman, J. F. Pinto, C.-H. Lin, P. C. Chang, S. J. Murry, and S.-S. Pei, "Type II mid-IR lasers operating above room temperature," *Electron. Lett.* **32**, 1593 (1996).

125. C. L. Felix, J. R. Meyer, I. Vurgaftman, C.-H. Lin, S. J. Murry, D. Zhang, and S.-S. Pei, "High-temperature 4.5 μm type-II quantum well laser with Auger suppression," *IEEE Photon. Technol. Lett.* **9**, 734 (1997).

126. C.-H. Lin, R. Q. Yang, S. J. Murry, S. S. Pei, C. Yan, D. L. McDaniel Jr., and M. Falcon, "Room-temperature low-threshold type-II quantum well lasers at 4.5 μm," *IEEE Phot. Tech. Lett.* **9**, 1573 (1997).

127. C.-H. Lin, S. J. Murry, R. Q. Yang, B. H. Yang, S. S. Pei, C. Yan, D. M. Gianardi Jr., D. L. McDaniel Jr., and M. Falcon, "Room-temperature mid-IR type-II quantum well lasers with high power efficiency," *J. Vac. Sci. Technol.* **B16**, 1435 (1998).

128. W. W. Bewley, C. L. Felix, E. H. Aifer, I. Vurgaftman, L. J. Olafsen, J. R. Meyer, H. Lee, R. U. Martinelli, J. C. Connolly, A. R. Sugg, G. H. Olsen, M. J. Yang, B. R. Bennett, and B. V. Shanabrook, "Above-room-temperature optically pumped mid-infrared W lasers," *Appl. Phys. Lett.* **73**, 3833 (1998).

129. M. E. Flatté, T. C. Hasenberg, J. T. Olesberg, S. A. Anson, T. F. Boggess, C. Yan, and D. L. McDaniel Jr., "III–V interband 5.2 µm laser operating at 185 K," *Appl. Phys. Lett.* **71**, 3764 (1997).

130. D. W. Stokes, L. J. Olafsen, W. W. Bewley, I. Vurgaftman, C. L. Felix, E. H. Aifer, J. R. Meyer, and M. J. Yang, "Type-II quantum-well 'W' lasers emitting at $\lambda = 5.4$–$7.3\,\mu m$," *J. Appl. Phys.* **86**, 4729 (1999).

131. M. E. Flatté, J. T. Olesberg, and C. H. Grein, "Ideal performance of cascade and noncascade intersubband and interband long-wavelength semiconductor lasers," *Appl. Phys. Lett.* **75**, 2020 (1999).

132. I. Vurgaftman and J. R. Meyer, "Optically pumped type-II interband terahertz lasers," *Appl. Phys. Lett.* **75**, 899 (1999).

133. H. Q. Le, C. H. Lin, and S. S. Pei, "Low-loss high-efficiency and high-power diode-pumped mid-infrared GaInSb/InAs quantum well lasers," *Appl. Phys. Lett.* **72**, 3434 (1998).

134. E. H. Aifer, W. W. Bewley, C. L. Felix, I. Vurgaftman, L. J. Olafsen, J. R. Meyer, H. Lee, R. U. Martinelli, and J. C. Connolly, "Cw operation of 3.4 µm optically-pumped type-II W laser to 220 K," *Electron. Lett.* **34**, 1587 (1998).

135. W. W. Bewley, C. L. Felix, I. Vurgaftman, D. W. Stokes, E. H. Aifer, L. J. Olafsen, J. R. Meyer, M. J. Yang, B. V. Shanabrook, H. Lee, R. U. Martinelli, and A. R. Sugg, "High-temperature continuous-wave 3–6.1 µm 'W' lasers with diamond-pressure-bond heat sinking," *Appl. Phys. Lett.* **74**, 1075 (1999).

136. C. L. Felix, W. W. Bewley, L. J. Olafsen, D. W. Stokes, E. H. Aifer, I. Vurgaftman, J. R. Meyer, and M. J. Yang, "Continuous-wave type-II 'W' lasers emitting at $\lambda = 5.4$–$7.1\,\mu m$," *IEEE Photon. Technol. Lett.* **11**, 964 (1999).

137. M. Beck, D. Hofstetter, T. Aellen, J. Faist, U. Oesterle, M. Ilegems, E. Gini, and H. Melchior, "Continuous wave operation of a mid-infrared semiconductor laser at room temperature," *Science* **295**, 301 (2002).

138. W. W. Bewley, C. L. Felix, E. H. Aifer, D. W. Stokes, I. Vurgaftman, L. J. Olafsen, J. R. Meyer, M. J. Yang, and H. Lee, "Thermal characterization of diamond-pressure-bond heat sinking for optically pumped mid-infrared lasers," *IEEE J. Quantum Electron.* **35**, 1597 (1999).

139. C. L. Felix, W. W. Bewley, I. Vurgaftman, L. J. Olafsen, D. W. Stokes, J. R. Meyer, and M. J. Yang, "High-efficiency midinfrared 'W' laser with optical pumping injection cavity," *Appl. Phys. Lett.* **75**, 2876 (1999).

140. W. W. Bewley, C. L. Felix, I. Vurgaftman, D. W. Stokes, J. R. Meyer, H. Lee, and R. U. Martinelli, "Optical-pumping injection cavity (OPIC) mid-IR 'W' lasers with high efficiency and low loss," *IEEE Photon. Technol. Lett.* **12**, 477 (2000).

141. H. K. Choi, A. K. Goyal, S. C. Buchter, G. W. Turner, M. J. Manfra, and S. D. Calawa, "High-power optically pumped GaInSb/InAs quantum well lasers with GaInAsSb

integrated absorber layers emitting at 4 μm," Conference on Lasers and Electro-Optics (San Francisco, May 7–12, 2000), *Tech Dig.*, p. 63.

142. W. W. Bewley, H. Lee, I. Vurgaftman, R. J. Menna, C. L. Felix, R. U. Martinelli, D. W. Stokes, D. Z. Garbuzov, J. R. Meyer, M. Maiorov, J. C. Connolly, A. R. Sugg, and G. H. Olsen, "Continuous-wave operation of λ = 3.25 μm broadened-waveguide W quantum-well diode lasers up to T = 195 K," *Appl. Phys. Lett.* **76**, 256 (2000).

143. H. Lee, L. J. Olafsen, R. J. Menna, W. W. Bewley, R. U. Martinelli, I. Vurgaftman, D. Z. Garbuzov, C. L. Felix, M. Maiorov, J. R. Meyer, J. C. Connolly, A. R. Sugg, and G. H. Olsen, "Room-temperature type-II W quantum well diode laser with broadened waveguide emitting at λ = 3.30 μm," *Electron. Lett.* **35**, 1743 (1999).

144. A. Joullié, E. M. Skouri, M. Garcia, P. Grech, A. Wilk, P. Christol, A. N. Baranov, A. Behres, J. Kluth, A. Stein, K. Heime, M. Heuken, S. Rushworth, E. Hulicius, and T. Simecek, "InAs(PSb)-based 'W' quantum well laser diodes emitting near 3.3 μm," *Appl. Phys. Lett.* **76**, 2499 (2000).

145. J. T. Olesberg, M. E. Flatté, B. J. Brown, C. H. Grein, T. C. Hasenberg, S. A. Anson, and T. F. Boggess, "Optimization of active regions in midinfrared lasers," *Appl. Phys. Lett.* **74**, 188 (1999).

146. J. T. Olesberg, M. E. Flatté, T. C. Hasenberg, and C. H. Grein, "Mid-infrared InAs/GaInSb separate confinement heterostructure laser diode structures," *J. Appl. Phys.* **89**, 3283 (2001).

147. J. Faist, F. Capasso, D. L. Sivco, C. Sirtori, A. L. Hutchinson, and A. Y. Cho, "Quantum cascade laser," *Science* **264**, 553 (1994).

148. J. Faist, F. Capasso, D. L. Sivco, C. Sirtori, A. L. Hutchinson, and A. Y. Cho, "Quantum cascade laser: An intersubband semiconductor laser operating above liquid nitrogen temperature," *Electron. Lett.* **30**, 865 (1994).

149. R. Q. Yang, "Infrared laser based on intersubband transitions in quantum wells," *Superlatt. Microstruct.* **17**, 77 (1995).

150. J. R. Meyer, I. Vurgaftman, R. Q. Yang, and L. R. Ram-Mohan, "Type-II and type-I interband cascade lasers," *Electron. Lett.* **32**, 45 (1996).

151. I. Vurgaftman, J. R. Meyer, and L. R. Ram-Mohan, "High-power/low-threshold type-II interband cascade mid-IR laser—Design and modeling," *IEEE Photon. Technol. Lett.* **9**, 170 (1997).

152. C. L. Felix, W. W. Bewley, I. Vurgaftman, J. R. Meyer, D. Zhang, C.-H. Lin, R. Q. Yang, and S.-S. Pei, "Interband cascade laser emitting >1 photon per injected electron," *IEEE Photon. Technol. Lett.* **9**, 1433 (1997).

153. R. Q. Yang, B. H. Yang, D. Zhang, C.-H. Lin, S. J. Murry, H. Wu, and S. S. Pei, "High power mid-infrared interband cascade lasers based on type-II quantum wells," *Appl. Phys. Lett.* **71**, 2409 (1997).

154. J. L. Bradshaw, R. Q. Yang, J. D. Bruno, J. T. Pham, and D. E. Wortman, "High-efficiency interband cascade lasers with peak power exceeding 4 W/facet," *Appl. Phys. Lett.* **75**, 2362 (1999).

155. L. J. Olafsen, E. H. Aifer, I. Vurgaftman, W. W. Bewley, C. L. Felix, J. R. Meyer, D. Zhang, C.-H. Lin, and S. S. Pei, "Near-room-temperature mid-infrared interband cascade laser," *Appl. Phys. Lett.* **72**, 2370 (1998).

156. H. F. Lockwood, K.-F. Etzold, T. E. Stockton, and D. P. Marinelli, "The GaAs p-n–p-n laser diode," *IEEE J. Quantum Electron.* **10**, 567 (1974).
157. R. Q. Yang, C.-H. Lin, P. C. Chang, S. J. Murry, D. Zhang, S. S. Pei, S. R. Kurtz, A.-N. Chu, and F. Ren, "Mid-IR interband cascade electroluminescence in type-II quantum wells," *Electron. Lett.* **32**, 1621 (1996).
158. R. Q. Yang, C.-H. Lin, S. J. Murry, S. S. Pei, H. C. Liu, M. Buchanan, and E. Dupont, "Interband cascade light emitting diodes in the 5–8 μm spectrum region," *Appl. Phys. Lett.* **70**, 2013 (1997).
159. C.-H. Lin, R. Q. Yang, D. Zhang, S. J. Murry, S. S. Pei, A. A. Allerman, and S. R. Kurtz, "Type-II interband quantum cascade laser at 3.8 μm," *Electron. Lett.* **33**, 598 (1997).
160. B. H. Yang, D. Zhang, R. Q. Yang, C.-H. Lin, S. J. Murry, and S. S. Pei, "Mid-IR interband cascade lasers with quantum efficiencies >200%," *Appl. Phys. Lett.* **72**, 2220 (1998).
161. R. Q. Yang, J. D. Bruno, J. L. Bradshaw, J. T. Pham, and D. E. Wortman, "High-power interband cascade lasers with quantum efficiency >450%," *Electron. Lett.* **35**, 1254 (1999).
162. J. D. Bruno, J. L. Bradshaw, R. Q. Yang, J. T. Pham, and D. E. Wortman, "Low-threshold interband cascade lasers with power efficiency exceeding 9%," *Appl. Phys. Lett.* **76**, 3167 (2000).
163. D. Zhang, E. Dupont, R. Q. Yang, H. C. Liu, C.-H. Lin, M. Buchanan, and S. S. Pei, "Long-wavelength IR (~10–15 um) electroluminescence from Sb-based interband cascade devices," *Opt. Expr.* **1**, 97 (1997).
164. E. Dupont, J. P. McCaffrey, H. C. Liu, M. Buchanan, R. Q. Yang, C.-H. Lin, D. Zhang, and S. S. Pei, "Demonstration of cascade process in InAs/GaInSb/AlSb mid-infrared light emitting devices," *Appl. Phys. Lett.* **72**, 1495 (1998).
165. I. Vurgaftman, J. R. Meyer, F. Julien, L. R. Ram-Mohan, "Design and simulation of low-threshold antimonide intersubband lasers," *Appl. Phys. Lett.* **73**, 711 (1998).
166. K. Ohtani and H. Ohno, "Mid-infrared intersubband electroluminescence in InAs/AlSb cascade structures," *Electron. Lett.* **35**, 935 (1999).
167. K. Ohtani and H. Ohno, "Mid-infrared intersubband electroluminescence in InAs/GaSb/AlSb type-II cascade structures," *Physica E* **7**, 80 (2000).
168. A. A. Allerman, S. R. Kurtz, R. M. Biefeld, and K. C. Baucom, "10-stage 'cascaded' InAsSb quantum well laser at 3.9 μm," *Electron. Lett.* **34**, 369 (1998).
169. S. R. Kurtz, A. A. Allerman, R. M. Biefeld, and K. C. Baucom, "High slope efficiency, 'cascaded' midinfrared lasers with type I InAsSb quantum wells," *Appl. Phys. Lett.* **72**, 2093 (1998).
170. C.-H. Lin, S. J. Murry, D. Zhang, P. C. Chang, Y. Zhou, S. S. Pei, J. I. Malin, C. L. Felix, J. R. Meyer, C. A. Hoffman, and J. F. Pinto, "MBE grown mid-infrared type-II quantum-well lasers," *J. Cryst. Growth* **175/176**, 955 (1997).
171. C. H. Kuo, C.-H. Lin, C. H. Thang, S. S. Pei, Y. C. Zhou, "High-power mid-IR type II quantum-well lasers grown on compliant universal substrate," *Electron. Lett.* **35**, 1468 (1999).
172. B. A. Matveev, N. M. Stus', and G. N. Talalakin, "Inverse defect formation during growth of epitaxial InAsSbP/InAs structures," *Sov. Phys. Cryst.* **33**, 124 (1988).

173. M. Aĭdaraliev, N. V. Zotova, S. A. Karandashev, B. A. Matveev, N. M. Stus', and G. N. Talalakin, "InAsSbP light emitting diodes for analysis of carbon oxides," *Pis'ma Zh. Tekh. Fiz.* **17**, 75 (1991) [*Sov. Tech. Phys. Lett.* **17**, 852 (1991)].
174. G. W. Turner, M. J. Manfra, H. K. Choi, and M. K. Connors, "MBE growth of high-power InAsSb/InAlAsSb quantum-well diode lasers emitting at 3.5 µm," *J. Cryst. Growth* **175/176**, 825 (1997).
175. M. J. Yang, W. J. Moore, B. R. Bennett, and B. V. Shanabrook, "Growth and characterisation of InAs/InGaSb/InAs/AlSb infrared laser structures," *Electron. Lett.* **34**, 270 (1998).
176. G. Tuttle, H. Kroemer, and J. H. English, "Effects of interface layer sequencing on the transport properties of InAs/AlSb quantum well: Evidence for antisite donors at the InAs/AlSb interface," *J. Appl. Phys.* **67**, 3032 (1990).
177. M. R. Kitchin, M. J. Shaw, E. Corbin, J. P. Hagon, and M. Jaros, "Optical properties of imperfect strained-layer InAs/Ga$_{1-x}$In$_x$Sb/AlSb superlattices with infrared applications," *Phys. Rev.* B**61**, 8375 (2000).
178. M. J. Shaw, E. A. Corbin, M. R. Kitchin, J. P. Hagon, and M. Jaros, "Qualitative theory of scattering in antimonide-based heterostructures with imperfect interfaces," *J. Vac. Sci. Technol.* B**18**, 2088 (2000).
179. J. Harper, M. Weimer, D. Zhang, C.-H. Lin, and S. S. Pei, "Cross-sectional scanning tunneling microscopy characterization of molecular beam epitaxy grown InAs/GaSb/AlSb heterostructures for mid-infrared interband cascade lasers," *J. Vac. Sci. Technol.* B**16**, 1389 (1998).
180. J. Harper, M. Weimer, D. Zhang, C.-H. Lin, and S. S. Pei, "Microstructure of the GaSb-on-InAs heterojunction examined with cross-sectional scanning tunneling microscopy," *Appl. Phys. Lett.* **73**, 2805 (1998).
181. A. L. Gray, T. C. Newell, L. F. Lester, and H. Lee, "High-resolution X-ray and transmission electron microscopic analysis of a GaInAsSb/AlGaAsSb multiple quantum well laser structure," *J. Appl. Phys.* **85**, 7665 (1999).
182. B. Z. Nosho, W. H. Weinberg, W. Barvosa-Carter, B. R. Bennett, B. V. Shanabrook, and L. J. Whitman, "Effects of surface reconstruction on III–V semiconductor interface formation: The role of III/V composition," *Appl. Phys. Lett.* **74**, 1704 (1999).
183. W. Barvosa-Carter, A. S. Bracker, J. C. Culbertson, B. Z. Nosho, B. V. Shanabrook, L. J. Whitman, H. Kim, N. A. Modine, and E. Kaxiras, "Structure of III-Sb(001) growth surfaces: The role of heterodimers," *Phys. Rev. Lett.* **84**, 4649 (2000).
184. J. D. Bruno, private communication.
185. J. S. Blakemore, *Semiconductor Statistics*, Wiley, New York, 1962.
186. J. R. Lindle, J. R. Meyer, C. A. Hoffman, F. J. Bartoli, G. W. Turner, and H. K. Choi, "Auger lifetime in InAs, InAsSb, and InAsSb-InAlAsSb quantum wells," *Appl. Phys. Lett.* **67**, 3153 (1995).
187. S. W. McCahon, S. A. Anson, D.-J. Yang, M. E. Flatté, T. F. Boggess, D. H. Chow, T. C. Hasenberg, and C. H. Grein, "Carrier recombination dynamics in a (GaInSb/InAs)/AlGaSb superlattice multiple quantum well," *Appl. Phys. Lett.* **68**, 2135 (1996).
188. W. W. Bewley, I. Vurgaftman, C. L. Felix, J. R. Meyer, C.-H. Lin, D. Zhang, S. J. Murry, S. S. Pei, and L. R. Ram-Mohan, "Role of internal loss in limiting type-II mid-IR laser performance," *J. Appl. Phys.* **83**, 2384 (1998).

189. A. Rakovska, V. Berger, X. Marcadet, B. Vinter, K. Bouzehouane, and D. Kaplan, "Optical characterization and room temperature lifetime measurements of high quality MBE-grown InAsSb on GaSb," *Semicond. Sci. Technol.* **15**, 34 (2000).

190. M. E. Flatté, C. H. Grein, T. C. Hasenberg, S. A. Anson, D.-J. Jang, J. T. Olesberg, and T. F. Boggess, "Carrier recombination rates in narrow-gap InAs/Ga$_{1-x}$In$_x$Sb superlattices," *Phys. Rev.* B**59**, 5745 (1999).

191. A. R. Beattie and P. T. Landsberg, "Auger effect in semiconductors," *Proc. Roy. Soc. London* **249 A**, 216 (1959).

192. H. P. Hjalmarson and S. R. Kurtz, "Electron Auger processes in mid-IR InAsSb/InGaAs heterostructures," *Appl. Phys. Lett.* **69**, 949 (1996).

193. A. R. Adams, "Band-structure engineering for low-threshold high-efficiency semiconductor lasers," *Electron. Lett.* **22**, 249 (1986).

194. E. Yablonovitch and E. O. Kane, "Correction to 'Reduction of lasing threshold current density by the lowering of valence band effective masses,'" *J. Lightwave Technol.* **4**, 504 (1986).

195. J. R. Meyer, C. L. Felix, W. W. Bewley, I. Vurgaftman, E. H. Aifer, L. J. Olafsen, J. R. Lindle, C. A. Hoffman, M.-J. Yang, B. R. Bennett, B. V. Shanabrook, H. Lee, C.-H. Lin, S. S. Pei, and R. H. Miles, "Auger coefficients in type-II InAs/Ga$_{1-x}$In$_x$Sb quantum wells," *Appl. Phys. Lett.* **73**, 2857 (1998).

196. B. L. Gel'mont, Z. N. Sokolova, and I. N. Yassievich, "Auger recombination in direct-gap *p*-type semiconductors," *Fiz. Tekh. Poluprovodn.* **16**, 592 (1982) [*Semiconductors* **16**, 382 (1982)].

197. A. Sugimura, "Band-to-band Auger effect in long wavelength multinary III-V alloy semiconductor lasers," *IEEE J. Quantum Electron.* **18**, 352 (1982).

198. M. Takeshima, "Auger recombination in InAs, GaSb, InP and GaAs," *J. Appl. Phys.* **43**, 4114 (1972).

199. A. Mozer, K. M. Romanek, O. Hildebrand, W. Schmid, and M. H. Pilkuhn, "Losses in GaInAsP/InP and GaAlSbAs/GaSb lasers—The influence of the split-off valence band," *IEEE J. Quantum Electron.* **19**, 913 (1983).

200. A. Haug, "Temperature dependence of Auger recombination in GaSb," *J. Phys.* C**17**, 6191 (1984).

201. G. Fuchs, S. Hausser, A. Hangleiter, G. Griffiths, H. Kroemer, and S. Subbanna, "Recombination in GaSb/AlSb multiple quantum wells under high excitation conditions," *Superlatt. Microstruct.* **10**, 361 (1991).

202. W. T. Cooley, R. L. Hengehold, Y. K. Yeo, G. W. Turner, and J. P. Loehr, "Recombination dynamics in InAsSb quantum-well diode lasers measured using photoluminescence upconversion," *Appl. Phys. Lett.* **73**, 2890 (1998).

203. C. H. Grein, P. M. Young, and H. Ehrenreich, "Minority carrier lifetimes in ideal InGaSb/InAs superlattices," *Appl. Phys. Lett.* **61**, 2905 (1992).

204. M. E. Flatté, C. H. Grein, H. Ehrenreich, R. H. Miles, and H. Cruz, "Theoretical performance limits of 2.1–4.1 µm InAs/InGaSb, HgCdTe, and InGaAsSb lasers," *J. Appl. Phys.* **78**, 4552 (1995).

205. M. E. Flatté, C. H. Grein, and H. Ehrenreich, "Sensitivity of optimization of mid-infrared InAs/InGaSb laser active regions to temperature and composition variations," *Appl. Phys. Lett.* **72**, 1424 (1998).

206. D.-J. Jang, M. E. Flatté, C. H. Grein, J. T. Olesberg, T. C. Hasenberg, and T. F. Boggess, "Temperature dependence of Auger recombination in a multilayer narrow-band-gap superlattice," *Phys. Rev.* B**58**, 13047 (1998).
207. S. A. Anson, J. T. Olesberg, M. E. Flatté, T. C. Hasenberg, and T. F. Boggess, "Differential gain, differential index, and linewidth enhancement factor for a 4 μm superlattice laser active layer," *J. Appl. Phys.* **86**, 713 (1999).
208. G. G. Zegrya and A. D. Andreev, "Mechanism of suppression of Auger recombination processes in type-II heterostructures," *Appl. Phys. Lett.* **67**, 2681 (1995).
209. A. D. Andreev and G. G. Zegrya, "Mechanism for a suppression of Auger recombination in type-II heterostructures," *Pis'ma Zh. Eksp. Teor. Fiz.* **61**, 749 (1995) [*JETP Lett.* **61**, 764 (1995)].
210. G. G. Zegrya and A. D. Andreev, "Theory of the recombination of nonequilibrium carriers in type-II heterostructures," *Zh. Eksp. Teor. Fiz.* **109**, 615 (1996) [*JETP* **82**, 328 (1996)].
211. E. E. Vdovkina, N. S. Baryshev, M. P. Shchetinin, A. P. Cherkasov, and I. S. Aver'yanov, "Photoelectric properties of HgCdTe at high temperatures," *Fiz. Tekh. Poluprovodn.* **10**, 183 (1976) [*Semiconductors* **10**, 109 (1976)].
212. J. Bajaj, S. H. Shin, J. G. Pasko, and M. Khoshnevisan, "Minority carrier lifetime in LPE HgCdTe," *J. Vac. Sci. Technol.* A**1**, 1749 (1983).
213. M. E. de Souza, M. Boukerche, and J. P. Faurie, "Minority-carrier lifetime in p-type (111)B HgCdTe grown by MBE," *J. Appl. Phys.* **68**, 5195 (1990).
214. J. Bonnet-Gamard, J. Bleuse, N. Magnea, and J. L. Pautrat, "Optical gain and laser emission in HgCdTe heterostructures," *J. Appl. Phys.* **78**, 6908 (1995).
215. J. L. Pautrat, E. Hadji, J. Bleuse, and N. Magnea, "Resonant-cavity IR optoelectronic devices," *J. Electron. Mat.* **26**, 667 (1997).
216. A. P. Mozer, S. Hausser, and M. H. Pilkuhn, "Quantitative evaluation of gain and losses in quaternary lasers," *IEEE J. Quantum Electron.* Optoelection. **21**, 719 (1985).
217. A. Haug and H. Burkhard, "Temperature dependence of threshold current of InGaAsP lasers with different compositions," *IEE Proc.-Optoelectron.* **134**, 117 (1987).
218. Y. Zou, J. S. Osinski, P. Grodzinski, P. D. Dapkus, W. Rideout, W. F. Sharfin, and F. D. Crawford, "Effect of Auger recombination and differential gain on the temperature sensitivity of 1.5 μm quantum well lasers," *Appl. Phys. Lett.* **62**, 175 (1993).
219. E. R. Youngdale, J. R. Meyer, C. A. Hoffman, F. J. Bartoli, C. H. Grein, P. M. Young, H. Ehrenreich, R. H. Miles, and D. H. Chow, "Auger lifetime enhancement in InAs-$Ga_{1-x}In_xSb$ superlattices," *Appl. Phys. Lett.* **64**, 3160 (1994).
220. T. C. Hasenberg, P. S. Day, E. M. Shaw, D. J. Magarrell, J. T. Olesberg, C. Yu, T. F. Boggess, and M. E. Flatté, "Molecular beam epitaxy growth and characterization of broken-gap (type II) superlattices and quantum wells for midwave-infrared laser diodes," *J. Vac. Sci. Technol.* B**18**, 1623 (2000).
221. H. Q. Le, C. H. Lin, S. J. Murray, R. Q. Yang, and S. S. Pei, "Effects of internal loss on power efficiency of mid-IR InAs-GaInSb-AlSb quantum well lasers and comparison with InAsSb lasers," *IEEE J. Quantum Electron.* **34**, 1016 (1998).
222. I. Vurgaftman and J. R. Meyer, "TE- and TM-polarized roughness-assisted free-carrier absorption in quantum wells at midinfrared and terahertz wavelengths," *Phys. Rev.* B**60**, 14294 (1999).

223. N. A. Gun'ko, V. B. Khalfin, Z. N. Sokolova, and G. G. Zegrya, "Optical loss in InAs-based long-wavelength lasers," *J. Appl. Phys.* **84**, 547 (1998).
224. A. A. Popov, V. V. Sherstnev, and Yu. P. Yakovlev, "Spectral and mode characteristics of InAsSbP/InAsSb/InAsSbP lasers in the spectral region near 3.3 µm," *Fiz. Tekh. Poluprovodn.* **32**, 1139 (1998) [*Semiconductors* **32**, 1019 (1998)].
225. M. Aĭdaraliev, N. V. Zotova, S. A. Karandashev, B. A. Matveev, M. A. Remennyĭ, N. M. Stus', and G. N. Talalakin, "Gain and internal losses in InGaAsSb/InAsSbP double-heterostructure lasers," *Fiz. Tekh. Poluprovodn.* **33**, 759 (1999) [*Semiconductors* **33**, 700 (1999)].
226. D. Z. Garbuzov, R. U. Martinelli, H. Lee, P. K. York, R. J. Menna, J. C. Connolly, and S. Y. Narayan, "Ultralow-loss broadened-waveguide high-power 2 µm AlGaAsSb/InGaAsSb/GaSb separate-confinement quantum-well lasers," *Appl. Phys. Lett.* **69**, 2006 (1996).
227. H. H. Chen, M. P. Houng, Y. H. Wang, and Y.-C. Chang, "Normal incidence intersubband optical transition in GaSb/InAs superlattices," *Appl. Phys. Lett.* **61**, 509 (1992).
228. F. Szmulowicz and G. J. Brown, "Calculation and photoresponse measurement of the bound-to-continuum infrared absorption in p-type GaAs/AlGaAs quantum wells," *Phys. Rev. B* **51**, 13203 (1995).
229. R. Kaspi, C. Moeller, A. Ongstad, M. L. Tilton, D. Gianardi, G. Dente, and P. Gopaladasu, "Absorbance spectroscopy and identification of valence subband transitions in type-II InAs/GaSb superlattices," *Appl. Phys. Lett.* **76**, 409 (2000).
230. M. Osinski and J. Buus, "Linewidth broadening factor in semiconductor lasers—An overview," *IEEE J. Quantum Electron.* **23**, 9 (1987).
231. A. N. Baranov, "Linewidth enhancement factor for GaInAsSb/GaSb lasers," *Appl. Phys. Lett.* **59**, 2360 (1991).
232. J. T. Olesberg, M. E. Flatté, and T. F. Boggess, "Comparison of linewidth enhancement factors in midinfrared active region materials," *J. Appl. Phys.* **87**, 7164 (2000).
233. W. W. Bewley, I. Vurgaftman, R. E. Bartolo, M. J. Jurkovic, C. L. Felix, J. R. Meyer, H. Lee, R. U. Martinelli, G. W. Turner, and M. J. Manfra, *IEEE J. Sel. Topics Quant. Electron.* **7**, 96 (2001).
234. B. W. Hakki and T. L. Paoli, "Gain spectra in GaAs double-heterostructure injection lasers," *J. Appl. Phys.* **46**, 1299 (1975).
235. C. E. Moeller, private communication.
236. A. Vicet, J.-C. Nicolas, F. Genty, Y. Rouillard, E. M. Skouri, A. N. Baranov, and C. Alibert, "Room temperature GaInAsSb/GaSb quantum well laser for tunable diode laser absorption spectroscopy around 2.35 µm," *IEE Proc. Optoelectron.* **147**, 172 (2000).
237. A. A. Popov, V. V. Sherstnev, and Yu. P. Yakovlev, "2.2 µm cw single-mode diode lasers with thermoelectric cooling," *Pis'ma Zh. Tekh. Fiz.* **23**, 39 (1997) [*Sov. Tech. Phys. Lett.* **23**, 591 (1997)].
238. A. Popov, V. Sherstnev, Y. Yakovlev, R. Mucke, and P. Werle, "Single-frequency InAsSb lasers emitting at 3.4 µm," *Spectrochim. Acta A* **52**, 863 (1996).
239. M. Aĭdaraliev, N. V. Zotova, S. A. Karandashev, B. A. Matveev, N. M. Stus', and G. N. Talalakin, "Emissive characteristics of mesa-stripe lasers (λ = 3.0–3.6 µm)

made from InGaAsSb/InAsSbP double heterostructures," *Pis'ma Zh. Tekh. Fiz.* **24**, 40 (1998) [*Sov. Tech. Phys. Lett.* **24**, 472 (1998)].

240. T. N. Danilova, O. I. Evseenko, A. N. Imenkov, N. M. Kolchanova, M. V. Stepanov, V. V. Sherstnev, and Yu. P. Yakovlev, "Influence of pumping uniformity on current tuning of the emission wavelength of InAsSb/InAsSbP diode lasers," *Pis'ma Zh. Tekh. Fiz.* **24**, 77 (1998) [*Sov. Tech. Phys. Lett.* **24**, 239 (1998)].

241. A. A. Popov, V. Sherstnev, Y. Yakovlev, P. Werle, and R. Mucke, "Relaxation oscillations in single-frequency InAsSb narrow band-gap lasers," *Appl. Phys. Lett.* **72**, 3428 (1998).

242. T. N. Danilova, A. P. Danilova, A. N. Imenkov, N. M. Kolchanova, M. V. Stepanov, V. V. Sherstnev, and Yu. P. Yakovlev, "InAsSb/InAsSbP heterostructure lasers with a large range of current tuning of the lasing frequency," *Pis'ma Zh. Tekh. Fiz.* **25**, 17 (1999) [*Sov. Tech. Phys. Lett.* **25**, 766 (1999)].

243. H. Q. Le, G. W. Turner, J. R. Ochoa, M. J. Manfra, C. C. Cook, and Y.-H. Zhang, "Broad wavelength tunability of grating-coupled external cavity midinfrared semiconductor lasers," *Appl. Phys. Lett.* **69**, 2804 (1996).

244. H. Yi, A. Rybaltowski, J. Diaz, D. Wu, B. Lane, Y. Xiao, and M. Razeghi, "Stability of far fields in double heterostructure and multiple quantum well InAsSb/InPAsSb/InAs midinfrared lasers," *Appl. Phys. Lett.* **70**, 3236 (1997).

245. T. N. Danilova, A. P. Danilova, O. G. Ershov, A. N. Imenkov, V. V. Sherstnev, and Yu. P. Yakovlev, "Spatial distribution of the radiation in the far zone of InAsSb/InAsSbP mesastrip lasers as a function of current," *Fiz. Tekh. Poluprovodn.* **32**, 373 (1998) [*Semiconductors* **32**, 339 (1998)].

246. A. P. Danilova, T. N. Danilova, A. N. Imenkov, N. M. Kolchanova, M. V. Stepanov, V. V. Sherstnev, and Yu. P. Yakovlev, "Spatial beam oscillations in stripe lasers utilizing InAsSb/InAsSbP heterojunctions," *Fiz. Tekh. Poluprovodn.* **33**, 1014 (1999) [*Semiconductors* **33**, 924 (1999)].

247. H. K. Choi, J. N. Walpole, G. W. Turner, S. J. Eglash, L. J. Missaggia, and M. K. Connors, "GaInAsSb-AlGaAsSb tapered lasers emitting at 2 μm," *IEEE Photon. Technol. Lett.* **5**, 1117 (1993).

248. H. K. Choi, J. N. Walpole, G. W. Turner, M. K. Connors, L. J. Missaggia, and M. J. Manfra, "GaInAsSb-AlGaAsSb tapered lasers emitting at 2.05 μm with 0.6-W diffraction-limited power," *IEEE Photon. Technol. Lett.* **10**, 938 (1998).

249. J. N. Walpole, H. K. Choi, L. J. Missaggia, Z. L. Liau, M. K. Connors, G. W. Turner, M. J. Manfra, and C. C. Cook, "High-power high-brightness GaInAsSb-AlGaAsSb tapered laser arrays with anamorphic collimating lenses emitting at 2.05 μm," *IEEE Photon. Technol. Lett.* **11**, 1223 (1999).

250. R. J. Lang, K. Dzurko, A. A. Hardy, S. DeMars, A. Schoenfelder, and D. F. Welch, "Theory of grating-confined broad-area lasers," *IEEE J. Quantum Electron.* **34**, 2196 (1998).

251. B. Pezeshki, M. Hagberg, M. Zelinski, S. D. DeMars, E. Kolev, and R. J. Lang, "400-mW single-frequency 660-nm semiconductor laser," *IEEE Photon. Technol. Lett.* **11**, 791 (1999).

252. R. E. Bartolo, W. W. Bewley, I. Vurgaftman, C. L. Felix, J. R. Meyer, and M. J. Yang, "Mid-infrared angled-grating distributed feedback laser," *Appl. Phys. Lett.* **76**, 3164 (2000).

253. W. Both, A. Bochkarev, A. Drakin, and B. Sverdlov, "Thermal resistivity of quaternary solid solutions InGaAsSb and GaAlSbAs lattice-matched to GaSb," *Electron. Lett.* **26**, 418 (1990).
254. T. Borca-Tasciuc, D. Achimov, W. L. Liu, G. Chen, H.-W. Ren, C.-H. Lin, and S. S. Pei, "Thermal conductivity of InAs/AlSb superlattices," *Proc. Conf. Heat Transfer and Transport Phenomena in Microsystems*, Canada, 2000.
255. S. A. Merritt, P. J. S. Heim, S. H. Cho, and M. Dagenais, "Controlled solder interdiffusion for high power semiconductor laser diode die bonding," *IEEE Trans. Comp., Packag., Manufact., Technol. B* **20**, 141 (1997).
256. C. Gmachl, A. M. Sergent, A. Tredicucci, F. Capasso, A. L. Hutchinson, D. L. Sivco, J. N. Baillargeon, S. N. G. Chu, and A. Y. Cho, "Improved cw operation of quantum cascade lasers with epitaxial-side heat-sinking," *IEEE Photon. Technol. Lett.* **11**, 1369 (1999).

CHAPTER 4

Lead-Chalcogenide-based Mid-Infrared Diode Lasers

UWE PETER SCHIEßL, JOACHIM JOHN, and PATRICK J. McCANN

4.1 INTRODUCTION

Lead-chalcogenide (IV-VI)-based mid-infrared (mid-IR) diode lasers are well known as key devices for high-sensitivity gas analysis for nearly a quarter of a century because they emit in the fundamental absorption region of molecules between 3 and 30µm. Their widespread commercial use, however, has been hindered until now because they operate in continuous wave (cw) mode only at cryogenic temperatures, requiring the use of expensive closed-cycle cryogenic coolers or liquid nitrogen.

Nowadays there is a growing worldwide demand for tunable mid-IR lasers. As a result an increasing number of groups are developing mid-IR lasers based on different material systems by employing powerful tools of modern semiconductor processing developed for near-infrared (NIR) lasers. This book gives an excellent overview of the different development strategies, which all have the objectives of making mid-IR lasers available for cw room-temperature operation.

Until now, however, lead-salt lasers are still leading the race for higher cw operating temperatures [1]. In addition lead-salt lasers have recently demonstrated pulsed operation above room temperature [1,2]. The high-temperature operation is possible in part because the Auger recombination rate is much lower in IV-VI systems than in other material systems such as antimonides [3]. Therefore the lead-chalcogenide-based lasers will continue to play an important role in high-resolution spectroscopy for gas analysis. A necessary prerequisite, however, is a continued effort in the development of lasers based on this interesting material system. The authors hope that this contribution will help motivate other groups worldwide to invest more efforts in this laser type

---

Long-Wavelength Infrared Semiconductor Lasers, Edited by Hong K. Choi
ISBN 0-471-39200-6 Copyright © 2004 John Wiley & Sons, Inc.

and to stimulate an open discussion on the potential of different mid-IR laser sources.

In this chapter we give an overview of the current state of the art of IV-VI lasers. We limit our discussions to the laser itself, since an excellent summary of general technology and preparation techniques as well as a collection of important material parameters has been given by Katzir et al. [4]. We review the current state of the art in the manufacturing of homostructure, double-heterostructure (DH), index-guided, quantum-well (QW), distributed feedback (DFB), and distributed Bragg reflector (DBR) lasers based on the IV-VI material systems. In addition a summary of the work on IV-IV materials epitaxy on BaF_2 and silicon is given. Growth on BaF_2, with its insulating and infrared transparent properties, has allowed unambiguous study of the electrical and optical characteristics of various IV-VI semiconductor materials. Results include demonstration of above room-temperature cw photoluminescence from multiple quantum-well materials grown on both BaF_2 and silicon. Future possibilities for laser fabrication involving substrate removal to enhance heat dissipation are also discussed.

4.2 HOMOSTRUCTURE LASERS

Homostructure lasers based on selected IV-VI materials are the most straightforward and easiest way to realize mid-IR laser emission. This laser type is still of importance for commercial laser production, especially in the longer wavelength region (>10µm). The first homostructure lasers were demonstrated in 1964 [5]. The major development of this simplest type of lasers was almost completed some 15 to 20 years ago. Since numerous review articles have documented the state-of-the-art of homostructure lasers [6–17], we limit our considerations to three prominent material systems: $Pb_{1-x}Sn_xTe$, $PbS_{1-x}Se_x$, and $Pb_{1-x}Sn_xSe$.

4.2.1 Material Properties

The binary compounds, PbSe, PbS, PbTe, and SnTe, crystallize in a cubic NaCl structure, while SnSe crystallizes in an orthorhombic B29 structure. Ternary compounds, $Pb_{1-x}Sn_xTe$ and $PbS_{1-x}Se_x$, exist for all x, while $Pb_{1-x}Sn_xSe$ with a cubic structure is limited to the lead-rich side ($x < 0.4$).

In Fig. 4.1 band-gap energy (emission wavelength), refractive index n, and lattice constant a of different ternary compounds at 77 K are plotted. A small difference in a is essential for the ability to form ternary compounds. For example, the difference in a between PbTe and PbS is quite large. In fact $PbTe_{1-x}S_x$ is very brittle and must be excluded from practical applications [16]. However, the lattice constant of $Pb_{1-x}Sn_xTe$, $PbS_{1-x}Se_x$, and $Pb_{1-x}Sn_xSe$ are comparable to their binary compounds, and are well suited for device fabrication. Most interesting is the fact that band inversion occurs in $Pb_{1-x}Sn_xSe$ as well as in

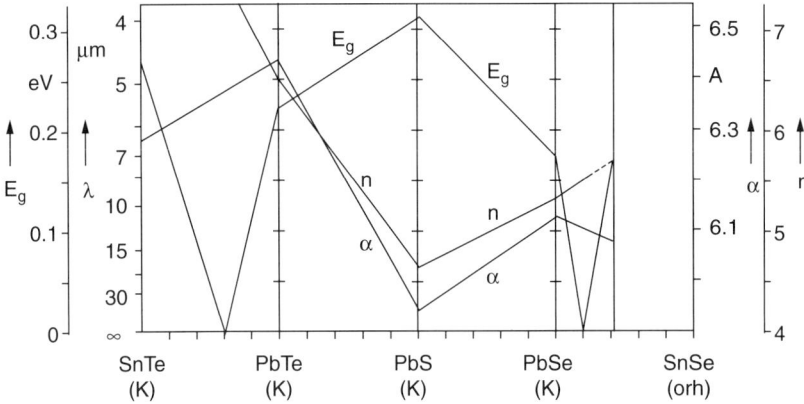

Fig. 4.1 Bandgap E_g, refractive index n and lattice constant a of $Pb_{1-x}Sn_xSe$, $PbTe_{1-x}S_x$, $PbS_{1-x}Se_x$ and $Pb_{1-x}Sn_xSe$ at 77 K as function of composition, taken from Ref. [16]

$Pb_{1-x}Sn_xTe$, which means that some compositions show a band gap with zero value [18]. In principle, lasers that emit at very long wavelengths could be made.

Functional dependencies for the band-gap energy of the different ternary compounds are given below. These formulas show very good agreement with low-temperature laser emission data and high-temperature optical absorption data [19].

$Pb_{1-x}Sn_xTe$ $(0 \leq x \leq 0.2)$
$$Eg(x,T) = 171.5 - 535x + \sqrt{(12.8)^2 + 0.19(T+20)^2}, \quad (4.1)$$

$PbS_{1-x}Se_x$ $(0 \leq x \leq 1)$ $\quad Eg(x,T) = 263 - 138x + \sqrt{400 + 0.265T^2}, \quad (4.2)$

$Pb_{1-x}Sn_xSe$ $(0 \leq x \leq 0.2)$ $\quad Eg(x,T) = 125 - 1021x + \sqrt{400 + 0.256T^2}. \quad (4.3)$

4.2.2 Device Fabrication

Fabrication of homostructure lasers is straightforward. First, a crystalline substrate based on a ternary material system is prepared by polishing at least one surface using mechanical and chemical techniques. The second step is to form a p-n junction by diffusion or ion implantation at a depth of several microns to ten microns. Then the backside of the wafer is polished to a typical thickness of 200 to 250 μm. Ohmic contacts are formed on both p- and n-sides. The wafer is cut into chips typically 250 × 360 × 220 μm. The chips can either be mounted on a copper heat sink by means of cold pressing the n-side into In, or by soldering the chip using a selected soft solder. Soft soldering has been demonstrated to be reliable for spectroscopic lasers after performing certain countermeasures to provide thermal cycling stability. For the p-side contact, a standard procedure is to use a rectangular-shaped cooper wire electroplated with In.

Formation of p-n Junction

Forming a p-n junction is one of the most critical steps. Typically diffusion is used for lead-salt lasers, but ion implantation was also applied. Since various materials system and diffusion sources were reviewed in [4], we will limit our considerations to the following basic systems.

p Diffusion in n Substrate by Chalcogenide. A surplus of chalcogenides forms acceptor-like states in lead salts. Because chalcogenides generally diffuse rapidly into the n-type material, it is necessary to use low diffusion temperatures to form shallow p-n junctions, leading to low carrier concentrations in the p-type layers. Generally, the low p-doping concentration is detrimental for the formation of good ohmic contacts, which makes this step more critical. In fact, to our knowledge, only one experimental result of chalcogenide diffusion of S in PbCdS [23] has been reported so far. This indicates the minor practical importance of this procedure. The p-n junction depth was 50 μm after a diffusion time of 20 minutes at 500°C.

n Diffusion in p Substrate by Lead. Due to the filling of lead vacancies, lead acts as a donor when it diffuses into a p-type material. Out-diffusion of chalcogenides at the same time creates chalcogenide vacancies, which also act as donors. It is intuitively clear that the diffusion coefficient of Pb should be a function of the hole concentration of the wafer. Typical diffusion coefficients at 500°C are of the order of $4–5 \times 10^{-12}$ cm^2/s [16] and 2.4×10^{-11} cm^2/s [24].

n Diffusion in p Substrate by Impurities. The use of Sb and Cd diffusion has been demonstrated to produce homostructure lasers. Sb diffusion in $Pb_{1-x}Sn_xTe$ was investigated in [25], and Sb-diffused lasers were reported in [26]. More popular is the application of Cd as a dopant. Cd diffusion in p-PbTe and p-PbSnTe was investigated in [27,28]. The solubilty of Cd in $Pb_{1-x}Sn_xTe$ for $0 \leq x \leq 0.25$ was independent of x. Saturation of the electron concentration was observed after filling the metal vacancies. Excess Cd is electrically inactive [28]. Cd diffusion in p-$Pb_{1-x}Sn_x$Se was investigated in [29]. In these material systems, Cd diffuses relatively fast by interstitial diffusion. Electrically Cd behaves in the same way as Pb.

Whereas the conventional method for Cd diffusion uses sealed ampoules, in which the Cd content has to be controlled in the submilligram range to avoid reactions such as Cd + PbSe ⇒ CdSe + Pb, a thermodynamically defined Cd diffusion source maintains a constant Cd pressure and prevents Cd chalcogenide formation [30] (Fig. 4.2). This system was successfully applied to PbCdS, PbSSe, and PnSnSe for the production of commercially available lead-salt lasers during the last 17 years. To obtain p-n junctions with depths of a few

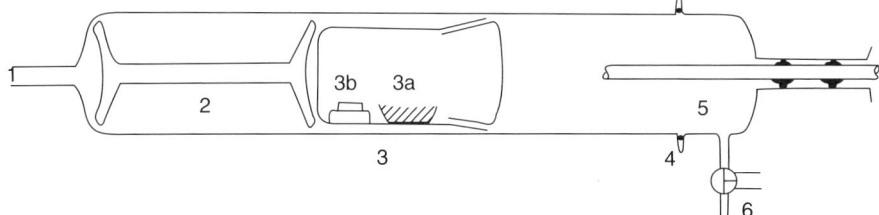

1. GAS INLET H$_2$/N$_2$
2. DISTANCE ROD for 3.
3. DIFFUSION CHAMBER with SOURCE MATERIAL CRU CIBLE, 3a, and QUARTZ PLATE with SUBSTRATE, 3b.
4. SEALING LID with O RING SEAL.
5. PUSH ROD for POSITIONING 3.
6. VACUUM / SPENT AIR CONNECTION [can be selected].

Fig. 4.2 Diffusion apparatus for the diffusion of Cadmium.

microns while keeping the diffusion time on the order of a few hours, the diffusion temperature has to be varied between 400°C and 550°C, depending on the band gap of the substrate material.

Ion Implantation. The use of ion implantation to incorporate a p-n junction into a lead-salt wafer was investigated in [31]. The authors implanted Ar$^+$ ions on p-PbSe wafers with carrier concentrations of around 10^{17} cm^{-3}. The energy, dose, and incident angle were 120 keV, 5×10^{16} cm^{-2}, and 7°, respectively. The ion implantation resulted in n-type doping with carrier concentrations on the order of 1×10^{18} cm^{-3}. The depth of the p-n junction was typically between 1 and 1.2 μm. The doping mechanism is considered to be a stoichiometric effect in PbSe since Ar$^+$ is neither a chemical donor nor acceptor. A possible explanation is that Se is driven out by the Ar$^+$ ions because Se has a high vapor pressure, which would leave Se vacancies and therefore conversion of conductivity. Measurements with FTIR indicated that the refractive index of the ion-implanted layer increased by 5%, whereas the energy gap decreased by 5%. Hence a quasi-single-heterostructure was formed.

Ohmic Contacts

The formation of ohmic contacts on lead salts is also one of the most critical technological steps. Laser degradation, low maximum operating temperature, fast tuning rates, and low yields are the consequences of poor contacts. The contact problem is minor for small gap semiconductors such as PbSnSe and PbSnTe. However, for higher band-gap materials such as PbSSe and PbCdSSe, a reliable contact technology is the key for good laser characteristics and low degradation.

On the n-side, the use of nonprecious materials with low work functions such as In was successful in achieving good ohmic contacts. However, nonprecious materials tend to react with the lead chalcogenide surface, leading to the formation of new phases. For example, the use of In leads to the formation of $InSe_2$, resulting in an enhanced series resistance. To avoid this problem, the use of a proper metal, which does not form chalcogenide layers like Pb, has to be taken into account. Another possibility is the application of special surface treatment procedures to form a Pb-rich surface, such as bombardment with Ar ions. Here the interaction with the big chalcogenide atoms leads to an excess of Pb on the surface. Excess Pb forms an n^+-doped surface—a well-known strategy for the formation of ohmic contacts on semiconductors.

On the p-side, ohmic contacts can be achieved by the use of noble metals with large work functions. However, noble metals do not easily form compounds, and therefore do not adhere well to the semiconductor surface. To improve the adhesion, sputtering techniques have been used with care, since investigations have shown that a p-type surface can be easily converted into an n-type surface [30]. Big chalcogenide atoms easily interact with high-energy collision partners, and therefore leave Pb atoms on the surface, resulting in an n-type surface. The formation of an excess chalcogenide layer on the surface by means of evaporation techniques followed by an annealing step is a possible way to preserve the high-chalcogenide concentration on the surface.

4.2.3 Device Characterization

Linden and Mantz [17] described a fast measurement setup in which 24 lasers could be simultaneously mounted onto a cold finger of a closed-cycle cooler. The test consisted of an electrical screening test in which resistance, threshold current, and output power of the lasers were evaluated. The lasers are mounted in a circular pattern, and each laser emits toward the center of the setup in which a Ge:Cu photoconductor was exploited to determine threshold current and a relative laser output.

Another way to achieve a fast function test is described in Fig. 4.4. Here the laser (see Fig. 4.3) is cooled by a liquid nitrogen cooling system, which is designed to achieve a constant cooling power over the entire operation temperature range between 77 and 300 K. With this setup, it is possible to cool a laser from room temperature down to 77 K in less than 1 minute. The time to heat up the laser from 77 to 300 K was optimized by several measures to be <5 minute. The laser can be tested either in pulsed mode (pulse lengths between 5 µs and 500 µs at a repetition rate of 1 kHz) or in cw mode. The system is able to monitor automatically the voltage versus current and the current versus output power characteristics at several operator-defined temperatures. A $\frac{1}{4}$-m monochromator, which can be easily placed between the cooler and the detector, allows the operator to manually determine the first spectroscopical features of the laser such as the emission wavelength at threshold.

After performing several quality measurements, which will be described later, the laser is tested for more detailed optical features in a measurement

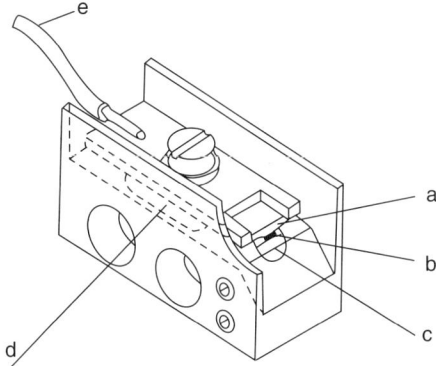

Fig. 4.3 Standard lead salt laser package showing (a) indium coated cooper wire, (b) laser chip, (c) indium coated cooper heat sink, (d) epoxy slab, (e) current lead.

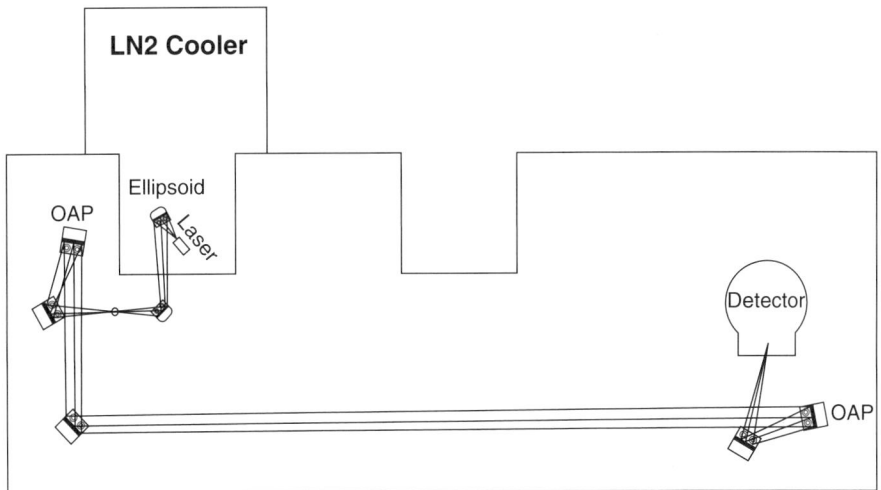

Fig. 4.4 Fast measurement setup at Laser Components.

setup like the one shown in Fig. 4.5. In this setup, five lasers can be cooled down to any operation temperature between 20 and 300 K. The temperature stability is better than 2 mK. The laser can be operated with a commercially available low-noise current source. The laser beam is collimated by using an off-axis ellipsoidal mirror objective (1:2 aperture) and an off-axis paraboloid. With this objective and the use of an alignment ocular in the focal position of the off-axis ellipsoid, the alignment procedure for the infrared laser radiation can be easily reduced to a visible alignment of the laser onto the cross hair of the ocular. This procedure generally takes no longer than 10 s. The collimated beam is then fed through a mirror chopper in one half-phase of the chopper

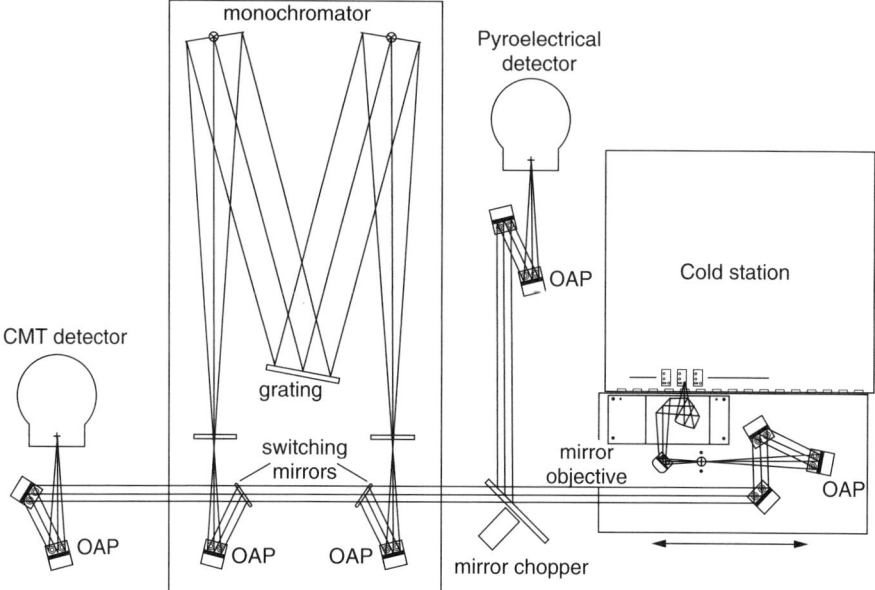

Fig. 4.5 End test ("lambda") measurement setup for monitoring the modal behavior of laser diodes.

onto a pyroelectric detector from Infratec to perform the measurement of the output power. The detector is calibrated using a commercially available thermopile detector system from Molectron. The other half-phase of the beam is fed through a commercial monochromator from Mütek Infrared and then collimated onto a two-color detector from PE Judson. This detector employs an InSb photodiode for the wavenumber region above $2000\,\text{cm}^{-1}$ and an HgCdTe photoconductive detector for the spectral region between 2000 and $500\,\text{cm}^{-1}$. With this system, the current versus voltage, current versus output power characteristics, the threshold dependence on operation temperature, and the spectral behavior at chosen temperatures and at variable injection currents can be monitored. After performing a software reduction of the spectra, modal charts are generated.

Modal charts are an extremely valuable tool to give the laser diode operator hints for finding the operation parameters for a particular application. On the left side of the mode charts are two columns of data providing important information about the quality of the laser modes. The first column gives the percentage of the total output power in the dominant mode. For example, 93% means that only 7% of the output power goes into other modes. This information is more valuable to the user than the sidemode suppression ratio. Even a high sidemode suppression ratio can result in a small percentage of power in the dominant mode, since a great number of small modes can add up to a

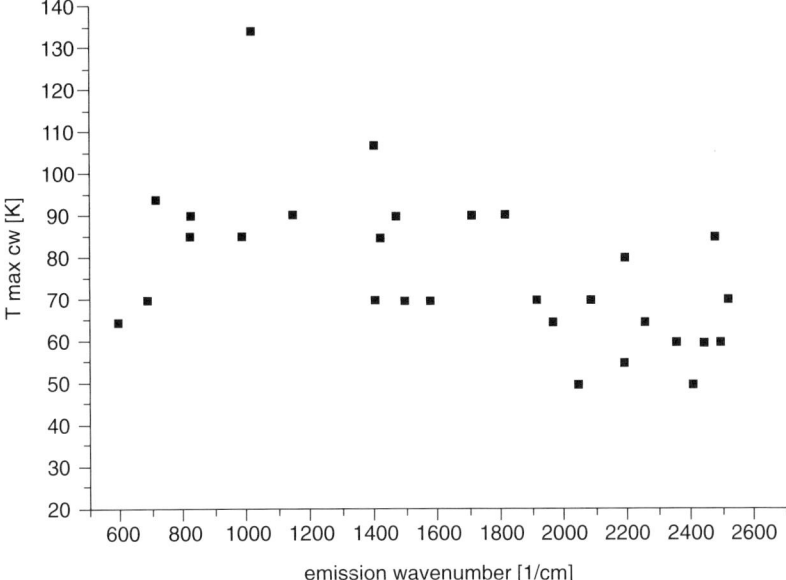

Fig. 4.6 Maximum operation temperature of homostructure lasers based on the material systems PbSnSe and PbSSe manufactured at the Fraunhofer Institute for physical measurement techniques in Freiburg and at Laser Components in Olching (both Germany).

considerable amount of the total output power of the laser diode. The second column gives the total power of the laser at the indicated current and operation temperature.

The fine-tuning range between two mode hops is typically 4 cm^{-1} for 5-μm diodes, 2 to 3 cm^{-1} for 10-μm lasers, and 1 to 2 cm^{-1} for 16-μm lasers. Typical tuning rates are 0.01 to 0.07 cm^{-1}/mA.

Figure 4.6 summarizes the current state of the maximum operating temperatures of homostructure lasers developed at the FHG IPM (Fraunhofer Institute for physical measurement techniques) and Laser Components (LC). The highest cw operating temperature is 135 K [30]. Figure 4.7 summarizes the maximum output power of nearly 30 lasers produced from $Pb_{1-x}S_xSe$ wafers. In the spectral range >1900 cm^{-1}, maximum output power >10 mW into a 1:2 aperture could be recorded. The maximum output power from a homostructure laser without antireflection coatings applied to the front facet was 61.5 mW at 1.54 A and 20 K. This result corresponds to a total efficiency of 14.3% per facet. The maximum differential quantum efficiency was 25.84% per facet [32]. All lasers were processed with 20-μm-wide stripe contacts. The maximum pulsed output powers of homostructure lasers are given in [4].

Ion-implanted lasers exhibited a maximum cw operating temperature of 115 K and maximum single-mode powers of 0.9 mW at 100 K at 7.8 μm.

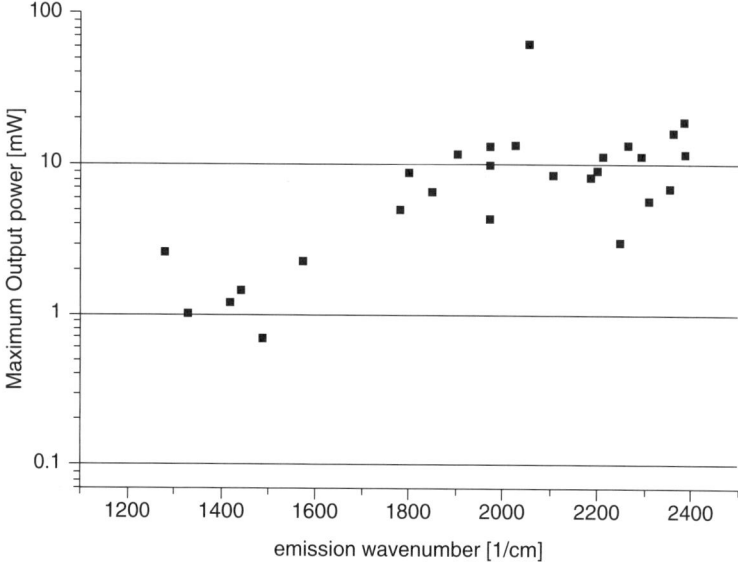

Fig. 4.7 Maximum output power of nearly 30 homostructure wafers produced from the $Pb_{1-x}S_xSe$ system at the Fraunhofer Institute for physical measurement techniques in Freiburg and at Laser Components in Olching (both Germany).

The modal tuning rate of these lasers varied between 0.05 and 0.01 cm^{-1}/mA [31].

Linewidth measurements on several homostructure lasers were performed. Current noise and vibrations from the cooling equipment often limit the linewidth of the laser in practical experiments [33]. Figure 4.8 shows the result of a self-beat experiment with a $Pb_{0.75}S_{0.25}Se$ laser. A white cell [34] was exploited for delaying one beam of the beat experiment. The experiment indicated that linewidths less than 10 MHz can be achieved with free-running lasers without using additional stabilizing measurements. Additional reduction of the linewidth by one to two orders of magnitude was demonstrated by means of optical stabilization [35].

4.3 DOUBLE-HETEROSTRUCTURE LASERS

Although the maximum operating temperature of some homostructure lasers is higher than 77 K, in practice this is uncommon. To improve this situation, a logical step was the development of DH lasers. This structure improves the electrical and optical confinement in the laser by employing two outer cladding layers with higher band-gap energy and lower refractive index than the inner active layer. Different material systems, epitaxial techniques including liquid

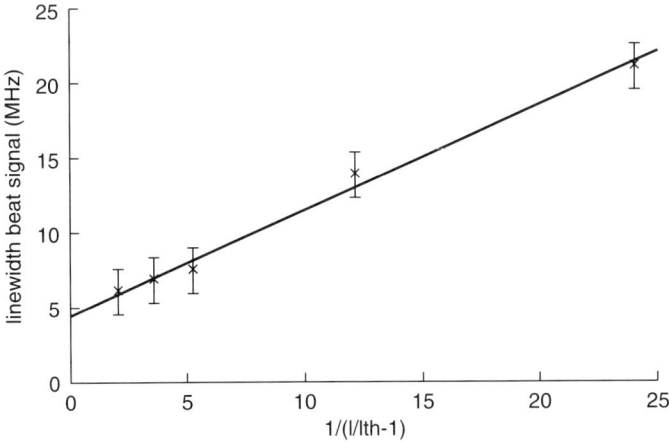

Fig. 4.8 Linewidth measurement of a $Pb_{0.75}S_{0.25}Se$ laser employing a self beat experiment, taken from Ref. [33].

phase epitaxy (LPE), hot-wall epitaxy (HWE), and molecular beam epitaxy (MBE) [16,36], and different philosophies regarding the necessity of lattice matching have been discussed in the past.

Material systems that were previously employed are $Pb_{1-x}Ca_xTe$ and $Pb_{1-x}Eu_xSe_yTe_{1-y}$ for the wavenumber range >1500 cm^{-1}, $PbS_{1-x}Se_x$, $Pb_{1-x}Eu_xSe$, and $Pb_{1-x}Sr_xSe$ for the wavenumber range >1200 cm^{-1}, and $Pb_{1-x}Sn_xSe$, $Pb_{1-x}Sn_xSe_{1-y}Te_y$, and $(Pb_{1-y}Sn_y)_{1-x}Yb_xTe$ for the longer wavenumber region [37].

Since only $Pb_{1-x}Eu_xSe$, $Pb_{1-x}Eu_xSe_yTe_{1-y}$, $Pb_{1-x}Sn_xSe$, and $Pb_{1-x}Sn_xSe_{1-y}Te_y$ are used nowadays, we will limit our discussions to these systems. A review of the current state of the art regarding other systems has been given in [4].

4.3.1 $Pb_{1-x}Eu_xSe_yTe_{1-y}$ Lasers

This material system was employed for laser production at Laser Analytics. They use a Varian Gen II MBE system (Fig. 4.9), which consists of a growth chamber, a preparation and analysis chamber, and a load lock assembly. Due to the isolation between the chambers, parallel activities without influencing neighboring chambers are possible. The employment of ultra-high vacuum (UHV) pumping equipment results in a background vacuum of typically 1–2 $\times 10^{-10}$ mbar. The growth chamber contains eight thermally isolated sources, each equipped with a shutter. The source isolation is achieved using a liquid nitrogen cryoshroud. The substrate holder is rotatable and its temperature can be kept stable during growth. Typical growth temperatures are 390°C [37]. The pre-cleaned substrates are loaded into the load lock, degassed, and transferred into the preparation chamber. After the substrates are baked at a temperature

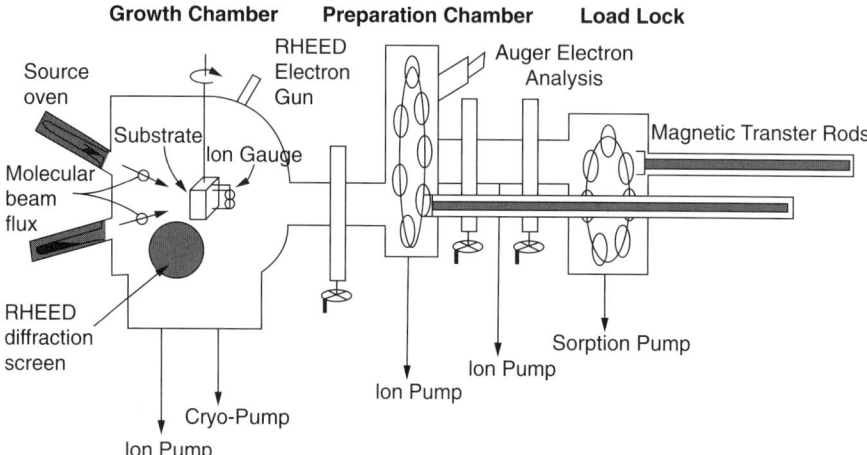

Fig. 4.9 Schematic of the Varian Gen II Molecular Beam Epitaxy (MBE) System, taken from Ref. [38], with kind permissions from Kluwer Academic Publisher.

slightly higher than the growth temperature, they are transferred into the growth chamber. Once the substrate temperature is stabilized at the growth temperature, the growth starts by opening the appropriate shutters of the sources. The molecular flux can be controlled via the source temperature, and the flux can be measured with an ion gauge, which can be rotated into the molecular beam.

The system is equipped with various in situ measurement tools. Reflection high-energy electron diffraction (RHEED) can be used as an in situ monitor of the growth rate and quality of the epitaxial layers. A residual gas analyzer can be employed for analyzing the residual gases during and between growth and of the molecular beam emerging from the sources. Auger surface analysis equipment mounted on the analysis chamber can be used for the measurement of surface composition. The surface of the wafer can be monitored before and after growth to verify the degree of contamination and compositional uniformity [38].

An example of a Laser Analytics' standard device structure is given in Fig. 4.10. The PbEuSeTe active and cladding quarternaries are lattice matched to PbTe and designed to be near the corner of the pseudobinary phase diagram. A typical contact stripe width is 20 μm.

A schematic of the carrier concentration, band-gap energy and the index of refraction profile is shown in Fig. 4.11. The maximum operating temperatures of PbEuSeTe DH lasers are summarized in Fig. 4.12. The highest operating temperature found for binary PbTe active layers was 175 K. With increasing Eu concentration in the active layer, the maximum temperature decreases gradually. Above 2.5% Eu, the decrease in maximum temperature

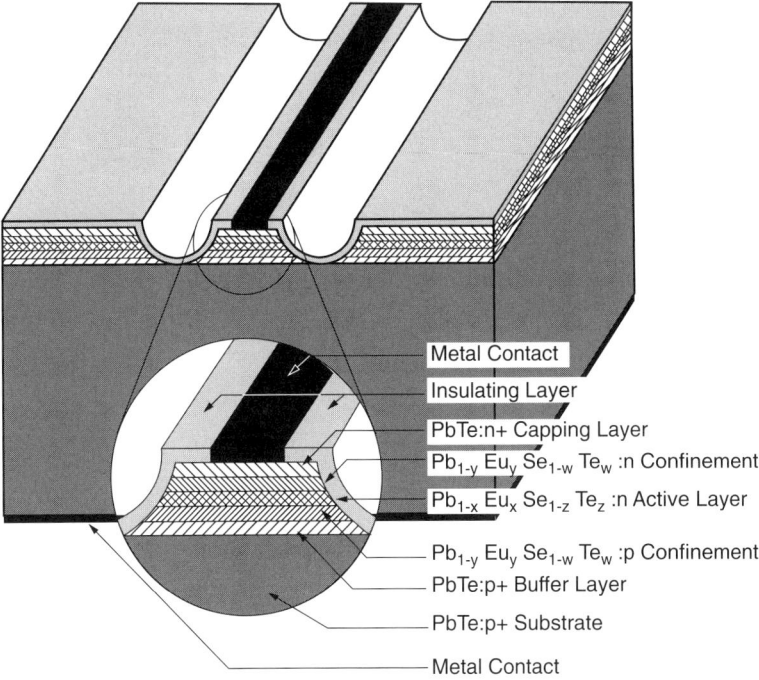

Fig. 4.10 Device structure of an MBE grown PbEuSeTe-PbTe DH laser, taken from Ref. [38], with kind permissions from Kluwer Academic Publisher.

is more steep due to inferior mobility and thermal conductivity of the compounds [39].

4.3.2 $Pb_{1-x}Eu_xSe$ and $Pb_{1-x}Sr_xSe$ Lasers

In early German development, PbS-PbS$_{1-x}$Se$_x$-PbS DH lasers operated up to 96 K cw [40], and PbS-PbSe-PbS DH lasers up to 120 K cw and 230 K with 200-ns pulses [41]. These lasers were made by HWE. The material system suffered from poor lattice match, probably leading to stress-induced dislocation near the active layer [42] and higher nonradiative recombination rates.

A straightforward approach to improve the maximum operating temperatures was to employ material systems with better lattice match. $Pb_{1-x}Eu_xSe$ is well suited for this case since the lattice mismatch between PbSe and EuSe is only 1.16% (Fig. 4.13). In fact, by carefully designing the active and confinement layers in a way that thicknesses of the active layer are smaller than the critical thickness for the formation of dislocations, one can avoid mismatch-induced dislocations [42]. Mixed crystals can be grown by MBE with gap energies equivalent to the mid-IR and doping levels up to the order of $10^{19}\,cm^{-3}$ [43]. Photoluminescence studies of the $Pb_{1-x}Eu_xSe$ system [44] and earlier

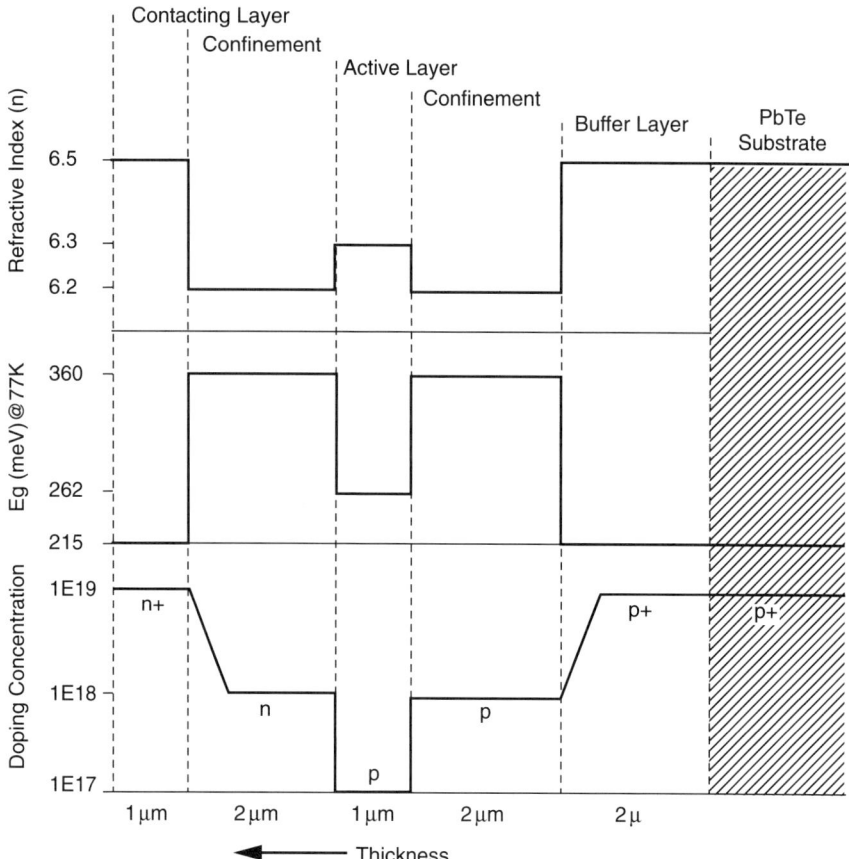

Fig. 4.11 Example for the design of an MBE fabricated lattice matched PbEuSeTe-PbTe laser structure, taken from Ref. [38], with kind permissions from Kluwer Academic Publisher.

studies [45] result in the following formula for the energy gap of $Pb_{1-x}Eu_xSe$, which should be valid for $x < 0.1$:

$$Eg(x,T) = 0.125 + 3.5x + \sqrt{2.65 \cdot 10^{-7} \cdot T^2 + 4 \cdot 10^{-4}}. \qquad (4.4)$$

For impurity doping, Bi and Ag are used. Both metals are expected to take Pb sites, Ag being an acceptor and Bi a donor. Figure 4.14 shows the carrier concentration of PbEuSe with a 1.5% Eu concentration. The flux unit corresponds to 0.07% Ag and 0.05% Bi. The layers were made with a 12% Se flux so that the nominally undoped material is p-type. The sticking coefficient of Bi was found to be nearly unity, making it an effective donor. Ag, however, has a saturated carrier concentration of $2 \times 10^{18}\,cm^{-3}$ at 77 K. A possible explanation is that at higher flux values Ag becomes incorporated in interstitial sites

DOUBLE-HETEROSTRUCTURE LASERS   159

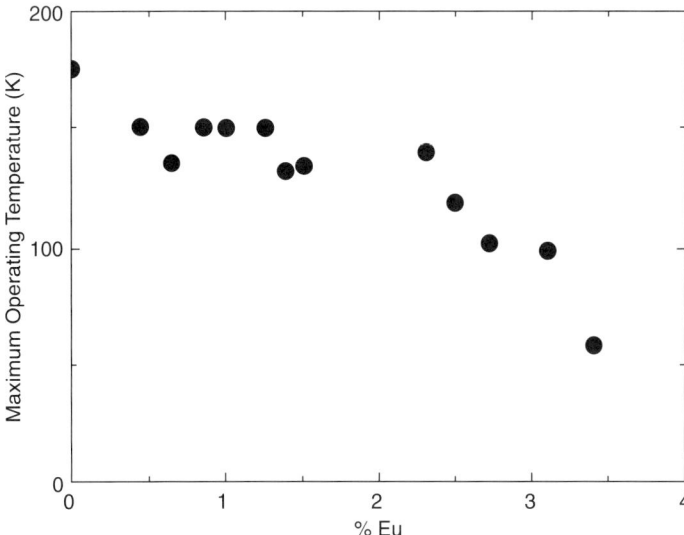

Fig. 4.12 The maximum operation temperature of MBE grown devices as a function of their Europium content, taken from Ref. [38], with kind permissions from Kluwer Academic Publisher.

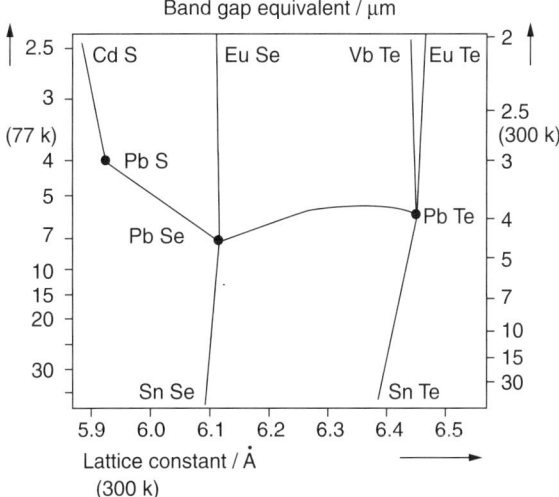

Fig. 4.13 Schematic diagram of the absorption edge wavelength at 77 K and 300 K as a function of the lattice constant within the lead salt based mixed crystal system, taken from Ref. [43], with permissions from Elsevier Science.

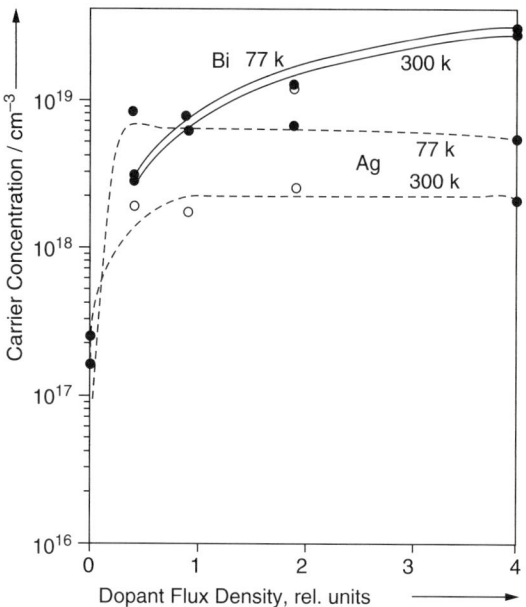

Fig. 4.14 Carrier concentration of PbEuSe containing 1.5% Eu. The flux unit corresponds to 0.07% Ag and 0.05% Bi. Open circles denote 300 K data, closed circles 77 K data. Broken lines connect data for Ag full lines for Bi. Reprinted from Ref. [43], with permissions from Elsevier Science.

[43]. This behavior is known from PbTe:Ag, where interstitial Ag was found to be a donor coexisting with substituional Ag, which acts as an acceptor [46].

The mobilities with dopant fluxes larger than 0.5 units are of the order of 10^3 cm^2/V-s at 77 K and 200 cm^2/V-s at 300 K for both n- and p-type. The carrier concentration as a function of the Eu content is plotted in Fig. 4.15.

Whereas the electron concentration changes only slightly with Eu content and is temperature independent in accordance with a shallow or resonant impurity state, the hole concentration behaves differently. At intermediate Eu contents, there is a pronounced difference between the concentration at 77 and 300 K. The authors explained this behavior with a localized state resonant with the valence band [43].

$Pb_{1-x}Eu_xSe$ diode lasers have been grown on (100)-oriented PbSe substrates using a PbSe and a Se effusion cell for PbSe, a Eu cell for the ternary material, and a Ag and a Bi source for p-type and n-type doping. The active layer was not intentionally doped and results in p-type carrier concentrations near 1×10^{17} cm^{-3}. The p-n junction was either located at the active layer or somewhat remote. DH structures as well as graded structures were realized. The Eu content of the confinement layer was usually 1.6% larger than that for the active layer. The thickness of the active layer was usually below 1 µm, with confinement layers of 2 to 4 µm thick. A stripe contact (20 µm) near the active

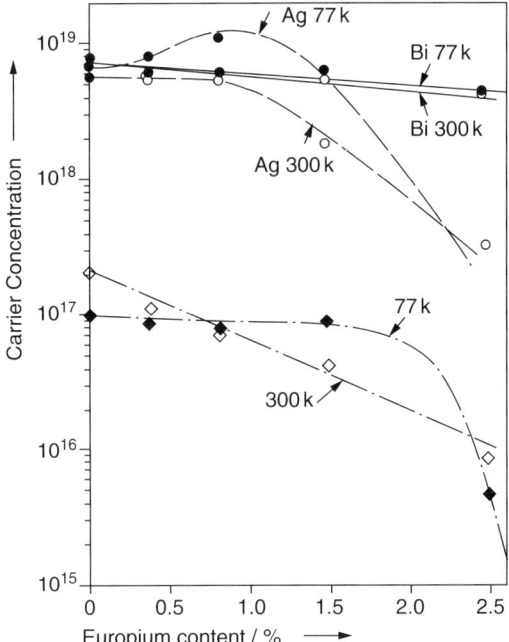

Fig. 4.15 Carrier concentration of samples that are doped with the unit dopant flux of Fig. 4.14 as a function of the Eu content. The undoped sample (bottom curves) has p-type. Reprinted from Ref. [43], with permissions from Elsevier Science.

layer and a full contact on the opposite side were formed by means of electroplating. The wafers were cleaved to chips with typical dimensions of $250 \times 230 \times 360\,\mu m$, cold pressed onto a cooper heat sink electroplated with In, and mounted in a standard tunable laser package (Fig. 4.3).

In Fig. 4.16, the threshold current versus operation temperature of a PbEuSe/PbSe/PbEuSe laser is plotted. A maximum operating temperature of 174 K, comparable to the values for the quaternary lattice-matched material, has been achieved, which indicates that the more complex lattice-matched growth is not necessary. However, the lattice mismatch results in a blue shift of the emission wavelength due to mismatch-induced strain in the active layer [42].

PbEuSe/PbSe/PbEuSe DH lasers with wavelength as short as $2.88\,\mu m$ were processed at IPM. Different tuning curves of lasers containing $Pb_{1-x}Eu_xSe$ active layers, with x increasing from 0 to 4.3% are plotted in Fig. 4.17. It was found that the yield decreased considerably with increasing Eu content in the active layer, and the maximum operating temperatures were reduced. This behavior was believed to be resulting from transitions involving the 4f levels of Eu.

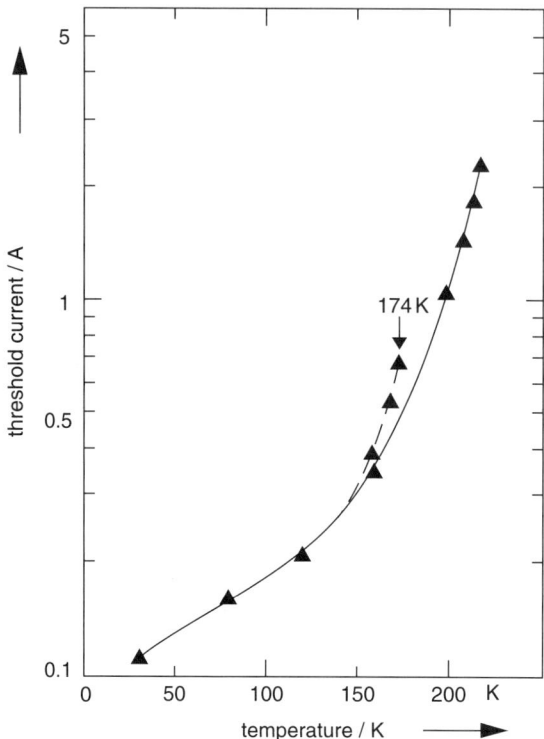

Fig. 4.16 Threshold current data for a PbEuSe/PbSe/PbEuSe DH laser as a function of the heat sink temperature. Broken lines connect cw data that are not coincident with pulsed threshold current. Taken from Ref. [63].

Figure 4.18 shows a calculated band structure for EuSe. A second valence band located several hundred meV from the regular within the forbidden band is formed by the Eu 4f levels. In studies of the photocurrent spectra in $Pb_{1-x}Eu_xSe$ samples, contributions to "band-like states" were found [47]. In optical spectra of photoluminescence and photocurrent, well seperated peaks, which were attributed to transitions between the conduction band and both valence bands, were observed [48].

Finally in a study of magnetoluminescence, it was possible to identify these additional lines as contributions from 4f levels in $Pb_{1-x}Eu_xSe$ [49]. In the same study the authors showed that the band parameters (energy gap, effective masses, and interband absorption edge) of PbSe are modified by alloying with EuSe and SrSe in a very similar manner, besides the fact that in $Pb_{1-x}Sr_xSe$ layers no evidence for 4f level transitions could be observed as was already stated from the band structure calculations.

Figure 4.19 shows a comparison of the band-gap energy and the temperature tuning coefficient of $Pb_{1-x}Eu_xSe$ and $Pb_{1-x}Sr_xSe$. Good agreements were

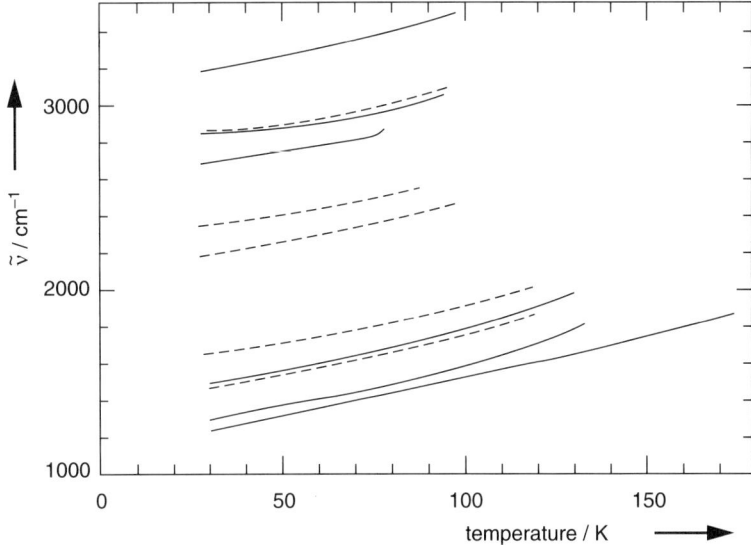

Fig. 4.17 Tuning curves of different lasers made with active layers of $Pb_{1-x}Eu_xSe$, with x increasing from 0% (bottom) to 4.3% (top). Data were taken at threshold for cw operation. Broken lines indicate graded structures, full lines indicate DH structures. Taken from Ref. [63].

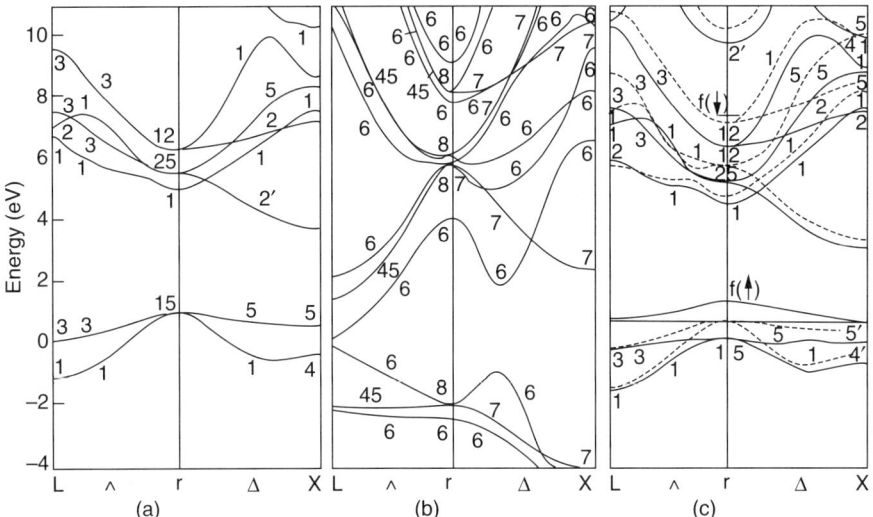

Fig. 4.18 Band structure of SrSe (a), PbSe (b) and EuSe (c) at T = 0 K. In (c) full curves denote the spin-down electrons. The numbers denote the irreducible representation at a given **k** vector. Taken from Ref. [49].

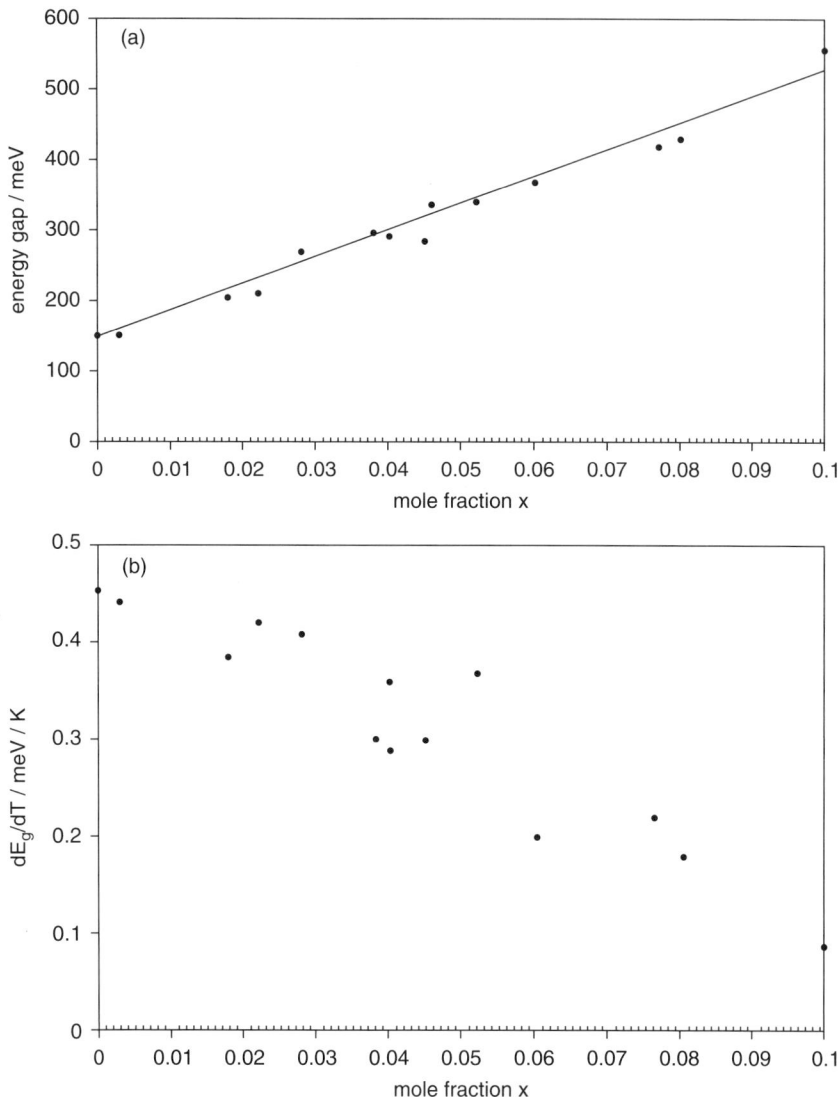

Fig. 4.19 (a) Energy gap of PbSe(••) as well as composition dependent gaps of $Pb_{1-x}Sr_xSe$ (*) and $Pb_{1-x}Eu_xSe$ (••) as determined from photoluminescence experiments at 10 K. (b) Temperature coefficient of the energy gap in PbSe (••), $Pb_{1-x}Sr_xSe$ (*) and $Pb_{1-x}Eu_xSe$ (••) for temperatures 70 K ≤ T ≤ 300 K. Taken from Ref. [49].

observed. Since the lattice mismatch of SrSe and PbSe amounts to only 1.73%, only slightly more than for EuSe (1.16%), the interfacial recombination for PbSrSe/PbSe/PbSrSe is presumably the same order as for PbEuSe/PbSe/PbEuSe. Moreover lattice constant measurements of PbSrSe indicate a very low lattice mismatch between PbSe and $Pb_{1-x}Sr_xSe$ as shown

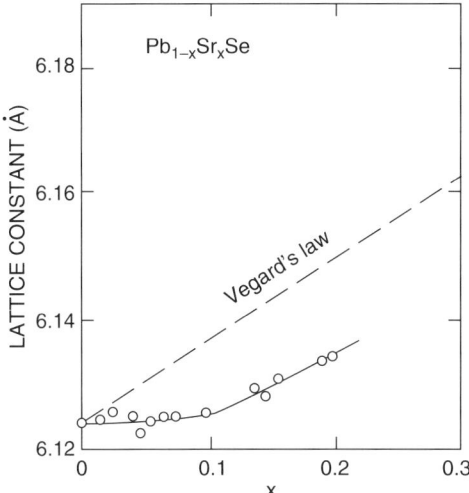

Fig. 4.20 Dependence of lattice constants on alloy composition in PbSe based alloys. Taken from Ref. [50].

in Fig. 4.20 [50]. Therefore it is no surprise that very good PbSrSe heterostructures can be prepared. Figure 4.21 shows the dispersion of several MBE-grown $Pb_{1-x}Sr_xSe$ layers with different Sr contents. The dispersion behavior is similar to that of PbEuSe materials.

DH diode lasers with $Pb_{1-x}Sr_xSe$ cladding layers and PbSe active layer have been grown at the FHG IPM . PbSe and Se effusion cells were used for the growth of the PbSe active layer, resulting in a p-type doping level of around $10^{17}\,cm^{-3}$. A Sr source was used for the ternary, with Ag and Bi_2Se_3 sources for p- and n-type doping. Ag and Bi doping of the confinement layers resulted in carrier concentrations of $3-4 \times 10^{18}\,cm^{-3}$. The Sr content of the confinement layer was 4%. The preparation of the high-purity Sr necessary for MBE is described in [51]. The mobilities of the p- and n-type layers were 320 and $2700\,cm^2\,V^{-1}s^{-1}$, respectively. The thicknesses of the layers were 4 μm (bottom p-type doped confinement layer), 0.5 μm (active layer), 2 μm (top confinement layer), and finally 0.5 μm (capping layer). Stripe-geometry lasers with 20 μm stripes were fabricated by the method described previously. Figure 4.22 shows the threshold current and the wavenumber at threshold under pulsed operation conditions (200 ns, 1 kHz repetition rate) as a function of temperature. A maximum operating temperature of 290 K was achieved [52].

Far-field data of DH lasers are given in Ref. [53]. Internal losses measured for mid-IR lead-salt lasers [54] could be qualitatively interpreted by assuming higher structural losses than losses calculated from free-carrier absorption and mobility [55].

The linewidth of DH lasers was measured with a Fabry-Perot interferometer and special feedback isolation techniques [56]. Free-running lasers were

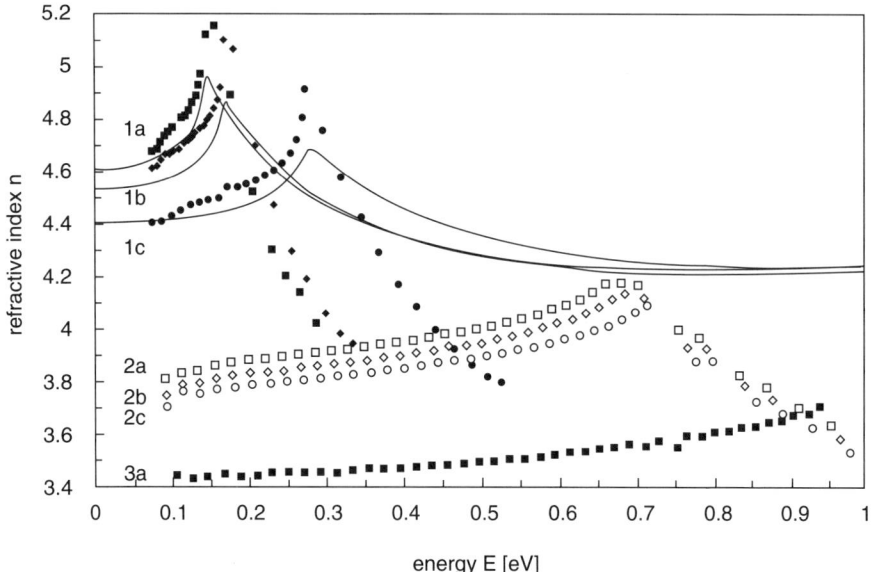

Fig. 4.21 Dispersion of $Pb_{1-x}Sr_xSe$ MBE layers with different composition. Experiment 1a,b,c,x = 0 (a,T = 10 K; b, T = 80 K; c, T = 300 K); 2a,b,c,x = 0.149; 3a, x = 0.214. The full curves have been calculated with $E_o = 10$ meV; $Eg = 0.149$ eV, $\beta = 4 \times 10^4 \text{cm}^{-1} \text{eV}^{-1/2}$ (1a), $Eg = 0.173$ eV, $\beta = 3.7 \times 10^4 \text{cm}^{-1} \text{eV}^{-1/2}$ (1b), $Eg = 0.278$ eV, $\beta = 3.6 \times 10^4 \text{cm}^{-1} \text{eV}^{-1/2}$ (1c). Taken from Ref. [48].

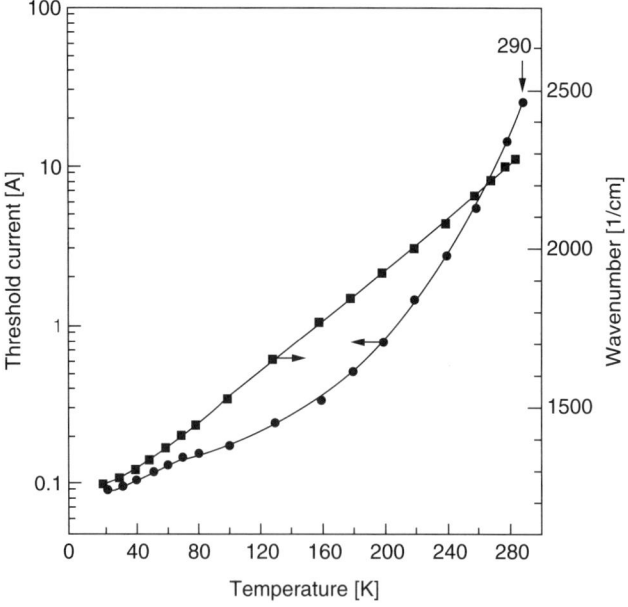

Fig. 4.22 Threshold current of a PbSrSe/PbSe/PbSrSe DH laser (●) in pulsed mode (left) and emission wave number (■) near threshold (right). Taken from Ref. [52].

often found to suffer from large line widths of the order of 100 MHz due to mode competition noise, except for isolated regions where they operated in predominantly a single mode. In such carefully selected emission ranges, the lasers had linewidths on the order of 10 MHz [55,56]. This is one of the main reasons why a carcful selection process is necessary to ensure good operational characteristics.

4.3.3 $Pb_{1-x}Sn_xTe$ and PbSnSeTe/PbSe Lasers

On the long-wavelength region, progress was made using PbSnSeTe and PbSnSe material systems. Ishida et al. reviewed the state of the art of PbSnTe lasers [50]. An interesting aspect was their proposal that $PbTe/Pb_{1-x}Sn_xTe$ ($x < 0.24$) has a type-I band structure, while $PbTe/Pb_{1-x}Sn_xTe$ ($x > 0.24$) becomes type-II, where the conduction band minimum of the PbTe is located below the valence band maximum of PbSnTe. Hence the injection efficiency of carriers into the active layer decreases with increasing difference between the Sn content in the active and confinement layers. This proposal was confirmed by various experiments [57–58]. In fact it delivered a reasonable explanation for the observation that the manufacturing of long-wavelength lasers based on the PbSnTe material system becomes very difficult with increasing difference in Sn contents between the confinement and active layers. The reduction of this difference has been demonstrated to deliver successful results [50].

Similar to the PbSnSe material system, the yield of (MBE-grown) PbSnTe lasers dropped dramatically when a Sn content corresponding to the emission wavelength beyond 10 µm was used. We assume that there is a problem of band offsets, as has been observed by Ishida et al. However, it was possible to grow PbSnSeTe/PbSe lasers by LPE [60], with emission wavelength up to 20 µm and operating temperatures up to 130 K. The reason why LPE produced superior lasers is not known. A possible explanation might be that an n-type substrate was used for the LPE compared to p-type substrates typically used for MBE.

4.3.4 Alternative Cladding Layer Materials

The prospect of extremely low-loss flouride glas fibers had stimulated some development effort in IV-VI material systems in the short wavelength region too. These material systems were based on the PbS system, which has a bandgap energy of 0.42 eV at 300 K. Cladding layers based on the compound PbCdS and PbSeS were prepared by two Japanese groups [61,62]. Pulsed laser operation up to 245 K at 2.97 µm and cw operation of up to 174 K were observed [63]. Since Eu becomes a deep donor in PbEuS, it could not be used as a cladding layer. With the $Pb_{1-x-y}Cd_xSr_yS/PbS$ system, a maximum operating temperature of 240 K with 1-µs pulses and 13-Hz repetition rate was observed [64]. They found that the maximum operating temperature was increased by using a tilted substrate, which reduced the lattice mismatch between the substrate and the confinement layer. A good discussion regarding the lattice mismatch of different material systems is given in [50].

Another interesting material system is PbCaTe. CaTe has a wide band gap (>4 eV) at room temperature. It has a lattice mismatch of only 1.5% with respect to PbTe, which is smaller than for PbEuTe and PbSrTe. The material is believed to play a useful role as an infrared laser material, especially for unipolar quantum cascade lasers, because it is possible to obtain high and thin quantum barrier layers that are favored for tunneling devices. The first results with this material system were presented in [65].

4.3.5 Quality Control Programs at Laser Components

After the technology for $Pb_{1-x}Eu_xSe$, $Pb_{1-x}Sr_xSe$ and $Pb_{1-x}Sn_xSe$ DH lasers was transferred from IPM to LC in 1993, considerable efforts have been made to improve the reliability of lead-salt lasers. This was motivated by the need to have single-mode lasers available for the community. With single-mode lasers, no expensive mode filters are necessary in the spectroscopical setup. Additionally the noise level for lasers operating in a single mode regime is considerably reduced due to the lack of mode competition between the different longitudinal modes [66,67]. Since single-mode emission is not an inherent feature of the DH lasers, it was necessary to select lasers from normally multimode wafers. This selection process was performed using an automatic measurement setup already described above.

The main problem to guarantee mode stability for the user was the need to cool the lasers from 300 K to their operation temperature in the vicinity of 80 K. During their lifetime, lasers will suffer numerous thermal cycling processes. The mechanical stress induced by different thermal expansion coefficients of semiconductor, contacts, and heat sinks can induce small changes in the microstructure of the thermal interface between the contact and chip, leading to changes in the laser characteristics due to different cooling profiles. We found these changes to be considerably pronounced during the first few cycles, as is demonstrated in Fig. 4.23 [32]. After monitoring the modal behavior of laser 322-hv-1-35 at 100 K, we performed 20 thermal cycles on this laser in the following way: The laser was mounted into a vacuum-tight box containing up to 24 additional lasers. By simply throwing this box into a vessel containing liquid nitrogen, we cooled the laser down to 77 K in less than 1 minute, and followed this by a warm-up phase of 4 minutes. The result of a particularly clear case is presented in the figure. The laser changed its modal behavior considerably during the first 20 cycles but remained stable after the additional 20 thermal cycles. This procedure has been a standard quality control measure at LC since 1994.

A second quality control step was the introduction of an aging step, which was introduced because changes were found in laser operational characteristics after storage for several months to years. A standard hypothesis in such cases is that thermally activated processes are responsible for the observed

DOUBLE-HETEROSTRUCTURE LASERS   169

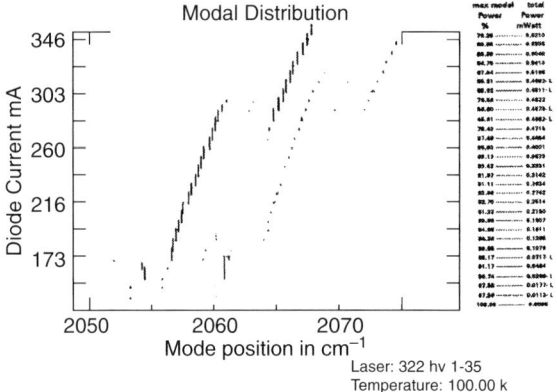

(a) Mode chart without temperature cycles

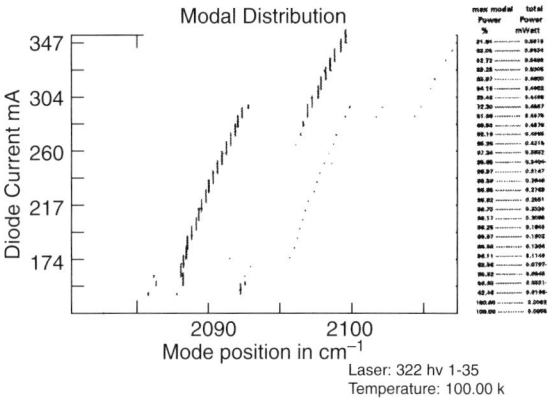

(b) Mode chart after 20 temperature cycles

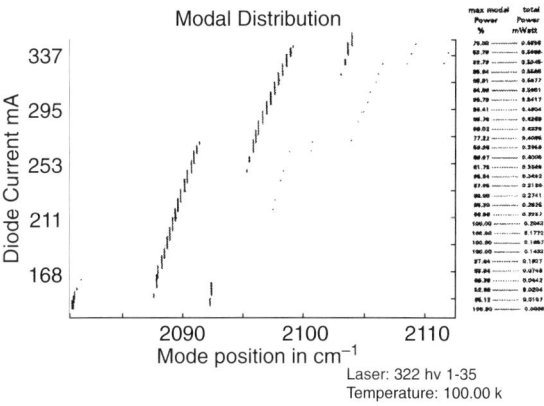

(c) Mode chart after 40 temperature cycles

Fig. 4.23 (a) mode chart without temperature cycles, (b) after 20 temperature cycles between 77 K and 300 K and (c) after 40 temperature cycles. Taken from Ref. [32]

behavior. Therefore a straightforward countermeasure is the aging of the devices at elevated temperatures. LC has used thermal aging as a quaility control measure since 1995.

Previous investigations of aging resulted in reduced threshold current, enhanced output power, and a reduced series resistance [68]. This behavior was explained by the annealing of crystal defects, which act as recombination centers in the active laser region. This model works qualitatively well for short periods (a few days) at 100°C. However, the behavior was totally reversed after storing the laser for much longer periods (weeks to months) at 100°C. The threshold currents were found to increase by 25–30%, the output power decreased by 50–75%, and the series resistance increased by 100–350% [32]. It was interesting to note that the series resistance increased much more in lasers stored in vacuum at 100°C. This behavior was explained with the interdiffusion of In from the heat sink into the Au contact and with the hypothesis that an InO layer might act as a diffusion barrier for this process [32].

As a final quality control measure, a room-temperature burn-in step was established at LC in 1997. During this step a constant current of the order of the 80-K threshold current is injected into a laser while it is at room temperature and above. Figure 4.24 shows the history of laser 404-hv-1-3. Here the threshold current and output power were monitored as a function of the integrated heat load applied to the laser during several room-temperature burn-ins. After the initial quality control measures, cycling, and aging (between 1 and 2), the output power increased dramatically. Performing several burn-ins

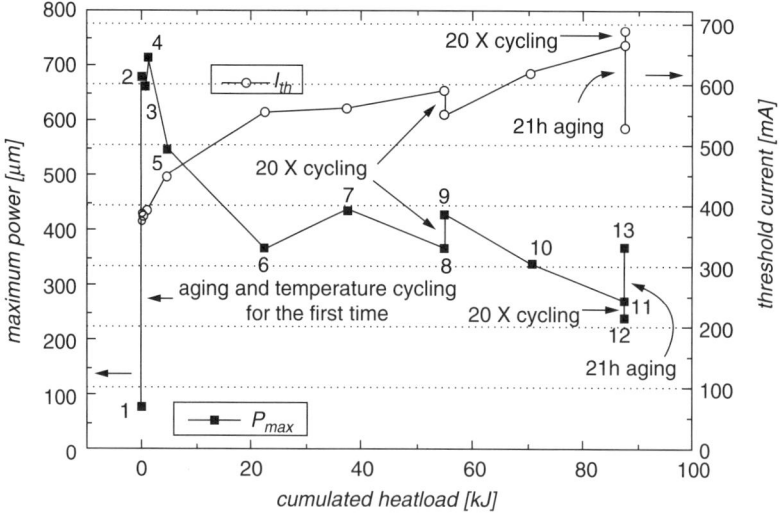

Fig. 4.24 Burn-in, aging and temperature cycling effects on laser 404-HV-1-3 (PbSnSe), taken from Ref. [32].

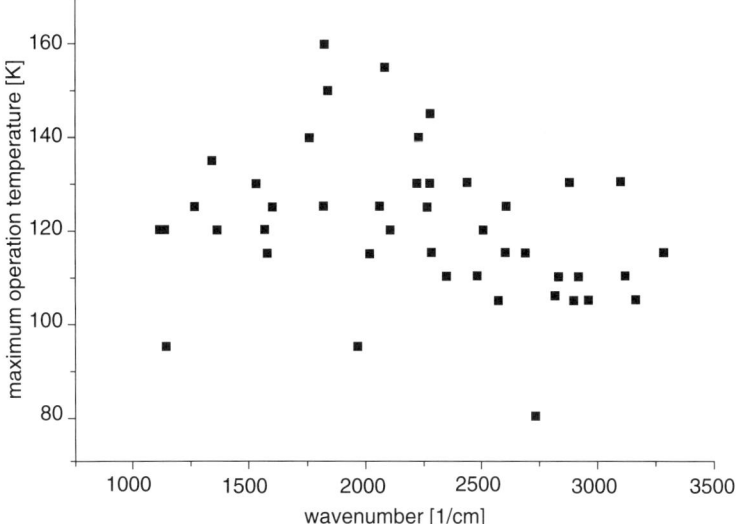

Fig. 4.25 Overview of the maximum operation temperatures typical for DH lasers produced at Laser Components (LC) since 1993. The design goal of LC was the manufacturing of wafer with strong single mode tendency. The achievement of high maximum operation temperatures was not a design criterion.

(3–8) resulted in an initial drop of output power followed by a stable behavior [32]. Naturally such behavior is a prerequisite for a successful burn-in.

Figure 4.25 gives an overview of the maximum operating temperatures typical for DH lasers produced at LC during the last few years. The design goal of LC was the manufacturing of wafers with strong single mode tendency. The achievement of high maximum operating temperatures was not a design criterion. Maximum output powers of LC DH lasers without additional AR coatings on the facet are shown in Fig. 4.26. Typically maximum output powers were in the range of 1 to 10 mW.

4.3.6 High-Temperature Operation of Double-Heterostructure Lasers

Figure 4.27 summarizes the current status of maximum cw operating temperatures of all semiconductor lasers. One can see that the maximum cw operating temperatures in the mid-IR are still dominated by the lead-salt lasers. In the range >3000 cm^{-1}, type-I QW and DH lasers operate cw at temperatures very close to the theoretical limit calculated by Horikoshi [69]. Optically pumped lasers can operate cw above 200 K. However, it is hard to compare optically and electrically pumped lasers directly.

A new effort to measure the maximum operating temperature of lead-salt lasers was motivated by an analysis of the Auger recombination rates in lead

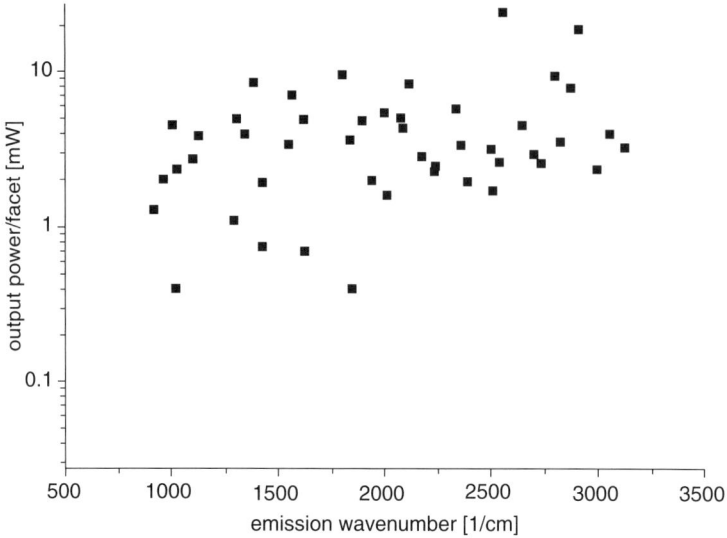

Fig. 4.26 Maximum output powers of LC DH lasers without additional AR coatings on the facet. Typical maximum output powers are situated in the range between 1–10 mW.

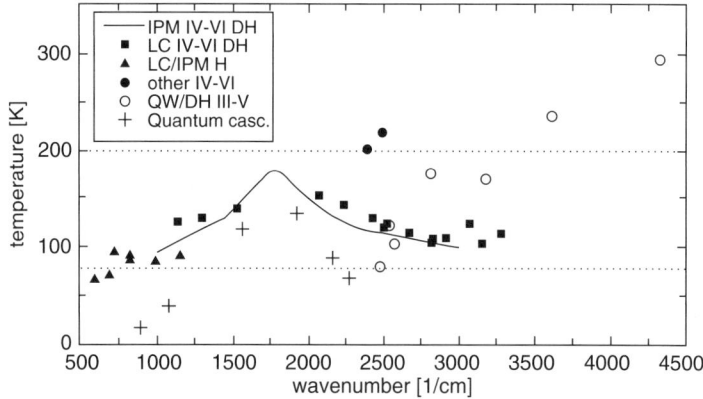

Fig. 4.27 Comparison of maximum operation temperatures for electrically pumped IV-VI, III-V and intersubband devices. Reprinted from Ref. [1], with permissions from Elsevier Science.

salts. Findlay et al. [3] found the Auger coefficient C of PbSe to be approximately constant between 300 and 70 K with a value of 8×10^{-28} cm^{6}s^{-1}. These values are between one and two magnitudes lower than for $Hg_{1-x}Cd_xTe$ alloys with a comparable band gap [3].

Intersubband QC lasers have demonstrated maximum pulsed operating temperatures up to 400 K [70]. The comparison of QC lasers with lead-salt

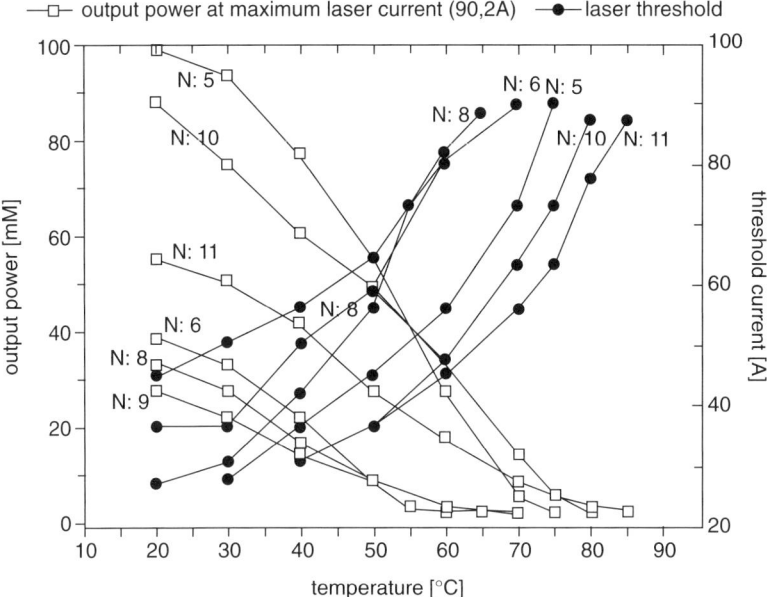

Fig. 4.28 Maximum output power /facet and threshold current of several lead salt lasers with binary active layer (pulse width 10 ns, repetition rate 1 kHz). Taken from Ref. [2].

lasers, however, was difficult until recently because the record 290-K-pulsed operation for lead-salt lasers achieved at the IPM in 1988 had employed relatively long current pulses (200 ns and 1 kHz repitition rate) [52].

Recent room-temperature results on lead-salt lasers were obtained using a pulser capable of injecting 10-ns pulses at a 1-kHz repetition rate. The very first test delivered a maximum operating temperature of 60°C [1]. Additional experiments on different lasers from other wafers have since been performed. The setup used for these tests is described in [2]. The results for lasers from wafer 430-HV-1 are plotted in Fig. 4.28 [2]. A maximum operating temperature of 85°C and a maximum output power of 100 mW were achieved. The wafer 430-HV-1 was grown by high-vacuum MBE on a p-type PbSe substrate. The active and cladding layers were PbSe and PbSrSe, respectively. The lasers have 20-μm Au stripe contacts on the n-side. A more detailed description of the design is given in [52,2]. The spectra of these lasers showed a pronounced multimode behavior [1]. The far-field behavior was typical for a lateral gain-guided laser with one- and two-lobe emission [2], since the lack of lateral index step allows multi-lobe emission.

Lasers with ternary active layers were also tested. No laser with Eu in the confinement layer operated near room temperature. However, laser emission at 30°C was obtained from wafer 441-HV-1 employing an active layer of

$Pb_{0.977}Sr_{0.023}Se$, with an emission wavelength of 2.7 µm. The output power at 20°C was about 16 mW [2]. For high-temperature operation it is necessary to minimize the carrier leakage in the structure by employing a band-gap difference between cladding and active layers as high as possible. Lasers with relatively weak electrical and optical confinement, which often show preferable spectral features at 80 K, could not be operated at room temperature. In wafer 441-HV-1, however, the band-gap difference was on the order of 1000 cm^{-1} due to an "accident" during the growth.

In the longer wavelength region, LC is working with PbSnSe. It demonstrated laser emission at nearly 5 µm up to 7°C by employing a $PbSe/Pb_{0.97}Sn_{0.03}Se$ system. The output power at 6°C was about 2 mW [2]. However, the performance of PbSnSe lasers degrades rapidly for longer wavelengths because this material system suffers from a negative band offset similar to the findings with PbSnTe [50]. For wavelengths longer than 10 µm, it has experienced difficulties in manufacturing lasers with maximum cw operating temperatures >80 K.

As a first application example, the measurement of CO_2 was demonstrated. The minimum detectable concentration was found to be around 100 parts per million for a 1-m path [2]. This relatively poor sensitivity is due to broad spectral width resulting from multimode operation. As a result these lasers could not be used for high-sensitivity spectroscopy applications, but for lower sensitivity applications such as detector test, IR alignment, detection of liquids, range finding, and security applications. By applying the distributed feedback (DFB) concept [71], however, single-mode emission can be obtained for room-temperature pulsed lead-salt lasers.

A new automatically controlled setup for routine characterizations of lead-salt lasers driven at room temperature under pulsed conditions has been completed. LC was working on the development of a DIL module, with an integrated lead-salt laser, Peltier cooler, thermistor and pulsed electronic source.

For a further increase in the cw operating temperature, an interesting proposal has recently been made by Shi [72]. He proposed a GaSb/PbSe/GaSb laser that should work cw at room temperature. GaSb has a number of advantages in comparison to PbEuSe such as a large difference in band gap, a big difference in refractive index, much better thermal conductivity, very high mobility, and commercially available substrates. However, realization of GaSb/PbSe lasers has not been demonstrated up to now. Some questions such as band offsets and lattice mismatch are still remaining. A successful heteroepitaxy of both materials, however, may lead to a breakthrough in the development of mid-IR lasers that can operate cw at room temperature.

Other hints to overcome the current problems with low cw operating temperatures are given by Tomm et al. [73]. They proposed the employment of lattice mismatch (strain) to lift the degeneracy of the carriers in the L valleys of the Brilloin zone. This could be theoretically achieved by growing the structure on (111)-oriented PbSe substrates. Problems, however, do arise when the resonator mirrors of the edge-emitting laser has to be formed. The realization of a vertical cavity surface emitting laser (VCSEL) structure would overcome these problems.

4.3.7 Index-Guided Double-Heterostructure Lasers

DH lasers employ a transversal index step, but only weak guidance resulting from the gain under the contact stripe is available in the lateral direction. Whereas these lasers show single-lobe far-field emission in the transverse direction, several higher order modes can often exist simultaneously in the lateral direction. Such a behavior is often detrimental for the use of the beam in optical systems, since undesired back reflection at the edges of optical elements may lead to so-called etalon noise. The lack of guidance can also result in self-pulsation and abrupt mode jumps, leading to enhanced noise levels due to mode competition.

In the past, several approaches to add an index step in the lateral direction have been investigated. Mesa structures, buried heterostructures (BH), and ridge-waveguide structrures have been attempted. Whereas mesa structures and BH employ relatively large index steps, requiring very small stripe widths for single fundamental mode propagation, this is not the case with ridge-waveguide lasers [74]. They operate in a single mode for considerably larger ridge widths, and therefore are favored for achieving higher single-mode output power. Current spreading in the top confinement layer, however, results in a considerably higher threshold current in ridge-waveguide devices compared to the mesa or BH lasers. Therefore ridge-waveguide lasers were under development at LC. First results indicate an improved far-field behavior that is competitive with BH lasers.

Mesa structure and BH lasers based on PbEuSe, PbSrSe, and PnSnSe, systems were investigated a few years ago at IPM. The lasers were made using MBE-grown DH wafers. Double-channel mesa structures were made by employing a wet chemical etch technique. The etch rate depended on the EuSe content. A 6% EuSe content led to a strong reduction of etch rate by about one order of magnitude compared to PbSe, which had been used for the active layer. This behavior led to a mushroom structure with an active layer width of only a few microns [75]. A threshold current as low as 1.5mA at 20K was observed for these lasers. The maximum operating temperature of these lasers was 150K. However, an inherent problem of mesa structure lasers is low yield due to the mechanically unstable structure. In addition a cycling stability and lifetime problems are expected from such structures, since the active layer is in contact with the atmosphere. Development of a reliable protective coating with excellent cycling stability would prevent this problem.

Further progress should be possible by employing the BH concept, which was first introduced in lead salts by Kasemet et al. [76]. They utilized the PbSeTe/PbSnTe lattice-matched system to prepare long-wavelength (10μm) diode lasers. BH lasers at IPM were fabricated in the following manner: The

MBE growth process was interupted after the active layer. By means of photolithographic and Ar^+ milling techniques, 5-μm mesa-shaped stripes [77] were realized from the active layer. The Ar^+ etching rate was found to be independent of the material composition. Nevertheless, it seems to have created surface defects that may influence MBE overgrowth and laser operation. Work by Palmetshofer [78] indicates that such defects can be annealed at a growth temperature of 400°C. An optimum shape for the overgrowth was achieved by an additional chemomechanical polishing step, which led to a significantly reduced formation of broad and deep cracks during the MBE overgrowth step [80].

BH lasers using PbSnSe for the active and PbEuSe for the confinement layers were fabricated with emission as long as $1028\,cm^{-1}$. For even longer wavelengths, PbSe was used as the confinement material. However, it was found that current leakage is a severe problem in such structures. For shorter wavelengths, PbSrSe was employed at active and confinement layers to achieve emission wavenumbers up to $2860\,cm^{-1}$ at 80 K. Due to rapidly degrading electrical parameters for larger gap materials, however, compromises had to be made, leading to structures with a lower band-gap difference between the confinement and active layers. This also resulted in larger leakage current.

A summary of the manufactured devices is presented in [80]. The authors also reported that the far-field behavior of BH lasers improved as expected. However, strong multimode behavior and low modal power generally resulted in larger linewidths in comparison to gain-guided lasers. A possible explanation for this behavior is the application of an Ar^+ milling process that seems to have enhanced the number of deep defects, leading to increased nonradiative recombination and therefore a reduced lasing efficiency. The application of a wet chemical etch technique together with the PbSrSe system, which does not show strong SrSe dependent etch rates, should improve this situation.

The BH concept was also developed at Laser Photonics Analytics Division (LPAD). The LPAD group was working with the lattice-matched PbEuSeTe material system. A typical LPAD standard design is described in [81]. The main difference in their concept compared to that of IPM was the application of a dilute 2% Br_2/HBr etch solution to obtain smaller stripe widths of the active layer. Additionally an electrochemical etching process was used to optimize the mesa shape and to clean the layers prior to MBE overgrowth. With 4-μm-wide active layers, single-mode operation was observed in about 30% of the fabricated devices for input currents ranging between $1.3 \times I_{th}$ and $4.85 \times I_{th}$. In some devices, single-mode tuning greater than $10\,cm^{-1}$ was demonstrated. With such structures, maximum cw operating temperatures of 195 and 183 K were demonstrated for 4-μm-wide BH lasers with 0.6-μm-thick PbTe and $Pb_{0.9976}Eu_{0.0024}Se_{0.0034}$ active layers, respectively [82]. An even higher cw operating temperature of 203 K was reported after further optimization [83].

Finally a separate-confinement buried-heterostructure laser (SCBH) was developed. The SC heterostructure consisted of a five-layer slab-waveguide design in which the carriers were confined in a small region located in a wider optical waveguide. The two outer regions served as confinement layers for the optical wave. The advantage of this concept was the ability to minimize the active layer thickness for lower threshold current without reducing the power-filling factor. With that concept, a SCBH diode laser was designed with a 0.15-μm and a 0.3-μm thick active layer and a 2-μm wide stripe width, leading to a maximum cw operating temperatures of 223 and 215 K. The threshold currents at 200 K were 103 and 253 mA, respectively [84].

At present, it is not clear if the effort to improve the far-field behavior in the lateral direction really contributes to spectral quality. To obtain a better far-field behavior, output power is sacrificed. The resulting larger linewidth is often detrimental to high-senitivity measurement of environmental gases. In fact investigations at Laser Components have shown that in most cases linewidth broadening and increased noise level of DH lasers can be attributed to poor confinement in the transverse direction, leading to multimode behavior in the transverse channel. Whereas lateral modes can normally be separated by means of a monochromator, this is not the case with transverse modes. The result of poor transverse confinement often leads to coexistence of several modes, which can only be detected with time-consuming linewidth measurements or with measurements using Doppler-limited gas absorption lines.

Laser Components as well as others have found a direct correlation of insufficient transverse confinement and a poor single-mode quality in such experiments. Careful planning of the transverse optical confinement should circumvent the occurrence of several transversal modes.

4.4 QUANTUM-WELL LASERS

Partin [85] published the first experimental results for IV-VI semiconductor quantum-well (QW) lasers in 1984. These devices, which were fabricated from PbTe QWs and PbEuSeTe barrier layers grown on (100)-oriented PbTe substrates, exhibited threshold current densities as low as 20 A/cm^2 at 12 K and maximum pulsed and cw operating temperatures of 241 and 174 K, respectively. A number of single and multiple-quantum-well (MQW) IV-VI semiconductor lasers have since been fabricated on (100)-oriented PbTe and PbSe substrates using QW/barrier materials combinations such as PbSnTe/PbTeSe [86] and PbSe/PbEuSe [87]. To date, the best performing IV-VI QW lasers have been made from PbSe wells and PbSrSe barrier layers [88]. Devices with seven-period MQW active regions exhibited maximum pulsed and cw operating temperatures of 282 and 165 K, respectively, with emission wavelengths of 4.3 and 5.0 μm. These maximum operating temperatures still fall short of non-QW lasers fabricated from IV-VI semiconductor materials. For example, a DH laser holds the record for highest pulsed operating temperature, 333 K [1], and a BH laser holds the record for highest cw operating temperature, 223 K [84]. Despite the lack of success in dramatically improving laser performance, it is useful to investigate further the use of IV-VI QW structures for laser fabrication, since new ways to exploit the quantum confinement effects in IV-VI semiconductors have yet to be fully explored.

IV-VI semiconductors have unique band structure properties that are quite different from III-V semiconductor materials. The direct band gap is at the four equivalent L-points in the Brillouin zone, the conduction and valence bands are nearly symmetric, and the valence band does not have a degenerate heavy-hole band. This last property is believed to be responsible for the much lower Auger recombination coefficients for bulk IV-VI materials [3] as compared to narrow band-gap III-V semiconductors [89]. The constant energy surfaces for electrons and holes in the conduction and valence bands are prolate ellipsoids of revolution characterized by longitudinal and transverse effective masses, m_L and m_T, respectively. For QW structures grown on (100)-oriented substrates, each of the four L-valleys along the $\langle 111 \rangle$ directions lies at the same angle with respect to the direction of potential variation, so electrons and holes in each valley have the same effective mass [90],

$$\frac{1}{m_{100}} = \frac{1}{3}\left[\frac{2}{m_T} + \frac{1}{m_L}\right]. \tag{4.5}$$

Quantum-confined energy levels are thus fourfold degenerate, and radiative recombination occurs between sub-bands with relatively large densities of states. This situation may explain why significant decreases in threshold currents have not been observed for QW laser structures grown on (100)-oriented substrates.

QW materials grown on (111)-oriented substrates, on the other hand, can provide a band structure that is more desirable for laser fabrication. In this case one of the L-valleys will be normal to the substrate surface, and the other three valleys will be at oblique angles to the substrate surface. Two different effective masses will thus exist for potential variation along the [111] direction [90],

$$m_{111}^{\text{longitudinal}} = m_L \tag{4.6}$$

and

$$\frac{1}{m_{111}^{\text{oblique}}} = \frac{1}{9}\left[\frac{8}{m_T} + \frac{1}{m_L}\right]. \tag{4.7}$$

The two different masses remove L-valley degeneracy, placing the longitudinal valley (normal to substrate surface) at a slightly lower energy than the three degenerate oblique valleys. With an approximate fourfold reduction in density of states, population inversion for a single L-valley can be achieved at much lower injection levels, enabling significant reductions in threshold currents. It is thus useful to consider the properties of IV-VI semiconductor QWs grown on (111)-oriented substrates.

The electronic band structures of IV-VI semiconductor QW materials grown on both (100)-oriented and (111)-oriented substrates have been studied using infrared transmission spectroscopy. Ishida et al. [91,92] measured infrared transmission through PbTe/PbEuTe MQW layer structures grown on (100)-oriented KCl substrates, and absorption edges corresponding to $n = 1$, 2, 3, and 4 sub-band transitions were observed in a sample with 30-nm-thick PbTe QWs. As the QW thickness decreased, the sub-band transition energies increased and their number decreased such that a 6-nm-thick QW sample had only one sub-band absorption edge of 314 meV at 77 K. No other absorption edges, which would indicate splitting of L-valley degeneracy, were observed. More recently Yuan et al. [93] performed Fourier transform infrared (FTIR) transmission spectroscopy measurements on PbTe/PbEuTe MQW structures grown on (111)-oriented BaF$_2$ substrates. Spectral fitting calculations using the Kramers-Kronig relationship yielded absorption coefficient versus photon energy plots that clearly showed removal of L-valley degeneracy. The 77-K spectrum for a 6.2-nm-thick PbTe MQW structure showed three absorption edges at 250 meV, 306 meV, and 327 meV, which corresponded respectively to sub-band transitions for the $n = 1$ longitudinal valley $(1-1)^l$, $n = 1$ oblique valleys, $(1-1)^o$, and $n = 2$ longitudinal valley, $(2-2)^l$. The absorption edge associated with the three degenerate $(1-1)^o$ oblique valley transitions was about five times larger than those associated with either the $(1-1)^l$ or $(2-2)^l$ longitudinal valley transitions, indicating the density of states ratio between these two sets of sub-bands.

In order to investigate more fully the electronic sub-band properties of (111)-oriented IV-VI MQW materials, a series of infrared transmission experiments employing NiCr anti-interference layers to suppress Fabry-Perot interference fringes was performed. Transmission spectra obtained from these measurements provide clear and direct evidence for removal of L-valley degeneracy. They also provide useful information for design and fabrication of tunable mid-IR lasers made from IV-VI semiconductor MQW materials. In this work, PbSe/PbSrSe MQWs were grown on freshly cleaved (111)-oriented BaF$_2$ substrates in an Intevac Mod Gen II MBE system. Structures with 15 PbSe QWs were directly deposited at 360°C on a 150-nm BaF$_2$ buffer layer grown at 500°C. PbSe QW and PbSrSe barrier layer thickness were in the range of 5–30 nm and 25–40 nm, respectively. A 3% Sr-to-PbSe flux ratio was used for the growth of PbSrSe, resulting in about 7.5% Sr in the ternary alloy and a room-temperature band gap of about 0.5 eV [94]. More details about PbSe/PbSrSe MQW growth on BaF$_2$ (111) can be found in [95]. High-resolution X-ray diffraction (HRXRD) measurements of the MBE-grown MQW structures showed numerous satellite peaks, and their spacings were used to calculate PbSe QW thicknesses. A Biorad (FTS-60) FTIR spectrometer was used to measure transmission through the MQW samples as a function of photon energy over a 700 to 6000 cm^{-1} spectral range for various sample temperatures between 4.2 and 300 K. Before measurement, each MQW layer was coated with an approximately 20-nm-thick layer of NiCr by thermal evap-

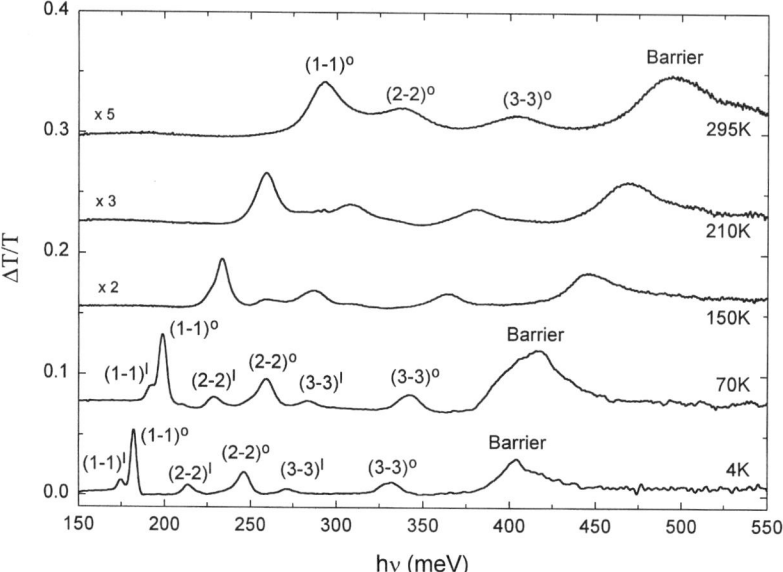

Fig. 4.29 FTIR differential transmission spectra measured at different temperatures from a PbSe/PbSrSe MQW sample with PbSe well width of 20.6 nm and PbSrSe barrier width of 35 nm. Three longitudinal valley sub-band transitions, $(1\text{--}1)^l$, $(2\text{--}2)^l$, and $(3\text{--}3)^l$, and three oblique valley sub-band transitions, $(1\text{--}1)^o$, $(2\text{--}2)^o$, and $(3\text{--}3)^o$, are unambiguously revealed.

oration. With proper impedance matching, this thin metal layer completely removes Fabry-Perot interference fringes from the transmission spectra, since light exiting the high index of refraction IV-VI material is not reflected back into the MQW layer [96]. FTIR transmission spectra were collected at two slightly different temperatures and subtracted from each other to produce a differential transmission ($\Delta T/T$) spectrum. This same technique, which enhances absorption features, has been used to study sub-band transitions in III-Sb quantum well structures [97].

Figure 4.29 shows differential transmission spectra for an MQW sample with 20.6-nm-thick PbSe QWs and 35-nm-thick PbSrSe barriers for five different sample temperatures. The 4-K spectrum exhibits seven peaks. The highest energy peak at 405 meV is associated with interband transitions in the PbSrSe barrier layers (the continuum), and the other six peaks are associated with inter-sub-band transitions between the longitudinal and oblique valleys in the PbSe QWs. The lowest energy peak at 173 meV is due to ground state longitudinal valley transitions, $(1\text{-}1)^l$, while the next peak at 180 meV is due to transitions between the three degenerate oblique valleys, $(1\text{-}1)^o$. A comparison of the integrated intensities for the $(1\text{-}1)^l$ and $(1\text{-}1)^o$ peaks reflects the difference in the joint density of states between these two transitions.

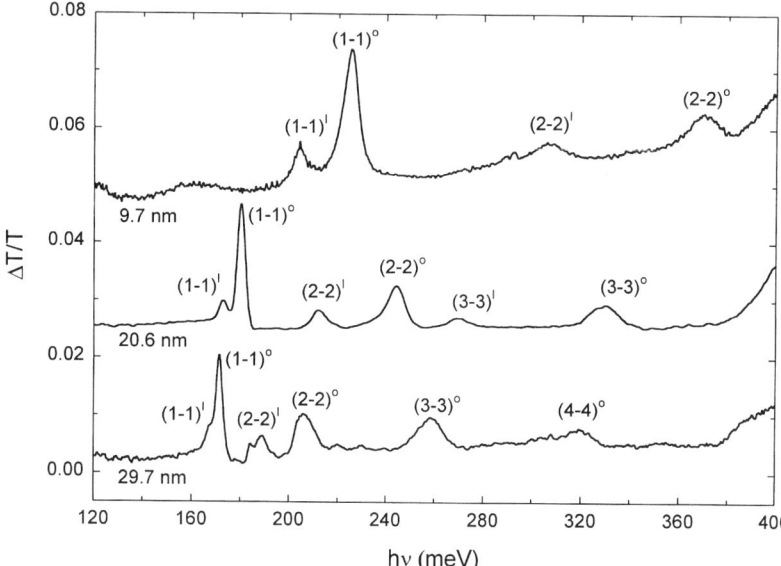

Fig. 4.30 FTIR differential transmission spectra for three MQW samples with well widths of 29.7 nm, 20.6 nm, and 9.7 nm. As well width decreases $(i–i)^l$ and $(i–i)^o$ move to higher energies and the number of confined states decreases.

Identification of the other allowed $(2-2)^l$, $(2-2)^o$, $(3-3)^l$, and $(3-3)^o$ transitions marked in Fig. 4.29 is assisted by comparing integrated intensities of the differential peaks. In each case, peaks for oblique valley transitions are larger than peaks for longitudinal valley transitions because of the larger transition matrix elements associated with the threefold degenerate oblique valleys. As temperature increases, all the sub-band and continuum transitions shift to higher energies and peaks broaden due to phonon scattering. At 210 K it is no longer possible to distinguish individual longitudinal valley transitions from oblique valley transitions.

Figure 4.30 shows differential transmission spectra for three MQW samples with different PbSe well widths. As the well width decreases from 29.7 to 9.7 nm, the splitting between the $(1-1)^l$ transition and the $(1-1)^o$ transition increases significantly, from 4 to 20 meV. The number of sub-bands involving both longitudinal and oblique valleys also decreases from 4 to 2. Again, in all three samples the relative peak intensities of the oblique-valley transitions $(i-i)^o$ compared to their corresponding longitudinal-valley transitions $(i-i)^l$ indicate the threefold degeneracy of the oblique valleys.

Energy level calculations for electrons and holes confined in IV-VI semiconductor QWs have been carried out by Kriechbaum [98], Geist [99], and Silva [100]. In PbSe/Pb(Eu,Sr,Mn)Se MQW systems a type-I band alignment was confirmed for low Eu, Sr, or Mn concentrations. Using barrier heights

TABLE 4.1 Band structure parameters used in the calculations of quantized states of electrons and holes in PbSe/PbSrSe MQW structures.

	PbSe		PbSrSe	
	4.2 K	295 K	4.2 K	295 K
E_g	0.150 eV	0.278 eV	0.405 eV	0.505 eV
m_e^l	$0.070\,m_0$	$0.113\,m_0$	$0.070\,m_0$	$0.113\,m_0$
m_e^o	$0.042\,m_0$	$0.069\,m_0$	$0.042\,m_0$	$0.069\,m_0$
m_h^l	$0.068\,m_0$	$0.108\,m_0$	$0.068\,m_0$	$0.108\,m_0$
m_h^o	$0.040\,m_0$	$0.054\,m_0$	$0.040\,m_0$	$0.054\,m_0$

obtained from the measured FTIR transmission spectra and the band structure parameters from [100] and [101] (see Table 4.1), sub-band transition energies were calculated using the envelope wave function approximation. Assumptions in the calculations included equal conduction- and valence-band edge discontinuities ($\Delta E_c : \Delta E_v = 1:1$), a reasonable assumption based on the work of Yuan et al. [93], and the same electron and hole effective masses in the barrier layers as those in PbSe. The solid and dotted lines in Fig. 4.31 show the results of these calculations along with the measured transition energies obtained from the differential transmission spectra. These calculations only consider the quantum size effect. Strain-induced shifts of transition energies in the PbSe well layers are not considered here. The lattice mismatch between BaF_2 (111) and the epilayer is around 1% at room temperature, and the total epilayer thickness is around 0.8 μm. It is thus expected that PbSe/PbSrSe epilayers on BaF_2 substrates are relaxed [102]. There is a 0.18% lattice mismatch between the PbSe wells and PbSrSe barriers, and this produces tensile strain in PbSe and compressive strain in the PbSrSe layers. Tensile strain in the PbSe QWs is expected to cause additional splitting of the degeneracy of the longitudinal and oblique valleys on the order of a few meV [98]. Availability of reliable shear deformation potential values would allow better quantification of this effect.

Comparing the calculations (lines) with the measured $(1-1)^l$ and $(1-1)^o$ transition energies at 4.2 and at 295 K shown in Fig. 4.31, it is clear that these calculations give a good description of the well width and temperature dependence for sub-band transitions in PbSe/PbSrSe MQW structures. In the room-temperature differential transmission spectra, due to phonon-scattered broadening, longitudinal and oblique sub-band transitions could not be resolved, so the experimental points represent the combination of ground state transitions from both longitudinal and oblique valleys.

The results described above clearly show that PbSe/PbSrSe MQW materials grown on (111)-oriented BaF_2 substrates have well-understood electronic properties, and it is expected that other IV-VI materials combinations will have similar electronic sub-band properties. These properties are desirable for

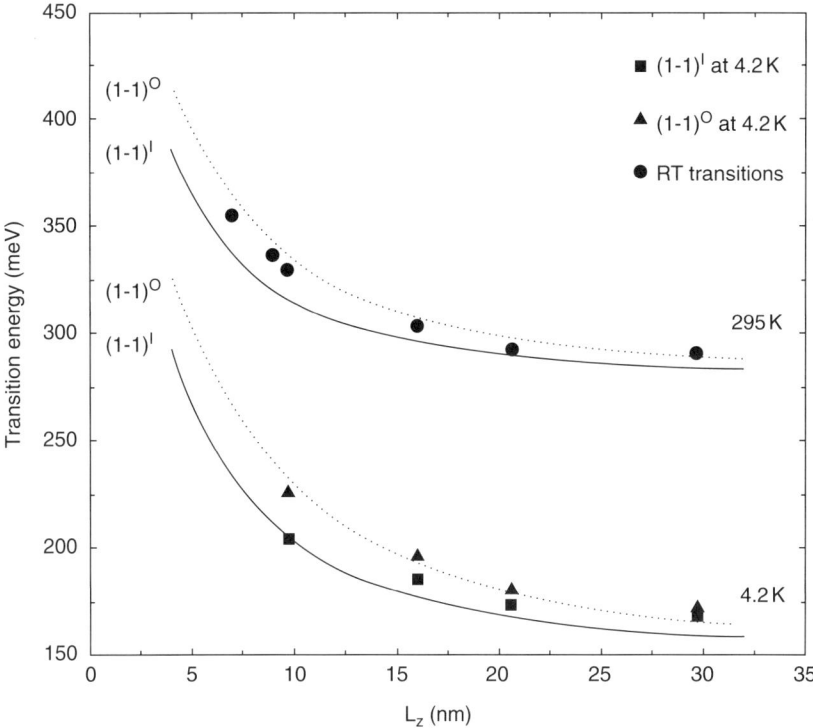

Fig. 4.31 Transition energies as a function of well width for PbSe/PbSrSe MQW structures at 4.2 K and 295 K. Calculations are plotted as lines, and experimental data from infrared transmission measurements are plotted as symbols: _ — $(1-1)^l$ at 4.2 K, ? — $(1-1)^0$ at 4.2 K, and • — at room temperature.

laser fabrication. The removal of L-valley degeneracy in which just one L-valley has a transition energy lower that the other three valleys facilitates population inversion; only one L-valley needs to be populated, whereas for QW materials grown on (100)-oriented substrates all four L-valleys need to be populated. Threshold currents should therefore be much lower in lasers fabricated from QW materials grown on (111)-oriented substrates. Unfortunately, it is difficult to fabricate IV-VI semiconductor lasers using (111)-oriented materials because IV-VI compounds cleave along the {100} planes and in-plane cleaved cavity lasers cannot be readily obtained. Other approaches to laser fabrication that do not require cleaving such as development of VCSEL structures are therefore worth pursuing.

New approaches to mid-IR laser fabrication have begun with the demonstration of two different IV-VI semiconductor VCSELs by Springholz et al. [103] and Shi et al. [104]. Both devices were grown on (111)-oriented BaF_2 and were optically pumped. The former device consisted of an active region with four 20-nm-thick PbTe QWs and PbEuTe/EuTe Bragg reflectors. Lasing occurred at temperatures where sub-band transition energies (the gain spectrum) coincided with cavity resonance modes. The maximum operating temperature of 82 K for this laser was thus determined by QW and cavity design. Much higher operating temperatures should be possible with cavity resonance modes designed for the correspondingly larger sub-band transition energies. The latter VCSEL consisted of a PbSe active region and PbSrSe/BaF$_2$ Bragg reflectors [105]. Lasing was observed at temperatures as high as 289 K. Again, the maximum operating temperature was determined by loss of overlap between the gain peak and the cavity resonance. These results demonstrate that it should be possible to fabricate IV-VI semiconductor lasers with higher maximum operating temperatures when VCSEL structures are optimized with properly matched QW sub-band energies and cavity resonance modes.

4.5 DFB AND DBR LASERS

4.5.1 Introduction

Fabry-Perot (FP) IV-VI lasers do not satisfy all conditions for heterodyne detection such as a continuously tunable longitudinal single mode beam or cw operating at temperatures above liquid nitrogen tempeature ($T_{cw} > 77$ K). Single-mode behavior and wider tuning ranges were obtained with corrugated structures in which frequency control is achieved by means of Bragg reflection [106]. In these structures, the lasing wavelength is well determined and can be controlled by the grating period (Fig. 4.32).

Distributed feedback (DFB) and distributed Bragg (DBR) lasers were first fabricated in IV-VI materials at MIT in the early 1970s [107,108,109]. These lasers used lateral gain-guided DH in combination with the DFB and DBR structures [110–118].

Monomode emission is desired in the three directions of wave propagation, in transversal, lateral, and longitudinal modes, respectively. After suppressing transversal multimode behavior successfully utilizing the gain-guided proper-

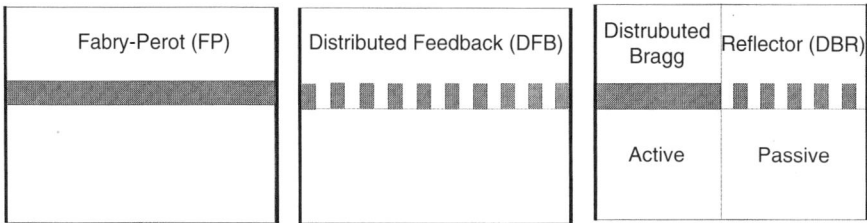

Fig. 4.32 Schematic describtion of the resonators in Fabry-Perot (FP), Distributed Feedback (DFB) and Distributed Bragg Reflector (DBR) Lasers.

ties of the DH, index-guided BH lasers for lateral monomode mode behavior were developed [119,120]. The next progress step was the implementation of corrugated DFB and DBR structures using an index-guided BH laser [121,122] (Fig. 4.33) or a technologically simpler index-guided single-heterostructure (SH) laser [123] (Fig. 4.34).

The theoretical background of laser light propagating in a waveguide and laser with Bragg reflectors is given by the mode coupling theory and the transfer matrix method, respectively. In the early 1970s, Kogelnik and Shank [106] developed a model for periodic modulated waveguides. The model describes the coupling of two waves with contrary propagation. For the special case of a periodic modulated waveguide with length L, solutions could be calculated under the condition that there is no reflection on the right side of the modulated part.

Figure 4.36 shows the calculated reflectivity and the transmissivity for different coupling coefficients versus the relative distance from the central Bragg frequency. The following conclusions could be made:

- The spectral width of the reflection is linear with the coupling coefficient κ, which means with the variation of the effective index of refraction.
- The reflection decreases with increasing absorption.
- For a larger κ, the influence of the absorption on the reflection is smaller.
- With increasing strength of κ, the reflection near the Bragg frequency and width of the reflection curve increases.

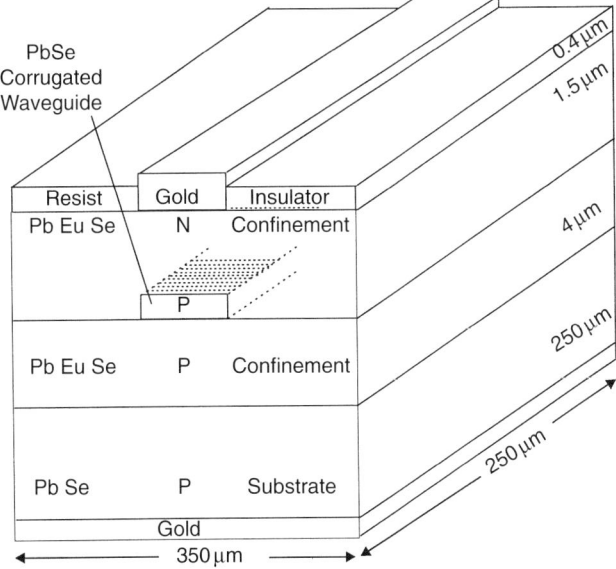

Fig. 4.33 Schematic description of a buried heterostructure (BH-DFB) laser.

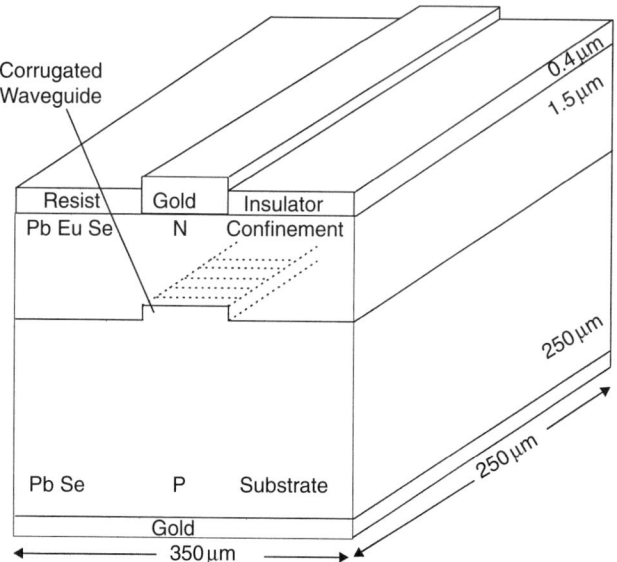

Fig. 4.34 Schematic description of an index guided single heterostructure (SH-DFB) laser.

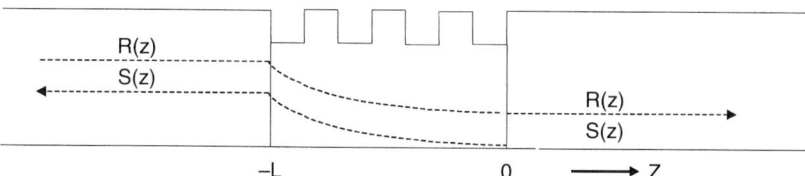

Fig. 4.35 Waveguide with periodic modulated area (Bragg-reflector). The incident wave R(z) from the left side is splitted into a partly reflected part, the wave S(z) propagating from right to left, and a transmitted part R(z).

The mode coupling theory gives a good description of the nature and the strength of the coupling and the behavior of the modes. It is less suited for the description of real laser structures because of the difficult integration of the influence of laser facets and disturbances in the theory.

An alternative theory to describe modulated waveguides is the transfer matrix analysis [124]. This theory considers the waveguide as an optical multilayer system. The axial distribution of the index of refraction is approximated by a constant index of refraction in a small area (Fig. 4.37). Transmission and reflection of a multilayer system can be calculated due to a recursive procedure [122].

In each layer a combination of an incident and a reflected wave in the axial direction determines the solution of the wave equation. Indexes of refraction

DFB AND DBR LASERS **187**

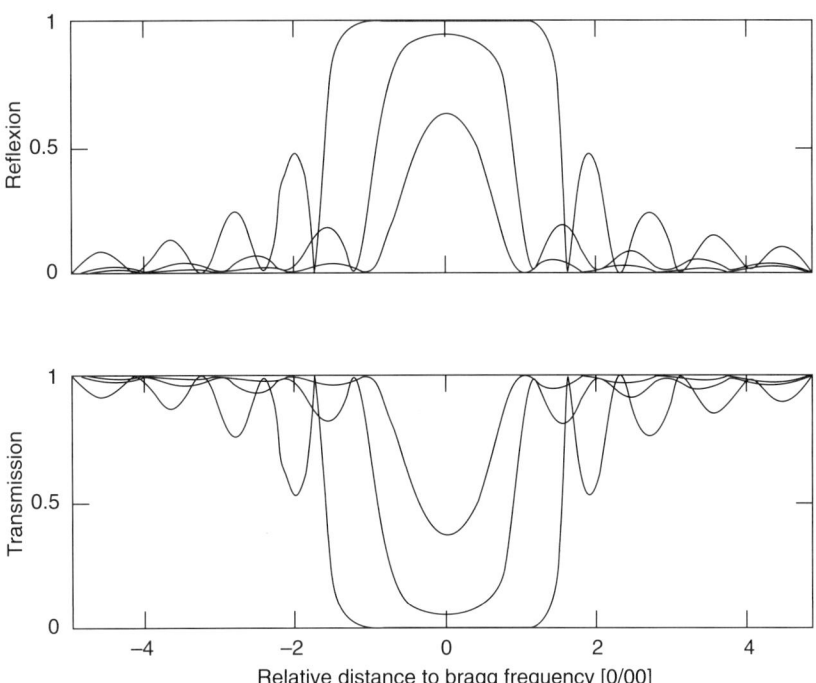

Fig. 4.36 Calculated reflection and transmission curves vs. the relative distance from the Bragg frequency for three different values of the coupling coefficient. Taken from Ref [121]

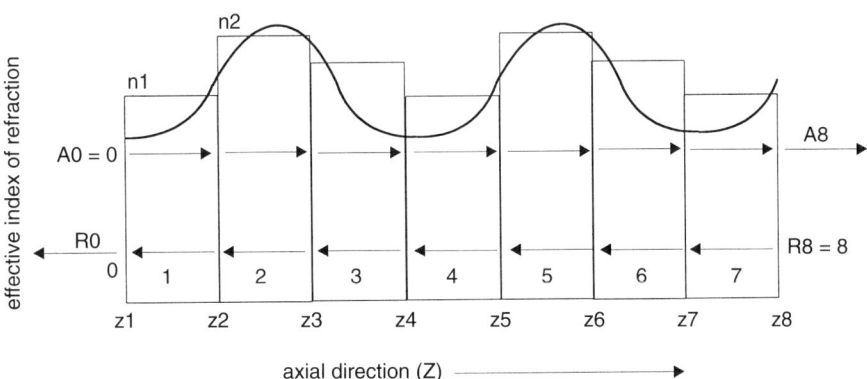

Fig. 4.37 Waveguide as an optical multilayer. The axial distribution of the index of refraction is approximated by a constant index of refraction in a small area.

and thickness are given for each layer. The boundary condition is a steady differential y-component of the electrical field E_y at the transition from one layer with given index of refraction and thickness to another. Thus the field coefficients must be chosen in a way that E_y is polarized in the transverse direction. From this condition it has been derived that the field coefficients of one layer can be recursively calculated from the field coefficients of the previous layer.

The correlation between to neighboring layers is given by the transfer matrix T.

$$\begin{bmatrix} A_{\ell+1} \\ R_{\ell+1} \end{bmatrix} = t_{\ell+1} \begin{bmatrix} A_\ell \\ R_\ell \end{bmatrix}. \tag{4.8}$$

The spectral position and threshold gain of the modes are determined by the roots of the matrix element t_{11}. It can be calculated numerically by varying the wave vector k_0 and the threshold gain g_{th}. The aim is to find the roots of the function $t_{11}(\beta)$ in the complex plane spreaded by the propagation constant β.

From the propagation constant, the value of the effective index of refraction n_{eff} can be calculated. This value is needed to determine the grating period Λ that should satisfy the condition

$$\Lambda = \frac{1/2 \lambda(T)}{n_{eff}(T)} \tag{4.9}$$

for a first-order grating, where λ is the wavelength that corresponds to the gain peak. The Bragg wavelength λ_B is not allowed as a resonator mode λ. In a first-order approximation, one gets

$$\lambda_q = \lambda_B + [(q + 1/2)a] \quad \text{with } a = \Lambda \lambda_B / L; q = 0, \pm 1, \pm 2, \ldots. \tag{4.10}$$

Both modes $q = 0$ and $q = -1$ are symmetric to λ_B at $\lambda = \lambda_B \pm \frac{1}{2}a$. Their reflection and gain are the same for symmetric reasons.

4.5.2 Experimental Work

Fabrication of Index-Guided DFB/DBR Laser Diodes
The PbSe/PbEuSe-based lasers are grown on (100) oriented PbSe substrates by MBE [125]. The mesa structure of the waveguide is formed by photolithography and ion-milling processes [122] before a second MBE overgrowth. The most delicate technological step is the transfer of the submicrometer grating into the mesa waveguide. This is conventionally done by holographic laser beam photolithography and dry etching techniques directly on the laserchip. This technology involves several processing steps, and the resulting grating is not suitable for the MBE overgrowth due to the surface condition.

A second method is the transfer of the grating using an embossing technique. This is a well-known process for integrated waveguides [117,71]. The embossing master submicrometer grating is performed in a (111)-oriented

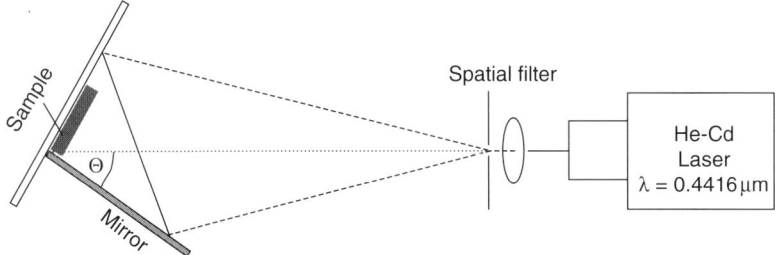

Fig. 4.38 Setup for holographic laser beam photolithography for the grating fabrication.

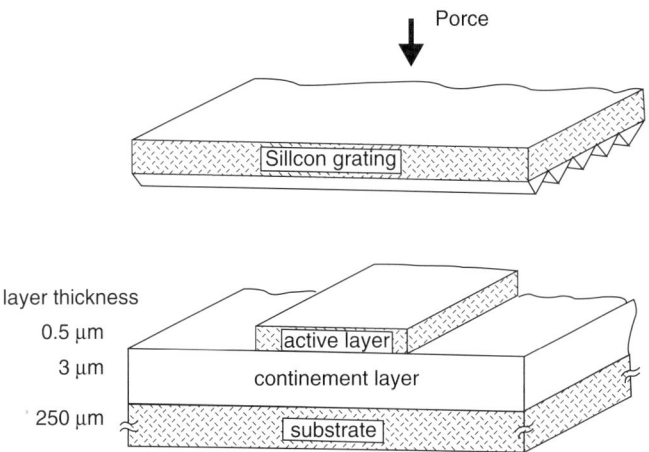

Fig. 4.39 Schematic description of the embossing.

silicon wafer (Fig. 4.38). As a radiation source for the holographic photolithography, an HeCd-laser is used that emits a linearly polarized beam at $\lambda_0 = 4416\,\text{Å}$. The angle Θ between the beams determines the grating period Λ. The correlation is given by the formula

$$\Lambda = \frac{\lambda}{2\sin\Theta}. \qquad (4.11)$$

A typical exposure time for a 0.5-µm-thick Shipley photoresist is 3 minutes. After a hard cure (1 h, 150°C) of the resist, the grating is transferred into the silicon by ion milling. The sample is mounted under 30° on a water-cooled rotatable stage. The etch depth is between 0.3 and 0.5 µm with a grating period varying from 0.5 µm up to 1 µm (Fig. 4.40).

The embossing can be performed with low force due to the thin (5–20 µm) ridge-waveguide structures. The overall pressure used is about $2\,\text{N/mm}^2$ for the whole PbSe wafer (Fig. 4.39). The silicon sample is made to be smaller than

190 LEAD-CHALCOGENIDE-BASED MID-INFRARED DIODE LASERS

Fig. 4.40 SEM picture of a silicon embossing master. Taken from Ref [123]

Fig. 4.41 SEM picture of a waveguide before embossing. Taken from Ref [126]

the IV-VI wafer to allow direct comparison between embossed and not embossed parts of one wafer. In Fig. 4.41 and Fig. 4.42 [126,127] SEM pictures of a waveguide before and after embossing are shown [71].

After the embossing treatment the sample is overgrown with the upper confinement layer of PbEuSe by MBE. Figure 4.43 shows a cross section of a DFB-BH structure. The corrugated waveguide is embedded between the two confinement layers. The material contrast is realized by a selective wet etch.

A standard photolithography step defined openings for the stripe contacts. Low-resistance p- and n-contacts were made simultaneously by electroplating gold. The wafer was cleaved into bars of about 500 μm length and mounted

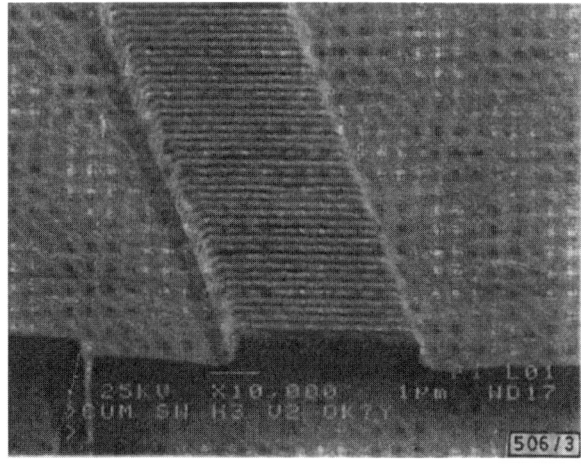

Fig. 4.42 SEM picture of a waveguide with embossed grating. Taken from Ref [123]

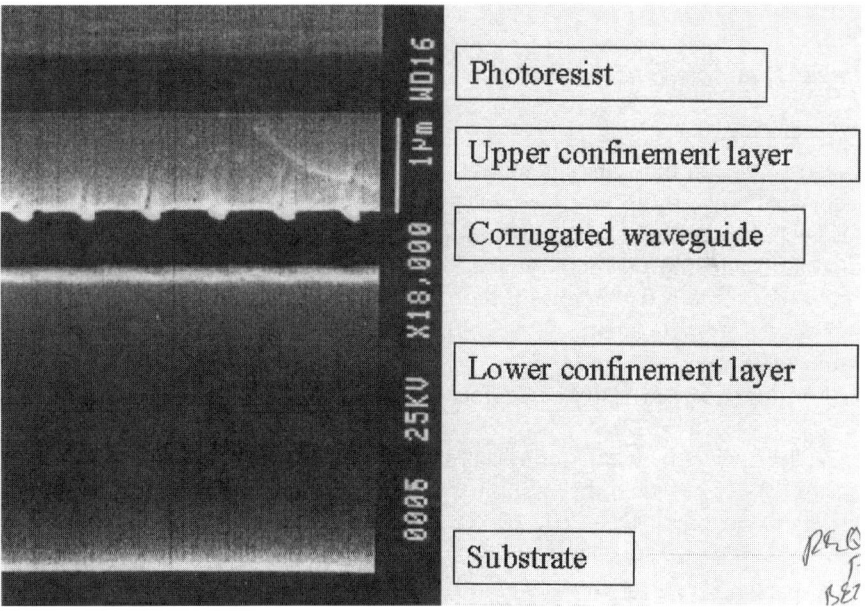

Fig. 4.43 SEM picture of a cross-section of a DFB-BH structure after a material contrast etch. Taken from Ref [127].

Fig. 4.44 Mounted laser chip on the indium-coated copper heat sink. Taken from Ref [121].

in a standard package using an In-coated copper heat sink as shown in Fig. 4.44.

Laser Characteristics

DFB-BH Laser. The lasers were tested in a closed-cycle helium cryostat. Mode spectra with dc bias current were taken at different temperatures. The BH lasers without grating on the same substrate were used a reference for the DFB/DBR lasers. At a certain temperature the current were successively increased and the positions of the laser modes were measured. The results show that there is no significant deterioration of electrical or optical properties caused from damaging between the grated and the nongrated lasers of the same substrate.

For the DFB lasers, the mode behavior could be classified into three types:

- The devices have no grating-supported mode, but the mode spacing in the feedback wavenumber region is different from that of the plain Fabry-Perot reference laser (Fig. 4.45).
- The devices show typically two or three modes with different mode spacing in the feedback region (Fig. 4.46).
- The devices show the desired monomode behavior (Fig. 4.47). The supported mode around $1500\,cm^{-1}$ fits to the grating period of 0.71 µm. The total tuning range of the DFB mode is about $7\,cm^{-1}$. Typical values for Fabry-Perot modes of this laser type are 2 to $3\,cm^{-1}$.

From the theory of DFB lasers, two symmetric modes could be preferred. The most elegant method to select one mode is the insertion of a $\lambda/4$ phase shift

DFB AND DBR LASERS **193**

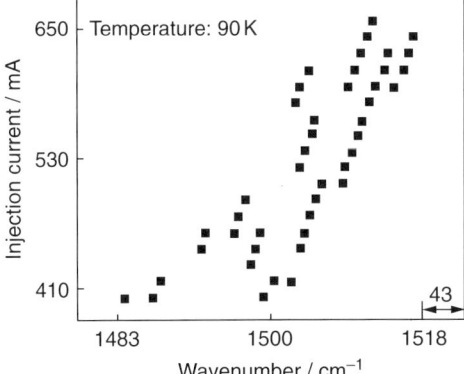

Fig. 4.45 Mode spectrum of a DFB laser showing no feedback mode behaviour. Taken from Ref [121].

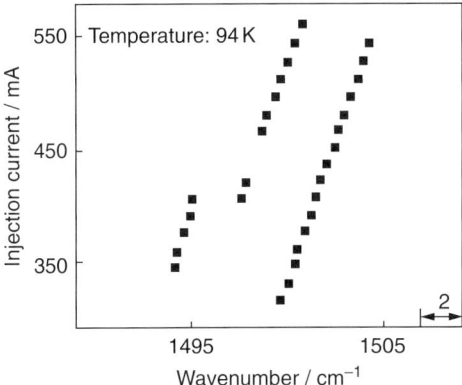

Fig. 4.46 Mode spectrum of a DFB laser showing two feedback modes. Taken from Ref [121].

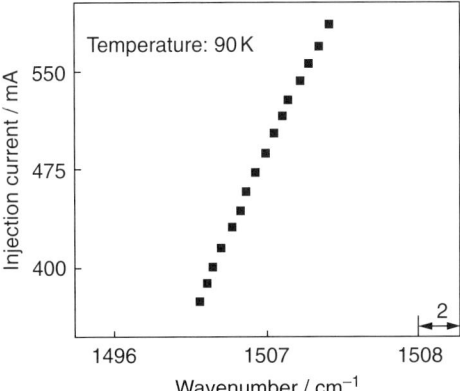

Fig. 4.47 Mode spectrum of a DFB laser showing monomode behaviour at the feedback wavelength. Taken from Ref [121].

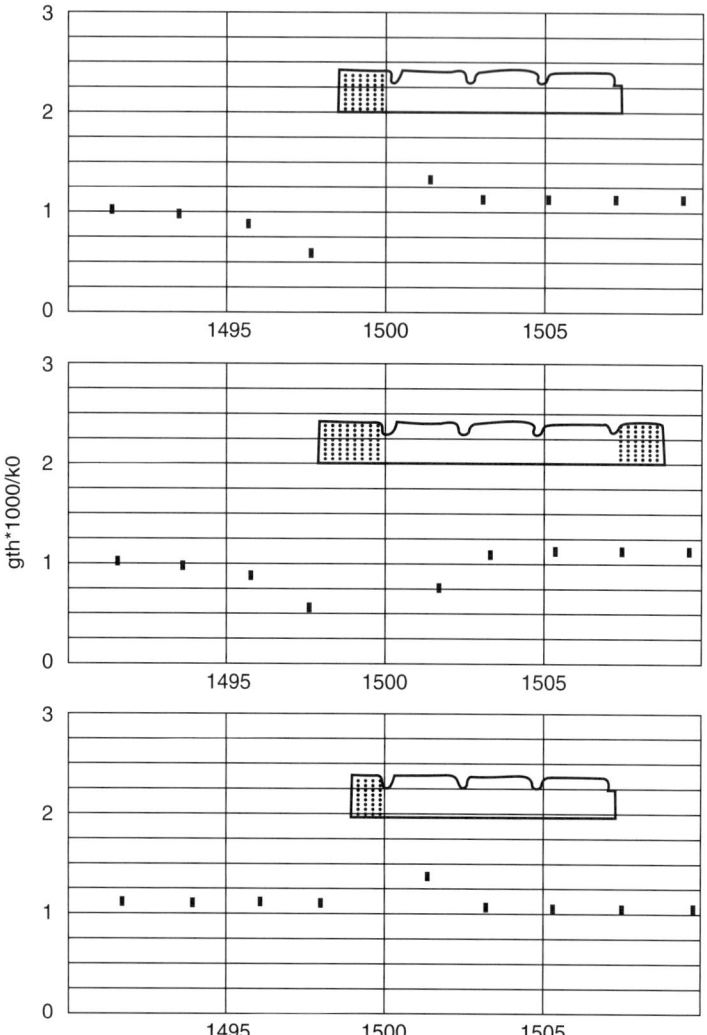

Fig. 4.48 Calculated threshold gain and schematic facet position. Taken from Ref [121].

in the grating. This is not very practical for technical reasons. Another method is the correct position of the laser facets that can influence the gain curve in a way that only one feedback mode is preferred [121].

Figure 4.48 shows three different possible facet positions and the corresponding calculated threshold gain. The resonator lengths of these three lasers are distinguished only by different fractions of a complete grating period, but with the totally different mode behaviors as shown in Figs. 4.45 to 4.47. Thus it is possible to calculate the mode behavior of randomly cleaved laser chips.

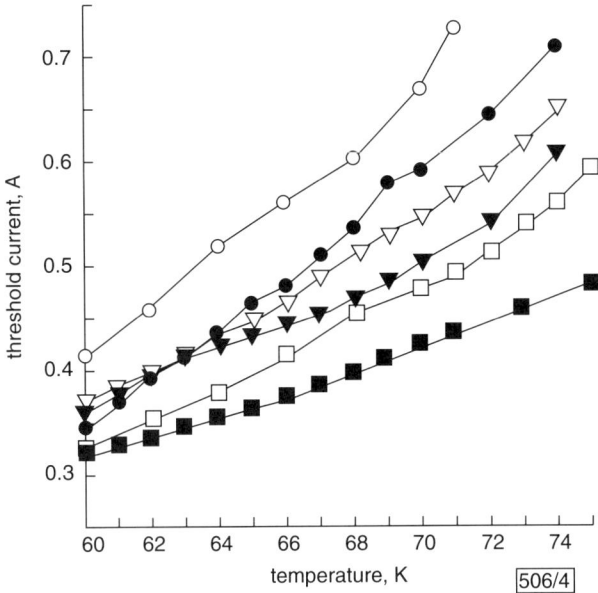

Fig. 4.49 Threshold current (cw) vs Temperature of embossed (filled symbols) and not embossed (open symbols) waveguides of the same wafer. Taken from Ref [123].

Unfortunately, the facet position cannot be defined before cleaving. So this process always will give a statistic distribution of monomode and multimode lasers.

DFB-SH-Laser. Single-heterostructure PbSe/PbEuSe DFB lasers were fabricated with the embossing technique [123]. To compare the lasers with embossed waveguide and those without treatment, the relation of threshold current to temperature was measured (Fig. 4.49). Although the semiconductor material of the active layer is mechanically deformed by the embossing procedure, the threshold current characteristic of these layers did not increase significantly. The emission mode behavior of SH lasers without grating shows the expected FP mode distribution with a maximum power of 0.1 mW per mode (Fig. 4.50). Figure 4.51 shows the mode behavior of a SH-DFB laser with a grating period of 0.74 µm. This corresponds to a Bragg frequency of 1350 cm^{-1}. This laser shows monomode mode behavior within a tuning range of 4 cm^{-1} and a power of 0.7 mW/mode. The lasers of this series achieved maximum operation temperatures of 100 K in cw mode.

DBR-DH Laser. Shani et al. [112] fabricated MBE stripe-geometry PbSnSe/PbEuSnSe DBR-DH lasers with a cw single mode behavior in the heat sink temperature range from 25 to 77 K. Figure 4.52 shows the schematic

196 LEAD-CHALCOGENIDE-BASED MID-INFRARED DIODE LASERS

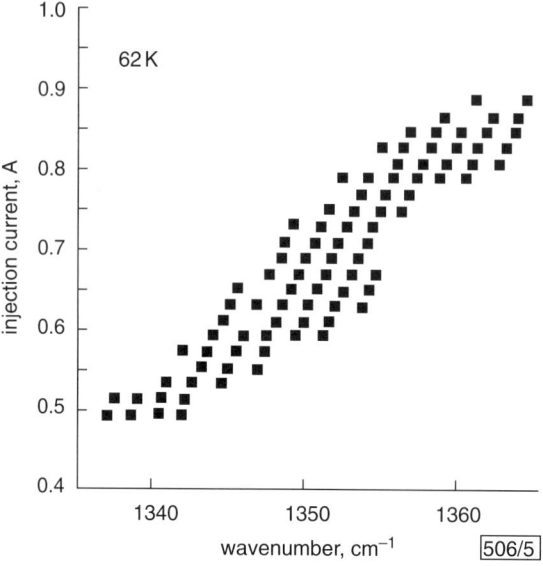

Fig. 4.50 Mode behavior of a SH laser without grating for different injection currents at 62 K. Taken from Ref [123].

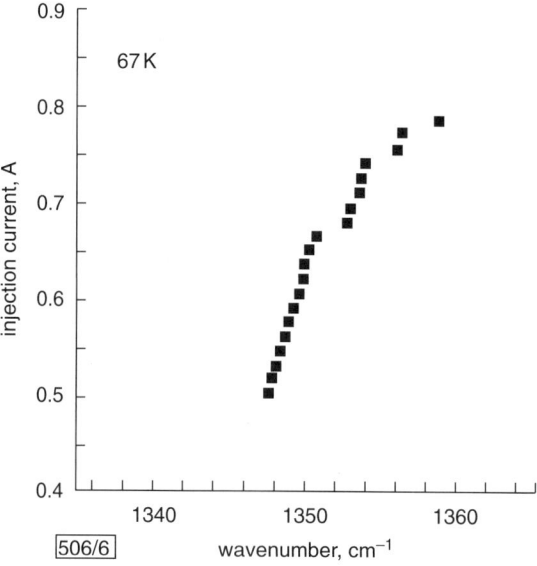

Fig. 4.51 Mode behavior of an embossed SH-DFB laser with feedback frequency of 1350 cm^{-1} and a grating periodicity of 0.74 Ïm. Taken from Ref [123].

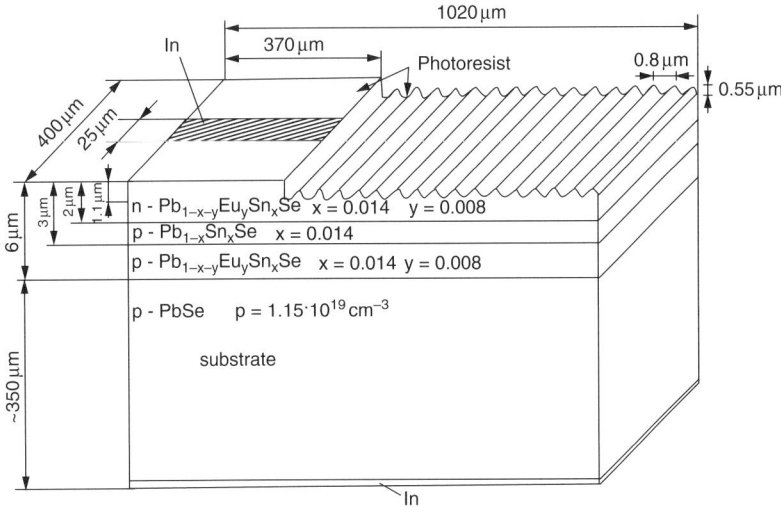

Fig. 4.52 Schematic describtion of a DBR laser after Shani et al. Taken from Ref [112].

description of this DBR laser, and its mode behavior at 77 K is given in Fig. 4.53.

A complete covered cw single mode tuning range of $24\,\mathrm{cm}^{-1}$ near $1280\,\mathrm{cm}^{-1}$ was also achieved with PbSnTe/PbTe DBR-DH lasers by Hsieh and Fonstad [109]. Schlereth [117] fabricated PbSe/PbEuSe DBR-DH lasers and found monomode behavior at 105 K and a continious tuning range of $15\,\mathrm{cm}^{-1}$ at about $1590\,\mathrm{cm}^{-1}$.

DBR-SH Laser. LPE broad-area lattice-matched PbSnTe/PbSeTe SH DBR lasers were reported by Kapon et al. [110,111]. Figure 4.23 shows a schematic illustration of the DBR laser. In Figs. 4.54 and 4.55, emission spectra of these lasers are shown. They operated in a single mode between heat sink temperatures from 8.5 to 38 K with a continuous tuning range of $6\,\mathrm{cm}^{-1}$ at $775\,\mathrm{cm}^{-1}$.

4.6 IV-VI EPITAXY ON BaF$_2$ AND SILICON

4.6.1 Introduction

IV-VI semiconductor epitaxy on BaF$_2$ substrates has been performed by a number of research groups since the first report of PbSnTe growth on (111)-oriented BaF$_2$ by Holloway and Logothetis [128]. In addition IV-VI epitaxy on Si substrates has been performed by a number of groups since the first report of PbSe growth on (111)-oriented Si with CaF$_2$ and BaF$_2$ buffer layers by Zogg et al. [129]. In both cases the close thermal expansion and lattice parameter matches between IV-VI materials and BaF$_2$ have allowed growth of high

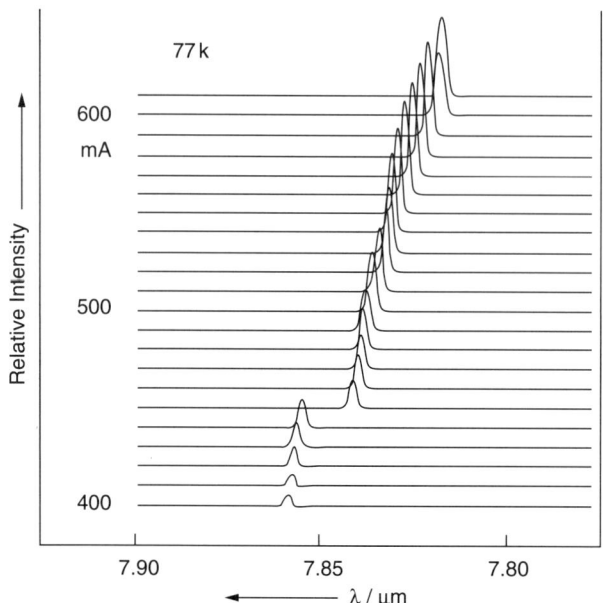

Fig. 4.53 Emission spectrum of a DBR laser, operated in cw mode, as a function of injection current (10 mA steps) at 77 K after Shani et al. Taken from Ref [112].

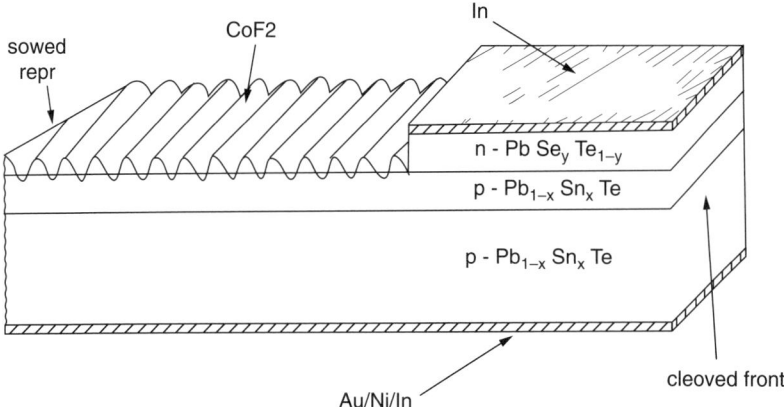

Fig. 4.54 Schematic illustration of the DBR PbSnSe/PbSeTe diode laser after Kapon and Katzir [111].

crystalline quality IV-VI epitaxial layers. Growth on BaF_2, with its insulating and infrared transparent properties, has allowed unambiguous study of the electrical and optical characteristics of various IV-VI semiconductor materials. This work provides useful data such as doping and band-gap information for fabrication of mid-IR devices based on these materials.

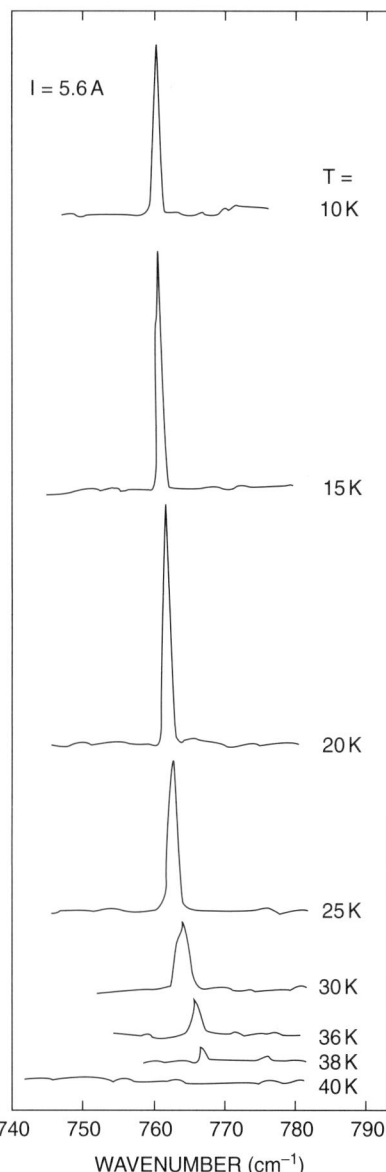

Fig. 4.55 Emission spectrum of a $Pb_{0.817}Sn_{0.183}Te/PbSe_{0.08}Te_{0.92}$ DBR laser at an injection current of 5.6 mA and for various heat sink temperatures after Kapon and Katzir [111].

Laser fabrication using conventional cleaved cavity techniques, however, is not possible for IV-VI layers on BaF_2 or silicon because IV-VI materials cleave along the {100} planes while BaF_2 and silicon cleave along the {111} planes. The water solubility of BaF_2, though, has created an opportunity to develop a new laser fabrication method that involves cleaving the epitaxial layers after removing the growth substrate by dissolving a BaF_2 buffer layer [130]. A major benefit of this new laser fabrication technique is that it enables mounting of epitaxially grown laser structures only between two thermally conductive materials. Finite element thermal modeling shows that a more than 60 K improvement in maximum operating temperature should be possible using these new methods [131]. Another approach to IV-VI laser fabrication on BaF_2 substrates involves development of vertical cavity structures. In this case the large difference in refractive indices between IV-VI materials and BaF_2 results in high reflectivities for only a three-pair $PbEuSe/BaF_2$ distributed Bragg reflector (DBR) [132]. These materials have been used to fabricate the first IV-VI laser on BaF2 [133] since the early work of Holloway et al. [134].

4.6.2 Growth and Characterization of IV-VI Layers on BaF_2

Hot wall epitaxy (HWE), MBE, and LPE have been used to grow IV-VI compounds and alloys on both (111)- and (100)-oriented BaF_2 substrates. Reviews of HWE and MBE techniques have been written by Lopez-Otero [135] and Holloway and Walpole [36], respectively, and techniques for LPE growth of IV-VI layers on BaF_2, which have helped to further IV-VI materials development, will be reviewed here. Results from this LPE work have provided new insights into IV-VI/BaF_2 interface chemistry, nucleation and growth mechanisms, lattice matching, doping, and band-gap dependence on the allowed composition and temperature. In addition recent results from MBE growth and characterization of PbSrSe/PbSe MQW structures on BaF_2 substrates will be reviewed. These new MQW materials have exhibited above-room-temperature cw photoluminescence (PL) in the 3 to 4μm spectral range, a result suggesting that cw room-temperature mid-IR lasers might be made from these new materials in the near future.

BaF_2 surface chemistry experiments were performed with an LPE apparatus to expose BaF_2 surfaces to selenium vapor at atmospheric pressure in a purified hydrogen ambient. Auger electron spectroscopic (AES) analysis of the exposed surfaces showed that a BaSe reaction layer forms [136], and that at temperatures near 620°C BaSe completely covers the BaF_2 surface. Continuous layer growth of PbSe on both (111)- and (100)-oriented BaF_2 substrates also occurs near 620°C in the LPE apparatus [137]. These experimental results suggest strongly that the BaSe reaction layer is epitaxial and that it promotes PbSe epitaxy on BaF_2 by catalyzing PbSe nucleation [138]. Additional LPE experiments also showed that controlling Se partial pressure through LPE graphite boat design was an important factor in obtaining good PbSe epitaxy on BaF2 [139]. This provided further support for the proposed epitaxy-enabling substrate surface reaction theory.

LPE was also used to grow ternary PbSe$_{0.78}$Te$_{0.22}$ [140] and quaternary Pb$_{1-x}$Sn$_x$Se$_{1-y}$Te$_y$ [141] pseudobinary alloys that are lattice matched with BaF$_2$. Lattice strain effects were clearly seen in the liquid-solid phase equilibria. For example, a high concentration of Te in the liquid growth solution, about 60% of the group-VI elements, was required to obtain the lattice-matched ternary alloy with 22% Te. A large Te chemical potential (i.e., concentration in the liquid) is necessary to overcome the lattice strain associated with incorporation of relatively large Te atoms in the growing solid. This effect was less in quaternary alloys, since incorporation of smaller Sn atoms reduced the lattice strain associated with Te incorporation [142]. Doping studies of lattice-matched PbSe$_{0.78}$Te$_{0.22}$ ternary layers grown on (100) BaF$_2$ showed that undoped layers were n-type, as expected due to growth from Pb-rich liquid solutions, which results in formation of group-VI vacancies, and that Th was an effective acceptor impurity [143]. These lattice-matching and doping data were used to grow full-fledged DH laser structures on (100)-oriented BaF$_2$ substrates with different Sn contents in the middle quaternary layer [144].

Infrared transmission spectra collected using FTIR spectroscopy have been obtained for various IV-VI layer structures grown on BaF$_2$ substrates. Absorption edges of LPE-grown lattice-matched three-layer DH laser structures on (100) BaF$_2$ at 130 K were observed to shift from 195 meV (6.4 µm) to 143 meV (8.7 µm) as Sn content in the growth solution for the Pb$_{1-x}$Sn$_x$Se$_{1-y}$Te$_y$ quaternary middle layer was increased from 0% to 12% [144]. This absorption edge energy dependence on Sn content is similar to what has been observed for laser emission energy measurements from DH lasers made with LPE-grown Pb$_{1-x}$Sn$_x$Se$_{1-y}$Te$_y$ quaternary alloys lattice matched with PbSe substrates [145]. Room-temperature FTIR transmission spectra for Pb$_{1-x}$Sr$_x$Se layers grown by MBE on (111) BaF$_2$ substrates (see Fig. 4.56) showed an absorption edge shift from 277 meV to 716 meV as the Sr/PbSe flux ratio was increased from 0% to 10% [146]. The spacings of below band-gap Fabry-Perot interference fringes were also used to determine refractive indexes, which decreased from 4.8 to 3.8 as the Sr/PbSe flux ratio was increased from 0% to 10%.

Both low-temperature and room-temperature PL have been observed from IV-VI layers grown on BaF$_2$ substrates. Comparison of low-temperature absorption edge energies with PL energies of LPE-grown PbSe$_{0.78}$Te$_{0.22}$ and Pb$_{0.95}$Sn$_{0.05}$Se$_{0.80}$Te$_{0.20}$ layers on (100) BaF$_2$ showed significant Burstein-Moss shifts consistent with Fermi energy locations 20 and 9 meV above the bottom of the conduction band for ternary and quaternary alloys, respectively [147]. Such degenerate materials are expected due to the high group-VI vacancy concentrations of these LPE-grown materials [143].

Recently, above room temperature, cw PL between 3 and 4 µm has been observed from PbSe/PbSrSe MQW structures grown on (111) BaF$_2$ substrates by MBE [148]. Fabry-Perot interference fringes dominated the spectra, indicating that luminescence was primarily due to stimulated emission processes.

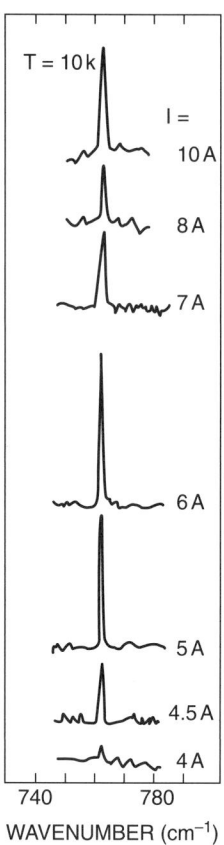

Fig. 4.56 Emission spectrum of a $Pb_{0.817}Sn_{0.813}Te/PbSe_{0.08}Te_{0.02}$ DBR laser at a heat sink temperature of 10 K and for various injection currents after Kapon and Katzir [111].

Gaussian fits to the spectra showed that emission energies at 25°C decreased from 402 to 312 meV as QW thickness increased from 4 to 20 nm. These PL measurements have also been used to quantify lattice heating effects due to the pump laser. Figure 4.57 shows measured PL spectra for a 40-period 10-nm MQW layer structure on BaF_2 for different diode laser pump currents. The inset shows Gaussian-fitted PL peak energies shifting from 341 to 352 meV as the pump laser current increases from 300 to 800 mA. Using the temperature tuning coefficient of 0.40 meV/K for this material [148], localized additional heating due to increasing the pump laser current is found to be 28°C.

4.6.3 Growth and Characterization of IV-VI Layers on Silicon

Despite the large thermal expansion and lattice parameter mismatches between IV-VI materials and Si, about 12% and 700%, respectively, high-

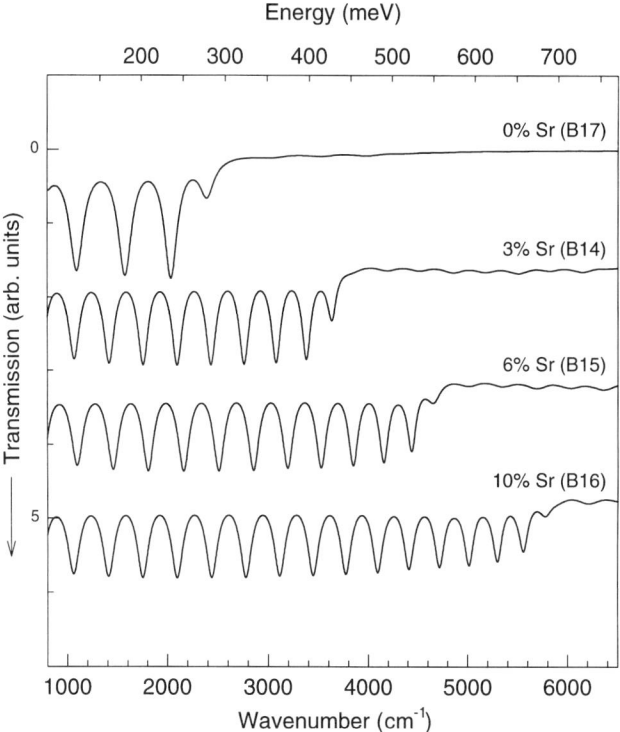

Fig. 4.57 Fourier transform infrared (FTIR) transmission spectra for $Pb_{1-x}Sr_xSe$ layers grown by MBE on (111)-oriented BaF_2.

quality IV-VI epitaxial layers can be grown on (111)-oriented silicon substrates using MBE. Initial work in this area focused on using a stacked BaF_2/CaF_2 buffer layer [129], but recent work has shown that just a thin CaF_2 layer produces good PbSe epitaxy [149]. This fluoride buffer layer, even when as thin as 50 Å, does appear to be necessary for good PbSe epitaxy, since attempts to grow PbSe directly on (111) Si have resulted in poor layer quality [150]. $PbSe/CaF_2$ interface chemistry has been studied using in situ X-ray photoelectron spectroscopy (XPS), and the interface was found to be abrupt and dominated by Pb-F and Ca-Se bonds [151]. Other fundamental studies include observation of RHEED oscillations, indicating layer-by-layer growth modes [152], doping [153], and band-gap studies for MBE-grown PbEuSe [154]. In addition DBR structures consisting of $PbEuSe/BaF_2$ pairs have been grown and characterized [133]. Reflection spectra show that broadband reflectivity of greater than 95% between 3 and 5 µm can be obtained with only three pairs of $PbEuSe/BaF_2$.

Plastic deformation of IV-VI materials has been shown to occur by movement of dislocations along {100} planes in the <110> directions [155]. This primary slip system allows plastic deformation of (111)-oriented layers grown on Si substrates. In-plane tensile strain in (100)-oriented layers, however, cannot be relieved by this primary dislocation glide system because there are no resolved shear stresses along the {100} planes in the <110> directions. MBE growth of PbSe on (100)-oriented Si at 280°C therefore results in a high crack-density, which is indicative of the large tensile strain that occurs in IV-VI layers when cooled to room temperature [150]. Combining MBE with LPE of PbSe, however, has enabled growth of entirely crack-free PbSe layers on $1 \times 1\,cm^2$ Si (100) substrates [156]. High-resolution X-ray diffraction (HRXRD) measurements also showed that these layers had good crystalline quality as indicated by full-width half-maxima (FWHM) values of below 200 arcsec.

Smaller band-gap $Pb_{1-x}Sn_xSe$ ternary layers were also grown by LPE on Si (100) substrates, but unlike binary PbSe layers they were not crack free [157]. HRXRD characterization showed that PbSnSe layers had a smaller elasticity, likely due to solid solution hardening, as exhibited by residual in-plane tensile strains of about 0.06% following growth, much lower than the 0.21% obtained for PbSe layers [158]. Crack densities in PbSnSe layers were observed to decrease when LPE growth temperatures were increased. In addition HRXRD FWHM data showed that PbSnSe crystalline quality improved as growth temperature was increased [159]. These results show that layers grown at higher temperatures are more able to plastically deform and that slip on one or more secondary glide systems, such as the <110> {110} system, increasingly occurs as growth temperature is raised from 420 to 480°C. High Se vacancy concentration, which increases with LPE growth temperature, can enhance layer plasticity if vacancies agglomerate and form edge dislocations along one or more higher order slip planes. This mechanism may explain why it is possible to obtain crack-free PbSe layers on (100) Si by LPE and not by MBE, where growth occurs at much lower temperatures and vacancy concentrations are smaller. FTIR transmission spectra for $Pb_{1-x}Sn_xSe$ layers grown on (100) Si substrates nevertheless showed expected band-gap dependence on temperature and alloy composition where $Pb_{1-x}Sn_xSe$ alloys with Sn contents between 0% and 10% could cover the $5\,\mu m$ ($2000\,cm^{-1}$) to $16\,\mu m$ ($625\,cm^{-1}$) spectral range.

Above room temperature, cw PL has been demonstrated from IV-VI MQWs grown by MBE on (111) Si substrates with CaF_2 and BaF_2/CaF_2 buffer layers. Figure 4.58 shows measured emission spectra at 25°C obtained with different diode laser pump currents for a sample with 20 periods of 10-nm-thick PbSe QWs. As with the PL spectra from IV-VI MQWs grown on BaF_2 substrates, there are Fabry-Perot resonance peaks, but they are not as pronounced because of weaker confinement of photons in the layer due to the closer refractive index match of PbSe ($n = 4.8$) to Si ($n = 3.4$) than that of BaF_2 ($n = 1.5$). The temperature tuning coefficient of 0.38 meV/K is similar to what was observed for MQW layers grown on BaF_2 substrates, but increasing pump laser power resulted in a much smaller blue shift. Gaussian-fitted PL peak energies shifted from 337 to 340 meV as the pump laser current was increased from 300 to 800 mA. This corresponds to an additional heating effect of only 7.8°C, more

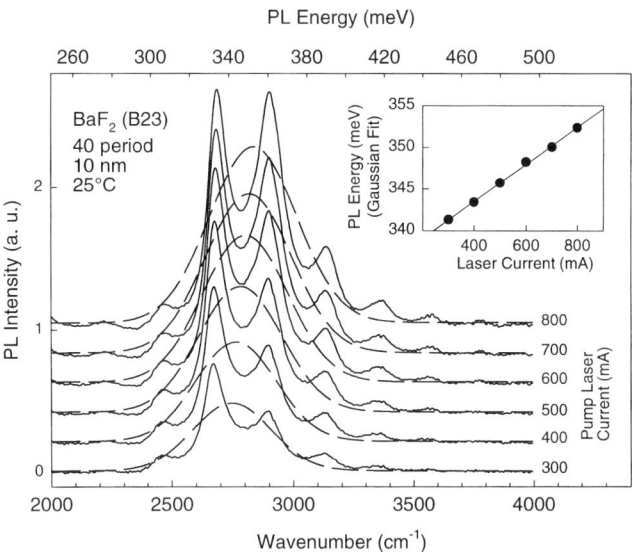

Fig. 4.58 Photoluminescence spectra at 25°C from a 40-period 10 nm thick PbSe/PbSrSe multiple quantum well structure grown by MBE on BaF$_2$ (111).

than 20°C less than the additional heating in IV-VI layers on BaF$_2$. Clearly, the higher thermal conductivity of the Si substrate (1.41 W/cm K) compared to that of BaF$_2$ (0.12 W/cm K) improves heat dissipation from the IV-VI layer.

Work has also been performed on removing Si growth substrates from MBE- and LPE-grown PbSe layers by dissolving BaF$_2$ buffer layers in water. This technique was successful in transferring 3-μm-thick MBE-grown PbSe epilayers from Si to Cu by using PbSn [160] and AuIn [161] bonding metallurgy. Transferred PbSe layers had mirror-like surfaces, and profilometer characterization showed a typical peak roughness of about 22 nm for the growth interface, not much higher than the roughness of the as-grown PbSe surface, about 12 nm. Nomarski microscopy did reveal threefold slip lines, indicating dislocations at the growth interface, whereas no slip lines were observed on the as-grown PbSe surface. The FWHM of the transferred PbSe layer's HRXRD peak was 160 arcsec, only slightly larger than the 153 arcsec FWHM for the as-grown layer, demonstrating that the transfer process preserves PbSe crystalline quality. Surface chemical analysis of transferred layers by XPS also indicated complete removal of BaF$_2$ buffer layer. PbSe layers grown by combination MBE/LPE on (100) Si have been removed from growth substrates and then cleaved along {100} planes using a novel cleaving tool [162]. Further development of these new substrate removal techniques is an active area of research. Benefits include a significant increase in laser active-region heat dissipation if epitaxially grown IV-VI laser structures can be sandwiched between two thermally conductive materials such as copper. The combination of MBE-

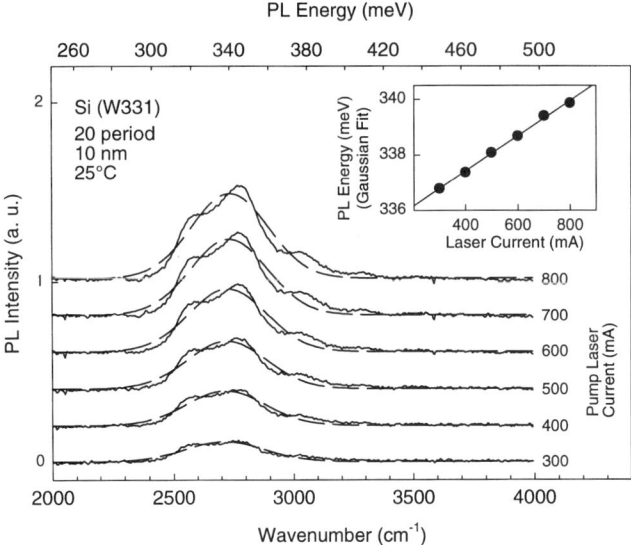

Fig. 4.59. Photoluminescence spectra at 25°C from a 20-period 10 nm thick PbSe/PbSrSe multiple quantum well structure grown by MBE on Si (111).

grown quantum-confined structures with new device packaging technologies as described here offers a viable way to achieve room-temperature cw mid-IR laser devices.

4.7 CONCLUSION

In this chapter we tried to give an overview of the current state of the art of lead-salt-based tunable diode laser technology. These types of laser diodes at present are the only available family that covers the complete fingerprint absorption range of molecules. Lead-salt lasers are commercially available. Their maximum operation temperatures in cw are still leading the race, despite the fact that to date only very limited effort is being spent on R&D activities in this interesting field. New strategies to improve the maximum operation temperatures were discussed, and with very limited effort pulse operation above room temperature was demonstrated successfully. The implementation of quantum wells as well as in edge-emitting and vertical surface-emitting devices combined with improved packaging techniques has an interesting potential for further improvements of the maximum cw operation temperature.

From a commercial point of view, however, the major breakthrough of this technology as well as others will be achieved only when tunable laser diodes over a broad wavelength region are available with operation temperatures well into the region that can be easily achieved by means of thermoelectric cooling methods. Lead-salt lasers have a good and promising potential to reach this goal. However, R&D activities and investment in this field has to be significantly enhanced in the future.

ACKNOWLEDGMENTS

The authors are grateful for helpful comments and correction of the manuscript by M. Tacke and H. Böttner.

REFERENCES

1. U. P. Schießl and J. Rohr, "60°C lead salt laser emission near 5-µm wavelength, " *Infrared Phys. Technol.* **40**, 325 (1999).
2. U. P. Schießl, M. Birle, and H.-E. Wagner, "Recent results about room temperature operation of lead salt laser diodes," *Optische Analysentechnik in Industrie und Umwelt—Heute und morgen*, Tagung Düsseldorf, March 20–21, 2000/VDI/VDE-Gesellschaft Mess- und Automatisierungstechnik. VDI Verlag, Disseldary 2000, pp. 113–118.
3. P. C. Findlay, C. R. Pidgeon, R. Kotitschke, A. Hollingworth, B. N. Murdin, C. J. G. M. Langerak, A. F. G. van der Meer, C. M. Ciesla, J. Oswald, A. Homer, G. Springholz, and G. Bauer, "Auger recombination of lead salts under picosecond free-electron-laser-excitations," *Phys. Rev.* B **58**, 12908 (1998).
4. A. Katzir, R. Rosman, Y. Shani, K. H. Bachem, H. Böttner, and H. M. Preier, "Tunable lead salt lasers," in P. K. Cheo, ed., *Handbook of Solid State Lasers*, 1989, pp. 227–347.
5. J. F. Butler et al., *Solid State Comm.* **2**, 303 (1964).
6. T. C. Harman, in D. L. Carter and R. T. Bate, eds. *Physics of Semimetals and Narrow—Gap Semiconductors*, Pergamon Press, New York, 1971, p. 363.
7. A. R. Calawa, "Small bandgap lasers and their uses in spectroscopy," *J. Luminescence* **7**, 477 (1973).
8. J. F. Butler, "Pb-salt tunable diode lasers," First European Electro Optics Markets and Technology Conference, Sept. 13–15, 1972, Geneva IPC Sci. & Technol. Press, London, 1973, p. 99.
9. A. P. Shotov, Suppl. *J. Jpn. Soc. Appl. Phys.* **42**, 282 (1973).
10. T. C. Harman and I. Melngailis, "Narrow gap semiconductors," Applied Solid State Sciences 4, Ed. R. Wolf (Academic Press, New York 1974), p. 1.
11. J. Hesse and H. Preier, "Lead salt laser diodes," in H. J. Queisser, ed., *Festkörperprobleme XV*, Pergamon/Vieweg, Braunschweig, 1975, p. 229.
12. J. Hesse, "IV-VI narrow gap semiconductors and devices," *Jpn. J. Appl. Phys.* **16**, Suppl. 16-1, 297 (1977).
13. I. Hayashi, "Recent progress in semiconductor lasers—Cw GaAs lasers are ready for new applications," *Appl. Phys.* **5**, 25 (1974/75).

14. J. Kuhl and W. Schmidt, *Appl. Phys.* **3**, 251 (1974).
15. A. Mooradian, T. Jaeger, and P. Stokseth, eds., *Tunable Lasers and Application*, Vol. 3, Springer, Berlin, 1976.
16. H. Preier, "Recent advances in lead-chalcogenide diode lasers," *Appl. Phys.* **20**, 189 (1979).
17. K. J. Linden, A. W. Mantz, "Tunable diode lasers and laser systems for the 3 to 30 μm infrared spectral region," in *SPIE Adv. Infrar. Fibers II* **320**, 109–120 (1982).
18. J. O. Dimmock, I. Melngailis, and A. J. Strauss, "Band structure and laser action in $Pb_xSn_{1-x}Te$," *Phys. Rev. Lett.* **16**, 1193 (1966).
19. R. Grisar, private communications.
20. L. N. Kurbatov et al., *Soviet Phys-Semicond.* **2**, 1008 (1969).
21. C. E. Hurwitz et al., "Electron beam pumped lasers of PbS, PbSe, and PbTe," *IEEE Trans. Qunatum. Electron.* **1**, 102 (1965).
22. J. F. Butler et al., *Solid State Comm.* **2**, 303 (1964).
23. K. W. Nill, A. J. Strauss, and F. A. Blum, "Tunable cw $Pb_{0.98}Cd_{0.02}S$ diode lasers emitting at 3.5 μm: applications to ultrahigh-resolution spectroscopy," *Appl. Phys. Lett.* **22**, 677 (1973).
24. W. Lo and D. Swets, "Diffused homojunction lead-sulfide-selenide diodes with 140 K laser operation," *Appl. Phys. Lett.* **33**, 938 (1978).
25. L. R. R. L. Guldi and G. A. Antcliffe, "Antimony diffusion into p-type $Pb_{1-x}Sn_xTe$," *J. Electrochem. Soc.* **121**, 1523 (1974).
26. W. Lo, *IEEE J. Quantum Electron.* **13**, 591 (1977).
27. E. Silberg, Y. Sternberg, and A. Yellin, "Cd diffusion into PbTe," *J. Solid State Chem.* **39**, 100 (1981).
28. E. Silberg and A. Zemel, "Cadmium diffusion studies of PbTe and $Pb_{1-x}Sn_xTe$ crystals," *J. Electr. Mater.* **8**, 99 (1979).
29. R. Grisar, W. J. Riedel, H. M. Preier, "Properties of diffused PbSnSe homojunction diode lasers," *IEEE J. Quantum Electron.* **17**, 586 (1981).
30. H. M. Preier, K.-H. Bachem, H. Böttner, D. Ball, W. J. Riedel, A. Jakubowicz, and A. Eisenbeiss, "Cd-diffused lead salt diode lasers and their application in multicomponent gas analysis systems," *Proc. SPIE* **483**, 10 (1983).
31. J. Xu, A. Lambrecht, and M. Tacke, "Lead chalcogenide implanted diode lasers in cw operation above 77 K," *Electron. Lett.* **30**, 7 (1994).
32. U. P. Schießl and H. E. Wagner, "New improvements in IV-VI tunable laser diodes at Laser Components," *Proc. 5th International Symposium on Gas Analysis by Tunable Diode Lasers*, pp. 251–260, VDI Verlag, Düsseldorf, 1998.
33. U. Lambrecht, "Abstimmung und linienbreite von diodenlasern in der infrarotspektroskopie," Diploma thesis, IPM and Universtity of Freiburg, 1991.
34. W. J. Riedel, M. Knothe, W. Kohn, R. Grisar, "An anastigmatic white cell for IR diode laser spectroscopy," in *Proc. Int. Symp. Monitoring of Gaseous Pollutants by Tunable Diode Lasers*, Freiburg, 1988, Kluwer Academic, Dordrecht, 1989.
35. M. Mürtz, M. Schaefer, M. Schneider, J. S. Wells, W. Urban, U. Schiessl, and M. Tacke, "Stabilization of 3.3 and 5.1 um lead-salt diode lasers by optical feedback," *Optics Commun.* **94**, 551 (1992).

36. H. Holloway and J. Walpole, "MBE Techniques for IV-VI optoelectronic devices," *Prog. Cryst. Growth Charac.* **2**, 49–84, Pergamon Press, Great Britain, 1979.
37. D. L. Partin, "Wavelength coverage of lead-europium-selenide-telluride diode lasers," *Appl. Phys. Lett.* **45**, 193 (1984).
38. H. Preier, Z. Feit, J. Fuchs, D. Kostyk, W. Jalanak, and J. Sproul, "Status of lead salt diode laser development at Spectra Physics," Proceedings of the International Symposium, Freiburg, 1988, Kluwer Academic, Dordrecht, 1989, pp. 85–102.
39. M. Tacke, "Recent results in lead salt laser development at the IPM," *Proc. Int. Symp.*, Freiburg, 1988, Kluwer Academic, Dordrecht, 1989, pp. 103–118.
40. H. Preier, M. Bleicher, W. Riedel, and H. Maier, "Double heterostructure PbS-PbS$_{1-x}$Se$_x$-PbS laser diodes with CW operation up to 96K," *Appl. Phys. Lett.* **28**, 669 (1976).
41. H. Preier, M. Bleicher, W. Riedel, H. Pfeiffer, and H. Maier, "Peltier cooled PbSe double-heterostructure lasers for IR-gas spectroscopy," *Appl. Phys.* **12**, 277 (1977).
42. H. Böttner, U. Schießl, and M. Tacke, "Dependance of lead chalkogenide diode laser radiation on lattice misfit induced stress," *Superlattice Microstruct.* **7**, 97 (1990).
43. P. Norton and M. Tacke, "MBE of Pb$_{1-x}$Eu$_x$Se for the use in IR devices," *J. Cryst. Growth* **81**, 405 (1986).
44. J. W. Tomm, K. H. Herrman, H. Böttner, M. Tacke, and A. Lambrecht, "A Luminescence study in the Pb$_{1-x}$Eu$_x$Se system," *Phys. Stat. Sol.* **A119**, 711 (1990).
45. B. D. Schwartz, C. A. Huber, and A. V. Nurmikko, *Lecture Notes Phys.* **152**, 163 (1982).
46. A. J. Strauss, "Effect of Pb and Te saturation on carrier concentration in impurity-doped PbTe," *Electron. Mater.* **2**, 553 (1973).
47. K. H. Herrmann, K.-P. Möllmann, J. W. Tomm, H. Böttner, M. Tacke, and J. Evers, "Some band structure related optical and photoelectrical properties of Pb$_{1-x}$Eu$_x$Se ($0 < x < 0.2$)," *J. Appl. Phys.* **72**, 1399 (1991).
48. K. H. Herrmann, U. Müller, and V. Meizer, "Interband and intraband contribution to refractive index in the new PbSe-based narrow-gap semiconductors," *Semicond. Sci. Technol.* **8**, 333 (1993).
49. J. W. Tomm, K.-P. Möllmann, F. Peuckert, K. H. Herrmann, H. Böttner, and M. Tacke, "Valence band hybridizing in europium-alloyed lead selenide," *Semicond. Sci. Technol.* **9**, 1033 (1994).
50. A. Ishida and H. Fuyiyasu, "Recent progress in lead-salt lasers," in Proc. SPIE **2682**, *Laser Diodes and Applications II*, San Jose, CA, 1996.
51. J. Evers, G. Oehlinger, A. Weiss, C. Probst, M. Schmidt, and P. Schramel, "Preparation and characterization of high purity calcium, strontium and barium," *J. Less-Common Metals*, 15 (1981).
52. B. Spanger, U. Schießl, A. Lambrecht, H. Böttner, and M. Tacke, "Near room-temperature operation of Pb$_{1-x}$Sr$_x$Se infrared diode lasers using molecular beam epitaxy growth techniques," *Appl. Phys. Lett.* **53**, 26 (1988).
53. M. Agne, A. Lambrecht, U. Schießl, and M. Tacke, "Guided modes and far-field patterns of lead chalcogenide buried heterostructure laser diodes," *Infrared Phys. Technol.* **35**, 47 (1994).

54. J. John, "Untersuchungen für Diodenlaser mit verteilter Rückkopplung für das mittlere Infrarot," Thesis, University of Freiburg, Germany, 1993.
55. M. Tacke, "New developments and applications of tunable IR lead salt lasers," *Infrared Phys. Technol.* **36**, 447 (1995).
56. G. Spilker, R. Daddato, U. Schießl, A. Lambrecht, and M. Tacke, "Linewidth and noise of lead chalcogenide diode lasers," *Proc. Int. Symp. Monitoring of Gaseous Pollutants by Tunable Laser Diodes*, Freiburg, 1991 Kluwer Academic, London, 1992.
57. A. Ishida and H. Fujiyasu, "Burnstein-Moss effect of PbTe-Pb$_{1-x}$Sn$_x$Te superlattice," *Jpn. J. Appl. Phys.* **24**, L956 (1985).
58. Y. Nishiyima, "PbSnTe double-heterostructure lasers and PbEuTe double-heterostructure lasers by hot-wall epitaxy," *J. Appl. Phys.* **65**, 935 (1989).
59. A. Ishida, M. Aoki, and H. Fujiyasu, "Semimetallic Hall properties of PbTe-SnTe superlattice," *J. Appl. Phys.* **58**, 1901 (1985).
60. Z. Feit, J. Fuchs, D. Kostyk, and W. Jalenak, "Liquid phase epitaxy grown PbSnSeTE/PbSe double heterostructure diode lasers," *Infrared Phys. Technol.* **37**, 439 (1996).
61. N. Koguchi, T. Kiyosawa, and S. Takahashi, "Double heterostructure Pb$_{1-x-y}$Cd$_x$S$_y$Se/PbS/Pb$_{1-x-y}$Cd$_x$Sr$_y$Se lasers grown by molecular beam epitaxy," *J. Cryst. Growth* **81**, 400 (1987).
62. A. Ishida, K. Muramatsu, H. Takashiba, and H. Fujiyasu, "Pb$_{1-x}$Sr$_x$S/PbS double-heterostructure lasers prepared by hot wall epitaxy," *Appl. Phys Lett.* **55**, 430 (1989).
63. M. Tacke, B. Spanger, A. Lambrecht, P. R. Norton, and H. Böttner, "Infrared double heterostructure diode laser made by molecular beam epitaxy of Pb$_{1-x}$Eu$_x$Se," *Appl. Phys. Lett.* **53**, 430 (1989).
64. N. Koguchi and S. Takahashi, "Double-heterostructure Pb$_{1-x-y}$Cd$_x$Sr$_y$S/PbS/Pb$_{1-x-y}$Cd$_x$Sr$_y$S lasers grown by molecular beam epitaxy," *Appl. Phys. Lett.* **58**, 799 (1991).
65. A. Ishida, T. Tsuchiya, N. Yoshioka, K. Ishino, and H. Fujiyasu, "PbCaTe films and PbCaTe/PbTe superlattices prepared by hot-wall epitaxy," *Jpn. J. Appl. Phys.* **38**, 4652 (1999).
66. H. Fischer, "Amplitudenrauschen von bleisalzlaserdioden im frequenzbereich von DC bis 500 MHz," FHG-IPM Forschungsbericht, Nr. 23, 1990.
67. G. Spilker, R. Daddato, U. Schiessl, A. Lambrecht, and M. Tacke, "Linewidth and noise of lead chalcogenide lasers," *Proc. Int. Symp.*, Freiburg, 1991, Kluwer Academic, Dordrecht, 1992, pp. 85–91.
68. T. Beyer, "Die Analyse der Alterung von Diodenlasern," Diploma theses, Freiburg i.Br., 1995.
69. Y. Horikoshi, *Semicond. Semimetal* 22 C, Academic Press, San Diego, 1985.
70. C. Sirtori et al., "Low-threshold InGaAsP ridge waveguide lasers at 1.3 μm," European Semiconductor Laser Workshop 1999, Paris, Sept. 24–25, 1999.
71. K. H. Schlereth and H. Böttner, "Embossed grating lead chalcogenide distributed-feedback lasers," *J. Vac. Sci. Technol.* **B1**, 114 (1992).
72. Z. Shi, "GaSb-PbSe-GaSb double heterostructure nidinfrared lasers," *Appl. Phys. Lett.* **72**, 1272 (1998).

73. J. W. Tomm, M. Mocker, T. Kelz, T. Elsaesser, R. Klann, B. V. Novikov, V. G. Talalaev, V. E. Tudorovskii, and H. Böttner, "Threshold of stimulated emission in multivalley lead salts," *J. Appl. Phys.* **78**, 7247 (1995).
74. I. P. Kaminow, L. W. Stulz, J. S. Ku, A. G. Dentai, R. E. Nahory, J. C. DeWinter, and R. L. Hartmann, *IEEE J. Quantum Electron.* **19**, 1312 (1983).
75. K. H. Schlereth, H. Böttner, and M. Tacke, "Mushroom double-channel double-heterostructure lead chalcogenide lasers made by chemical etching," *Appl. Phys.* **56,** 22 (1990).
76. D. Kasemset, S. Rotter, and C. G. Fonstadt, "$Pb_{1-x}Sn_xTe/PbTe_{1-y}Se_y$ lattice-matched buried heterostructure lasers with cw single mode output," *IEEE Elect. Dev. Lett.* **1**, 75 (1980).
77. K. H. Schlereth, B. Spanger, H. Böttner, A. Lambrecht, and M. Tacke, "Buried waveguide DH-PbEuSe-lasers grown by MBE," *Infrared Phys.* **30**, 449 (1989).
78. L. Palmetshofer, "Ion-implanted IV-VI semiconductors," *Appl. Phys. A* **34**, 139 (1984).
79. A. Lambrecht, H. Böttner, M. Agne, R. Kurbel, A. Fach, B. Halford, U. Schießl, and M. Tacke, "Molecular beam epitaxy of laterally structured lead chalcogenides for the fabrication of buried heterostrcuture lasers," *Semicond. Sci. Technol.* **8**, 334 (1993).
80. A. Lambrecht, H. Böttner, and M. Tacke, "Development of infrared tunable diode lasers specifically suited for spectroscopical applications," in *Transport and Chemical Transformation of Pollutants in the Troposphere*, Vol. 8, Springer, Berlin, 1994, pp. 237–243.
81. Z. Feit, D. Kostyk, R. J. Woods, and P. Mak, "PbEuSeTe buried heterostructure lasers grown by molecular-beam epitaxy", *J. Vac. Sci. Technol.* B**8**, 200 (1990).
82. Z. Feit, D. Kostyk, R. J. Woods, and P. Mak, "Molecular beam epitaxy grown PbEuSeTe buried-heterostructure lasers with continuous wave operation at 195 K," *Appl. Phys. Lett.* **57**, 2891 (1990).
83. Z. Feit, D. Kostyk, R. J. Woods, and P. Mak, "Single-mode molecular beam epitaxy grown PbEuSeTE/PbTe buried heterostructure diode lasers for CO_2 high-resolution spectroscopy," *Appl. Phys. Lett.* **58,** 343 (1991).
84. Z. Feit, M. McDonald, R. J. Woods, V. Archambault, and P. Mak, "Low threshold PbEuSeTe/PbTe separate confinement buried heterostructure diode lasers," *Appl. Phys. Lett.* **68**, 738 (1996).
85. D. L. Partin "Single quantum well lead-europium-selenide-telluride diode lasers," *Appl. Phys. Lett.* **45**, 487 (1984).
86. A. Ishida, H. Fujiyasu, H. Ebe, and K. Shinohara, "Lasing mechanism of type-I' PbSnTe-PbTeSe multiquantum well laser with doping structure," *J. Appl. Phys.* **59**, 3023 (1986).
87. D. Partin, "Lead salt quantum effect structures," *IEEE J. Quantum Electron.* **24**, 1716 (1988).
88. Z. Shi, M. Tacke, A. Lambrecht, and H. Böttner, "Midinfrared lead salt multi-quantum-well diode lasers with 282 K operation," *Appl. Phys. Lett.* **66**, 2537 (1995).
89. J. R. Meyer, C. L. Felix, W. W. Bewley, I. Vurgaftman, E. H. Aifer, L. J. Olafsen, J. R. Lindle, C. A. Hoffman, M. J. Yang, B. R. Bennett, B. V. Shanabrook, H. Lee,

C. H. Lin, S. S. Pei, and R. H. Miles, "Auger coefficients in type-II InAs/Ga$_{1-x}$In$_x$Sb quantum wells," *Appl. Phys. Lett.* **73**, 2857 (1998).

90. M. F. Khodr, P. J. McCann, and B. A. Mason, "Effects of nonparabolicity on the gain and current density in EuSe/PbSe$_{0.78}$Te$_{0.22}$/EuSe IV-VI semiconductor quantum well lasers," *IEEE J. Quantum Electron.* **32**, 236 (1996).

91. A. Ishida, Y. Sase, and H. Fujiyasu, "Optical properties of PbTe/Pb$_{1-x}$Eu$_x$Te superlattices prepared by hot wall epitaxy," *Appl. Surf. Sci.* **33/34**, 868 (1988).

92. A. Ishida, S. Matsuura, M. Mizuno, and H. Fujiyasu, "Observation of quantum size effects in optical transmission spectra of PbTe/Pb$_{1-x}$Eu$_x$Te superlattices," *Appl. Phys. Lett.* **51**, 478 (1987).

93. S. Yuan, H. Krenn, G. Springholz, G. Bauer, and M. Kriechbaum, "Large refractive index enhancement in PbTe/Pb$_{1-x}$Eu$_x$Te multiquantum well structures," *Appl. Phys. Lett.* **62**, 885 (1993).

94. A. Lambrecht, N. Herres, B. Spanger, S. Kuhn, H. Bottner, M. Tacke, and J. Evers, "Molecular beam epitaxy of Pb$_{1-x}$Sr$_x$Se for use in IR devices," *J. Cryst. Growth* **108**, 301 (1991).

95. X. M. Fang, K. Namjou, I. Chao, P. J. McCann, N. Dai, and G. Tor, "Molecular beam epitaxy of PbSrSe and PbSe/PbSrSe multiple quantum well structures for use in mid-infrared light emitting devices," *J. Vac. Sci. Technol.* **18**, 1720 (2000).

96. S. W. McKnight, K. P. Stewart, H. D. Drew, and K. Moorjani, "Wavelength-independent anti-interference coating for the far-infrared," *Infrared Phys.* **27**, 327 (1987).

97. N. Dai, F. Brown, P. Barsic, G. A. Khodaparast, R. E. Doezema, M. B. Johnson, S. J. Chung, K. J. Goldammer, and M. B. Santos, "Observation of excitonic transitions in InSb quantum wells," *Appl. Phys. Lett.* **73**, 1101 (1998).

98. M. Kriechbaum, K. E. Ambrosch, E. J. Fantner, H. Clemens, and G. Bauer, "Electronic structure of PbTe/Pb$_{1-x}$Sr$_x$Te superlattices," *Phys. Rev.* B**30**, 3394 (1984).

99. F. Geist, H. Pascher, M. Kriechbaum, N. Frank and G. Bauer, "Magneto-optical properties of diluted magnetic PbSe/Pb$_{1-x}$Mn$_x$Se superlattices," *Phys. Rev.* B**54**, 4820 (1996).

100. E. A. de Andrada e Silva, "Optical transition energies for lead-salt semiconductor quantum wells," *Phys. Rev.* B**60**, 8859 (1999).

101. Landolt-Börnstein, "Numerical data and functional relationships in science and technology," in O. Madelung, ed., *Non-Tetrahedrally Bonded Compounds*, Vol. 17, Springer, Berlin, 1987.

102. J. W. Matthews and A. E. Blakeslee, "Defects in epitaxial multilayers," *J. Cryst. Growth* **27**, 118 (1974).

103. G. Springholz, T. Schwarzl, M. Aigle, H. Pascher, and W. Heiss, "4.8 µm vertical emitting PbTe quantum-well lasers based on high-finesse EuTe/Pb$_{1-x}$Eu$_x$Te microcavities," *Appl. Phys. Lett.* **76**, 1807 (2000).

104. Z. Shi, G. Xu, P. J. McCann, X. M. Fang, N. Dai, C. L. Felix, W. W. Bewley, I. Vurgaftman, and J. R. Meyer, "IV-VI compound mid-infrared high-reflectivity mirrors and vertical-cavity surface-emitting lasers grown by molecular beam epitaxy," *Appl. Phys. Lett.* **76**, 3688 (2000).

105. X. M. Fang, H. Z. Wu, N. Dai, Z. Shi, and P. J. McCann, "Molecular beam epitaxy of periodic BaF$_2$/PbEuSe Layers on Si (111)," *J. Vac. Sci. Technol.* B**17**, 1297 (1999).

106. H. Kogelnik and C. V. Shank, "Coupled-wave theory of distributed feedback lasers," *J. Appl. Phys.* **43**, 2327 (1972).
107. J. N. Walpole, A. R. Calawa, S. R. Chinn, S. H. Groves, and T. C. Harman, "Distributed feedback $Pb_{1-x}Sn_xTe$ double-heterostructure lasers," *Appl. Phys. Lett.* **29**, 307 (1976).
108. J. N. Walpole, A. R. Calawa, S. R. Chinn, S. H. Groves, and T. C. Harman, "cw operation of distributed feedback $Pb_{1-x}Sn_xTe$ lasers," *Appl. Phys. Lett.* **30**, 524 (1977).
109. H. Hsieh and C. G. Fonstad, "Liquid-phase epitaxy grown PbSnTe distributed feedback laser diodes with broad continuous single-mode tuning range," *IEEE J. Quantum Electron.*, **16**, 1039 (1980).
110. E. Kapon, A. Zussmann, and A. Katzir, "Distributed Bragg-reflector lattice-matched $Pb_{1-x}Sn_xTe$/ $PbSe_yTe_{1-y}$ diode lasers," *Appl. Phys. Lett.* **44**, p. 275 (1984).
111. E. Kapon and A. Katzir, "Distributed Bragg-reflector lattice-matched $Pb_{1-x}Sn_xTe$/ $PbSe_yTe_{1-y}$ diode lasers," *IEEE J. Quantum Electron.*, **21**, 1947 (1985).
112. Y. Shani, A. Katzir, K.-H. Bachem, P. Norton, M. Tacke, and H. M. Preier, "77 K cw operation of distributed Bragg reflector $Pb_{1-x}Sn_xSe$/ $Pb_{1-x-y}Eu_ySn_xSe$ diode lasers," *Appl. Phys. Lett.* **48**, 1178 (1986).
113. Y. Shani, R. Rosman, A. Katzir, P. Norton, M. Tacke, and H. M. Preier, "Distributed Bragg reflector $Pb_{1-x}Sn_xSe$/$Pb_{1-x-y}Eu_ySn_xSe$ diode lasers with a broad single-mode tuning range," *J. Appl. Phys.* **63**, 11, 5603 (1988).
114. Y. Shani, A. Katzir, M. Tacke, and H. M. Preier, "Metal clad $Pb_{1-x}Sn_xSe/Pb_{1-x-y}Eu_ySn_xSe$ distributed feedback lasers," *IEEE J. Quantum Electron.* **24**, 2135 (1988).
115. Y. Shani, A. Katzir, M. Tacke, and H. M. Preier, "Highly collimated laser beams from grating coupled emission $Pb_{1-x}Sn_xSe/Pb_{1-x-y}Eu_ySn_xSe$ diode lasers," *Appl. Phys. Lett.* **53**, 462 (1988).
116. Y. Shani, A. Katzir, M. Tacke and H. M. Preier, "$Pb_{1-x}Sn_xSe/Pb_{1-x-y}Eu_ySn_xSe$ corrugated diode lasers," *IEEE J. Quantum Electron.* **25**, 1828 (1989).
117. K. H. Schlereth, H. Böttner, and M. Tacke, "Mushroom' double-channel double-heterostructure lead chalcogenide lasers made by chemical etching," *Appl. Phys. Lett.* **56**, 2169 (1992).
118. K. H. Schlereth, B. Spanger, H. Böttner, A. Lambrecht, and M. Tacke, "Buried waveguide DH-PbEuSe-Lasers grown by MBE," *Infrared Phys.* **30**, 449 (1990).
119. Z. Feit, D. Kostyk, R. J. Woods, and P. Mak, "Molecular beam epitaxy-grown PbSnTe-PbEuSeTe buried heterostructure diode lasers," *IEEE Photon. Technol. Lett.* **2**, 860 (1990).
120. Z. Feit, D. Kostyk, R. J. Woods, and P. Mak," Single-mode molecular beam epitaxy grown PbEuSeTe/PbTe buried-heterostructure diode lasers for CO_2 high-resolution spectroscopy," *Appl. Phys. Lett.* **58**, 343 (1991).
121. M. A. Fach, H. Böttner, K. H. Schlereth, and M. Tacke," Embossed-grating lead chalcogenide buried-waveguide distributed-feedback lasers," *IEEE J. Quantum Electron.* **30**, (1994).
122. K. H. Schlereth, "Diodenlaser mit verteilter Rückkopplung für das mittlere Infrarot," Ph.D. thesis, University of Wuerzburg, 1990.
123. J. John, A. Fach, H. Böttner, and M. Tacke, "Embossed monomode single heterostructure distributed feedback lead chalcogenide diode lasers," *Electron. Lett.* **28**, 2180 (1992).

124. T. Makino and J. Glinski, "Transfer matrix analysis of the amplified spontaneous emission of DFB semiconductor laser amplifiers," *IEEE J. Quantum Electron* **24**, 1507 (1988).
125. A. Lambrecht, A. Fach, R. Kurbel, B. Halford, H. Böttner, and M. Tacke, "Epitaxial growth of laterally structured lead chalcogenide lasers," *Proc. Int. Symp. Monitoring of Gaseous Pollutants by Tunable Lasers*, Freiburg, 1991, Kluwer Academic, Dordrecht, 1992.
126. J. John, "Untersuchungen für diodenlaser mit verteilter rückkopplung für das mittlere infrarot," Thesis, University of Freiburg, 1993.
127. A. Fach, "Diodenlaser mit verteilter Rückkopplung für das mittlere Infrarot," Thesis, University of Freiburg, 1991.
128. H. Holloway and E. M. Logothetis, "Epitaxial growth of lead tin telluride," *J. Appl. Phys.* **41**, 3543 (1970).
129. H. Zogg and M. Huppi, "Growth of high quality epitaxial PbSe onto Si using a $(Ca, Ba)F_2$ buffer layer," *Appl. Phys. Lett.* **47**, 133 (1985).
130. P. J. McCann, "New fabrication method promises higher Mid-IR laser operating temperatures," *Comp. Semicond.* **5**, 57 (1999).
131. K. R. Lewelling and P. J. McCann, "Finite element modeling predicts possibility of thermoelectrically-cooled lead-salt diode lasers," *IEEE Photon. Technol. Lett.* **9**, 297 (1997).
132. X. M. Fang, H. Z. Wu, N. Dai, Z. Shi, and P. J. McCann, "Molecular beam epitaxy of periodic BaF_2/PbEuSe layers on Si(111)," *J. Vac. Sci. Technol. B* **17**, 1297 (1999).
133. Z. Shi, G. Xu, P. J. McCann, X. M. Fang, N. Dai, C. L. Felix, W. W. Bewley, I. Vurgaftman, and J. R. Meyer, "IV-VI compound mid-infrared high-reflectivity mirrors and vertical-cavity surface-emitting lasers grown by molecular beam epitaxy," *Appl. Phys. Lett.* **76**, 3688 (2000).
134. H. Holloway, W. H. Weber, E. M. Logothetis, A. J. Varga, and K. F. Yeung, "Injection luminescence and laser action in epitaxial PbTe diodes," *Appl. Phys. Lett.* **21**, 5 (1972).
135. A. Lopez-Otero, "Hot wall epitaxy," *Thin Solid Films* **49**, 3 (1978).
136. P. J. McCann and C. G. Fonstad, "Auger electron spectroscopic analysis of barium fluoride surfaces exposed to selenium vapor," *J. Electron. Mater.* **20**, 915 (1991).
137. P. J. McCann and C. G. Fonstad, "Liquid phase epitaxy growth of PbSe on (111) and (100) barium fluoride," *J. Cryst. Growth* **114**, 687 (1991).
138. P. J. McCann, "The role of substrate surface reactions in heteroepitaxy of PbSe on BaF_2," Heteroepitaxy of Dissimilar Materials, *Mat. Res. Soc. Symp. Proc.* **221**, 289 (1991).
139. P. J. McCann and S. Aanegola, "The role of graphite boat design in liquid phase epitaxial growth of $PbSe_{0.78}Te_{0.22}$ on BaF_2," *J. Cryst. Growth* **141**, 376 (1994).
140. P. J. McCann and C. G. Fonstad, "Growth of $PbSe_{0.78}Te_{0.22}$ lattice matched with BaF_2," *Thin Solid Films* **227**, 185 (1993).
141. P. J. McCann and D. Zhong, "Liquid phase epitaxy growth of $Pb_{1-x}Sn_xSe_{1-y}Te_y$ alloys lattice matched with BaF_2," *J. Appl. Phys.* **75**, 1145 (1994).
142. P. J. McCann, "The effect of composition dependent lattice strain on the chemical potential of tellurium in $Pb_{1-x}Sn_xSe_{1-y}Te_y$ quaternary alloys," *Mat. Res. Soc. Symp. Proc.* **311**, 149 (1993).

143. P. J. McCann, S. Aanegola, and J. E. Furneaux, "Growth and characterization of thallium and gold doped PbSe$_{0.78}$Te$_{0.22}$ layers lattice matched with BaF$_2$ substrates," *Appl. Phys. Lett.* **65**, 2185 (1994).
144. I. Chao, P. J. McCann, W. Yuan, E. A. O'Rear, and S. Yuan, "Growth and characterization of IV-VI semiconductor heterostructures on (100) BaF$_2$," *Thin Solid Films* **323**, 126 (1998).
145. Z. Feit, J. Fuchs, D. Kostyk, and W. Jalenak, "Liquid phase epitaxy grown PbSnSeTe/PbSe double heterostructure diode lasers," *Infrared Phys. Technol.* **37**, 439 (1996).
146. X. M. Fang, K. Namjou, I. Chao, P. J. McCann, N. Dai, and G. Tor, "Molecular beam epitaxy of PbSrSe and PbSe/PbSrSe multiple quantum well structures for use in mid-infrared light emitting devices," *J. Vac. Sci. Technol.* (October 1999).
147. P. J. McCann, L. Li, J. Furneaux, and R. Wright, "Optical properties of ternary and quaternary IV-VI semiconductor layers on (100) BaF$_2$ substrates," *Appl. Phys. Lett.* **66**, 1355 (1995).
148. P. J. McCann, K. Namjou, and X. M. Fang, "Above-room-temperature continuous wave mid-infrared photoluminescence from PbSe/PbSrSe quantum wells," *Appl. Phys. Lett.* **75**, 3608 (1999).
149. H. Zogg, A. Fach, J. John, J. Masek, P. Müller, and C. Paglino, "Epitaxy of IV-VI materials on Si with fluoride buffers and fabrication of IR-sensor arrays," in *Narrow Gap Semiconductors 1995*, Institute of Physics, London, 1995, p. 160.
150. P. Müller, A. Fach, J. John, A. N. Tiwari, H. Zogg, and G. Kostorz, "Structure of epitaxial PbSe grown on Si(111) and Si(100) without a fluoride buffer layer," *J. Appl. Phys.* **79**, 1911 (1996).
151. X. M. Fang, W. K. Liu, P. J. McCann, B. N. Strecker, and M. B. Santos, "XPS study of the PbSe/CaF$_2$ (111) interface grown on Si by MBE," *Mat. Res. Soc. Symp. Proc.* **450**, 457 (1997).
152. P. J. McCann, X. M. Fang, W. K. Liu, B. N. Strecker, and M. B. Santos, "MBE growth of PbSe/CaF$_2$/Si(111) heterostructures," *J. Cryst. Growth* **175/176**, 1057 (1997).
153. X. M. Fang, I-Na Chao, B. N. Strecker, P. J. McCann, S. Yuan, W. K. Liu, and M. B. Santos, "Molecular beam epitaxial growth of Bi$_2$Se$_3$- and Tl$_2$Se-doped PbSe and PbEuSe on CaF$_2$/Si(111)," *J. Vac. Sci. Technol.* B**16**, 1459 (1998).
154. X. M. Fang, I. Chao, B. N. Strecker, P. J. McCann, S. Yuan, W. K. Liu, and M. B. Santos, "MBE growth of PbEuSe on CaF$_2$/Si(111)," in S. C. Shen, D. Y. Tang, G. Z. Zheng, and G. Bauer, eds., Proc. Eighth Int. Conf. Narrow Gap Semiconductors, World Scientific, Singapore, 1998, p. 101.
155. H. Zogg, S. Blunier, A. Fach, C. Maissen, and P. Müller, "Thermal-mismatch-strain relaxation in epitaxial CaF$_2$, BaF$_2$/CaF$_2$, and PbSe/BaF$_2$/CaF$_2$ layers on Si(111) after many temperature cycles," *Phys. Rev.* B**15**, 10801 (1994).
156. B. N. Strecker, P. J. McCann, X. M. Fang, R. J. Hauenstein, M. O'Steen, and M. B. Johnson, "LPE growth of crack-free PbSe layers on (100)-oriented silicon using MBE-grown PbSe/BaF$_2$/CaF$_2$ buffer layers," *J. Electron. Mater.* **26**, 444 (1997).
157. H. K. Sachar, I. Chao, P. J. McCann, and X. M. Fang, "Growth and characterization of PbSe and Pb$_{1-x}$Sn$_x$Se on Si (100)," *J. Appl. Phys.* **85**, 7398 (1999).

158. H. K. Sachar, P. J. McCann, and X. M. Fang, "Strain relaxation in IV-VI semiconductor layers grown on silicon (100) substrates," *Mat. Res. Soc. Symp. Proc.* **505**, 185 (1998).
159. C. P. Li, P. J. McCann, and X. M. Fang, "Strain relaxation in PbSnSe and PbSe/PbSnSe layers grown by liquid phase epitaxy on (100)-Oriented silicon," *J. Cryst. Growth* **208**, 423 (2000).
160. H. Z. Wu, X. M. Fang, R. Salas, D. McAlister, and P. J. McCann, "Transfer of PbSe/PbEuSe Epilayers grown by MBE on BaF_2-coated Si (111)," *Thin Solid Films* **352**, 277 (1999).
161. H. Z. Wu, X. M. Fang, D. McAlister, R. Salas, Jr., and P. J. McCann, "Molecular beam epitaxy growth of PbSe on BaF_2-coated Si (111) and observation of the PbSe growth interface," *J. Vac. Sci. Technol.* B**17**, 1263 (1999).
162. D. W. McAlister, P. J. McCann, H. Z. Wu and X. M. Fang, "Fabrication of thin film cleaved cavities using a bonding and cleaving fixture," *IEEE Photon. Technol. Lett.* **12**, 22 (2000).

CHAPTER 5

InP and GaAs-based Quantum Cascade Lasers

JÉRÔME FAIST and CARLO SIRTORI

5.1 INTRODUCTION

5.1.1 Quantum Engineering

In quantum engineering of electronic energy states and wavefunctions, ultrathin layers (0.5–100 nm) of semiconductor compounds used with different compositions allow the design of quantum phenomena that are typically observed in atomic structures [1]. This approach is the basis for modifying, in a unique way, the optical and transport properties of semiconductors, opening avenues to artificial materials and the creation of useful devices. The quantum cascade (QC) laser is one the best examples of quantum devices where fundamental properties such as the emission wavelength are not an intrinsic property of the semiconductor but a result of the *design* of the epitaxial layers. Similarly population inversion is not caused by some intrinsic physical property of the system but must be designed by a suitable engineering of the wavefunctions.

5.1.2 Organization of the Chapter

This chapter is divided as follows: In Section 5.2, the key features of the QC laser are discussed. In Section 5.3, the fundamentals of the population inversion between subbands are derived and a general description of the three-quantum-well active region is given. Waveguide and technology are discussed in Section 5.4. The high-power, room-temperature performance of this design is discussed in Section 5.5. In Section 5.6, the development of QC lasers based on GaAs/AlGaAs is discussed along with the new waveguide associated

---

Long-Wavelength Infrared Semiconductor Lasers, Edited by Hong K. Choi
ISBN 0-471-39200-6   Copyright © 2004 John Wiley & Sons, Inc.

with this material system. Section 5.7 discusses the relation between the conduction-band discontinuity and performances in a wide wavelength range. The spectral characteristics of QC lasers are discussed in Section 5.8. In Section 5.9, we discuss the nanofabrication of QC lasers in order to decrease their threshold current and control their emission properties. Section 5.10 discusses the implementation of distributed feedback (DFB) grating in QC lasers. These proved to be essential for obtaining the single-mode operation needed for gas spectroscopy. Finally, in Section 5.11, we discuss some of the outlooks for active region designs, base materials and operating wavelengths.

5.2 QUANTUM CASCADE LASER FUNDAMENTALS

5.2.1 History

The idea of a unipolar laser based on intersubband transitions in a semiconductor heterosctructure can be traced to the seminal work of R. F. Kasarinov and R. A. Suris in 1972 [2,3], who gave a theoretical treatment of the light amplification in a superlattice under a strong applied electric field. On the experimental side, the advent of molecular beam epitaxy (MBE) in the 1970s allowed for the first time the fabrication of heterostructures based on III–V compounds with very sharp interfaces and excellent compositional control, and created a great interest in the study of multi-quantum-well structures [4], especially for their *interband* transition properties. In 1985, there was the first observation of intersubband absorption in a GaAs/AlGaAs multi-quantum well [5]. Early attempts were made to implement experimentally the proposal of Kasarinov and Suris in a GaAs/AlGaAs superlattice. They led to the observation of intersubband luminescence pumped by resonant tunneling by M. Helm in 1989 [6]. Kasarinov's paper spurred large theoretical activities, which yielded a large number of proposals for intersubband lasers. They were based on resonant tunneling in superlattices [7] or superlattices with two quantum-well thicknesses [8]. Later they included also resonant tunneling structures with injection in the first excited state [9,10,11], injection in a superlattice miniband [12] or even above-barrier Bragg-confined states [13].

The first successful intersubband semiconductor laser was achieved in 1994 in Bell Laboratories using InGnAs/InAlAs quantum wells on InP substrates (or InP-based quantum wells) [14]. In the following years the Bell Labs group made a strong effort to improve the performance and illustrated the potential of these devices as a revolutionary light source for molecular spectroscopy [15,16,17]. Many important milestones for semiconductor lasers with emission wavelength in the mid infrared (3–15μm), such as room-temperature operation and high cw output power ($\approx 1\,\text{W}$) at cryogenic temperatures, were demonstrated by this group in the following years [18,19,20,21].

Nevertheless, the development of QC laser was carried out in other research groups. At Northwestern University, gas-source MBE was successfully used to

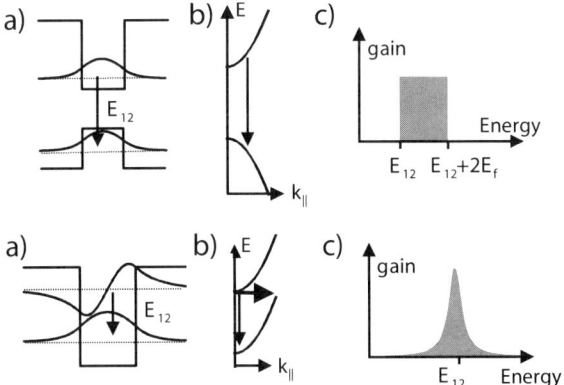

Fig. 5.1 Interband (*top*) versus intersubband (*bottom*) transitions. (*a*) real-space drawing of the potential profile, (*b*) dispersion of the electronic subbands in the plane and (*c*) joint density of state.

grow QC lasers with high performances [22,23]. The same growth technique was also used by a group from the Shanghai Insitute of Metallurgy [24].

Groups at Thomson-CSF, TU-Vienna, University of Sheffield, and University of Glasgow worked on the development of GaAs-based QC lasers. The achievement of the first GaAs-based QC laser [25] was the proof that QC lasers concepts could be developed in different material systems. This development will be discussed further in this chapter.

5.2.2 Unipolarity and Cascading

The intersubband QC laser differs in many fundamental ways from diode lasers. All of the differences are consequences of two main features that are unique to QC lasers and distinguish them from conventional semiconductor light emitters: *unipolarity* (electrons only) and a *cascading scheme* (electron recycling). These two features, shown schematically in Fig. 5.1 and Fig. 5.2, are independent, and can be used separately as it has been already demonstrated experimentally. In 1997, Garcia et al. [26] demonstrated a cascade interband laser at nearly 830nm, and Gmachl et al. [27] in 1998 demonstrated an intersubband laser with a one-period active region. Recent applications of the cascading scheme to near-IR vertical cavity surface-emitting lasers are very promising [28,29].

5.2.3 Intersubband Transitions

The unipolarity in QC lasers is a consequence of the nature of the optical transitions that occur between conduction-band states (subbands) arising from size quantization in semiconductor heterostructures. These transitions are commonly denoted as intersubband transitions. As shown in Fig. 5.1, their initial and final states are in the conduction band, and therefore they have the same

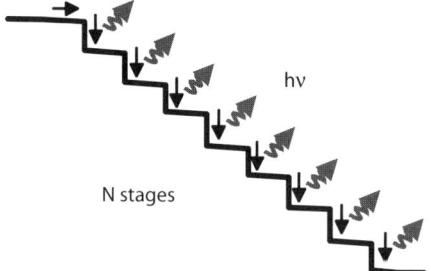

Fig. 5.2 Operating principle of a cascade structure: An electron cascade down an potential staircaise, emitting photons at each step.

curvature in the reciprocal space. If one neglects nonparabolicity, the joint density of states is very sharp and similar to the one of atomic transitions. In contrast to interband transitions, the gain linewidth is now indirectly dependent on temperature through collision processes and many-body effects. Moreover, for these devices the emission wavelength is not dependent on the band gap of constituent materials but can be tuned over a wide range ($\lambda = 4.28$ to $24\,\mu m$ using InGaAs/AlInAs lattice-matched materials) on InP by tailoring the layer thicknesses.

The largest achievable photon energy is ultimately set by the conduction-band discontinuity of the materials, while on the long-wavelength side there are no fundamental limits preventing the fabrication of QC devices emitting in the far infrared. On the other hand, semiconductor diode lasers, including quantum-well lasers, rely on transitions between energy bands in which conduction-band electrons and valence-band holes are injected into the active layer through a forward-biased p-n junction and radiatively recombine across the material band gap [30]. As shown in Fig. 5.1, the latter essentially determines the emission wavelength. As shown schematically in Fig. 5.2, the other fundamental feature of QC lasers is the multi-stage cascade scheme, where electrons are recycled from period to period, contributing each time to the gain and the photon emission. Thus, in an ideal case, each electron injected above threshold generates N_p laser photons where N_p is the number of stages, leading to a differential efficiency and therefore an optical power proportional to N_p [20].

5.3 FUNDAMENTALS OF THE THREE-QUANTUM-WELL ACTIVE-REGION DEVICE

An essential feature of the QC laser is its architecture. As shown in Fig. 5.3, each stage comprises an active region followed by an injection/relaxation region. The active region comprises a ladder of electronic states engineered such as to maintain population inversion between two levels exhibiting a large oscillator strength. We will discuss here the implementation of a ladder of

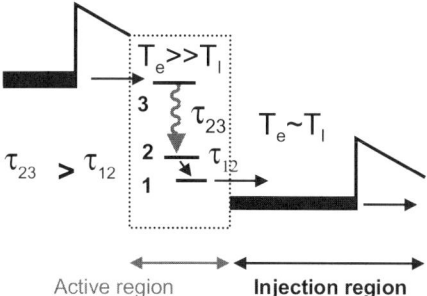

Fig. 5.3 Design principle of a quantum cascade laser structure. Each repeat of a unit cell period consists of an active region followed by an injection/relaxation region.

three states. Assuming a simplified model with no nonparabolicity (atomic picture) and a 100% injection efficiency in the state $n = 3$, the population inversion condition is simply $\tau_{32} > \tau_2$, where τ_{32}^{-1} is the nonradiative scattering rate from level 3 to level 2 and τ_2^{-1} is the total rate out of level $n = 2$. In coupled-well structures, the nonradiative channel $3 \to 1$ is usually not negligible, and therefore the condition above on the lifetimes is acutally less stringent than the condition $\tau_3 > \tau_2$ between the total lifetimes of states $n = 3$ and $n = 2$.

5.3.1 Active Region

This ladder of electronic states can be obtained in a variety of multiple quantum-well structures based on two or three quantum wells. We will distinguish these structures by the character of the 3 to 2 (laser) transition. The transition is said to be *vertical* when the two wavefunctions of states $n = 3$ and $n = 2$ have a large overlap, and *diagonal* when this overlap is significantly reduced. The implementation of such an active region based on a *diagonal* transition and where the population inversion is maintained between the second and third state is shown in Fig. 5.4 for two devices emitting at about the same wavelength ($\lambda \approx 9\text{–}10\,\mu\text{m}$) but based on either the InGaAs/AlInAs or the GaAs/AlGaAs materials.

Discrete electronic states are created by quantum confinement in the direction of the growth. Parallel to the layers, however, these states have plane-wave-like energy dispersion. Such a dispersion for a three-level structure is depicted in Fig. 5.5. Under the proper applied bias, the structure is designed so that the energy separation between the $n = 1$ and $n = 2$ subbands is equal to an optical phonon energy (34 meV in InGaAs, 36 meV in GaAs). The optical phonon transition rate is therefore resonantly enhanced, and the scattering time from the $n = 2$ level drops to subpicosecond values ($\tau_2 \cong 0.2\,\text{ps}$). On the other hand, the optical phonon scattering between the $n = 3$ and $n = 2$ subbands leads to longer lifetimes because it is associated with a much larger wavevector transfer. The population inversion condition is therefore satisfied even for long-wavelength lasers [31].

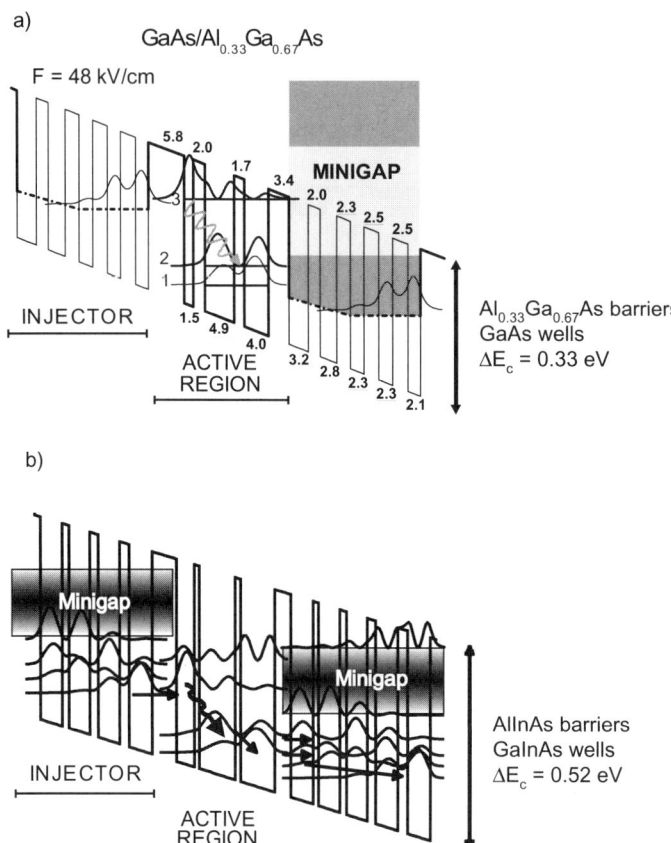

Fig. 5.4 Schematic band structure of a quantum cascade laser based on a three-quantum-well active region, for two different matierals. (*a*) GaAs/Al$_{0.33}$Ga$_{0.67}$As lattice-matched on a GaAs substrate. (*b*) AlInAs/GaInAs lattice-matched on an InP substrate.

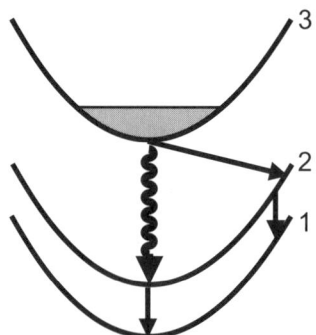

Fig. 5.5 Schematic representation of the dispersion of the $n = 1, 2$, and 3 states parallel to the layers; $k_\parallel$ is the corresponding wavenumber. The straight arrows represent the intersubband optical-phonon processes, and the wavy arrow the radiative transitions.

By varying the thickness of the coupling barrier separating the first two quantum wells of the active region, as well as the width of the first quantum well (counting from the injection barrier), the magnitude of the overlap between the wavefunctions of the $n = 3$ and $n = 2$ states can be strongly varied, and the character of the transition changed from *vertical* to *diagonal*.

For very thin barriers and the first quantum well, the wavefunction of the $n = 3$ state is strongly localized in the last two quantum wells, and the overlap between the wavefunctions of the $n = 3$ and $n = 2$ states is maximized. In this regime the transition has a *vertical* character.

A vertical transition was adopted for the short-wavelength ($\lambda = 3$–$5\,\mu m$) QC devices [32,33,34]. Its advantage is the maximization of the oscillator strength between the two states and the minimization of the broadening of the intersubband luminescence. However, the disadvantage of such a design is that the $n = 3$ to $n = 2$ lifetime decreases as the emission wavelength increases, from $\tau_{32} \cong 2.1\,ps$ at $\lambda \cong 5\,\mu m$ to $\tau_{32} \cong 1.3\,ps$ at $\lambda \cong 11\,\mu m$ (at $T = 0\,K$), leading to an unfavorable ratio of lifetimes between the upper and lower states of the laser transition at long wavelengths.

On the other hand, the lifetime of the $n = 3$ state can be significantly increased by a thick barrier (at the cost of a reduced oscillator strength between the two states) and the population inversion further enhanced. This approach was used very early at both short ($\lambda = 4.3\,\mu m$ [14]) and long ($\lambda = 8.4\,\mu m$ [35]) wavelengths. For a long-wavelength laser operating at $\lambda = 11\,\mu m$ and using a diagonal transition, the resulting lifetime between level 3 and 2 is $\tau_{32} = 2\,ps$ [36]. While this is accompanied by a concomitant decrease in optical matrix element to $z_{32} = 2.4\,nm$ (instead of $z_{32} = 3.1\,nm$ for lasers based on vertical transitions [31]), threshold current densities and slope efficiencies benefit from a more favorable ratio of population between states 3 and 2 given by the ratio of lifetimes $\tau_{32}/\tau_2 = 5$, instead of $\tau_{32}/\tau_2 = 3.4$ for the vertical device.

As compared to three-level active region based on a coupled well [37,35], the three-well active-region also improves the injection efficiency by reducing the overlap between the ground state wavefunction of the injection/relaxation region and the states $n = 1$ and $n = 2$ of the active region, while maintaining a good injection efficiency into state 3. This feature is very important to maintain a high-injection efficiency even as the temperature is raised. We believe it is the main reason why high-performance room-temperature operation was obtained in three-quantum-well active region devices [18,38,36,39], even if room-temperature operation has also been demonstrated in a double-quantum-well active region device [19].

5.3.2 Doping and Injection/Relaxation Region

It is usually beneficial to leave the active region not intentionally doped, since, as we have shown experimentally, the presence of dopants in the active region significantly broadens the laser transition [40]. The relaxation/injection region is the section of the period where the electrons cool down to the lattice temperature and are re-injected in the next period. Because this region provides the electron population that will be injected, it has to be doped to prevent the strong space-charge build-up if the electron population was injected from the contact. In the most recent structure, this doped region was restricted to the center of the injection region in order to have a modulation doping of the ground state of the injector.

In all of these designs the injection/relaxation region includes a minigap facing level 3 to prevent electron escape from the latter into the injector and with a miniband facing levels $n = 2$ and $n = 1$ to enhance the extraction from lower levels of the active region. Moreover it was found beneficial to have the width of the lower miniband decrease toward the injection barrier in order to "funnel" electrons into the ground state of the relaxation/injection region.

5.3.3 Threshold Current Density

From a simple rate equation analysis, one derives easily the threshold current density of a QC laser based on the generic three level arrangement shown in Fig. 5.3:

$$J_{th} = \frac{1}{\tau_3(1-\tau_2/\tau_{32})}\left[\frac{\varepsilon_0 n L_p \lambda(2\gamma_{32})}{4\pi q_0 \Gamma_p N_p z_{32}^2}(\alpha_m + \alpha_w) + q_0 n_2^{therm}\right]. \quad (5.1)$$

In Eq. (5.1) α_w is the waveguide losses and α_m the mirror losses, given by the expression $\alpha_m = -\ln(R)/L_{cav}$, where R is the (intensity) facet reflectivity and L_{cav} the laser cavity length. z_{32} is the optical dipole matrix element, λ is the emission wavelength, L_p is the length of each stage, Γ_p is the overlap between the mode and one period of the structure, N_p is the number of repeat periods, n is the mode refractive index, ε_0 the vacuum permittivity, and q_0 the electron charge.

To describe the finite temperature behavior, we added a term due to the thermal population n_2^{therm} of the state $n = 2$, which can be approximated by an activated behavior $n_2^{therm} = n_g \exp(-\Delta/kT)$, where n_g is the sheet doping density of the injector and Δ the energy difference between the Fermi level of the injector and the $n = 2$ state of the laser transition. A slightly more accurate model is obtained by computing the electron distribution of the $n = 2$ state directly as a function of (electron) temperature using the Fermi distribution of the electron in the injector and a step two-dimensional density of states. However, measurements [41] and Monte Carlo simulations [42] have suggested that depending on the band structure and the drive conditions, the electron distribution and temperature might depart significantly from thermal equilibrium with the lattice. To model the high-temperature behavior, we also assume that the lifetimes τ_i are limited by optical phonon emission and decrease with temperature according to the Bose-Einstein factor for stimulated absorption and emission of optical phonons:

$$\tau_i^{-1}(T) = \tau_i^{-1}(0)\left(1 + \frac{2}{\exp(h\omega_{LO}/kT) - 1}\right), \quad (5.2)$$

where $h\omega_{LO}$ is the phonon energy.

In Eq. (5.1) two quantities are difficult to compute. The waveguide loss can be computed assuming optical losses dominated by free carrier absorption, computed using a Drude conductivity model. Although a good agreement was found in other experiments [43,44] (see also data on GaAs QC laser further in the text), this approach sometimes leads to values lower by a factor of 2 to 3 than the one obtained by the technique based on the analysis of the sub-threshold luminescence (Hakki-Paoli) [45,46]. This issue is still not resolved and is the subject of intense research. A technique based on an analysis of multisection devices also demonstrated good agreement with the waveguide loss computed by a Drude conductivity model [47].

It is now accepted that the linewidth of intersubband transitions in the mid-infrared is limited, to a major extent, by interface roughness [48]. As a consequence the broadening parameter $2\gamma_{32}$ entering Eq. (5.1) cannot be computed, and is assumed to have a value given by the experimental FWHM of the luminescence measured on devices processed without an optical cavity as to suppress amplification of the spontaneous emission.

Using the same simple rate equation approach, the slope efficiency (in mW/A) of the laser can be derived:

$$\frac{dP}{dI} = N_p \frac{h\nu}{q_0} \frac{\alpha_m}{\alpha_m + \alpha_w} \frac{\tau_{32} - \tau_2}{(\tau_{32} + \tau_{32}\tau_2/\tau_3 - \tau_2)}, \quad (5.3)$$

where h is Plank's constant and v is the laser operating frequency. An analysis of Eq. (5.1) and (5.3) leads to the conclusion that a low-threshold, high-efficiency laser will be obtained in a structure that achieves the following:

- A large ratio of upper-state to lower-state lifetimes τ_{32}/τ_2
- A low waveguide loss α_w
- Narrow transition linewidth γ_{32}
- Long upper state lifetime τ_3

The multiple QC designs that have appeared in the literature simply reflect the complicated trade-off that have to be made when trying to fulfill these conflicting requirements.

In the rate equations described above, we assumed implicitly that the injection efficiency into state 3 is unity, that is, each electron leaving the injector is going into state $n = 3$. This assumption may be wrong, for example, in the presence of a strong elastic or inelastic scattering, which may scatter this electrons directly to state $n = 2$, or when the current is larger than the resonant tunneling point maximum and the injector is not anymore well aligned energetically to state $n = 3$. In the latter case we have to assume that only a fraction of the electron η_{in} is injected into the state $n = 3$, the remaining fraction $(1 - \eta_{in})$ being injected into states $n = 2$ and $n = 1$. Such processes (i.e., that yield $\eta_{in} < 1$) are strongly detrimental to the laser performances by strongly increasing the threshold current density and decreasing the slope efficiency. In practice, the difficulty of taking into account such nonideal injection condition comes from the poor knowledge of η_{in}, since both theoretical computation or direct mesurement are extremely cumbersome.

5.3.4 Effect of Cascading on the Performances of QC Lasers

In QC lasers the threshold current density and the slope efficiency (dP/dI) are a function of the number of periods in the structure. This is a direct consequence of the cascade scheme and can be simply put in evidence in our formalism:

Threshold current density. The total overlap factor between the optical mode and the gain region, Γ, is a function of the number of periods of the structure,

$$\Gamma = \sum_p^{N_p} \Gamma_p. \qquad (5.4)$$

This expression can be simplified to $\Gamma = \Gamma_p N_p$, when Γ covers only a central part of the optical mode, namely when the Γ_p are almost identical. The approximation is valid in practice in all the cases with the exception of the waveguides based on surface plasmon modes [49]. In the general case it becomes apparent that the threshold current density (Eq. 5.1) is inversionally proportional to the number of periods and can be written as

$$J_{th} = \frac{\alpha_{tot}}{g N_p \Gamma_p}. \qquad (5.5)$$

The reduction of the current density relies simply on a geometrical increase of the gain region interacting with the optical mode. In the cascade scheme the laser active regions are electrically pumped in a series configuration, and the amount of carriers injected per period therefore does not vary with their number. Adding more period does not affect the population inversion in each single stage, but it increases the voltage, as it will be explained later. This situation is very different from that of conventional diode lasers, where by increasing the number of quantum wells in the active region, one has also to increase the current to reach threshold.

Slope efficiency. The slope efficiency is the quantity that most directly benefits from the cascade scheme. In the case of an ideal QC laser, an electron that crosses one stage of the structure is "recycled" in the next adjacent stage. Above threshold, each electron injected into the active region of the device creates N_p emitted photons. This makes the slope efficiency of a QC lasers directly proportional to N_p, typically around 30 to 50 times higher than what would be obtained without the cascade scheme, as shown directly in Eq. (5.3).

As is clear from Eqs. (5.1) and (5.3), N_p acts favorably by decreasing the threshold and increasing the slope efficiency. The price to pay is on the voltage applied to the structure, which is also directly proportional to the number of periods. Nevertheless, it is easy to demonstrate that the contributions to the voltage coming from parasitic resistances are inversely proportional to N_p. This is a net advantage that makes the cascade scheme very attractive as a mean for increasing the wall plug efficiency of this devices.

The low-temperature dependence of the threshold current density and the slope efficiency are studied in detail in Ref. [50].

5.4 WAVEGUIDE AND TECHNOLOGY

5.4.1 Waveguide

In order to obtain laser action, the active region must be inserted in an optical waveguide. Waveguide design of semiconductor lasers emitting in the mid- to far-infrared (4–20 µm) has to contend with much higher internal cavity losses and much thicker cladding layers than near-IR diode lasers [21,49,51].

For wavelengths larger than 10 µm, the cladding thickness is at the limit of epitaxial growth. Optical losses can reach very high values (>100 cm^{-1}) and are at present one of the major limiting factors on the performance of mid-IR lasers [21,49]. This problem is more severe for narrow-gap semiconductor lasers [51] than for QC lasers. For the latter, α_w is predominantly controlled by free-carrier absorption, whereas for the former, there is also a strong contribution from intraband absorption associated with subband resonances, especially in the valence band [51]. Both of these effects increase with the carrier concentration, and are therefore very sensitive to the thermally activated population across the band gap. This is a fundamental advantage of QC lasers, which are based on wide band-gap semiconductors where the thermally activated population at room temperature is negligible. This is obviously not the case for lasers based on narrow-gap semiconductors, for which the thermal population at room temperature can be a significant fraction of the carriers at threshold. This important difference on the temperature dependence of the waveguide losses is a major asset of the QC laser technology.

An optimum QC laser waveguide must fulfill the requirements of low optical losses for a TM propagating mode, as required by the optical selection rules for intersubband tranistions. At the same time a minimum total thickness is very important to optimize the thermal transport across the device, and to minimize the growth time and the number of defects.

TABLE 5.1 Waveguide layer structure.

Material	Thickness (nm)	Doping (cm^{-3})
GaInAs	1050	7×10^{18}
AlInAs	1050	1.5×10^{17}
GaInAs	800	6×10^{16}
35x Active	1700	4×10^{16}
GaInAs	700	6×10^{16}
InP		$1–3 \times 10^{17}$

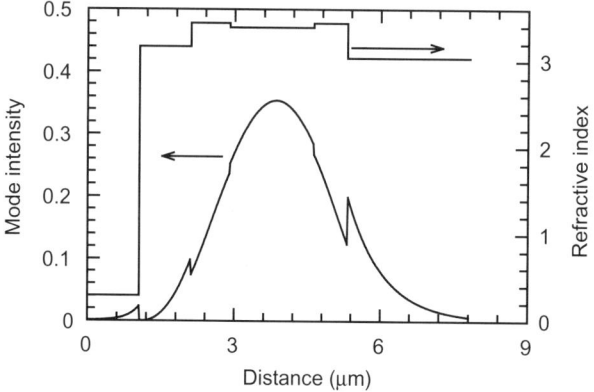

Fig. 5.6 Calculated refractive index and mode profile of a waveguide for a $\lambda = 10\,\mu m$ device.

Waveguide technologies are very dependent on the base material chosen. For InGaAs/AlInAs devices grown on InP substrates, the latter can be used as a cladding material because its lower refractive index. This approach is not possible for devices based on GaAs substrates, and other approaches must be developed.

The complete layer sequence of a waveguide for a laser operating at $\lambda \approx 10\,\mu m$ is shown in Table 5.1; the corresponding refractive index profile and calculated mode profile are shown in Fig. 5.6. The mode profile, effective index and optical loss are computed by solving the wave equation for a planar waveguide with a complex propagation constant, modeling each layer with its complex refractive index. On both sides of the 35-period active region/injector region, the waveguide is comprised of two 800-nm- and 700-nm-thick GaInAs guiding layers that enhance the optical confinement by increasing the average refractive index difference between the core and cladding regions of the waveguide. The bottom cladding is the InP substrate. The top cladding consists of a 1050-nm-thick AlInAs layer followed by a 1050-nm-thick heavily doped GaInAs cladding region. The purpose of this heavily doped GaInAs layer is to decouple the lossy ($\alpha = 140\,cm^{-1}$) metal contact—semiconductor interface plasmon mode from the laser mode by enhancing the difference of the effective refractive indexes of the two modes.

In all of our waveguide designs, the transitions between the low and high refractive index materials are graded (either with compositional or digital graded region) over a distance of about 30 nm. This grading has the purpose of reducing the series resistance by forming a smooth band profile, preventing the formation of a barrier between the cladding layers. Another important feature of the design is to minimize the optical losses due to free carriers. This is obtained by reducing to its minimum value the doping level around the waveguide region while maintaining low resistivity.

5.4.2 Processing

Typically the samples were processed into mesa-etched wide waveguides by wet etching through the active region to about 1-µm deep into the substrate. An insulating layer is then grown by chemical vapor deposition to provide insulation between the contact pads and the doped InP substrate. For minimum losses, SiO_2 is chosen for short wavelengths ($\lambda < 5\,\mu m$) and long wavelengths ($\lambda = 11\,\mu m$), and Si_3N_4 is chosen for the intermediate ones. Windows are etched through the insulating layer by plasma etching, exposing the top of the mesa. Ti/Au nonalloyed ohmic contacts are provided to the top layer and the substrate. The devices are then cleaved in 0.5- to 3-mm-long bars, soldered to a copper holder, wire-bonded and mounted either in the cold head of a temperature-controlled He flow cryostat or onto a Peltier-cooled laser head.

5.5 HIGH-POWER, ROOM-TEMPERATURE OPERATION OF THREE-QUANTUM-WELL ACTIVE REGION DESIGNS

5.5.1 High Power at Room Temperature

High-power, room-temperature operation was demonstrated for the first time in a three-quantum-well active region structure with a vertical transition operating at $\lambda \cong 5\,\mu m$ [18]. The same design also demonstrated high-performance operation at shorter ($\lambda \cong 3.5\,\mu m$ [33]) and longer wavelengths in the second atmospheric window ($\lambda \cong 8.5\,\mu m$ [38,23] and $\lambda \cong 11\,\mu m$ [36]). This design was the basis for the first successful demonstration of a GaAs-based QC laser [25]. The flexibility and the high performances achieved over a very wide wavelength range by this design makes it an excellent benchmark against which other QC laser designs can be compared.

In this section we describe QC lasers based on a three-quantum-well active region that operate at about 10 µm. The schematic conduction-band profile with the relevant wavefunction is displayed in Fig. 5.4. The structure is grown by MBE using InGaAs and AlInAs alloys lattice matched on an InP substrate, and consists of a 35-period active region embedded into the optical waveguide described in the previous section, and for which the refractive index profile is displayed in Fig. 5.6.

For the electrical and optical testing, the lasers are mounted on the cold finger of a He-cooled, temperature-controlled cryostat. The device is driven with 1 00-ns-long pulses at a 5-kHz repetition rate. The light is collected and colli-

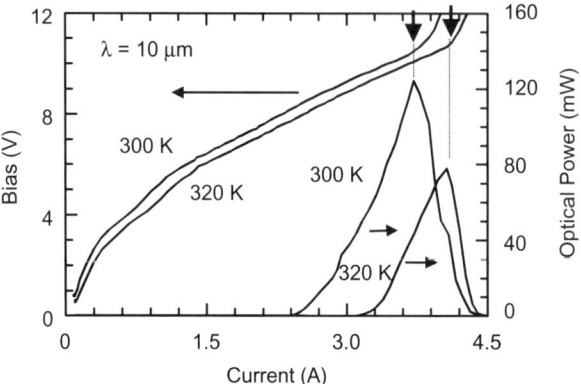

Fig. 5.7 Voltage (*left curve*) and peak optical power measured at 300 and 320 K displayed as a function of injected current. The pulse duration is 100 ns and the repetition rate 5 kHz. The abrupt increase in the voltage and the concomitants decrease in optical power are due to a resonant tunneling effect.

mated by a high-numerical-aperture (f/0.8) aspheric Ge lens. The collimated beam is then focused on a fast (subnanosecond rise time), room-temperature, calibrated HgCdTe detector. This optical arrangement is necessary to obtain high optical collection efficiency (60–80% of the light beam exiting from the facet of the device) owing to the large divergence, especially in the plane perpendicular to the layers, of the light beam. The peak current, voltage, and light intensity are measured by a digital oscilloscope and a boxcar averager. The calibration of the responsivity of the HgCdTe detector is an important step in order to obtain a reliable power readings from the measurement. This calibration was performed with an FTIR for the relative spectral response followed by a thermopile calibration at a reference wavelength.

Figure 5.7 shows the peak optical power collected from a single facet versus drive current for temperatures of $T = 300$ K and 320 K. The average slope efficiency is $dP/dI = 100$ mW/A at $T = 300$ K, with a maximum power of 90 mW at this temperature. This large slope efficiency is mainly a consequence of the cascaded geometry, which allows electrons to be recycled from period to period. The current is limited by resonant tunneling injection, as shown by the abrupt increase in voltage from 10 to more than 12 V at the maximum injection current (3.2 A) accompanied by a concomitant abrupt decrease in the optical output power. This behavior has been extensively discussed in a previous work [52]. It originates from an abrut decrease in the injection efficiency as the injector ground state becomes misaligned with the $n = 3$ state of the laser transition. The transition occurs in a very narrow range of drive current and is an attractive (although never tried, to our knowledge) means of modulating the laser at a very high speed. Such modulation experiment has already been performed on the *linear* part of the light versus current characteristics,

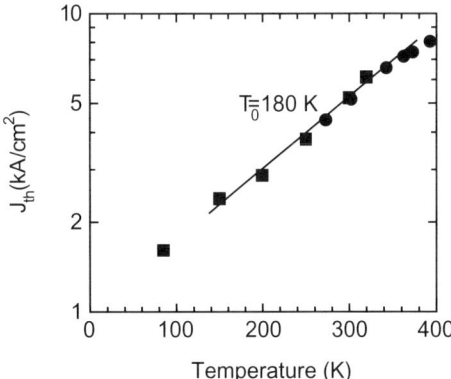

Fig. 5.8 Threshold current density as a function of temperature for the three-quantum-well active-region QC laser. The weak temperature dependence of the threshold current density is a very attractive feature of QC lasers.

demonstrating that pulses as short as 200 ps can be produced in a cryogenically cooled devices [53].

In Fig. 5.8 the threshold current density J_{th} is plotted as a function of temperature. It has a value of only 5.2 kA/cm² at 300 K. The data between $T \approx$ 200 K and 320 K can be described by the usual exponential phenomenological behavior $J_{th} = J_0 \exp(T/T_0)$ with an average $T_0 \approx 170$ K. This value of T_0 is much larger than the value measured typically in interband lasers operating in the mid-infrared ($T_0 \approx 20$–50 K) and is comparable to high-performance near-infrared interband lasers. The weak temperature dependence of the QC laser threshold current density can be ascribed to the following fundamental reasons:

- The material gain is insensitive to the thermal broadening of the electron distribution in the excited state, since the two subbands of the laser transition are nearly parallel.
- Auger intersubband recombination rates are negligible compared to the optical phonon scattering rates.
- The variation of the excited state lifetime with temperature is small, being controlled by the Bose-Einstein factor for optical phonons.
- The measured luminescence linewidth is weakly temperature dependent.

These points are well illustrated in Fig. 5.9 where the electroluminescence of this three-quantum-well active-region sample is plotted as a function of temperature. For these experiments, the sample was processed into a square (180 μm side mesa) and the light was extracted through a polished 45-degree wedge in the substrate. As shown in Fig. 5.9 and for the reasons mentioned above, neither the luminescence efficiency nor its linewidth varies by more than a factor of two between cryogenic and room temperature.

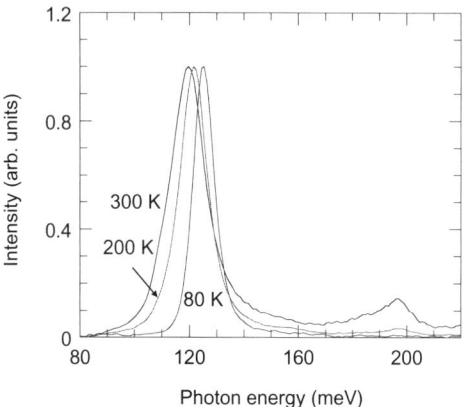

Fig. 5.9 Luminescence spectra for the three-quantum-well active-region design for various temperatures, as indicated.

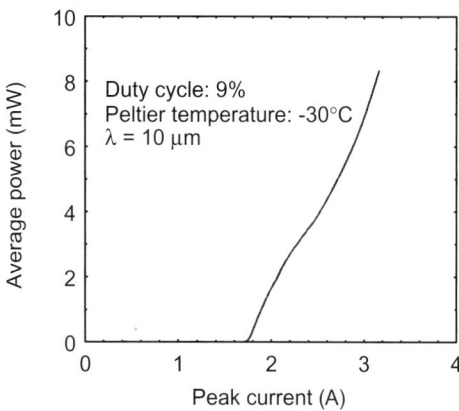

Fig. 5.10 Average power as a function of peak injected current for a three-quantum-well device operated on a Peltier-cooled laser housing.

5.5.2 High Room-Temperature Average Power

The high peak powers reported in Fig. 5.7 translate into large average powers near room temperature. In Fig. 5.10 the average optical power as a function of injected current is reported for a device operated at a fairly high duty cycle of 9%. In these experiments the device was mounted in a commercial temperature-controlled Peltier-cooled laser housing fitted with a low-impedance line. The average power was measured directly with a thermopile powermeter. Average powers up to 10 mW were obtained in this experimental arrangement. These high output powers make this device attractive for spectroscopy techniques such as photoacoustic spectroscopy that require fairly large amount of optical power.

Optical power in the 2 to 10 mW range around room temperature was also achieved in devices operating at λ ≈ 5 μm [18]. Even larger average powers of 14 mW were achieved using a chirped superlattice active region structure by Tredicucci and coworkers [54].

5.5.3 Continuous-Wave Operation

QC laser devices are only operated in pulsed mode at room temperature because the heat dissipation in the active region (about 100 kW/cm^2) is too large to be removed in a continuous manner. As the temperature is lowered, however, the concomitant decrease in threshold current density and increase in heat conductivity of the claddings enable continuous-wave operation of most of the QC lasers at cryogenic (typically 20–150 K) temperatures. Continuous-wave performances depend critically on the heat dissipation through the device. For this reason, for devices mounted junction up and for a fixed threshold current density, devices operating with narrow waveguide stripes will perform better than wider stripe devices simply because of a lower heat dissipation.

Because the stripe width scales with wavelength, it is therefore easier to operate shorter wavelength QC lasers in continuous wave. High optical power and cw operation of three-quantum-well devices operating at λ ≈ 5 μm have been obtained and are shown in Fig. 5.11. These devices displayed extremely large optical powers [20]. Figure 5.11 shows the optical power versus driving current from a single facet of one such laser. In this measurement the light is collected by a parabolic aluminum cone acting as a nonimaging energy concentrator with near unity collection efficiency, and the intensity is measured with a broad-band laser power meter.

A maximum power of more than 700 mW is achieved at 22 K for a drive current of about 2.1 A, corresponding to a current density $J \approx 6.2$ kA/cm^2. The slope efficiency dP/dI in the single mode portion of the L-I is 582 mW/A, with a lasing threshold of approximately 1.3 kA/cm^2. This large value of slope efficiency is a direct consequence of the cascade geometry of the structure, which allows more than one photon emitted per electron above threshold. The value of the slope efficiency agrees very well with the value predicted by Eq. (5.3), 624 mW/A, using the measured value of the waveguide loss $\alpha = 13.8$ cm^{-1} [20]. The "wall plug" efficiency (conversion efficiency from electrical power to optical power) is plotted in the inset of Fig. 5.11 as a function of injected current: a maximum of about 8.5% is obtained at around 1.7 A. A clear feature in the L-I characteristic emerges around 1.2 A (400 mW) The spectral analysis of the emission shows that it is related to the laser switching from single-mode to multimode operation.

5.6 GaAs-BASED QC LASERS

5.6.1 Active Region Design

In general, the active region of the GaAs-based QC lasers follows the same design rules as the one based on InGaAs/AlInAs. Most of the devices demon-

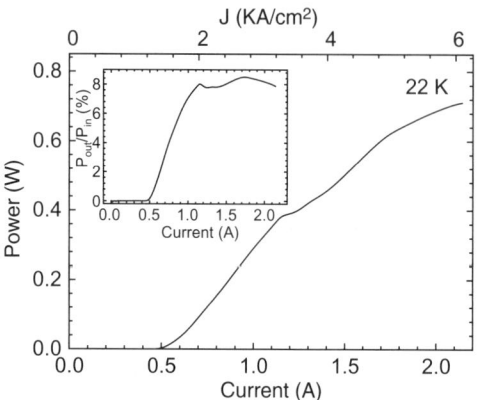

Fig. 5.11 Continuous-wave power versus injected current for a 5 μm QC laser at cryogenic temperature (20 K). Very high optical power (700 mW) and slope efficiency (560 mW/A) were achieved because the architecture of the active region is cascaded.

strated so far [25,44] followed the three-quantum-well architecture described in Section 5.3 and are shown in Fig. 5.4, although the first superlattice design has now been reported [55]. As shown in Fig. 5.4, the smaller conduction-band discontinuity of $GaAs/Al_{0.33}Ga_{0.67}As$ ($\Delta E_c = 0.33\,eV$) compared to lattice-matched InGaAs/AlInAs ($\Delta E_c = 0.52\,eV$) means that for a given emission wavelength, the upper state of the laser transition will be closer to the discontinuity in GaAs-based QC laser. The physical implications of this material parameter will be discussed in more detail in the next section. In addition GaAs exhibits a larger effective mass ($m^*/m_0 = 0.067$) than InGaAs ($m^*/m_0 = 0.043$). This has the effect of increasing the optical phonon scattering rates by approximately

$$\sqrt{\frac{m^*(GaAs)}{m^*(InGaAs)}} = 1.2.$$

5.6.2 Waveguide Design

The higher refractive index of the GaAs substrate than that of AlGaAs has an important impact on the design of QC laser waveguide for this material, as the substrate cannot be used as a cladding layer. The first demonstrations of GaAs-based QC lasers had waveguides that were based on thick $Al_xGa_{a1-x}As$ claddings with a large Al content x.

However, in recent GaAs QC lasers the planar optical confinement relies on a novel Al-free design [44]. In these waveguides the variations of the refractive index are exclusively based on a change in the doping concentration in the GaAs rather than on alternating materials. Apart from the lower losses, which will be discussed in detail later, there are several advantages in fabricating mid-IR waveguides without AlGaAs cladding layers.

In AlGaAs layers, the ionized carrier concentrations have significant variations as a function of different environmental conditions (typically stray light illumination and temperature). These variations produce strong instabilities on device thresholds and slope efficiencies [56]. In addition the growth of thick, high-quality GaAs layers is much easier than AlGaAs layers, which are anyway limited to a maximum thickness of approximately 1.5 μm, due to residual strain. Also the ternary alloys such as AlGaAs have very poor thermal conductivity compared with the binary crystals. A simple calculation shows that by using GaAs cladding layers, the thermal resistance of the lasers is reduced by a factor of 2. Finally, there are no heterobarriers between the contacts and the active region, that may increase the differential resistance across the device.

Figure 5.12 presents all the relevant parameters and the resulting mode profile of the waveguide we are using. The active region is sandwiched between two thick layers (4.5 μm) of GaAs, and no AlGaAs is used as the cladding material. In our waveguides the decrease of the refractive index, needed for planar optical confinement, is achieved by increasing the doping concentration in the last micron of the GaAs layers (Fig. 5.12). The doping (6 × 10^{18} cm^{-3}) is chosen to shift the plasma frequency of the n^{++} layers close to that of the laser [44]. This strongly depresses the real part of the refractive index n (middle part of Fig. 5.12) but also increases its imaginary part k, thus the absorption coefficient of these layers (lower part of Fig. 5.12) [44].

To avoid very high waveguide losses, it is necessary to minimize the overlap factor between the n^{++} layers and the optical mode. This is readily achieved by separating these layers from the active region by two thick low-doped (4 × 10^{16} cm^{-3}) GaAs layers. The calculated overlap factor with the lossy cladding layers, Γ_{n++}, is only 0.008, but it accounts for 90% of the total waveguide losses. In fact, in these layers the absorption coefficient α_{n++} imposed by free carriers at around 9.5 μm wavelength is 1740 cm^{-1}, which multiplied by Γ_{n++} gives 14 cm^{-1}. Note that these waveguides are not based on a pseudometallic optical confinement, since the real part of the complex dielectric function of the heavily doped layers is still the dominant factor ($n \gg k$) [44].

The waveguide losses α_w in QC lasers are typically characterized using two independent methods: one based on a plot of the threshold current density versus reciprocal cavity length (1/L) and the other on an analysis of subthreshold emission spectra for different injection currents. The former method is based on the threshold condition which is a mere rewriting of Eq. (5.1):

$$J_{th} = \frac{\alpha_w + \alpha_m}{\Gamma_{ac} g_d} \qquad (5.6)$$

Recall that the mirror loss is written as $\alpha_m = -ln(R)/L$ and the overlap factor for the whole active region is $\Gamma_{ac} = N_p \Gamma_p$.

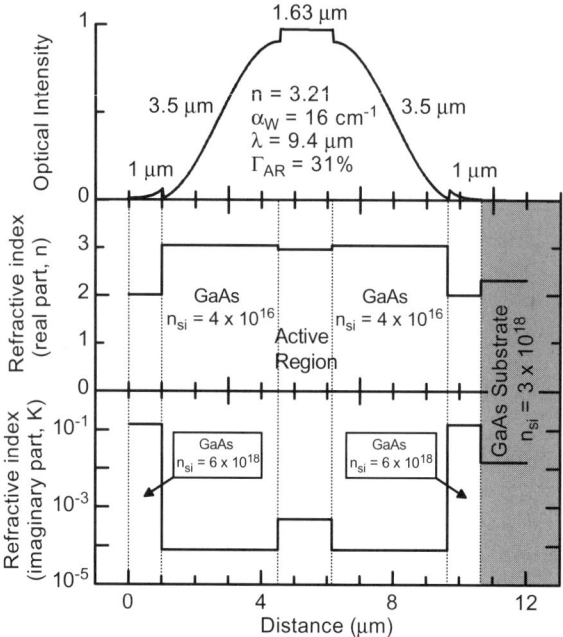

Fig. 5.12 (*Upper part*) Calculated intensity distribution of the fundamental TM mode of the waveguide. Also indicated are the relevant thicknesses of the structure and the important parameters characterizing the waveguide. (*Middle part*) Real part of the refractive index profile in the direction perpendicular to the layers. (*Lower part*) Imaginary part of the refractive index profile in the direction perpendicular to the layers. Note that the latter increases more than three orders of magnitude in plasmon confinement cladding layers.

Figure 5.13 shows the plot of the measured threshold versus $1/L$ at 77 K for characteristic laser with an Al-free waveguide and emission wavelength at $\lambda \approx 9\,\mu m$. The data can be correctly fitted with Eq. (5.1) only up to current densities of the order of 7 to $8\,kA/cm^2$; above this value they lose their linear dependence, a sign of gain saturation. From the slope and the intercept of the straight line below the saturation point, one gets α_w and the modal gain coefficient $g_d\Gamma_{ac}$, expressing the linearity between the modal gain G_M and injected current density J:

$$G_M = g_d \Gamma_{ac} J. \tag{5.7}$$

For the latter, we find $g_d = 4.9\,cm/kA$ and the corresponding value for the waveguide losses is $19\,cm^{-1}$.

The second method is based on a measurement of the fringe contrast of the Fabry-Perot modes of the cavity below threshold. The net modal gain ($G_M(\lambda) - \alpha_w$) is extracted from the fringe contrast using numerical Fourier analysis of the subthreshold spectra in a modified version of the Hakki-Paoli technique [46].

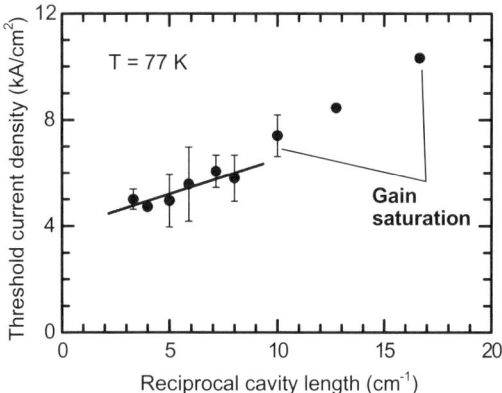

Fig. 5.13 Measured threshold current density versus reciprocal cavity length at 77 K. For current densities above 7 kA/cm², the data show evidence of gain saturation. The solid line is the least square fit to the data below saturation. From its slope and intercept, the modal gain and the waveguide loss are determined. For this measurement, 35 devices were tested.

In Fig. 5.14, the peak net modal gain ($G_M(\lambda) - \alpha_w$) is plotted versus drive current density. We can still observe the linear dependence up to 7.5 kA/cm². After this value the gain saturates, in accordance with the previous measurement. From the data in the linear regime, we derive $g_d \Gamma_{ac} = 4.4$ cm/kA and $\alpha_w = 21$ cm^{-1}, in good agreement with the $1/L$ method. The measured values of the waveguide losses are also in good agreement with the predicted value of 16 cm^{-1}, which is calculated by solely taking into account the free-carrier absorption of the different layers. This value is approximately half of the previously measured internal losses in AlInAs/InGaAs/InP waveguides in the same wavelength range. This is consistent with the assumption that free-carrier absorption is the dominant factor in the waveguide losses, since a higher effective mass corresponds a lower absorption ($m*(GaAs) = 0.067 m_0$; $m*(GaInAs) = 0.0427 m_0$). By optimizing the doping in the n^{++} layers, it should be possible to obtain structures with $\alpha_w = 10$ cm^{-1} for a wavelength of $\lambda \approx 10$ μm.

The power versus current characteristics (P-I) of these devices operating in pulsed mode at 77 K is shown in Fig. 5.15. Note that the power rises up to 1.5 W, the highest peak power ever reported for a GaAs-based QC laser.

5.7 ROLE OF THE CONDUCTION-BAND DISCONTINUITY

Beside the effective mass, the next most important material parameter affecting the performances of QC lasers is the size of the conduction-band discontinuity ΔE_c between the two materials. The influence of this parameter has been difficult to evaluate in the early work as only one composition (and therefore only one value of the conduction band discontinuity) satisfies the lattice matched condition for the InGaAs/InAlAs on the InP substrate. The achieve-

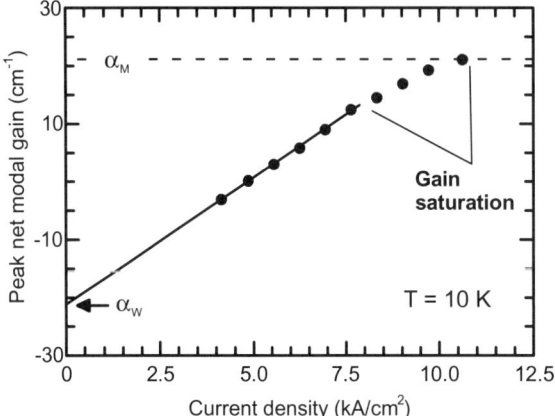

Fig. 5.14 Net peak modal gain $(G_M(\lambda) - \alpha_w)$ of a 0.6-mm-long laser as deduced from the subthreshold emission spectra plotted as a function of the current density J. It is apparent that gain saturation is occurring above 7.5 kA/cm^2, in agreement with the data of Fig. 5.13. The laser threshold is reached when $(G_M(\lambda) - \alpha_w)$ exactly compensate the mirror losses $\alpha_m = 21$ cm^{-1} (*dashed line*). The slope and intercept of the line fit to the data, prior to saturation, give G_M and α_w.

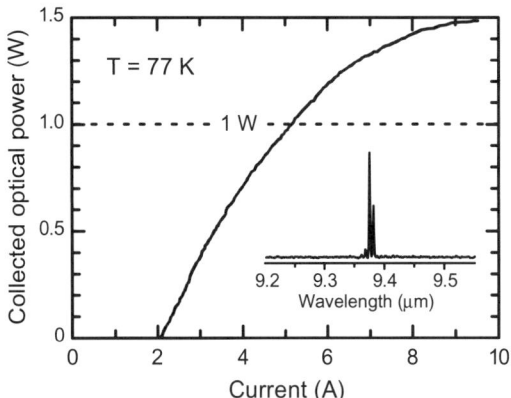

Fig. 5.15 Power-current *(P-I)* characteristic, at 77 K, as recorded using a f/0.8 optics and a calibrated room-temperature HgCdTe detector from a single facet with approximately 60% collection efficiency. The device (2 mm long, 30 μm wide) is driven in pulsed mode (100 ns width, 1 kHz repetition rate), with the measurement performed using an adjustable gate integrator. The high-resolution pulsed spectrum of this device is shown in the inset.

ment of Al$_x$Ga$_{1-x}$As-based QC lasers brought a very important contribution to this field by allowing to study the influence of ΔE_c on device performance systematically for the first time, as lattice-matched material can be achieved for all Al contents.

5.7.1 Strain-Compensated InGaAs/AlInAs Lasers for 3–5 µm Operation

The size of ΔE_c limits, on the high-energy side, the maximum value of the emitted photon energy. For the $In_{0.53}Ga_{0.47}As/In_{0.52}Al_{0.48}As$ material lattice matched to InP where $\Delta E_c = 0.52\,eV$, cryogenic temperature operation was reached ($T < 140\,K$) up to a photon energy of $h\nu = 290\,meV$, but room-temperature operation was limited to $h\nu = 250\,meV$, corresponding to a wavelength of $\lambda > 5\,\mu m$.

Therefore, to provide a good coverage of the 3 to 5 µm atmospheric window, a material with a larger conduction-band discontinuity has to be used. Strain-compensated InGaAs/AlInAs is an almost ideal material system with high conduction-band discontinuity. In these strain-compensated materials, the compressive strain introduced by a pseudomorphic layer with a larger lattice constant (in our case $In_xGa_{1-x}As$ with $x > 53\%$) is compensated by a strain of equal magnitude but opposite sign provided by a material of a smaller lattice constant (in our case $In_yAl_{1-y}As$). In this way the growth of very thick stacks is possible as long as the thickness of the individual layers remains smaller than a value given by the physics of the growth mechanism (critical layer thickness).

The fact that its active region comprises only very thin layers (<5 nm) of barrier and well materials in approximately equal amounts makes this approach especially well suited for QC lasers design [33]. This approach, which adds flexibility to the quantum design by allowing a selection of the desired discontinuity, includes the constraint that the tensile strain balances the compressive one in each of the 16 periods of the structure.

In the design presented here, a 1.2% compressive strain in the $In_{0.7}Ga_{0.3}As$ layers, which represents 40% of the thickness, is compensated by the 0.8% tensile strain in the $In_{0.4}Al_{0.6}As$ layers, which themselves represent 60% of the active layer thickness. This combination of materials yields of $\Delta E_c = 0.74\,eV$, significantly larger than $0.52\,eV$ achieved in lattice-matched systems. The conduction-band discontinuity was evaluated using Van der Walle's model-solid theory [57].

A structure with a 16-period active region based on the $In_{0.7}Ga_{0.3}As/In_{0.4}Al_{0.6}As$ alloy pair was grown by MBE on an n-doped InP substrate [33]. The design of the active region, shown in Fig. 5.16, used the approach developed for room-temperature operation based on a three-well vertical transition and a funnel injector.

The calculated laser transition energy is $E_{23} = 392\,meV$, which corresponds to an emission wavelength of $\lambda = 3.16\,\mu m$. The waveguide is grown using lattice-matched $In_{0.53}Ga_{0.47}As$ and $In_{0.52}Al_{0.48}As$ materials.

Figure 5.17 shows the optical power versus drive current for various temperatures between $T = 200\,K$ and $280\,K$. The slope efficiency decreases from $dP/dI = 80\,mW/A$ at $T = 200\,K$ to $dP/dI = 10\,mW/A$ at $T = 280\,K$, with a maximum power of 4 mW at this temperature. The decrease in slope efficiency at the maximum operation temperature is attributed to saturation of the

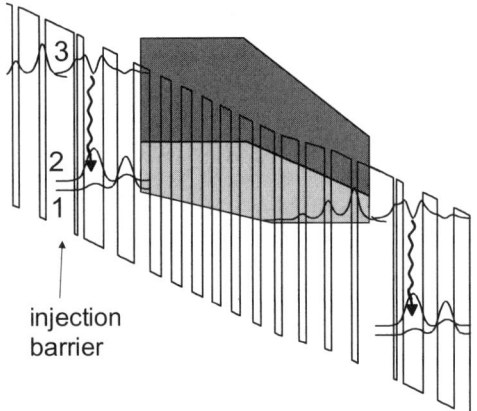

Fig. 5.16 Conduction band diagram of one stage (active plus injector region) of the strain-compensated structure under an applied electric field of 9.6×10^4 V/cm. The wavy line indicates the laser transition designed for $\lambda = 3.15\,\mu m$. The moduli squared of the relevant wavefunctions are shown. The layer sequence of one period of the structure, in nm, from left to right (starting from the injection barrier) is **4.5**/0.5/**1.2**/3.5/**2.3**/3.0/**2.8**/2.0/**1.8**/1.8/**1.8**/<u>1.9</u>/**1.8**/<u>1.5</u>/**2.0**/<u>1.5</u>/**2.3**/<u>1.4</u>/**2.5**/<u>1.3</u>/**3.0**/1.3/**3.4**/1.2/**3.6**/1.1 where $In_{0.4}Al_{0.6}As$ layers are in bold, $In_{0.7}Ga_{0.3}As$ in roman and underlined numbers correspond to Si-doped layers with $N_d = 2.5 \times 10^{17}\,cm^{-3}$.

current supplied through the injection barrier. The emission wavelength of this device is $\lambda = 3.49\,\mu m$ at $T = 10\,K$ and $\lambda = 3.58\,\mu m$ at $T = 270\,K$. We attribute the difference between the calculated and measured wavelength to an uncertainty in the conduction-band discontinuity and to underestimation of the nonparabolicity effects.

The dashed curve in Fig. 5.17 is the current-voltage characteristics, and the operating voltage is between 6.5 V ($T = 10\,K$) to 8.5 V (at $T = 275\,K$). The threshold current density J_{th} between 200 and 270 K can be described by the usual exponential behavior $J_{th} \approx \exp(T/T_0)$ with an average $T_0 = 85\,K$. As will be shown in the next section, the likely explanation for the lower T_0 of these devices, compared to the ones operating at long wavelength, is related to the proximity of the upper state of the laser transition with the conduction-band discontinuity.

The use of strain-compensated active layers clearly enables QC lasers to be fabricated across most of the $\lambda = 3$–$5\,\mu m$ wavelength. Using this approach, Kohler and coworkers [34] obtained excellent room-temperature performances at $\lambda = 4.6\,\mu m$ with hundreds of mw of single-mode power and maximum operating temperature larger than 320 K.

5.7.2 Role of ΔE_c on the High-Temperature Performances of GaAs QC Lasers

The peak output power obtained with GaAs QC lasers up to a temperature $T = 77\,K$ is comparable or better than that of InP/GaInAs/AlInAs lasers. However, GaAs lasers exhibit still higher threshold current densities

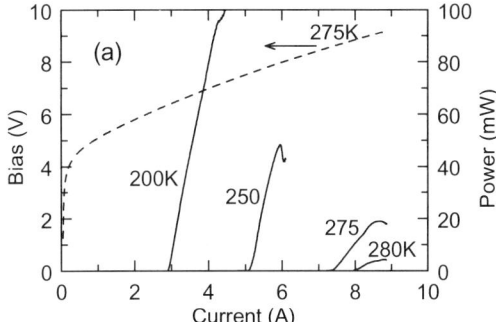

Fig. 5.17 Collected pulsed optical power from a single facet versus injection current for various heat sink temperatures. The collection efficiency is estimated to be 50%. The dashed curve is the current-voltage characteristic at $T = 275\,\text{K}$.

(in 3–5 kA/cm^2) and a much poorer high-temperature performance, with the maximum operating temperature being limited to below $T < 280\,\text{K}$ for the best devices [58]. At high temperatures the devices are characterized by a larger temperature sensitivity of the threshold, namely a lower value of T_0 parameter. These device characteristics are very reminiscent of the performances short-wavelength ($\lambda < 4.6\,\mu\text{m}$) InGaAs/AlInAs-based QC lasers.

Recent studies on the gain mechanism in GaAs lasers show a non-unity injection efficiency into the upper level of the laser transition [60]. This is due to the thermal activation of electrons from the injector directly into the continuum states (Fig. 5.18) and is the principal reason why lower thresholds and room-temperature operation is hindered. To suppress or at least reduce this parasitic current path, it is necessary to increase the energy separation between the injector and the continuum states ΔE_{act}. Furthermore band-structure calculations show that this escape probability increases with the applied electric field, which gives rise to more pronounced injector/excited-continuum coupling [61]. Finally, scattering-assisted injection of electrons into lateral valleys (X, L) above the Γ point of the GaAs and the AlGaAs alloy can also influence device performance. All of these mechanisms diminish the optical gain of the laser structure.

An increase of the energy separation between the injector state $n = 3$ and the continuum states ΔE_{act} can be readily achieved in two ways: the first is by increasing the wavelength of operation and therefore sinking the excited level of the laser transition inside the quantum wells (Fig. 5.18); the second is to use higher conduction-band offsets to deepen the levels inside the multi-quantum-well potential and hence have more freedom to increase the energy separation between the subbands (Fig. 5.19) [62]. Both these approaches have shown significant improvements in the thermal behavior of our lasers.

$Al_{0.33}Ga_{0.67}As/GaAs$ 11-μm Lasers

The purpose of the structure presented in this section is to reduce the parasitic electron transfer, while keeping the Al composition at 33% and hence the

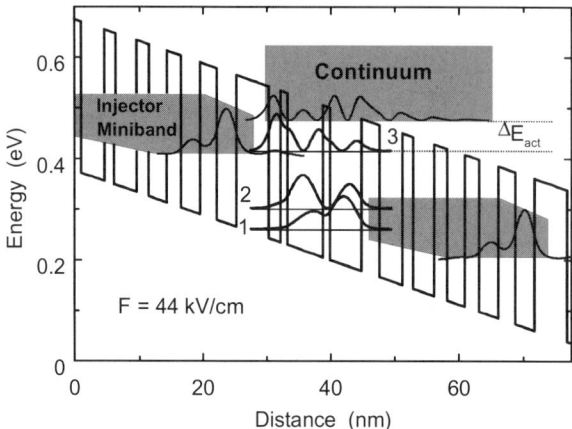

Fig. 5.18 Conduction-band energy diagram of a portion of an 11 μm QC laser under an applied electric field $F = 44$ kV/cm [59]. The subband alignment corresponds to the situation at the threshold. Shown are the moduli squared of the relevant wavefunctions. High-temperature operation is limited by activation of electrons above the energy barrier ΔE_{act}.

conduction-band discontinuity unchanged. The active region, which is shown in Fig. 5.18, consists of three coupled quantum wells, as described in detail in Section 5.3. The lifetime of the upper laser level $n = 3$ is $\tau_3 = (\tau_{31}^{-1} + \tau_{32}^{-1})^{-1} = 1.2$ ps ($\tau_{32} = 2$ ps). The optical matrix element is $z_{32} = 2.1$ nm. Under an external field of $F = 44$ kV/cm, we calculate an energy difference $E_{32} = 112$ meV ($\lambda = 11.1$ μm).

The following features characterize the basic design of the active region. Compared to previous QC lasers, we increased the thickness of each of the three strongly coupled quantum wells of the active region. This was to lower the laser transition energy (3–2) and allow a better confinement of the $n = 3$ state. At the same time we kept the energy difference between the two anti-crossed states $n = 1$ and $n = 2$ close to the LO phonon energy to ensure fast depopulation of the state $n = 2$. We achieved the necessary energy splitting by reducing the width of the barrier that couples the two thick wells of the active region. Under typical biasing conditions, the energy separation, ΔE_{act} between the injector/excited states and miniband-like continuum states is thus increased. With the $n = 3$ state as a reference point, band-structure calculations should give an energy difference of $\Delta E_{act} = 58$ meV to the lowest delocalized continuum state. This value is significantly higher than $\Delta E_{act} = 38$ meV obtained from a comparative calculation for our $\lambda = 9$ μm structure shown in Fig. 5.4 under appropriate biasing conditions (close to resonance). As a consequence the thermally activated leakage of electrons to the continuum should be significantly reduced in this laser structure. We believe that this design feature will notably improve the high-temperature operation of the laser.

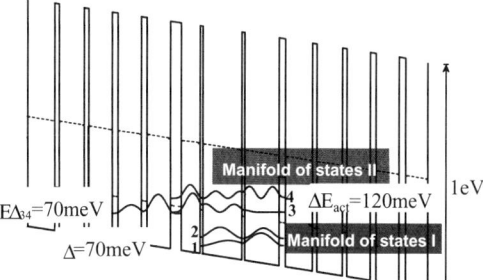

Fig. 5.19 Conduction-band diagram of an injector/active region/injector sequence of a AlAs/GaAs QC laser with emission wavelength at $\lambda = 11\,\mu m$ [62]. Note that the manifold of states II is equivalent to the continuum states, located above the barriers of the active region in Fig. 5.18. Here the height of the AlAs barriers enables the confinement of these states and therefore increases their energy separation ΔE_{act}. The dashed line represents the X-valley conduction-band profile.

In these devices, collected output power in excess of 100 mW at $T = 250\,K$ was demonstrated, with a maximum operating temperature reaching 280 K. The threshold current density as a function of the temperature followed the typical exponential dependence ($\exp(T/T_0)$) with $T_0 = 130\,K$ up to 250 K and $T_0 = 90\,K$ thereafter. The estimated waveguide losses measured using the modified Hakki-Paoli method were 44 cm^{-1}. This rather high value is a consequence of the nonoptimized waveguide design but also of the longer emission wavelength, which always leads to higher losses.

AlAs/GaAs Lasers

Another solution to increase the energy separation between the injector and the continuum states, located above the $n = 3$ state, is to increase the barrier height, therefore their Al content. To this end we used AlAs as the barrier material to fully exploit the Γ point conduction-band discontinuity of $\approx 1\,eV$ with respect to GaAs. The bulk effective band offset in the AlAs/GaAs heterostructure is obviously much lower and is determined by the Γ–X indirect discontinuity of $\approx 195\,meV$. In our device we took advantage of the quantum confinement in the X-valley, which pushes the ground state of the very thin AlAs layers approximately to the level of the GaAs X minimum. In addition we had the GaAs layers act as thick tunneling barriers, to hinder the perpendicular transport in the X valley. Consequently electrons scattered into the X valleys do not contribute to the transport and cannot directly influence the optical gain. In view of these considerations and the results we are going to present, it is safe to state that QC active regions in the AlAs/GaAs heterostructure can be designed by taking into account only the Γ discontinuity if the ratio of the AlAs to GaAs layer thickness is $d_{AlAs}/d_{GaAs} < 0.3$ ($d_{AlAs}/d_{GaAs} = 0.17$ in our structure). In the AlAs/GaAs active regions (Fig. 5.19), thermal activated processes are practically suppressed because the $n = 4$ state is more than 70 meV above the excited state of the laser transition. This state is also well confined within the three wells that define the active region and is not likely to contribute to the transport. In addition the bottom of the manifold of extended states II, which we believe is the main channel for the thermally activated escape of electrons, is situated 120 meV above the state $n = 3$ (ΔE_{act} in Fig. 5.19). In QC lasers this quasi-miniband is usually placed at roughly the same energy as the $n = 4$ state, which therefore becomes the coupling state between the injector and the extended states responsible for carrier leakage. Finally, as in previous QC lasers (based on GaAs or InP substrate), we avoid electron thermal back-filling into the $n = 2$ state from the injector ground state by separating these states by an energy $\Delta > 70$ meV.

The electron lifetime on excited subbands is controlled by the electron-longitudinal optical phonon interaction. Due to the large elastic discontinuity between the two materials, the AlAs/GaAs heterostructure cannot be correctly described using a model for bulk phonons. The scattering rates of the subbands $n = 1, n = 2$, and $n = 3$, which permit an estimation of the population inversion, were calculated within the framework of the macroscopic dielectric model [63]. Our calculations, which take into account both confined and interface phonons, show that the relaxation time from the $n = 3$ state into $n = 2$ state is $\tau_{32} = 1.6$ ps and from $n = 3$ into $n = 1$ is $\tau_{31} = 4.5$ ps.

The collected peak power from a single facet versus injected current densities, at two different heat sink temperatures, is displayed in Fig. 5.20, for a 3-mm-long device. At $T = 10$ K the threshold current density is 5.2 kA/cm^2, and the maximum peak power is in excess of 350 mW, limited by the occurrence of negative differential resistance (NDR) at an applied bias of $U = 8$ V.

The temperature dependence of the threshold current density does not follow the usual exponential behavior but has, between 130 and 230 K, a linear dependence. Nevertheless, an exponential fit of the data over this range would give a extremely large value of the characteristic temperature $T_0 = 320$ K. In addition the differential slope efficiency is nearly constant over the whole temperature range of operation, which is more evidence for the suppression of thermally activated leakage.

In this structure, laser action at temperatures higher than $T = 230$ K is prevented by the occurrence of NDR when the voltage reaches 8 V (Fig. 5.20). This is due to the breaking of the alignment between the injector ground state and the excited state of the laser transition, which dramatically reduces the amount of carriers entering into the active region. In addition the differential resistance of the device increases with temperature, thus reducing the current density needed to reach the limiting voltage. The available current range before the occurrence of the NDR is therefore reduced when the temperature is increased.

Figure 5.21 summarizes the improvements in temperature dependence of GaAs lasers using the arguments presented above. The device with the poorest thermal behavior and the lowest operating temperature (200 K), emits at ≈9 μm and is based on a 33% concentration Al heterostructure. The highest

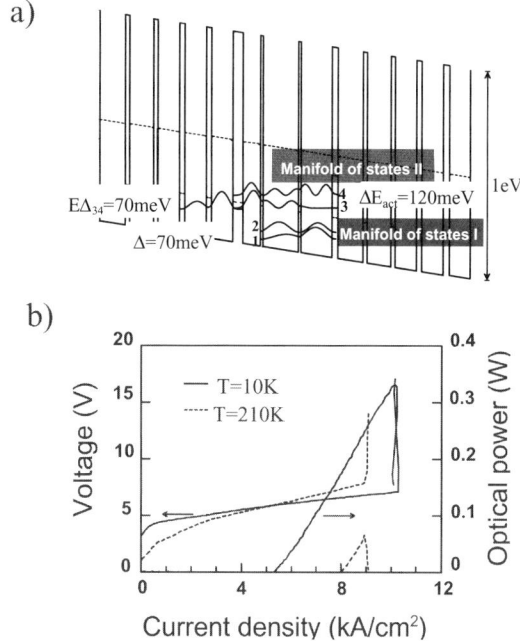

Fig. 5.20 Voltage and output power vs current density of a 3-mm-long and 20-μm-wide device for two different operating temperatures (*solid lines*: $T = 10$ K; *dashed lines*: $T = 210$ K). Devices were driven with 100 ns pulses at 1 kHz repetition frequency.

operating temperature to date, 280 K, has been achieved with a heterostructure of similar composition but a longer emission wavelength ($\approx 11\,\mu$m). The best temperature dependence has been achieved using an AlAs/GaAs heterostructure with a Γ–Γ band offset of ≈ 1 eV (cf. 295 meV for an AlAs mole fraction $x = 33\%$). Although the maximum operating temperature is limited to 230 K by NDR effects, a device of this type is clearly a good candidate for room-temperature operation.

5.8 SPECTRAL CHARACTERISTICS OF QC LASERS

5.8.1 Pulsed Operation

In the preceding sections we described QC lasers operating in pulsed mode with high optical powers at and above room temperature. Under pulsed operation, these lasers exhibit a relatively broad-band, multimode operation, as expected for devices based on Fabry-Perot cavities. The spectra depend not only on the type of the intersubband transition (diagonal versus vertical) but also of the current and temperature.

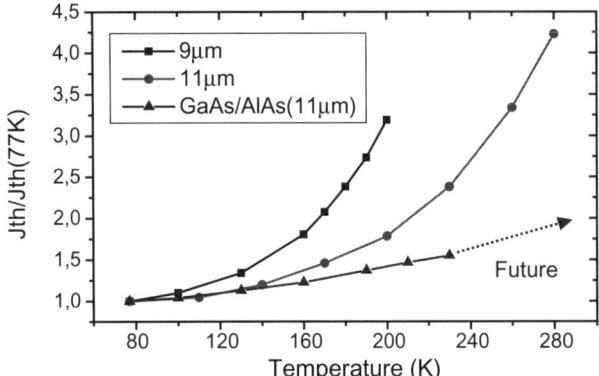

Fig. 5.21 Comparison of the relative temperature dependence of the threshold current density for the devices discussed in the text. The improvements in temperature dependence and maximum operating temperature arise from successively increasing the lasing wavelength and ΔE_{act}. Squares show a typical characteristic for a structure emitting at $\lambda \approx 9\,\mu m$ with an AlAs mole fraction $x = 33\%$, which has the maximum operating temperature less than 200 K. The circles are for a similar heterostructure emitting at $\approx 11\,\mu m$ showing superior temperature dependence and the maximum operating temperature of $T = 280\,K$. The triangles are for an AlAs/GaAs heterostructure with the greatest carrier confinement.

In general, two common features may be observed with increasing current (and voltage) for devices based on diagonal transitions: a *broadening* and a *blue shift* of the laser transition. These two features are apparent in Fig. 5.22. In this figure the emitted laser spectrum at $T = 250\,K$ is displayed as a function of applied bias for a three-quantum-well active region device emitting around $\lambda = 10\,\mu m$. From this measurement it is clear that the spectral features are associated with the change in the position of the gain maximum as a function of bias.

As already mentioned, the blue shift is observed only in devices based on *diagonal* transitions, and the magnitude of this shift is roughly proportional to the change in applied voltage and to the distance between the center of mass of the upper ($n = 3$) and lower ($n = 2$) state wavefunctions [39]. Strong broadening of the emitted spectrum at room temperature is also observed when driving high-performance devices based on other active region designs much above threshold, with a typical broadening of $\Delta v/v = 4\%$ at the maximum operating power point.

The tunability of the gain spectrum with voltage in three-quantum-well diagonal active-region QC devices enables the fabrication of electrically tunable mid-infrared lasers. To enable separate control of emission wavelength and optical power, the top contact metallization of such a device consists of two segments of equal length (750 μm), separated by a small gap. The resistance between the two sections (20 Ω) is much larger than the differential resis-

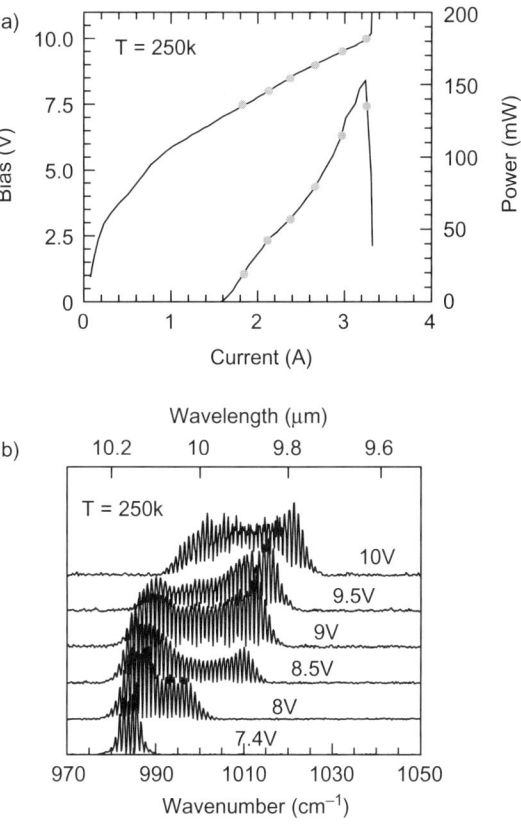

Fig. 5.22 Spectral behavior of Fabry-Perot devices measured at 250 K. (a) Light and bias voltage versus current characteristics of the device. The filled disks indicate the position at which the spectra are taken. (b) High-resolution spectra of the device taken at various voltages.

tance of a single section (2Ω), allowing the injection of different current densities J_1 and J_2 in the two sections (Fig. 5.23) [39]. The total gain spectrum is now the sum of the two gain spectra of the individual sections. In this way the value of the peak gain *and* its location can be controlled continuously.

In Fig. 5.23 the tunable operation of a 1.5-mm-long laser with two sections of equal lengths is displayed. The measurement was performed in pulsed operation, with 100-ns-long pulses at a 5-kHz repetition rate. The temperature was set at $T = 260$ K. For this measurement, the current I_1 in the first section was set to some value. The current I_2 in the second section was then increased until a fixed amount of power (5 mW in Fig. 5.23) was obtained. A few representative spectra for various injected current pairs are displayed in Fig. 5.23. For these low powers the spectra are limited to a few (1 or 2) longitudinal modes of the Fabry-Perot cavity. The mode jumps between the different spectra

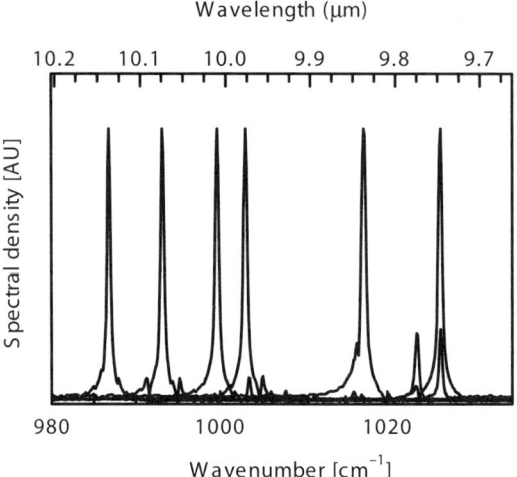

Fig. 5.23 Laser tuning at at $T = -10°C$ and constant optical power $P = 5\,mW$. A few representative spectra for various current pairs. The longest wavelength spectrum is obtained with equal injected current densities in both sections; the one with the shortest wavelength by current injection in one of the two sections only.

reflect the continuous tuning of the gain curve. Some longitudinal modes do not appear, however. This behavior, already observed in the tuning of Fabry-Perot QC lasers operating in continuous wave, is attributed to defects in the waveguide, which force a mode selection by slightly varying the loss of each individual mode. The total tuning range is $40\,cm^{-1}$, which represents a relative tuning $\Delta v/v$ of about 4%.

5.8.2 Continuous-Wave Operation

When QC lasers are operated in continuous wave, single-mode operation has been observed in many instances without the help of external stabilization [64,65,20]. For this reason the following discussion is also valid for distributed feedback (DFB) QC laser devices. The tuning, modal behavior, and linewidths of such a device as a function of temperature and current were studied [64,65,20]. In those works it was found that single-mode operation can be observed in a large temperature and optical power range. The range for single-mode operation was found to decrease almost linearly from $P = 200\,mW$ at $T = 30\,K$ to a few mw close to the maximum operating temperature $T = 100$ K [20].

These characteristics are summarized in Fig. 5.24 and 5.25, where the emission wavelength is plotted as a function of injection current at a fixed temperature and as a function of temperature at a fixed current. The tuning of the single mode QC laser may be understood by assuming that the emission wavelength tunes with the temperature dependence of the refractive index of the active region. For a device operated with an injection current I and at bias U, the active region temperature T_{act} can be related to the temperature of the heat sink T_h by a single thermal resistance R_{th}:

$$T_{active} = T_h + R_{th}UI. \tag{5.8}$$

For InP-based devices, the change of refractive index with temperature at frequency ν yields a tuning of $\frac{1}{\nu}\frac{\Delta\nu}{\Delta T} = 6 \times 10^{-5}\,\text{k}^{-1}$ for $T > 100\,\text{K}$. This value decreases at $T < 100\,\text{K}$ and is only $4 \times 10^{-5}\,\text{K}^{-1}$ at $T = 50\,\text{K}$. The same coefficients are also valid for DFB operating in pulsed and continuous-wave mode.

A high-resolution spectrum of a device operating in continuous wave, measured with a Fabry-Perot interferometer, is displayed in Fig. 5.26. We believe that the measured linewidth ($\Delta\nu = 70\,\text{MHz}$) is limited by the experimental conditions. This value is very close to the the resolution of the spectrometer (about 50 MHz), and corresponds to a current fluctuation of about 1 mA, or a temperature fluctuation of 50 mK. Given the fact that the device was held in a close-cycle He cryostat, it is very likely that the residual temperature fluctuations are responsible for the device linewidth. In fact the Schwalow-Townes limit for this kind of device should be in the KHz range. The fundamental limitation for the linewdith of a laser is given by the ratio of spontaneous to stim-

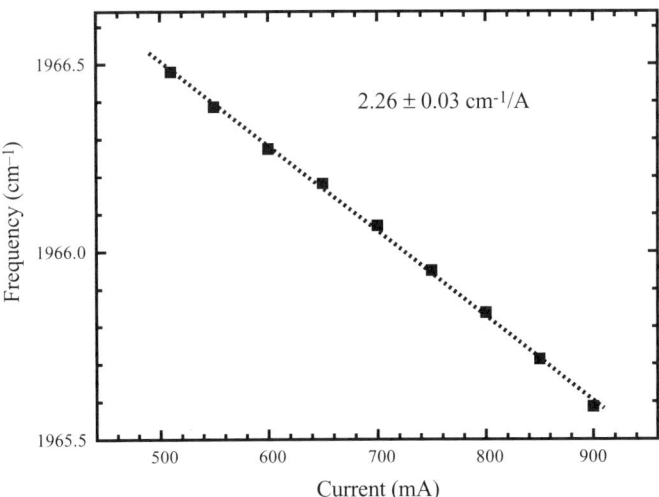

Fig. 5.24 Emission frequency tuning of a $\lambda = 5\,\mu\text{m}$ laser (whose L-I curve is shown in Fig. 5.11) as a function of driving current at a temperature of 20 K. The dashed line is the result of a linear fit.

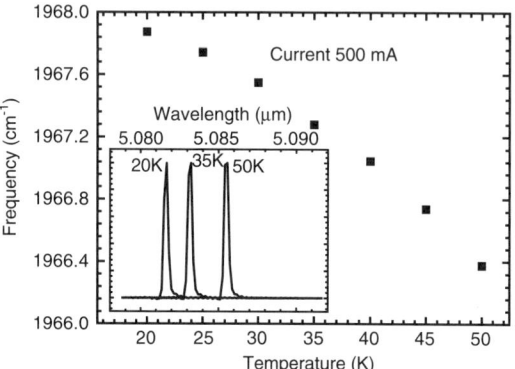

Fig. 5.25 Emission frequency of a laser as a function of heat sink temperature at a fixed current corresponding to an optical power of a few mW. (*Inset*): Measured spectra at three different temperatures, plotted in a linear scale.

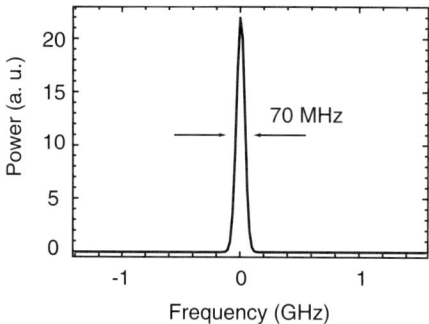

Fig. 5.26 High-resolution spectrum of a laser, taken with a Fabry-Perot interferometer at a current level corresponding to an optical power of a few mW.

ulated photon emission in the lasing mode. Extremely narrow fundamental linewdiths are expected for QC lasers for the following reasons:

- The ratio of spontaneous to stimulated emission coefficients decreases with the square of the energy, and is therefore much lower in the mid-infrared compared to the near-infrared.
- QC lasers exhibit very high optical power.
- The linewidth enhancement factor, related to the change in refractive index induced by a change in carrier concentration, is close to zero in QC lasers.

A beautiful experimental proof of these considerations was given in recent collaborative work of Bell Labs, Pacific Northwest, and NIST [66]. In these experiments, a QC laser was stabilized actively on the absorption line of NO molecule. Using this active feedback loop, a linewidth of only a few KHz was demonstrated.

5.9 DISTRIBUTED FEEDBACK QUANTUM CASCADE LASERS

Applications such as remote chemical sensing and pollution monitoring require a tunable source with a narrow linewidth. For most sensor applications, the linewidth must be narrower than the pressure-broadened linewidth of gases at room temperature, which is about one wavenumber, and the source must be tunable over a few wavenumbers.

As shown in Fig. 5.22, Fabry-Perot devices operating close to room-temperature exhibit spectra that are much too broad to enable such a fine spectroscopy. By incorporating a grating in a QC laser structure, the emission can be stabilized even during the short pulses (≈ 100 ns) that are typically used to drive the device. In fact, these DFB QC lasers do operate at and above room temperature with a linewidth sufficiently narrow for a large class of applications.

In DFB lasers, the coupling constant κ quantifies the amount of coupling between the forward and backward waves traveling in the cavity. In the coupled-mode theory of DFB lasers [67], κ is written as

$$\kappa = \frac{\pi}{\lambda_B} n_1 + i \frac{\alpha_1}{2}, \tag{5.9}$$

where n_1 is the amplitude of the periodic modulation of the real part of the effective index (n_{eff}) of the mode, induced by the grating of periodicity Λ. The corresponding modulation of the absorption coefficient has an amplitude α_1. For first-order gratings, the grating periodicity is fixed at the desired operating wavelength λ_B by the Bragg reflection condition $\Lambda = \lambda_B/2n_{eff}$.

For optimum performance in slope efficiency and threshold current, the product κL_{cav}, where L_{cav} is the cavity length, must be kept close to unity [67]. In the first DFB lasers, a simple design rule was followed. Room-temperature, high-performance devices were obtained with cavity length $L_{cav} \approx 2$–3 mm, so the coupling constant should be designed with a value $|\kappa| \approx 5$ cm^{-1} corresponding to $n_1 \approx 10^{-3}$ for an index-coupled or $\alpha_1 \approx 2.5$ cm^{-1} for a loss-coupled structure. In practice, however, it is found that a larger value of $|\kappa|$ is desirable, since it allows the fabrication of devices with shorter cavity lengths. Device operation in a regime where the product $\kappa L_{cav} \approx 2$–3 leads to better performances with lower threshold current densities and a more stable single-mode operation.

Different fabrication schemes have been demonstrated for the implementation of DFB-QC lasers. These can be separated roughly in three classes, as shown in Fig. 5.27, depending on the respective location of the grating and the active region. The first solution adopted is etching and metalizing the grating directly on top of the waveguide (Fig. 5.27a). The second is the conventional approach used in telecom lasers, which cosists of etching the grating close to the active region and then regrowing the top cladding (Fig. 5.27b). The third solution is to etch the grating in a special waveguide designed with air used as the top cladding region (Fig. 5.27c). In this geometry the current is injected in the device laterally.

We will discuss these three approaches separately in the following sections:

5.9.1 Metalized Top Grating

This configuration was used for the first demonstration of DFB QC lasers [68]. In this configuration, the grating is etched across the top plasmon-confining layer, removing it partially [68] or even completely, as in recent works [34,69]. The need for a strong coupling between the grating and the optical mode means that in this configuration, the waveguide must be designed with a thin top cladding so that the amplitude of the confined mode at the grating surface must be about 0.5–1% of its maximum value in the center of the guide.

The first DFB QC lasers were realized using a three-quantum-well active region designed for 8 and 5.4 μm. The first-order grating of periodicity (850 nm for the $\lambda = 5.4$ μm and 1250 nm for the $\lambda = 8$ μm devices) was exposed by contact photolithography and subsequently etched to a depth of nearly 250 nm by wet chemical etching as the first processing step. In the grating grooves the thickness of the heavily doped plasmon-confining layer is reduced, and the guided mode interacts more strongly with the metal contact, therefore locally increasing the loss. For this reason we expect the coupling constant to be complex, namely to exhibit both real and imaginary parts. In these samples, $n_1 \approx 5 \times 10^{-4}$ and $\alpha_1 \approx 0.5\text{--}2\,\text{cm}^{-1}$. Complex-coupled DFB lasers exhibit in general better single-mode yield because the loss component lifts the degeneracy between the two modes on each side of the stopband [67]. In addition, since the loss component originates from the metal, we do not expect any saturation behavior with increasing optical intensity.

The optical power versus drive current for a 2.25-mm-long DFB laser operating at $\lambda \approx 8$ μm is displayed in Fig. 5.28. At $T = 300$ K, the measured slope efficiency ($dP/dI = 10$ mW/A) and maximum optical power ($P_{max} = 12$ mW) are somewhat lower than the values obtained in the Fabry-Perot devices ($dP/dI = 45$ mW/A and $P_{max} = 15$ mW).

The laser spectra were measured with a Fourier Transform Infrared Spectrometer (FTIR) with a maximum resolution of $0.125\,\text{cm}^{-1}$. The spectrum (Fig. 5.29) is single mode with no observable side lobes in the temperature range of 80 to 315 K. This DFB laser is continuously wavelength tunable from 7.78 μm to 7.93 μm by changing its operating temperature between 80 *and* 315 K.

From these data, the presence of the DFB grating seems to reduce the slope efficiency by a factor of 3 to 10 as compared to Fabry-Perot devices processed

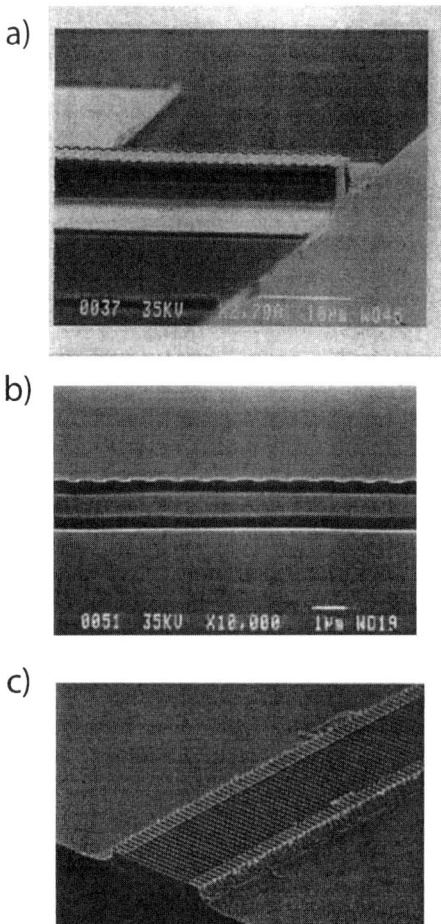

Fig. 5.27 Scanning electron micrograph of a device from each of the major classes of DFB-QC devices. (*a*) metalized top grating, (*b*) grating in the active region, and (*c*) top grating with lateral injection.

from the same epilayer but leaves the threshold current density essentially unchanged. We attribute this reduction in slope efficiency to underestimation of the coupling coefficient κ, implying that our devices operate in the regime of $\kappa L > 1$. In the latter case the differential efficiency is strongly quenched by spacial hole burning. Assuming a simple rate equation model with a constant injection current density, we write the (total) differential quantum efficiency for one period of the active region as:

$$\eta = \frac{G_{th} - \alpha_w}{\alpha_w} \int_{L_{cav}/2}^{-L_{cav}/2} I^2(z) dz, \qquad (5.10)$$

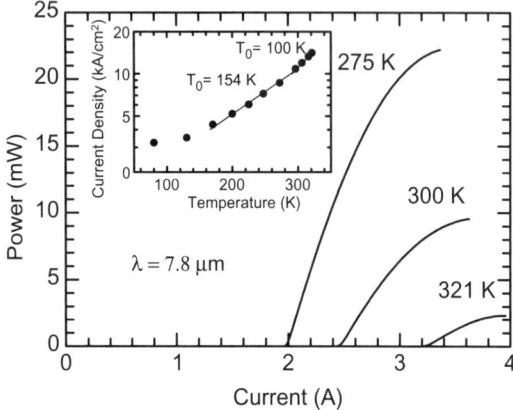

Fig. 5.28 Collected pulsed optical power from a single facet versus injection current at various heat sink temperatures for a device operating at $\lambda \approx 8\,\mu m$. *Inset:* Threshold current density in pulsed operation as a function of temperature. The solid lines indicate the range over which the T_0 parameter is derived.

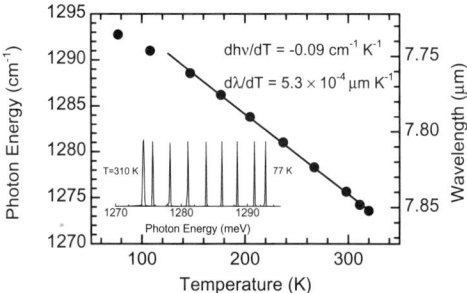

Fig. 5.29 Tuning range of the single-mode emission. A few selected single-mode high-resolution spectra are shown in the inset.

where G_{th} is the threshold modal gain, α_w is the waveguide loss, and $I(z)$ is the (average) intensity along the DFB cavity of length L_{cav}, normalized such that

$$\frac{1}{\sqrt{L_{cav}}} \int_{-L_{cav}/2}^{L_{cav}/2} I(z)dz = 1. \tag{5.11}$$

Equation (5.9) predicts a strong reduction of slope efficiency compared to the Fabry-Perot case when $\kappa L \gg 1$.

At $T = 300\,K$, the QC lasers dissipate too much power ($\approx 80\,kW/cm^2$) to operate in continuous wave. This large amount of thermal power heats the device during the electrical pulse, causing its emission wavelength to drift. For this reason the spectra of these devices was found to be very narrow, limited by our spectrometer's resolution ($0.125\,\text{cm}^{-1}$) for very short pulses (5–10 ns) just above threshold and increased with pulse length up to $1\,\text{cm}^{-1}$ for a 100-ns-long pulse.

To quantify this effect and check that the broadening was only caused by the thermal drift, we performed time-resolved spectra of a DFB laser emitting at $\lambda = 5.4\,\mu\text{m}$. To this end this laser was excited with 100-ns-long electrical pulses with current $I = 3.14\,\text{A}$. The peak optical power at this current is $P \approx 50\,\text{mW}$. The laser emission is detected with a room-temperature mercury cadmium telluride (MCT) detector with a sub–ns time constant. The signal is sampled by a boxcar integrator, whose output is fed back into the FTIR. The resulting spectra taken at 10 ns interval with a 3-ns-long gate are shown on Fig. 5.30. From these spectra it is clear that the laser keeps a narrow emission line, which drifts with time at a rate of $0.03\,\text{cm}^{-1}/\text{ns}$. From this we extrapolate a temperature increase of 20 K during a 100-ns-long pulse. The rate of wavelength shift is proportional to the power dissipated in the active region therefore will decrease with operating point and threshold current density.

These first devices were fabricated by etching only part of the plasmon-confining layer of the waveguide. In a recent work [34] the etching was deepened across the whole plasmon layer. This had the effect of not only changing the effective index of the mode, but the complete mode shape. The authors described therefore their coupling constant with an additional term coming from the change in overlap between the mode and the active region. However, the description of the coupling constant in terms of the effective index approximation becomes inadequate. A new theoretical work describing the interaction correctly [70] will be outlined briefly a little further in the text. At any rate, the coupling constant for such a laser is expected to be large. Devices fabricated with this technique at $\lambda = 4.6\text{–}4.7\,\mu\text{m}$ [34] and $\lambda = 9\text{–}10\,\mu\text{m}$ [71] showed excellent single-mode performance. In these devices the active region was also based on the vertical transition three-quantum-well architecture. The room-temperature performances of the 4.6 μm devices were especially impressive and are displayed in Fig. 5.31. Single-mode peak optical powers above 100 mW at 300 K were demonstrated. In the work cited above, the theoretical approach used to predict the value of the coupling coefficient is based on an effective index approximation. This approximation is known to be inaccurate when the optical mode is in contact with a metal layer and therefore there are abrupt local changes in the electric field direction and magnitude. For this reason a more accurate model was developed by N. Finger and coworkers at the Technical University of Vienna [70]. This model is based on an expansion of the exact solution into a two-dimensional Floquet-Bloch basis. The result of a calculation of the real and imaginary part of the coupling constant as a function of etch depth for a first-order square grating is shown in Fig. 5.32. The computation was performed assuming an Al-free waveguide very similar to the one shown in Fig. 5.12. On such a grating geometry, the effective index approximation would fail to predict even the correct order of magnitude of the coupling constant for most of the etch depth range.

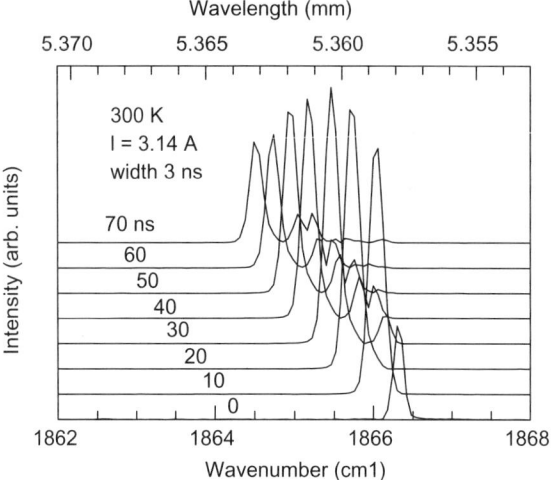

Fig. 5.30 Time-resolved spectra of a $\lambda = 5.4\,\mu m$ device at $T = 300\,K$. The length of the electrical pulse is 100 ns. The spectra are taken 10 ns apart with a 3-ns-long gate. The peak optical power is 50 mW.

As shown in Fig. 5.32, the results obtained from such a model have some startling features:

- Depending on the depth of the etching, the coupling can either be index- or loss-dominated.
- Loss coupling appears as a resonance-like behavior for a given etch depth.
- There is an etch depth for which the coupling is essentially index-dominated and for which the coupling coefficient almost does not depend on this depth.

The last feature is especially useful for a stable fabrication technique. GaAs-based DFB-QC lasers processed using these design rules, where the top grating was etched by reactive ion etching, displayed excellent single-mode and threshold properties [69].

5.9.2 Index-Coupled Lasers

Complex-coupled DFB lasers with an etched grating on the surface have some very advantageous features: because the coupling coefficient has also a loss component, they are much more likely to be monomode regardless of the position of the cleaved facet. Most important, their fabrication is very simple and requires only one growth step at the beginning. The grating periodicity can be adjusted a posteriori by trial and error on the grown layer. Although this grating strongly influences the QC laser through its plasmon-enhanced waveguide, it is located away from the active region and the region of maximum intensity of the laser mode. For relatively shallow gratings obtained by wet

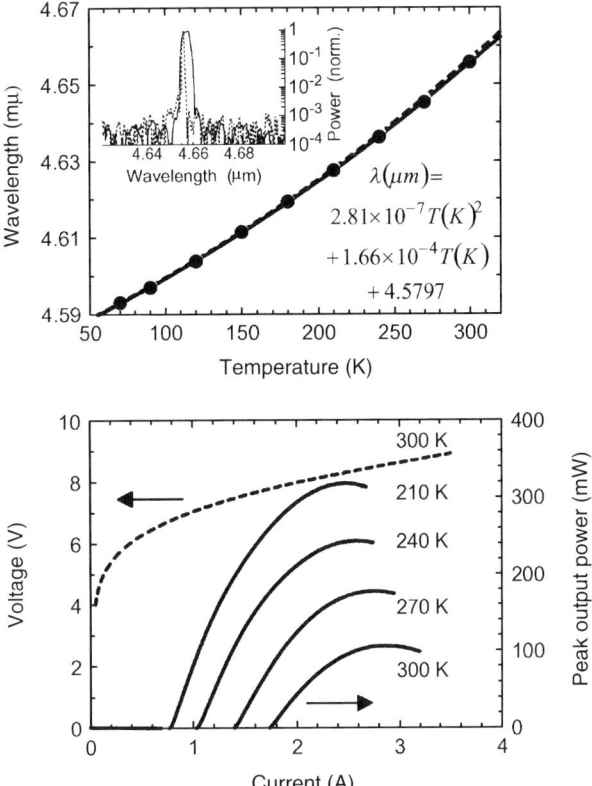

Fig. 5.31 *Top*: Emission wavelength as a function of the heat sink temperature of a DFB-QC laser device operated in pulsed mode (50 ns pulse width and 5 kHz repetition rate). The closed circles are obtained from spectra close to the threshold, the open circles at approximately twice the threshold current. The lines represent a quadratic fit through the data. *Inset*: Two characteristic single-mode emission spectra obtained at room temperature at 3.2 A (*solid line*) and 1.8 A (*dashed*). The line broadening at the higher current is due to wavelength chirp induced by the pulsed nature of the drive current. *Bottom*: Voltage (V, *dashed*) and light (L, *solid*) versus current (I) characteristics at various heat sink temperatures of the laser. (From Ref. [34], with permission)

chemical etching, strong coupling is only attainable with a reduced upper cladding layer thickness. This increases the waveguide loss through the absorbing top metal layer and in turn decreases the performance of the DFB laser. On the other hand, if high aspect ratio gratings obtained typically by reactive ion etching are used, the data shown in Fig. 5.32 demonstrate that strong, index-dominated coupling may be achieved by top metallized grating.

A more classical approach relies on a grating etched directly inside the active region that allows strong coupling with negligible additional loss. This approach was first demonstrated successfully in three-quantum-well active-region devices [72,73,74]. The devices were processed as follows: In the first

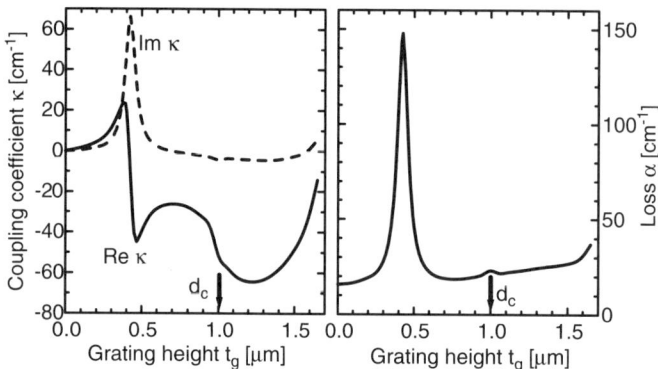

Fig. 5.32 Calculated complex coupling constant κ as a function of etch depth for an Al-free GaAs waveguide.

MBE step, the QC active region is grown embedded between two InGaAs layers. The upper InGaAs layer serves as the host region for the first-order grating. The latter is transferred by contact lithography and wet chemical etching. In the second MBE step, InP is epitaxially grown directly on top of the grating. The grating strength is controlled by the grating depth and duty cycle during grating fabrication (etching) and the reflow of material in the regrowth process. It has an approximatively trapezoidal shape with a duty cycle of 30–50%.

Continuously tunable single-mode operation is achieved from $\lambda = 8.38\,\mu m$ to $8.61\,\mu m$ ($\Delta\lambda = 230\,nm$) in the same wafer using two different grating periods: $\Lambda_1 = 1.35\,\mu m$ (8.38 to 8.49 µm) and $\Lambda_2 = 1.375\,\mu m$ (8.47 to 8.61 µm). The tuning mechanism is identical to all the other single-mode devices, namely the mode tunes with the temperature dependence of the refractive index. It was found experimentally that this tuning rate was about half the one experienced by the intersubband transition. As a consequence it is not possible to achieve perfect overlap between the gain spectra and Bragg reflection peak at both cryogenic and room temperature. The fact that the device can be operated with such a wide temperature range, while maintaining a dynamical single-mode operation with side-mode suppression ratio larger than 30 dB, demonstrates that strong index coupling has been achieved in this device. The modulation Δn is estimated as the difference between the modal refractive indexes of the "undisturbed" waveguides at the location of the grating grooves (InP) and plateaus (InGaAs). The deviation of the grating shape from a sinusoidal shape as well as a duty cycle other than 50% reduces the modulation amplitude of the grating. A correction factor of 0.8 (estimated for a trapezoidal grating with 50% duty cycle) finally results in $\Delta n = 1.79 \times 10^{-2}$. From this value, and using Eq. (5.8), we obtain a coupling coefficient of $\kappa_{in} = 33\,cm^{-1}$. A clear Bragg stopband with width $\Delta\lambda_{Bragg} = \Delta n \Lambda \approx 24\,nm$ follows from the strong index coupling of the DFB-QC laser. The two Bragg resonances on either side of the stopband—located on the slope of the narrow gain spectrum—experience a strong discrimination with respect to each other due to the large value of $\Delta\lambda_{Bragg}$.

Tunable single-mode lasing has also been achieved in continuous-wave operation. The lasing resonance has been tuned from 8.47 to 8.54 µm by changing the heat sink temperature from 20 to 120 K. The L-I characteristics of the DFB-QC devices operated in continuous wave at various heat sink temperatures is displayed in Fig. 5.33. As shown in this figure, single-mode continuous wave optical powers of 50 mW at liquid nitrogen temperature are achieved for 1.5-mm-long and 17-µm-wide devices.

5.9.3 Lateral-Injection Surface-Grating Laser

One possibility to prevent the top contact metal from introducing a large waveguide loss is to avoid the metal on top of the waveguide altogether [75,76]. Such a device consists basically of a waveguide with a semiconductor lower cladding layer and air acting as top cladding. The heavily n-doped InGaAs cap layer, which serves as host layer for the grating, is highly conducting to allow lateral current injection and distribution throughout the device. The very favorable consequences of such a design are as follows:

- A large refractive index step between the semiconductor and air, allowing a strong grating coupling of the device.
- Low calculated waveguide losses of 12 cm^{-1} for a $\lambda \approx 10$ µm laser device.

The resulting high coupling coefficient of the grating potentially allows the fabrication of short devices with a low threshold current.

The growth process started with the lower waveguide layers (InGaAs Si, 1×10^{17} cm^{-3}, total thickness 1.5 µm), followed by an active region (thickness 1.75 µm), and was finished by a thicker set of upper waveguide layers (thickness 2.2 µm), and a 0.5 µm thick highly n-doped cap layer. This cap layer was also the host layer for the grating, as mentioned in the introduction. The active region, which thus formed the central part of the waveguide, consisted of 35 periods of the $\lambda \approx 10$ µm three-quantum-well active region described in Fig. 5.4a.

The fabrication process started by holographically defining a grating with a 1.6378 µm period (assuming an effective index $n_{eff} = 3.163$), and wet chemical etching of the grating to a depth of 0.3 µm in a $H_3PO_4:H_2O_2:H_2O$ solution (4:1:10, etch rate 800 nm/min). We used a 488-nm Ar-ion laser and a 90-degree corner reflector mounted on a rotational stage for the grating exposure. Wet chemical etching in a $HBr:HNO_3:H_2O$ solution (1:1:10, etch rate 800 nm/min) was then used to define ridge waveguides with a width of 35 to 55 µm (etch depth = 4.5 µm) and a length of 1.5 mm. 300-nm-thick Si_3N_4 served as an electrical passivation layer, and Ti/Au (10/1000 nm) was used as top contact metal. Thinning, back contacting (Ge/Au/Ag/Au, 12/27/50/100 nm), and

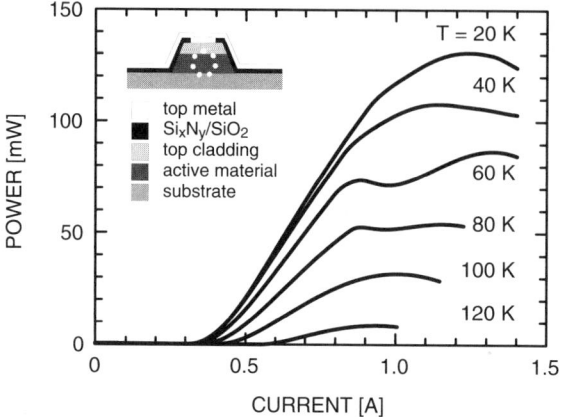

Fig. 5.33 Continuous wave operation of a buried grating DFB-QC laser at various temperatures.

cleaving completed the processing. As shown by the scanning electron micrograph in Fig. 5.27c), the contact metal covered only the edges (about 5 μm on each side) of the ridge to prevent large absorption losses in the waveguide, but still allow lateral current injection.

All devices were mounted ridge-side up on copper heat sinks, and then placed into a Peltier-cooled aluminum box with an antireflection-coated ZnSe window. In this box they were held at a constant temperature between −30°C and 60°C. A commercial pulse generator (Alpes Lasers, TPG 128) with power supply (Alpes Lasers, LDD 100) allowed us to deliver 22.5- or 45-ns-long current pulses at a variable repetition frequency of up to 5 MHz to the laser. For the measurement of L-I curves, we directly measured the average power with a calibrated thermopile detector.

Typical P-I and I-V curves of a 55-μm-wide and 1.5-mm-long device at 1.5% duty cycle are shown in Fig. 5.34. The emitted average power drops from 7.2 mW at −30°C to 2 mW at 60°C. These values correspond to peak powers of 480 and 135 mW at the respective temperatures. We observed threshold currents of 4.35 Å at −30°C and 5.8 Å at 60°C; these values are equivalent to threshold current densities of 5.3 kA/cm^2 and 7 kA/cm^2, respectively. Since the Bragg peak tuned toward the center of the gain curve at higher temperatures, we observed a relatively high T_0 of 310 K.

The maximum average output power was also measured at the same temperatures as in Fig. 5.34 but with a doubled duty cycle of 3%. The highest output power was achieved at −30°C; its value at the thermal rollover point was 13.6 mW. At room temperature we still observed 4 mW, and finally, at 60°C, the value decreased to 1.5 mW. The last number, which is smaller than the one for 1.5% duty cycle, indicates that at a temperature of 60°C the duty cycle with the best performance is smaller than 3%. Considering the high voltage nec-

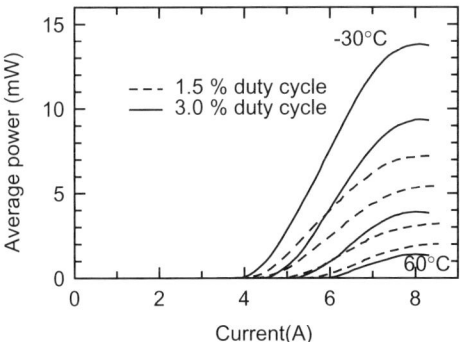

Fig. 5.34 Average optical power versus injected current characteristic of a lateral-injection DFB-QC laser at various temperatures. Dashed lines are measurements performed at a 3% duty cycle, and solid lines at a 1.5% duty cycle.

essary to achieve these output powers, it becomes clear that the device suffers from excessive heating, in particular, at higher temperatures.

This hypothesis is confirmed by Fig. 5.35, which shows emission spectra at maximum output power and different temperatures between 30°C and 60°C. A pulse length of 45 ns and a pulse repetition frequency of 667 kHz were used to drive the laser at a duty cycle of 3%. The single emission peak tunes from 968.6 cm^{-1} at 30°C to 961.4 cm^{-1} at 60°C. The temperature tuning is $\Delta h\nu/\Delta T = -0.08$ cm^{-1}/K, a value that is 27% larger than the one we reported earlier for devices running at a low duty cycle. Since the temperature tuning of a DFB laser is only due to a temperature-induced refractive index change, we are able to estimate the overheating of the laser at a heat sink temperature of 60°C. Under the assumption of a temperature difference ΔT between the active region and the heat sink at 30°C, we find at 60°C using the usually observed temperature tuning of DFB lasers a much larger temperature difference of $\Delta T = +25°C$. This is consistent with the observed increase in threshold current from 5.8 to 6.1 A at this temperature when going from a 1.5% to 3% duty cycle. Constant emission peak width of 0.5 cm^{-1} and side-mode suppression ratio >35 dB were seen for all temperatures. The linewidth could be further decreased (down to 0.25 cm^{-1}) by driving the device with shorter (22.5 ns long) pulses.

In laterally injected DFB-QC lasers, the metallization runs up the sidewalls of the mesa and on the top of the device with width of about 4 µm. The metal contact will therefore induce large lateral waveguide losses in narrow devices. On the other hand, the same loss will allow fairly large devices to run in a single transverse mode. The fundamental mode may differ from a Gaussian profile because the refractive index profile is affected by the nonuniform injection density across the device, as shown in Fig. 3.6.

Evidence for the strong coupling efficiency of the surface grating, lateral-injection devices can be obtained from a measurement of their electrolumi-

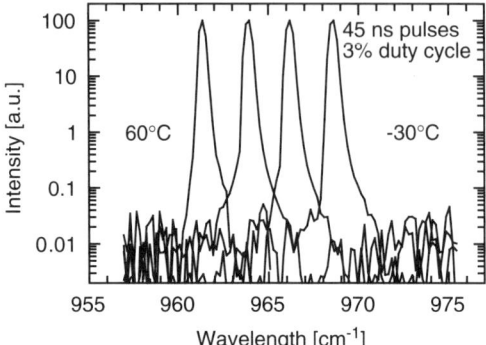

Fig. 5.35 Emission spectrum of a 1.5-mm-long, lateral-injection DFB-QC laser at various temperatures driven at a 1.5% duty cycle

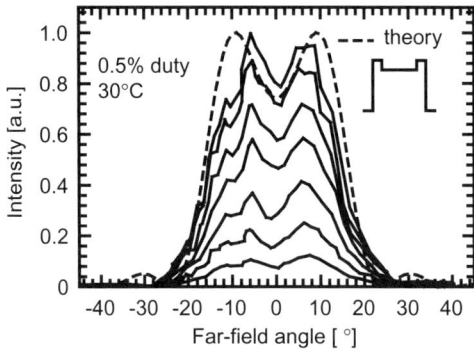

Fig. 5.36 Far-field emission profile of a 1.5-mm-long, lateral-injection DFB-QC laser as a function of drive current. The temperature is 30°C and the device is driven at a 1.5% duty cycle.

nescence spectrum below threshold. In Fig. 5.37 we present three luminescence spectra measured at 85, 105, and at 150 K. In all of them, the spontaneous emission peaks around 980 cm^{-1}, and there occur regular Fabry-Perot modes with a spacing of 1.7 cm^{-1} (cavity length: 850 μm). The Bragg reflector's stopband with a width of 2.5 cm^{-1} is clearly visible at 995.7 cm^{-1} (85 K), 994.6 cm^{-1} (105 K), and 992.3 cm^{-1} (150 K). From the stopband width, we determined the coupling coefficient of the grating to be $\kappa = \pi n_{eff} \Delta \lambda / \lambda^2 = 24$ cm^{-1}; this number agrees well with a value obtained from an estimation based on the effective refractive index difference of $\Delta n = 1.8 \times 10^{-2}$ between areas with and without the grating layer ($\kappa = \pi \Delta n_{eff}/2\lambda = 28$ cm^{-1}). Since the partial removal of the contact layer leads to a slight gain variation, a small amount of loss coupling might also be present in this device.

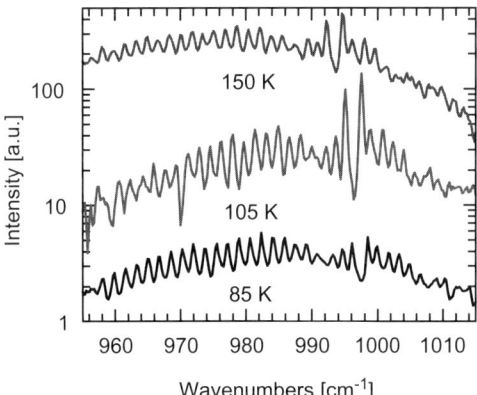

Fig. 5.37 Spontaneous emission spectra of a 45-µm-wide and 850-µm-long DFB-QC laser at 85, 105, and 150 K.

5.10 MICROSTRUCTURED QC LASERS

A strong limitation of QC lasers is the high room-temperature threshold current (>1 A). It implies that a large amount of power must be delivered to the device before light can be extracted from the device. More sophisticated QC laser processing could allow for more compact laser cavities with lower mirror and waveguide losses.

5.10.1 Microdisk QC Laser Resonators

Early work concentrated on the so-called whispering disk resonators [32,77], following an approach that had already been demonstrated in interband semiconductor lasers [78]. Whispering gallery QC lasers have also been demonstrated using GaAs-based QC lasers [79]. In these devices laser action occurs in a disk laser in the so-called whispering gallery modes [78], which correspond to light traveling close to the perimeter of the disk impinging onto its edge in angles larger than the critical angle of refraction ($\Theta > \arcsin(1/n_{eff})$, where n_{eff} is the two-dimensional effective refractive index of the disk). This results in a high quality-factor (Q) resonator, where the losses are due to (1) tunneling of light (which is usually negligible), (2) scattering from surface roughness, and (3) intrinsic waveguide loss. Since a disk laser consists mainly of a thin disk of active material free-standing in air waveguide, losses due to cladding layers and substrate are kept to a minimum.

These devices were fabricated by a two-step wet etching process. After patterning the wafer by optical lithography, cylinders are wet chemically etched in an aged $HBr:HNO_3:H_2O$ (1:1:10) solution at room temperature, an isotropic etchant for AlInAs, GaInAs and InP. In a second step, a rhomboidal InP pedestal is formed by etching in $HCl:H_2O$ (1:1) solution at 40°C, which selectively etches only InP. The active region of the device was based on a three-quantum-well active region optimized for operation at λ = 5.4 μm.

Figure 5.38 gives a scanning electron micrograph of a disk laser. Figure 5.39 shows the cryogenic L-I characteristic of a 17-μm-diameter device, which exhibits a threshold of I_{th} = 2.85 mA corresponding to a threshold current density of J_{th} = 1.25 kAcm^{-2}.

Unfortunately, these very low threshold currents were achieved at the cost of a geometry that prevents efficient heat extraction from the device. For this reason neither continuous-wave nor room-temperature operation were achieved in these devices.

Laser cavities with high Q and very low volume have been used in semiconductor diode lasers to modify the photon density of states and control the spontaneous emission. These systems have been the experimental support for spectacular demonstrations of very important effects long sought in semiconductors, such as the Purcell effect [80] and controlled single-photon emission (e.g., see [81]). Many researchers have devoted a much effort to the realization of very small-volume cavities in order to convey the light generated by spontaneous emission in only one mode, so to strongly reduce the threshold of the device and, at the limit, obtain almost thresholdless laser.

In this framework one quantity of interest is the fraction of spontaneous emission that can be channeled into a single-cavity mode. This parameter is commonly referred as the spontaneouse-mission coupling value or the β factor, and is by definition 0 < β < 1. In this section it will be shown that the threshold of a QC laser is unexpectedly independent from the value of the β factor due to the different physical mechanisms that govern the lifetime of the upper state in QC and diode lasers; completely controlled by nonradiative processes in the first case while dominated by the spontaneous emission lifetime in the second. The microcavity effect will be treated in a simple way by adding a term, proportional to β, in coupled rate equations. In this case, in

Fig. 5.38 Scanning electron microscope image of a 25-μm-diameter QC disk laser with total height of 15.5 μm.

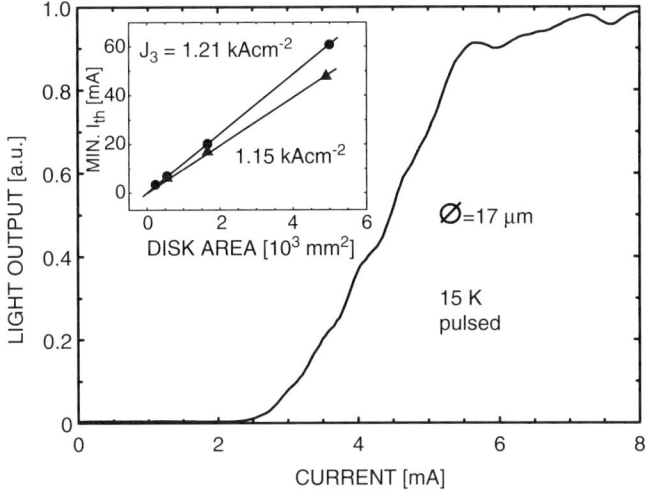

Fig. 5.39 Light output versus injection current (L-I) of a 17μm-diameter QC disk laser. The threshold current of $I_{th} = 2.85$ mA corresponds to a threshold current density of $J_{th} = 1.25$ kAcm^{-2}. *Inset*: threshold current density versus disk size.

order to focus on the physics of our interest, we will consider two levels and we will assume that the lifetime of the lower state of the laser transition τ_1 is vanishingly small compared to that of the excited state τ_2. This allows us to derive an analytical solution that includes as evidence the physical quantity of our interest.

We can therefore write

$$\frac{dS}{dt} = (\sigma N_2 - \alpha \bar{c})S + \frac{\beta N_2}{\tau_{rad}}, \quad (5.12)$$

$$\frac{dN_2}{dt} = J - N_2/\tau_2 - S\sigma N_2, \quad (5.13)$$

where N_2 is the electron density in the excited state ($n = 2$), S is the photon density, σ is the gain cross section per unity time, $J = J/q$ is pumping current density divided by the electronic charge, α is the total losses of the system, $1/\tau_{rad}$ is the reciprocal of the spontaneous emission rate, and $\bar{c}$ is the velocity of light divided by the refractive index. The solution of the rate equations in a steady state gives for the photon density,

$$S^2 \sigma \bar{c} \alpha - S\left(J\sigma - \frac{\bar{c}\alpha}{\tau_2}\right) - \frac{J\beta}{\tau_{rad}} = 0. \quad (5.14)$$

Since $S > 0$ by definition, we take the positive solution:

$$S(J) = \frac{1}{2\bar{c}\alpha} - \frac{1}{2\sigma\tau_2} + \sqrt{\left(\frac{1}{2\bar{c}\alpha} - \frac{1}{2\sigma\tau_2}\right)^2 + \frac{J}{2\bar{c}\alpha}\frac{2\beta}{\sigma\tau_{rad}}}. \qquad (5.15)$$

This curve represents the light versus current characteristic of the device. A family of these curves are plotted in Fig. 5.40 for a set of different values of β, from 0.01 to 1. In calculating the curves of Fig. 5.40, we have used a higher rate of spontaneous emission (1 ns) than the one is normally found for QC lasers in the mid-infrared (≈ 10–500 ns). This has been done to enhance the effect of β and make the graph easier to read by having a greater separation among the curves. From the figure two things are immediately apparent: the threshold is almost unchanged and all the curves have an asymptotic behavior at high current with a straight line. The asymptote can be easily calculated:

$$S_\infty(J) = \frac{J}{\bar{c}\alpha} + \frac{1}{\sigma}\left(\frac{\beta}{\tau_{rad}} - \frac{1}{\tau_2}\right). \qquad (5.16)$$

This straight line represents the behavior of the laser when light generation is completely dominated by stimulated emission. Its intercept with the density current axis is commonly used as a definition of threshold. It is straightforward to see that the changes in threshold, due to β, are proportional to the term and therefore are very small even when β ≈ 1, since the ratio τ_2/τ_{rad} is of the order of 10^{-4} to 10^{-5}, depending on the wavelength. The physical explanation of this result lays in the fact that in QC lasers the lifetime of the upper state of the laser transition is totally controlled by nonradiative processes, and the amount of losses due spontaneous emission coupled into other radiative modes has negligible contribution. This is not the case for diode lasers where the lifetime is controlled by the spontaneous emission time. For these devices the possibility of channeling all the radiation in one mode can be of great advantage.

An interesting improvement to this geometry was achieved by a quadrupolar deformation of these disk devices [82]. In the latter work the outcoupling efficiency of the lasers were improved by more than three order of magnitudes as the resonator was progressively deformed. Output powers up to a few mW were achieved at cryogenic temperatures. A similar approach was also used later in GaAs-based QC lasers [83].

5.10.2 High Reflectors

A more straightforward approach to the miniaturization of QC lasers is the fabrication of very short devices with high-reflectivity mirrors. This geometry allows an easier outcoupling and heat dissipation than the disk devices. In a recent work [84], focused-ion beam apparatus was used to etch in situ quarter-wave reflectors at the facet of the device.

Figure 5.41 shows an SEM picture of such an etched mirror, along with the computed reflectivity. Because of diffraction losses, the reflectivity of the

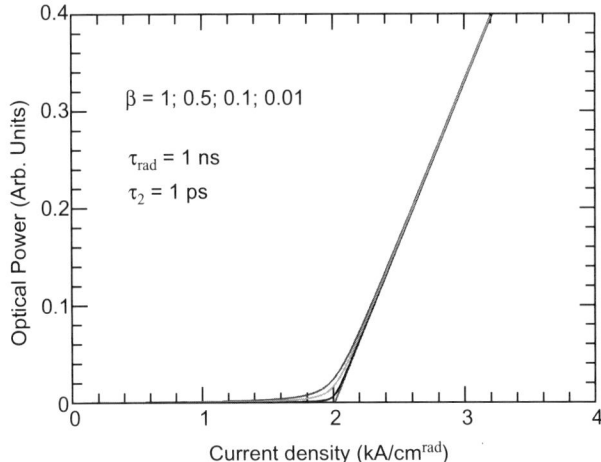

Fig. 5.40 Light versus current characteristics of QC lasers with different values of β. The only region that is sensibly modified is the one close to threshold. For very high values of β, the light has much stronger values below threshold. Note that in order to make the effect of β more evident, we chose the spontaneous emission rate 50 times larger than the actual value.

mirror saturates already after two periods. The threshold current of devices based on a Fabry-Perot cavity, a DFB cavity, and a distributed Bragg reflector (DBR) are compared in Fig. 5.42. The lower cavity losses of the DBR reflector allow for a lower threshold and higher operating temperatures. A 500-μm-long device fitted with such reflectors operates with the same mirror losses as a 3mm-long device. In principle, such a process could be optimized for lower diffraction losses by reducing the length of the air gaps responsible for the diffraction (going away from an exact λ/4 condition) or decreasing the refractive index step. In this way QC lasers with very short cavities should be feasible.

5.10.3 Buried-Heterostructure Lasers

The threshold current of a QC laser could, in principle, be strongly reduced by etching very narrow waveguide stripes. However, two phenomena strongly limit this option for the achievement of very low threshold currents. First of all, the lateral waveguide losses strongly increase as the lateral dimension of the waveguide is reduced. Second, simply reducing the lateral size of the waveguide also reduces the heat dissipation capability of the device.

However, lower lateral waveguide losses and improved lateral heat dissipation can both be achieved by substituting the Si_3N_4/SiO_2 films by InP. This has already been achieved using solid source MBE and in situ thermal Cl_2 etching [85]. However, this previous approach, although promising in principle, had the main disadvantage that it did not yield a planarized structure

268 InP AND GaAs-BASED QUANTUM CASCADE LASERS

a)

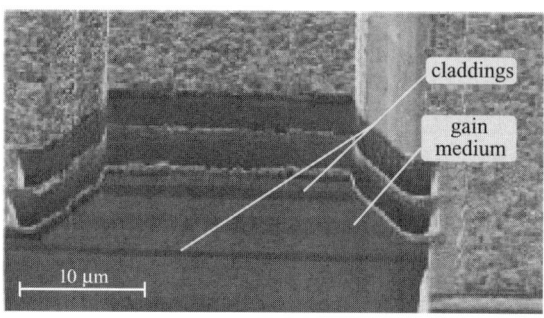

b)

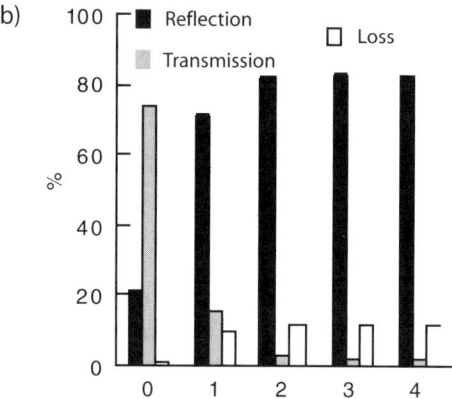

Fig. 5.41 *Top*: Scanning electron micrograph of Bragg reflector etched with a focused ion beam apparatus into a GaAs/AlGaAs QC laser. *Bottom*: Reflection, transmission, and loss coefficients computed as a function of the number of Bragg reflectors. Because of diffraction losses, the reflection coefficient saturates at 80% after two periods. (Reprinted from [84], with permission).

because the regrown material had to be etched on the top of the stripe. These devices also suffered from electrical instabilities due to dislocation defects originating from the regrowth interface, which prevented their use at a high duty cycle.

The approach described here is based on a selective lateral MOCVD regrowth of undoped InP on a ridge masked by SiO_2 [86]. Reference lasers and buried lasers were processed in a wafer from the same MBE growth in order to investigate the influence of the lateral InP layer on laser performance. Both samples were etched as 28-μm-wide and 6-μm-deep mesa ridge waveguides. Our standard process was then applied to the reference samples. Our buried laser samples were transferred into an MOCVD reactor after mesa etching to form the buried heterostructure, the top of the mesa ridge being

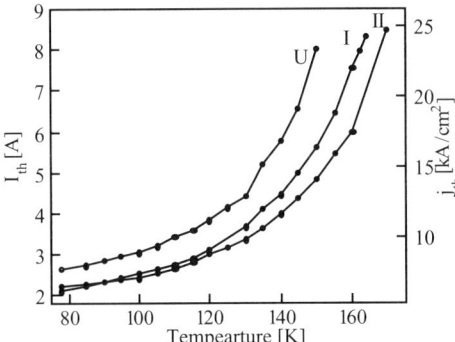

Fig. 5.42 Threshold current (*left axis*), threshold current density (*right axis*) as a function of temperature for Fabry-Perot devices (*U*), DFB devices (*I*), and Bragg reflector devices (*II*).

covered by SiO_2 inhibiting InP regrowth on top of the cladding. The SiO_2 film is removed chemically after deposition of 3 μm of nonintentionally doped InP, and nonalloyed Ti/Au ohmic contacts were finally evaporated on top of the highly doped cladding layer. The cross sections resulting from two processing steps are compared in Fig. 5.43.

After processing, the lasers were cleaved in bars and the facets left uncoated. They were then mounted epilayer up on a Peltier-cooled/controlled head of a commercially available laser housing. The lasers were driven by 60-ns current pulses at different repetition rates up to 3 MHz. The optical output power from one facet was measured using a thermopile power meter. The performance of two otherwise identical QC lasers with and without buried heterostructure is compared in Fig. 5.43, where the maximum average power from a single facet is plotted versus duty cycle at an operating temperature of 273 K.

As expected, the buried heterostructure process yields both lower optical losses, translating into a higher slope efficiency and higher thermal conductivity, allowing higher average powers and higher duty-cycle operation. The data of Fig. 5.43c is plotted along with the result of a simple model. In the latter the threshold current density is assumed to have the exponential dependence in the active region temperature $\cong \exp(T_{act}/T_0)$ with a (measured) T_0 value of 200 K, the active region temperature is related to the heat sink temperature by a single thermal resistance, and the average slope efficiency displays the measured temperature dependence modeled by a linear dependence. Specific thermal conductances of $G = 110 \, W \, cm^{-2} \, K^{-1}$ and $G = 60 \, W \, cm^{-2} \, K^{-1}$ are fitted from the data of Fig. 5.43 for the buried heterostructure and ridge process, respectively. These compare well with the calculated values of $130 \, W \, cm^{-2} \, K^{-1}$ and $90 \, cm^{-2} \, K^{-1}$ by a finite-element technique.

Wet oxidation of AlInAs is another attractive means of achieving lateral guiding with good current confinement, high thermal transport, and low losses.

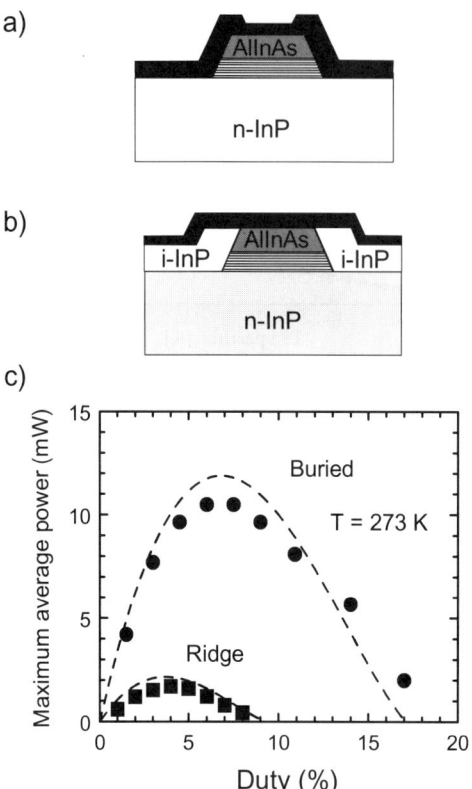

Fig. 5.43 Schematic cross section of a laser processed with an oxide insulating layer (*a*) and with a buried heterostucture layer (*b*). (*c*) Maximum average output power versus duty cycles for the reference (*lower curve*) and the buried-heterostructure (*upper curve*) samples. The results of the model are indicated by dashed lines.

This technique has been recently demonstrated in three-quantum-well active region QC lasers operating at $\lambda = 5\,\mu m$ [87]. High slope efficiency ($dP/dI = 0.69\,W/A$) and fundamental-mode operation in both the vertical and transverse directions were reported at cryogenic temperatures.

5.11 OUTLOOK ON ACTIVE REGION DESIGNS AND CONCLUSIONS

So far we have described work performed using the three-quantum-well active region design. As we noted, this design offers very good performance levels in a wide range of temperatures and wavelengths. It is by far not the only solution to the problem of achieving population inversion between subbands. Ever since its first demonstrations, a large number of other designs have been proposed and demonstrated. As shown in Fig. 5.44, these active regions have been based on the following:

- One-quantum-well, with the population inversion based on tunneling from the ground state of the quantum well and nonparabolicity [88].
- Two-quantum-well, with the population inversion now based on the resonance with the optical phonon [37,65,19].
- A photon-assisted tunneling transition across a barrier, with the population inversion solely based on tunneling [89].
- Transitions between minibands in a superlattice active-region, where the population inversion relies on a difference between the phase space to scatter into and out from the lower state of the laser transition [90].

With the exception of the single-quantum-well active-region device, all of the other designs demonstrated operation up to room temperature. Nevertheless, other than the three-quantum-well design, it is fair to say that the design based on superlattice active region has led to the most interesting developments. The initial devices, based on *doped* superlattice active regions, demonstrated very high output powers (1 W) at cryogenic temperature, but their threshold current density remained prohibitively high [90]. Great progress in performance was achieved by replacing the doping of the active region by a "chirped superlattice" to compensate for the applied electric field [91,54] and by separating the dopants from the electron reservoir [92]. Other groups have also demonstrated high performances using chirped superlattice devices [93,94,23].

The extremely fast electron lifetime from the lower state enables superlattice active region designs to operate at very long wavelengths down to 17 μm [21]. In another work, laser action on two different emission lines ($\lambda = 7$ and 9μm) was demonstrated [95]. Intersubband lasers with superlattice active regions have also demonstrated in the GaAs material system [55].

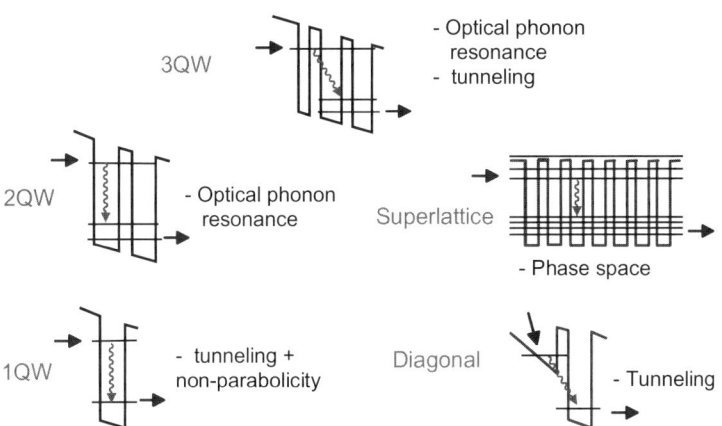

Fig. 5.44 Various active regions described in the literature. They are based on one-quantum-well [88], two-quantum-well [19], and three-quantum-well active regions as extensively discussed in the text; photon-assisted tunneling structures [89]; and superlattice active regions [90].

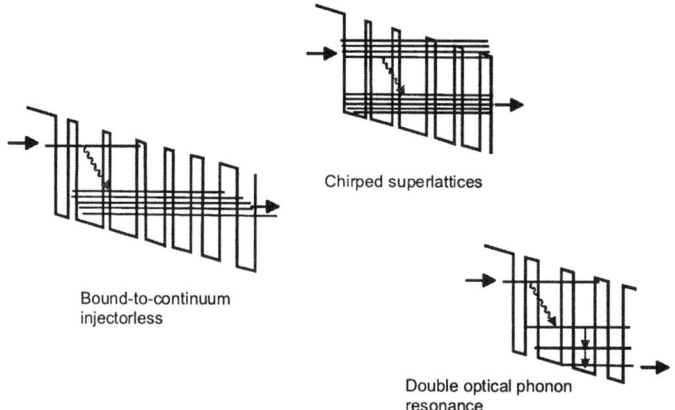

Fig. 5.45 Designs demonstrated recently. The active regions are based on a chirped superlattice [91], bound-to-continuum [98], or two-phonon resonances [99].

Three-quantum-well active region designs have also shown interesting developements. GaAs-based QC devices are particularly attractive because they enable devices based on more than one Al-content to be used in the active region. Specifically, the insertion of a high Al content barrier between the first two quantum wells has been studied [96]. Another attractive possibility is the use of InGaAs in the center of the wells to increase the emission wavelength [97].

New designs trying to combine the features of the three-quantum-well active regions and the superlattice active regions have recently been proposed and demonstrated. As shown in Fig. 5.45, these approaches rely either on a bound-to-continuum or on a two-phonon resonances. In both cases the high injection efficiency into the upper state of the laser transition was preserved by an upper-state wavefunction penetrating significantly into the injection barrier. For the case of the bound-to-continuum design [98], high extraction efficiency was obtained by having the lower state of the laser transition belonging to a miniband.

Excellent high-power, high-temperature operation with up to 90 mW peak power was obtained at 150°C. In the case of the two-phonon transition lasers [99], the high extraction efficiency was gained by spacing the first three levels by an optical phonon energy each, allowing very efficient electron extraction from level $n = 3$. Excellent high-temperature performances were also obtained in this work, with single-mode peak powers up to 1 W at room temperature and 70 mW at 120°C.

ACKNOWLEDGMENTS

JF would like to acknowledge Mattias Beck, Daniel Hofstetter, Stephane Blaser, Michel Rochat, Thierry Aellen from the University of Neuchâtel,

Antoine Muller, Yargo Bonetti, and Hege Bianchini from Alpes Lasers who contributed to parts of this work. CS is very grateful to Hideaki Page, Cyrille Becker, Valentin Ortiz, and Xavier Marcadet for their important contribution in the work that was performed on GaAs structures. Funding from the Swiss National Science Foundation and other EU science agencies is gratefully acknowledged.

REFERENCES

1. F. Capasso, "Bandgap and interface engineering for advanced electronic and photonic devices," *MRS Bulletin* **16**, 23 (1991).
2. R. Kazarinov and S. Suris, "Possibility of the amplification of electromagnetic waves in a semiconductor with a superlattice," *Sov. Phys. Semicond.* **5**, 707 (1971).
3. R. Kazarinov and S. Suris, "Electric and electromagnetic properties of semiconductors with a superlattice," *Sov. Phys. Semicond.* **6**, 120 (1972).
4. A. Cho, *Molecular Beam Epitaxy*, AIP Press, Woodbury, NW, 1994.
5. L. C. West and S. J. Eglash, "First observation of an extremely large-dipole infrared transition within the conduction band of a GaAs quantum well," *Appl. Phys. Lett.* **46**, 1156 (1985).
6. M. Helm, P. England, E. Colas, F. DeRosa, and Jr S. Allen, "Intersubband emission from semiconductor superlattices excited by sequential resonant tunneling," *Phys. Rev. Lett.* **63**, 74 (1989).
7. F. Capasso, K. Mohammed, and A.Y. Cho, "Resonant tunneling through double barriers, perpendicular quantum transport phenomena in superlattices, and their device applications," *IEEE J. Quantum Electron.* **22**, 1853 (1986).
8. H. C. Liu, "A novel superlattice infrared source," *J. Appl. Phys.* **63**, 2856 (1988).
9. A. Kastalsky, V. J. Goldman, and J. H. Abeles, "Possibility of infrared laser in a resonant tunneling structure," *Appl. Phys. Lett.* **59**, 2636 (1991).
10. S. I. Borenstain and J. Katz, "Evaluation of the feasibility of a far-infrared laser based on intersubband transitions in GaAs quantum wells," *Appl. Phys. Lett.* **55**, 654 (1989).
11. Q. Hu and S. Feng, "Feasibility of far-infrared lasers using multiple semiconductor quantum wells," *Appl. Phys. Lett.* **59**, 2923 (1991).
12. P. Yuh and K. L. Wang, "Novel infrared band-aligned superlattice laser," *Appl. Phys. Lett.* **51**, 1404 (1987).
13. G. N. Henderson, L. C. West, T. K. Gaylord, C. W. Roberts, E. N. Glytsis, and M. T. Asom, "Optical transitions to above-barrier quasibound states in asymmetric semiconductor heterostructures," *Appl. Phys. Lett.* **62**, 1432 (1993).
14. J. Faist, F. Capasso, D. L. Sivco, C. Sirtori, A. L. Hutchinson, and A. Y. Cho, "Quantum cascade laser," *Science* **264**, 553 (1994).
15. K. Namjou, S. Cai, E. A. Whittaker, J. Faist, C. Gmachl, F. Capasso, D. L. Sivco, and A. Y. Cho, "Sensitive absorption spectroscopy with a room-temperature distributed-feedback quantum-cascade laser," *Opt. Lett.* **23**, 219 (1998).

16. S. W. Sharpe, J. F. Kelly, J. S. Hartmann, C. Gmachl, F. Capasso, D. L. Sivco, J. N. Baillargeon, and A. Y. Cho, "High-resolution (Doppler-limited) spectroscopy using quantum-cascade distributed-feedback lasers," *Opt. Lett.* **23**, 1396 (1998).
17. B. A. Paldus, C. C.. Harb, T. G. Spence, R. N. Zare, C. Gmachl, F. Capasso, D. L. Sivco, J. N. Baillargeon, A. L. Hutchinson, and A. Y. Cho, "Cavity ringdown spectroscopy using mid-infrared quantum-cascade lasers,". *Opt. Lett.* **25**, 666 (2000).
18. J. Faist, F. Capasso, C. Sirtori, D. L. Sivco, J. N. Baillargeon, A. L. Hutchinson, S. Chu, and A. Y. Cho, "High power mid-infrared ($\lambda \sim 5\mu m$) quantum cascade lasers operating above room temperature," *Appl. Phys. Lett.* **68**, 3680 (1996).
19. C. Sirtori, J. Faist, F. Capasso, D. L. Sivco, A. L. Hutchinson, and A. Y. Cho, "Mid-infrared ($8.5\mu m$) semiconductor lasers operating at room temperature," *IEEE Photon. Technol. Lett.* **9**, 294 (1997).
20. J. Faist, A. Tredicucci, F. Capasso, C. Sirtori, D. L. Sivco, J. N. Baillargeon, A. L. Hutchinson, and A. Y. Cho, "High-power continuous-wave quantum cascade lasers," *IEEE J. Quantum Electron.* **34**, 336 (1998).
21. A. Tredicucci, C. Gmachl, F. Capasso, D. L. Sivco, A. L. Hutchinson, and A. Y. Cho, "Long wavelength superlattice quantum cascade lasers at $\lambda = 17\mu m$," *Appl. Phys. Lett.* **74**, 638 (1999).
22. S. Slivken, C. Jelen, A. Rybaltowski, J. Diaz, and M. Razeghi, "Gas-source molecular beam epitaxy growth of an $8.5\mu m$ quantum cascade laser," *Appl. Phys. Lett.* **71**, 2593 (1997).
23. S. Slivken, A. Matlis, C. Jelen, A. Rybaltowski, J. Diaz, and M. Razeghi, "High-temperature continuous-wave operation of $\lambda = 8\mu m$ quantum cascade lasers," *Appl. Phys. Lett.* **74**, 173 (1999).
24. A. L. Li, J. X. Chen, Q. K. Yang, and Y. C. Ren, "GSMBE grown infrared quantum cascade laser structures," *J. Cryst. Growth* **201/202**, 901 (1999).
25. C. Sirtori, P. Kruck, S. Barbieri, P. Collot, J. Nagle, M. Beck, J. Faist, and U. Oesterle, "GaAs/Al$_x$Ga$_{1-x}$As quantum cascade lasers," *Appl. Phys. Lett.* **73**, 3486 (1998).
26. J. C. Garcia, E. Rosencher, P. Collot, N. Laurent, J. L. Guyaux, B. Winter, and J. Nagle, "Epitaxially stacked lasers with Esaki junction: a bipolar cascade laser," *Appl. Phys. Lett.* **71**, 3752 (1997).
27. C. Gmachl, F. Capasso, A. Tredicucci, D. L. Sivco, A. L. Hutchinson, S. Chu, and A. Y. Cho, "Noncascaded intersubband injection lasers at $\lambda = 7.7\mu m$," *Appl. Phys. Lett.* **73**, 3830 (1998).
28. A. Korshak, Z. S. Gribnikov, and V. V. Mitin, "Tunnel-junction-connected distributed-feedback vertical-cavity surface-emitting laser," *Appl. Phys. Lett.* **73**, 1475 (1998).
29. J. K. Kim, S. Nakagawa, E. Hall, and L. A. Coldren, "Near-room-temperature continuous-wave operation of multiple-active-region $1.55\mu m$ vertical-cavity lasers with high differential efficiency," *Appl. Phys. Lett.* **77**, 3137 (2000).
30. A. Yariv, *Quantum Electronics*, John Wiley & Sons, New York, 1989.
31. C. Sirtori, J. Faist, F. Capasso, D. L. Sivco, A. L. Hutchinson, and A. Y. Cho, "Long wavelength infrared $\lambda \sim 11\mu m$ quantum cascade lasers," *Appl. Phys. Lett.* **69**, 2810 (1996).
32. J. Faist, C. Gmachl, M. Striccoli, C. Sirtori, F. Capasso, D. L. Sivco, and A. Y. Cho, "Quantum cascade disk lasers," *Appl. Phys. Lett.* **69**, 2456 (1996).
33. J. Faist, F. Capasso, D. L. Sivco, A. L. Hutchinson, S. Chu, and A. Y. Cho, "Short wavelength ($\lambda \sim 3.4\mu m$) quantum cascade laser based on strained compensated InGaAs/AlInAs," *Appl. Phys. Lett.* **72**, 680 (1998).

34. R. K. Köhler, C. Gmachl, A. Tredicucci, F. Capasso, D. L. Sivco, S. Chu, and A. Y. Cho, "Single-mode tunable pulsed, and continuous wave quantum-cascade distributed feedback lasers at λ ~ μm 4.6–4.7 μm," *Appl. Phys. Lett.* **76**, 1092 (2000).
35. C. Sirtori, J. Faist, F. Capasso, D. L. Sivco, A. L. Hutchinson, and A. Y. Cho, "Quantum cascade laser with plasmon-enhanced waveguide operating at 8.4 μm wavelength," *Appl. Phys. Lett.* **66**, 3242 (1995).
36. J. Faist, C. Sirtori, F. Capasso, D. L. Sivco, J. N. Baillargeon, A. L. Hutchinson, and A. Y. Cho, "High-power long-wavelength (λ = 11.5 μm) Quantum cascade lasers operating above room temperature," *IEEE Photon. Technol. Lett.* **10**, 1100 (1998).
37. J. Faist, F. Capasso, C. Sirtori, D. L. Sivco, A. L. Hutchinson, and A. Y. Cho, "Vertical transition quantum cascade laser with Bragg confined excited state," *Appl. Phys. Lett.* **66**, 538 (1995).
38. C. Gmachl, A. Tredicucci, F. Capasso, A. L. Hutchinson, D. L. Sivco, J. N. Baillargeon, and A. Y. Cho, "High-power λ ~ 8 μm quantum cascade laser with near optimum performance," *Appl. Phys. Lett.* **72**, 3130 (1998).
39. A. Müller, M. Beck, J. Faist, and U. Oesterle, "Electrically tunable, room-temperature quantum-cascade lasers," *Appl. Phys. Lett.* **75**, 1509 (1999).
40. J. Faist, F. Capasso, C. Sirtori, D. L. Sivco, A. L. Hutchinson, S. Chu, and A. Y. Cho, "Narrowing of the intersubband electroluminescent spectrum in coupled-quantum well heterostructures," *Appl. Phys. Lett.* **65**, 94 (1994).
41. M. Troccoli, G. Scamarcio, V. Spagnolo, A. Tredicucci, C. Gmachl, F. Capasso, D. L. Sivco, A. Y. Cho, and M. Striccoli, "Electronic distribution in superlattice quantum cascade lasers," *Appl. Phys. Lett.* **77**, 1088 (2000).
42. R. Claudia Iotti and F. Rossi, "Carrier thermalization versus phonon-assisted relaxation in quantum-cascade lasers: A monte carlo approach," *Appl. Phys. Lett.* **78**, 2902 (2001).
43. C. Sirtori, J. Faist, F. Capasso, D. L. Sivco, A. L. Hutchinson, and A. Y. Cho, "Pulsed and continuous-wave operation of long wavelength infrared (λ = 9.3 μm) quantum cascade lasers," *IEEE J. Quantum Electron.* **33**, 89 (1997).
44. C. Sirtori, P. Kruck, S. Barbieri, H. Page, J. Nagle, M. Beck, J. Faist, and U. Oesterle, "Low-loss al-free waveguides for unipolar semiconductor lasers," *Appl. Phys. Lett.* **75**, 3911 (1999).
45. B. W. Hakki and T. L. Paoli, "Gain spectra in gaas double-heterostructure injection lasers," *J. Appl. Phys.* **46**, 1299 (1975).
46. D. Hofstetter and J. Faist, "Measurement of semiconductor laser gain and dispersion curves utilizing fourier transforms of the emission spectra," *IEEE Photon. Technol. Lett.* **11**, 1372 (1999).
47. M. Rochat, M. Beck, J. Faist, and U. Oesterle, "Measurement of far-infrared waveguide loss using a single-pass technique," *Appl. Phys. Lett.* **78**, 1967 (2001).
48. K. L. Campman, H. Schmidt, A. Imamoglu, and A. C. Gossard, "Interface roughness and alloy-disorder scattering contributions to intersubband transition linewidths," *Appl. Phys. Lett.* **69**, 2554 (1996).
49. C. Sirtori, C. Gmachl, F. Capasso, J. Faist, D. L. Sivco, A. L. Hutchinson, and A. Y. Cho, "Long-wavelength (λ = 8–11.5 μm) semiconductor lasers with waveguides based on surface plasmons," *Opt. Lett.* **23**, 1366 (1998).
50. C. Gmachl, F. Capasso, A. Tredicucci, D. L. Sivco, R. Köhler, A. L. Hutchinson, and A. L. Cho, "Dependence of the device performance on the number of stages in quantum-cascade lasers," *IEEE J. Select. Topics Quantum Electron.* **5**, 808 (1999).

51. W. W. Bewley, C. L. Felix, E. H. Aifer, I. Vurgaftman, L. J. Olafsen, J. R. Meyer, H. Lee, R. U. Martinelli, J. C. Connolly, A. R. Sugg, G. H. Olsen, M. Yang, B. R. Bennett, and B. V. Shanabrook, "Above-room-temperature optically pumped mid-infrared W lasers," *Appl. Phys. Lett.* **73**, 3833 (1998).

52. C. Sirtori, F. Capasso, J. Faist, A. L. Hutchinson, D. L. Sivco, and A. Y. Cho, "Resonant tunneling in quantum cascade lasers," *IEEE J. Quantum Electron.* **34**, 1722 (1998).

53. R. Paiella, F. Capasso, C. Gmachl, C. G. Bethea, D. L. Sivco, J. N. Baillargeon, A. L. Hutchinson, and A. Y. Cho, "High-speed operation of gain-switched midinfrared quantum cascade lasers," *Appl. Phys. Lett.* **75**, 2536 (1999).

54. A. Tredicucci, F. Capasso, C. Gmachl, D. L. Sivco, A. L. Hutchinson, A. Y. Cho, J. Faist, and G. Scamarcio, "High-power inter-miniband lasing in intrinsic superlattices," *Appl. Phys. Lett.* **72**, 2388 (1998).

55. G. Strasser, S. Gianordoli, L. Hvozdara, W. Schrenk, K. Unterrainer, and E. Gornik, "GaAs/AlGaAs superlattice quantum cascade lasers at $\lambda \sim 13\,\mu m$," *Appl. Phys. Lett.* **75**, 1345 (1999).

56. C. Sirtori, S. Barbieri, P. Kruck, V. Piazza, M. Beck, J. Faist, U. Oesterle, P. Collot, and J. Nagle, "Influence of dx centers on the performance of unipolar semiconductor lasers based on GaAs-Al$_x$Ga$_{1-x}$As," *IEEE Photon. Technol. Lett.* **11**, 1090 (1999).

57. C. Walle, "Band linups and deformation potentials in the model-solid theory," *Phys. Rev.* B**39**, 1871 (1989).

58. H. Page, A. Robertson, C. Sirtori, C. Becker, G. Glastre, and J. Nagle, "Demonstration of $\lambda \sim 11.6\,\mu m$ GaAs-based quantum cascade laser operating on a Peltier cooled element," *IEEE Photon. Technol. Lett.* **13**, 556 (2001).

59. P. Kruck, H. Page, C. Sirtori, S. Barbieri, M. Stellmacher, and J. Nagle, "Improved temperature performance of Al$_{0.33}$Ga$_{0.67}$As/GaAs quantum-cascade lasers with emission wavelength at $\lambda \sim 11\,\mu m$," *Appl. Phys. Lett.* **76**, 3340 (2000).

60. S. Barbieri, C. Sirtori, H. Page, M. Beck, J. Faist, and J. Nagle, "Gain measurements on gaas-based quantum cascade lasers using a two-section cavity technique," *IEEE J. Quantum Electron.* **36**, 736 (2000).

61. S. Barbieri, C. Sirtori, H. Page, M. Stellmacher, and J. Nagle, "Design strategies for gaas-based unipolar lasers: Optimum injector-active region coupling via resonant tunneling," *Appl. Phys. Lett.* **78**, 282 (2001).

62. C. Becker, C. Sirtori, H. Page, G. Glastre, V. Ortiz, X. Marcadet, M. Stellmacher, and J. Nagle, "AlAs/GaAs quantum cascade lasers based on large direct conduction band discontnuity," *Appl. Phys. Lett.* **77**, 463 (2000).

63. K. Huang and B. Zhu, "Dielectric continuum model and frölich interaction in superlattices," *Phys. Rev.* B**38**, 13377 (1988).

64. J. Faist, F. Capasso, C. Sirtori, D. L. Sivco, A. L. Hutchinson, and A. Y. Cho, "Continuous wave operation of a vertical transition quantum cascade laser above T = 80 K," *Appl. Phys. Lett.* **67**, 3057 (1995).

65. C. Sirtori, J. Faist, F. Capasso, D. L. Sivco, A. L. Hutchinson, S. Chu, and A. Y. Cho, "Continuous wave operation of mid-infrared (7.4–8.6 μm) quantum cascade lasers up to 110k temperature," *Appl. Phys. Lett.* **68**, 1745 (1996).

66. R. M. Williams, J. F. Kelly, J. S. Hartman, S. W. Sharpe, M. S. Taubman, J. L. Hall, F. Capasso, C. Gmachl, D. L. Sivco, J. N. Baillargeon, and A. Y. Cho, "Kilohertz linewidth from frequency-stabilized mid-infrared quantum cascade lasers," *Opt. Lett.* **24**, 1844 (1999).

67. H. Kogelnik and C. Chank, "Coupled-wave theory of distributed feedback lasers," *J. Appl. Phys.* **43**, 2327 (1972).
68. J. Faist, C. Gmachl, F. Capasso, C. Sirtori, D. L. Sivco, J. N. Baillargeon, and A. Y. Cho, "Distributed feedback quantum cascade lasers," *Appl. Phys. Lett.* **70**, 2670 (1997).
69. W. Schrenk, N. Finger, S. Gianordoli, L. Hvozdara, G. Strasser, and E. Gornik, "GaAs/AlGaAs distributed feedback quantum cascade lasers," *Appl. Phys. Lett.* **76**, 253 (2000).
70. N. Finger, W. Schrenk, and E. Gornik, "Analysis of tm-polarized dfb laser structures with metal surface gratings," *IEEE J. Quantum Electron.* **36**, 780 (2000).
71. R. Köhler, C. Gmachl, F. Capasso, A. Tredicucci, D. L. Sivco, and A. Y. Cho, "Single-mode tunable quantum cascade lasers in the spectral range of the CO_2 laser at $\lambda = 9.5–10.5 \mu m$," *IEEE Photon. Technol. Lett.* **12**, 474 (2000).
72. C. Gmachl, J. Faist, J. N. Baillargeon, F. Capasso, C. Sirtori, D. L. Sivco, S. Chu, and A. Y. Cho, "Complex-coupled quantum cascade distributed-feedback laser," *IEEE Photon. Technol. Lett.* **9**, 1090 (1997).
73. C. Gmachl, F. Capasso, J. Faist, A. L. Hutchinson, A. Tredicucci, D. L. Sivco, J. N. Baillargeon, S. Chu, and A. Y. Cho, "Continuous-wave and high-power pulsed operation of index-coupled distributed quantum cascade laser at $\lambda \sim 8.5 \mu m$," *Appl. Phys. Lett.* **72**, 1430 (1998).
74. C. Gmachl, F. Capasso, A. Tredicucci, D. L. Sivco, J. N. Baillargeon, A. L. Hutchinson, and A. Y. Cho, "High power, continuous-wave, current-tunable, single-mode quantum-cascade distributed-feedback lasers at $\lambda \sim 5.2 \mu m$ and $\lambda \sim 7.95 \mu m$," *Opt. Lett.* **25**, 230 (2000).
75. D. Hofstetter, J. Faist, M. Beck, A. Müller, and U. Oesterle, "Demonstration of high-performance 10.6 μm quantum cascade distributed feedback lasers fabricated without epitaxial regrowth," *Appl. Phys. Lett.* **75**, 665 (1999).
76. D. Hofstetter, J. Faist, M. Beck, and U. Oesterle, "Surface-emitting 10.1 μm quantum-cascade distributed feedback lasers," *Appl. Phys. Lett.* **75**, 3769 (1999).
77. C. Gmachl, J. Faist, F. Capasso, C. Sirtori, D. L. Sivco, and A. Y. Cho, "Long-wavelength (9.5–11.5 μm) microdisk quantum cascade lasers," *IEEE J. Quantum Electron.* **33**, 1567 (1997).
78. R. Slusher, A. Levi, U. Mohideen, S. McCall, S. Pearton, and R. Logan, "Threshold characteristics of semiconductor microdisk lasers," *Appl. Phys. Lett.* **63**, 1310 (1993).
79. S. Gianordoli, L. Hvozdara, G. Strasser, W. Schrenk, K. Unterrainer, and E. Gornik, "Gaas/algaas-based microcylinder lasers emitting at 10 μm," *Appl. Phys. Lett.* **75**, 1045 (1999).
80. J. Gerard, B. Sermage, B. Gayral, B. Legrand, E. Costard, and V. Thierry-Mieg, "Enhanced spontaneous emission by quantum boxes in a monolithic optical microcavity," *Phys. Rev. Lett.* **81**, 1110 (1998).
81. P. Michler, A. Kiraz, C. Becher, W. Shoenfeld, P. Petroff, L. Zhang, and A. Imamoglu, "A quantum dot single-photon turnstile device," *Science* **290**, 2282 (2000).
82. C. Gmachl, F. Capasso, E. Narimanov, J. U. Nöckel, D. Stone, J. Faist, D. L. Sivco, and A. Y. Cho, "High-power directional emission from microlasers with chaotic resonators," *Science* **280**, 1556 (1998).

83. S. Gianordoli, L. Hvozdara, G. Strasser, W. Schrenk, J. Faist, and E. Gornik, "Long-wavelength ($\lambda = 10\,\mu m$) quadrupolar-shaped GaAs-AlGaAs microlasers," *IEEE J. Quantum Electron.* **36**, 458 (2000).
84. L. Hvozdara, A. Lugstein, N. Finger, S. Gianordoli, W. Schrenk, K. Unterrainer, E. Bertagnolli, G. Strasser, and E. Gornik, "Quantum cascade lasers with monolithic air-semiconductor Bragg reflectors," *Appl. Phys. Lett.* **77**, 1241 (2000).
85. M. Beck, J. Faist, C. Gmachl, F. Capasso, D. Sivco, J. Baillargeon, and A. Hutchinson, "Buried heterostructure quantum cascade lasers," *SPIE Proc.* **3284**, 231 (1998).
86. M. Beck, J. Faist, U. Oesterle, M. Ilegems, E. Gini, and H. Melchior, "Buried heterostructure quantum cascade lasers with a large optical cavity waveguide," *IEEE Photon. Technol. Lett.* **12**, 1450 (2000).
87. C. D. Farmer, P. T. Keithley, C. N. Ironside, C. R. Stanley, L. R. Wilson, and J. W. Cockburn, "A quantum cascade laser fabricated using planar native-oxide layers," *Appl. Phys. Lett.* **77**, 25 (2000).
88. J. Faist, F. Capasso, C. Sirtori, D. L. Sivco, A. L. Hutchinson, M. S. Hybertsen, and A. Y. Cho, "Quantum cascade lasers without intersubband population inversion," *Phys. Rev. Lett.* **76**, 411 (1996).
89. J. Faist, F. Capasso, C. Sirtori, D. L. Sivco, A. L. Hutchinson, and A. Y. Cho, "Laser action by tuning the oscillator strength," *Nature* **387**, 777 (1997).
90. G. Scamarcio, F. Capasso, C. Sirtori, J. Faist, A. L. Hutchinson, D. L. Sivco, and A. Y. Cho, "High-power infrared (8-µm wavelength) superlattice lasers," *Science* **276**, 773 (1997).
91. A. Tredicucci, F. Capasso, C. Gmachl, D. L. Sivco, A. L. Hutchinson, and A. Y. Cho, "High performance interminiband quantum cascade lasers with graded superlattices," *Appl. Phys. Lett.* **73**, 2101 (1998).
92. A. Tredicucci, F. Capasso, C. Gmachl, D. L. Sivco, A. L. Hutchinson, and A. Y. Cho, "High-performance quantum cascade lasers with electric-field-free undoped superlattice," *IEEE Photon. Technol. Lett.* **12**, 260 (2000).
93. A. Matlis, S. Slivken, A. Tahraoui, K. J. Luo, Z. Wu, A. Rybaltowski, C. Jelen, and M. Razeghi, "Low-threshold and high power $\lambda \sim 9.0\,\mu m$ quantum cascade lasers operating at room temperature," *Appl. Phys. Lett.* **77**, 1741 (2000).
94. A. Tahraoui, A. Matlis, S. Slivken, J. Diaz, and M. Razeghi, "High-performance quantum cascade lasers ($\lambda \sim 11\,\mu m$) operating at high temperature (T > 425 K)," *Appl. Phys. Lett.* **78**, 416 (2001).
95. A. Tredicucci, C. Gmachl, F. Capasso, D. L. Sivco, A. L. Hutchinson, and A. Y. Cho, "A multiwavelength semiconductor laser," *Nature* **396**, 350 (1998).
96. L. R. Wilson, P. T. Keightley, J. W. Cockburn, M. S. Skolnick, J. C. Clark, R. Grey, and G. Hill, "Controlling the performance of GaAs-AlGaAs quantum-cascade lasers via barrier height modifications," *Appl. Phys. Lett.* **76**, 801 (2000).
97. L. R. Wilson, J. W. Cockburn, M. J. Steer, D. A. Carder, M. S. Skolnick, M. Hopkinson, and G. Hill, "Decreasing the emission wavelength of GaAs-AlGaAs quantum cascade lasers by the incorporation of ultrahin InGaAs layers," *Appl. Phys. Lett.* **78**, 413 (2001).
98. J. Faist, M. Beck, T. Aellen, and E. Gini, "Quantum cascade lasers based on a bound-to-continuum transition," *Appl. Phys. Lett.* **78**, 147 (2001).
99. D. Hofstetter, M. Beck, T. Aellen, and J. Faist, "High-temperature operation of distributed feedback quantum-cascade lasers at $5.3\,\mu m$," *Appl. Phys. Lett.* **86**, 396 (2001).

CHAPTER 6

Widely Tunable Far-Infrared Hot-Hole Semiconductor Lasers

ERIK BRÜNDERMANN

6.1 INTRODUCTION

The idea of hot-hole lasers, and especially lasers using germanium material, has a long history that reaches back to the 1950s. Shockley [1] analyzed electron mobility data in germanium and its dependence on temperature and electric fields up to 40 kV/cm. He found that the significant scattering process of hot carriers is interaction with optical phonons, mainly optical phonon emission. This process has a threshold in carrier energy, the optical phonon energy which is 37 meV in germanium.

For high enough electric fields and at low temperature hot carriers accelerate without acoustical phonon interaction along the crystallographic direction in which the electric field is applied. These hot carriers reach the optical phonon energy and lose all their energy due to emission of optical phonons. They accelerate again, repeating this directional motion in momentum space. This motion is called streaming motion.

The idea of using such a hot carrier, nonequilibrium distribution for amplification of long wavelength radiation was presented in 1958 by Krömer [2]. He proposed a so-called negative mass amplifier and generator (NEMAG) that utilizes warping of the constant energy surfaces and nonelastic optical phonon scattering. It was thought that such an oscillator using p-type germanium could be operated with an electric field of a few thousand V/cm along a ⟨100⟩ crystallographic axis.

Kurosawa and Maeda used Monte Carlo simulations to investigate the properties of hole distributions in p-type germanium and in *perpendicular* electric and magnetic fields [3,4]. They found a region in momentum space where carriers are accumulated, since carriers oscillate at the cyclotron reso-

Long-Wavelength Infrared Semiconductor Lasers, Edited by Hong K. Choi
ISBN 0-471-39200-6   Copyright © 2004 John Wiley & Sons, Inc.

nance frequency with negligible scattering interaction. Within this work they proposed a laser mechanism that uses a transition from the accumulated region to the region where the carriers are not accumulated, since carriers frequently emit optical phonons and experience inelastic scattering in streaming motion.

In 1979 Andronov [5] proposed involving two bands in this mechanism, the valence bands of light and heavy holes, to form a population inversion. Finally, at the beginning of the 1980s several experiments demonstrated a population inversion in p-type germanium at 4.2 and 10 K [6,7,8,9,10]. Vorobjev et al. [10] found that a population inversion also exists at 80 K although less pronounced.

Traditionally the laser process is divided into two mechanisms: the first involves a population inversion between the light- and heavy-hole bands (so-called intervalence-band or IVB lasers) and the second is based on a population inversion between two light-hole Landau levels (so-called light-hole cyclotron resonance or LHCR lasers). This distinction is somewhat artificial because within the IVB laser a variety of simultaneous transitions occur that can be attributed to LHCR-type transitions. The LHCR laser is just a special case. All other possible transitions have very low gain, such as by a special choice of crystallographic axes and acceptor concentration, to only maintain laser action between two light-hole Landau levels. In general, the upper laser states are light-hole states.

Strong stimulated emission was observed in 1984. Laser emission was attributed to transitions between the valence bands of light and heavy holes [11,12] and between two light-hole Landau levels [13]. The emission was stimulated in the wavelength range from 100 to 300 μm with an output power up to 10 W. The acceptor concentration of these germanium lasers ranged from 6×10^{12} cm^{-3} to 5×10^{14} cm^{-3}.

Another topic in the 1980s was the emission of germanium crystals in *parallel* electric and magnetic fields. The emitted wavelength lies between 1 and 5 mm. The emission is based on cyclotron resonance transitions of heavy holes with negative effective mass. This device comes close to the initial proposal by Krömer. A line width of less than 6 MHz was found for those lasers that allowed spectroscopic applications [14].

Various review articles [15,16,17] give a detailed overview of the field until 1987. A compilation of papers on germanium lasers and masers can be found in special issues [18,19] published in 1991 and 1992.

Until 1995 all germanium lasers were doped with the hydrogenic acceptor gallium, except one that was doped with thallium [20]. These lasers operated with a low-duty cycle of 10^{-5}. The laser performance was dramatically improved with the discovery of germanium laser material doped with nonhydrogenic acceptors, such as beryllium [21].

This material has led to strong laser emission up to duty cycles of 5% [22], repetition rates up to 45 kHz [23], and laser pulse lengths up to 32 μs [23]. Recently nonthermal far-infrared emission was detected for continuous excitation [22].

Further technological improvements have led to a more convenient laser operation. Germanium lasers can be operated with small permanent magnets and in closed cycle refrigerators [24] at temperatures of up to 40 K [23] without the need for liquid helium.

6.1.1 Tunable Germanium Lasers

One key feature of the germanium laser is its large tuning range. The laser can be tuned by 60% in reference to the middle frequency of 2.5 THz or 83 cm^{-1}, respectively. In particular, lasers made from beryllium-doped germanium material can be tuned continuously over the spectral range from 30 to 140 cm^{-1} or 1 to 4 THz [25].

The peak of the gain spectrum, typically 10 cm^{-1} wide, is tuned by an external magnetic induction from 0.2 to 1.7 T. The rather wide gain spectrum can already be understood in a simple semiclassical model described in Section 6.2.1. The underlying quantum mechanical mechanism of Landau level splitting is discussed in Section 6.2.6 in detail. Within this gain spectrum a mechanical external resonator can be used to achieve single mode tunable far-infrared laser radiation. Germanium laser resonators will be discussed in Section 6.5.1.

Above magnetic inductions of 1.7 T the gain spectrum narrows significantly because only a limited number of Landau level cyclotron resonance transitions, mainly light- to light-hole level transitions, are allowed. Then the emission frequency is linear tunable with the magnetic induction with a ratio of approximately 20 cm^{-1}T^{-1}. The laser emits in a very narrow line (less than 0.1 cm^{-1}) even without an external mechanical resonator.

6.1.2 Motivation

The strong interest in a coherent, compact, powerful, tunable, continuous-wave, solid state, and far-infrared laser is driven by a multitude of potential applications. Far-infrared molecular spectroscopy is an important tool to investigate chemical processes that occur in astronomical objects and in the atmosphere. Far-infrared lasers would be valuable as local oscillators in heterodyne receivers for the study of quantized rotational states of molecules and fine structure lines of atoms in atmospheric research [26] and in starforming regions [27]. However, far-infrared radiation is strongly absorbed by water molecules at sea level, so most observations of the latter kind need to take place on airborne and spaceborne platforms. The limited space, power, and time of flight require compact size, continuous-wave operation, and low power consumption.

A far-infrared, compact, powerful, and tunable laser would also be very useful in laboratory-based applications. Many materials have signatures in the far-infrared or terahertz frequency region including dielectrics [28], semiconductors [28], superconductors [29], liquids [30], and gases [31].

Interesting physical phenomena also exist in this spectral region: van der Waals bonding energies, such as in soft molecular crystals, molecular clusters, biomolecules and organic semiconductors; bonding energies of molecules bound to surfaces and dust particles; molecular formation in flames [32]; protein vibrational states [33], and modes in DNA, such as base roll and propeller twist modes [34]; acoustical and optical phonons; hole energy states; and cyclotron resonance frequencies in three and lower dimensional systems. The germanium laser is a powerful, tunable, and very compact source operating in the terahertz frequency range which will allow us to harvest this wealth of scientific information.

The germanium laser system also offers a wide range of very interesting physically and technologically challenging problems. The laser depends on a rather large number of parameters. These parameters are impurity, acoustical and optical phonon scattering, band structure, band warping, crystal symmetry, crystallographic orientations, electric and magnetic fields, Landau level structure, Stark effect, the magnetically induced Hall effect, which in turn depends on the device geometry and device dimensions, photon-hole interaction, and temperature.

The near equal strength of the electric and magnetic field can lead to new physics because most textbook theories treat cases in which only one of the two fields is weak. The application of uniaxial stress gives another adjustable parameter to influence the band structure and scattering processes, mainly impurity scattering.

Gas laser heterodyne spectrometers have been used on an airborne platform for astronomical observations [35]. Continuous-wave emission of the local oscillator is desired to efficiently use the flight or mission time. A Schottky diode can be used as a mixer in such heterodyne receivers. The diode typically requires 1 mW/THz local oscillator power. Hot electron bolometers can lower this power demand by one or two orders of magnitude.

Gas laser heterodyne spectrometer have a frequency resolution of 1 MHz at terahertz frequencies. This resolution is necessary to analyze Doppler-shifted molecular emission that allows one to extract information on the dynamic processes in molecular clouds. This information is crucial to determine the precursors of star formation.

As will be shown in the following sections, the germanium laser is a very useful device that not only promises to deliver the high performance required in airborne applications but also high power and short pulses, attractive features in laboratory-based research. Short, several tens of picoseconds (ps) long pulses can be generated by mode-locked germanium lasers.

6.1.3 Applications

There have been several examples of terahertz imaging using short ps-pulses emitted from nonlinear crystals that show great promise for applications in diverse fields like production quality control and medical imaging. However, such sources cannot deliver the high spectral purity and sufficient power that is required for many spectroscopic applications in semiconductor physics and physical chemistry.

The germanium laser is on the verge of becoming a source for widespread spectroscopic applications due to a range of improvements like convenient laser operation in a closed cycle machine, tunable resonators, and improved duty cycle. However, there were already a few applications of the germanium lasers in spectroscopy: intracavity absorption measurements of quantum wells [36], spectroscopy of semiconductors [13,20], and spectroscopy of H_2O [37].

6.2 HOT-HOLE LASER MODEL

In this section the concept of the hot-hole laser is introduced. Qualitative and some quantitative results can already be derived from a very simple semiclassical model. This model will build a bridge to a quantum mechanical model and more detailed articles in the literature. The theoretical background of such lasers will be discussed in the light of experimental investigations. Carrier transport in a bulk semiconductor, band structure, effects due to the applied external electric and magnetic fields, and scattering mechanisms will be addressed. A few comments on the optical gain and on Monte Carlo simulations are given. The section concludes with thermal effects because device heating always plays a prominent role in the laser development from an initial pulsed to continuous-wave operation.

6.2.1 Semiclassical Model

Far-infrared laser emission cannot be generated in a diode laser. The diode laser is based on a population inversion between the conduction and valence bands and the laser emission results from recombination of electrons and holes. The available band gaps of semiconductors only allow near to mid-infrared emission. A diode laser cannot be realized in germanium because the minimum of the conduction band is found at the L point, whereas the maximum of the valence band is found at the Γ point in the center of the Brillioun zone (see Fig. 6.1).

However, in germanium a laser based on a hot-hole population inversion between two valence bands can be realized. The valence band is split into three bands: the band of heavy holes, the band of light holes, and the hole split-off band that is not degenerate like the other two bands at the Γ point due to spin-orbit interaction. By forming a population inversion between the light- and heavy-hole bands, a wide range of frequencies from very low frequencies up to a fundamental limit can be generated due to the degeneration of the bands at the Γ point. This fundamental limit is the optical phonon energy, which has a value of 37 meV in germanium corresponding to 9 THz. In prac-

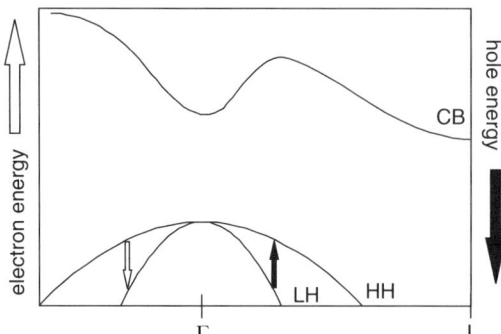

Fig. 6.1 Part of the germanium band structure (not to scale) with the conduction (CB) and two of the valence bands: heavy hole (HH) and light hole (LH) band. A population inversion can form between the light and heavy hole bands. For electrons the laser emission is due to transitions from the HH band to the LH band. For holes the reader may rotate the figure by 180° to obtain the LH band as the upper overpopulated band with vertical transitions to the HH band, the ground states.

tice, germanium lasers emit frequencies ranging from 1 to 4 THz or 30 to 140 cm^{-1} [25], respectively.

A population inversion between the two-hole bands will result in photon emission by vertical transitions of electrons from the heavy hole band to the light hole band. In Fig. 6.1 the y-axis is the electron energy. In the following discussion holes are chosen as the carriers of interest and the diagram can be rotated by 180° so that now vertical transitions of holes from top to bottom, from the upper light hole band to the lower heavy hole band, are obtained.

The split-off band is neglected because of its separation from the Γ point by 0.3 eV. Only processes that are less separated in energy from the valence band maximum up to a few multiples of the optical phonon energy are considered. In silicon all three bands are important because the split-off band separation of 44 meV is smaller than the optical phonon energy of 63 meV. The conduction band is also neglected, which leaves the two bands of interest, the light- and heavy-hole band. More details on silicon hot hole lasers can be found in Section 6.6.3.

Germanium is a very suitable material for the hot-hole laser mechanism because it can be grown in a very pure state and is easily doped with acceptors with good control of the donor concentration and, therefore, doping compensation. A bulk single crystal of p-type germanium is the laser material.

Two-Dimensional Model

The formation of the population inversion between the light- and heavy-hole bands can be described in a simple, semiclassical picture. At low temperature, acoustical phonon scattering is negligible. In practice, the laser would be cooled below 20 K, for example, by liquid helium. Impurity scattering can also be neglected for a low acceptor concentration. A further condition is an electric $\vec{E}$ field perpendicular to a magnetic induction $\vec{B}$. The latter is often called a crossed electric and magnetic field. Most carriers are at low energy with a small momentum before initiation of the electric field or pump pulse.

The equation of motion consisting of the combined electric field and Lorentz force is used:

$$\vec{F} = \hbar \frac{d\vec{k}}{dt} = m* \frac{d\vec{v}}{dt} = e(\vec{E} + \vec{v} \times \vec{B}), \tag{6.1}$$

with the charge e, the effective mass $m*$, and the velocity $\vec{v}$ of the holes. In a xyz-coordinate system the electric field is oriented parallel to the x-axis ($\vec{E} = (E, 0, 0)$), and the magnetic induction is parallel to the z-axis ($\vec{B} = (0, 0, B)$). The initial hole velocity is bound by $\vec{v}(t = 0) = (0, 0, 0)$. In solving Eq. (6.1), we obtain an orbit in velocity space:

$$v_x = \frac{E}{B} \sin(\omega_{CR} t),$$
$$v_y = \frac{E}{B} (\cos(\omega_{CR} t) - 1),$$
$$v_y = 0. \tag{6.2}$$

The kinetic hole energy W is

$$W = \frac{m*}{2} \vec{v}^2$$
$$= m* \left(\frac{E}{B}\right)^2 (1 - \cos(\omega_{CR} t))$$
$$= 2W_D (1 - \cos(\omega_{CR} t)), \tag{6.3}$$

and it uses the drift energy

$$W_D = \frac{m*}{2} \left(\frac{E}{B}\right)^2 \tag{6.4}$$

The hole oscillates at the cyclotron resonance (CR) frequency $\omega_{CR} = eB/m*$ through the origin in an orbit centered at the point $P_D = (0, v_D, 0)$. The drift velocity v_D is given by the applied fields

$$\vec{v}_D = \frac{\vec{E} \times \vec{B}}{B^2} \tag{6.5}$$

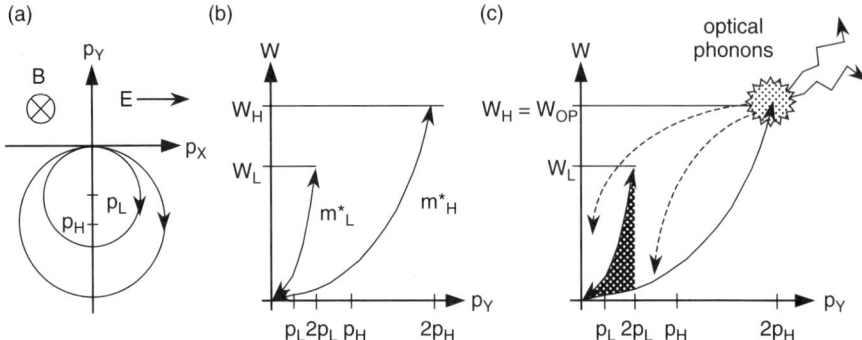

Fig. 6.2 (A) The main trajectories of light and heavy holes are centered at p_L and p_H passing through the origin in momentum space. The electric field $\vec{E}$ is parallel to the x-axis and the magnetic induction $\vec{B}$ parallel to the z-axis. (B) Hole motion in a parabolic band diagram with the maximum energies W_L and W_H for the light and heavy holes. Neither diagram is to scale. Note that the hole effective mass ratio m_H^*/m_L^* in germanium is about eight. (C) Population inversion formation by optical phonon emission of hot heavy holes transforming partially by a probability determined by the density of states into light holes which do not emit optical phonons.

Here this velocity reduces to a single component on the v_y-axis with a value of $v_D = -E/B$. In momentum space two circular trajectories, one light hole and the other heavy hole, are obtained due to their different effective mass (see Fig. 6.2a). The light- and heavy-hole trajectories are centered at the points $(0, p_y, 0)$ in momentum space with p_y components of $p_L = m_L^* v_D$ and $p_H = m_H^* v_D$, respectively.

This velocity or momentum oscillation corresponds to a kinetic energy oscillation between $W = 0$ and $W = 2m^* v_D^2 = 4W_D$. The light- and heavy-hole effective masses are $m_L^* = 0.043 m_{e^-}$ and $m_H^* = 0.35 m_{e^-}$ with the free electron mass m_{e^-}. The effective mass ratio m_H^*/m_L^* is approximately eight in germanium. Note that the effective mass values are given for a magnetic induction oriented along a $\langle 110 \rangle$ crystallographic axis. The different masses lead to different maximum energies. The maximum energy is $W_L = 2m_L^* v_D^2$ for the light and $W_H = 2m_H^* v_D^2$ for the heavy holes (see Fig. 6.2b).

A population inversion results when light holes do not scatter with optical phonons and heavy holes scatter into the light-hole band with a probability given by the ratio of the heavy- and light-hole band densities of states, namely $W_L < W_{OP} \leq W_H$ (see Fig. 6.2c). W_{OP} denotes the optical phonon energy (37 meV in germanium). If the heavy holes cannot reach the optical phonon energy, namely $W_L < W_H < W_{OP}$, pumping of heavy holes into the light hole band ceases and laser emission should not occur. The hole population inversion is destroyed if both heavy and light holes emit optical phonons, namely $W_{OP} \leq W_L < W_H$.

The limiting E/B-field ratios are 0.96 kV/(cm T) and 2.75 kV/(cm T) for $W_H = W_{OP}$ and $W_L = W_{OP}$, respectively (1 kV/(cm T) = 10^5 cm/s). This simple model seems to be appropriate for gallium-doped germanium (Ge:Ga) lasers

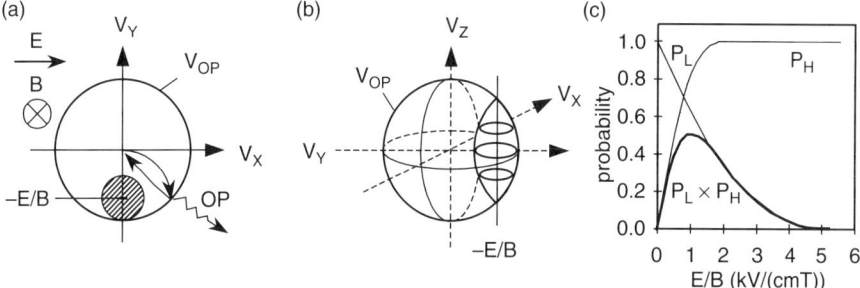

Fig. 6.3 Reprinted from Infrared Phys. Technol. 40(3), E. Bründermann et al., Novel design concepts of widely tunable germanium terahertz lasers, 141–151, Copyright 1999, with permission from Elsevier Science. (A) Two dimensional velocity space in crossed $\vec{E}$ and $\vec{B}$ fields for an electric field $\vec{E}$ parallel to the x-axis and a magnetic induction $\vec{B}$ parallel to the z-axis. The main hole trajectory starts at the origin. Oblique hatching: hole trajectories with a long lifetime, without optical phonon (OP) interaction, form a disc in the $v_z = 0$ plane. (B) Hole trajectories without optical phonon scattering form a spindle in three dimensions. (C) Probability of light holes P_L to be within the spindle volume, to be upper laser states, and the probability of heavy holes P_H to be outside of their spindle volume participating in optical phonon interaction and pumping of the upper laser states. Total probability $P_L \times P_H$ describing the formation of a population inversion.

because the onset of laser emission is typically found above 1 kV/(cmT) and the maximum emission around 1.4 kV/(cmT).

However, the simple model fails in the case of beryllium-doped germanium (Ge:Be) lasers. The onset of laser emission already occurs well below 0.96 kV/(cmT), for example, at 0.7 kV/(cmT). The maximum emission is typically found at $E/B = 0.96$ kV/(cmT). The latter is reasonable. If the heavy-hole energy W_H coincides with the optical phonon energy W_{OP}, then the heavy hole can scatter into the light-hole band with zero remaining kinetic energy. But why do Ge:Be lasers show laser emission at lower E/B-field ratios?

Three-Dimensional Model

An answer can be given by studying the problem in a three dimensional spherical model. Figure 6.3a shows a disc that encloses all trajectories without optical phonon interaction in the $v_z = 0$ plane. In three dimensions this area becomes a spindle like volume (see Fig. 6.3b). The disc or spindle is also called the accumulation area. Only light holes that are within this volume will have a long lifetime. Heavy holes have to be outside this volume because for them optical phonon interaction is necessary to transform into light holes. These light holes can be accumulated in the light hole spindle. From the parameter

$$v_m = \sqrt{v_{OP}^2 - \left(\frac{E}{B}\right)^2}, \qquad (6.6)$$

the spindle volume follows as

$$V = 2\pi\left(\upsilon_{OP}^2\left(\upsilon_m - \frac{E}{B}\arccos\left(\frac{E/B}{\upsilon_{OP}}\right)\right) - \frac{\upsilon_m^3}{3}\right). \tag{6.7}$$

The probability of light holes being within the light hole spindle is $P_L = V_L/V_B$, the ratio of the light hole spindle volume V_L and the total volume of the ball $V_B = 4\pi\upsilon_{OP}^3/3$. The probability of heavy holes being outside of the heavy hole spindle of volume V_H is $P_H = 1 - V_H/V_B$. The velocity υ_{OP} is defined by the equation

$$\frac{m^*}{2}\upsilon_{OP}^2 = W_{OP}. \tag{6.8}$$

This velocity differs for light and heavy holes by a factor, the square root of the effective mass ratio.

The product $P_L P_H$ can describe the quality of the population inversion as a function of the applied E/B-field ratio (see Fig. 6.3c). Ge:Be lasers operate from 0.7 to 2 kV/(cmT) or for $P_L P_H > 0.4$. Note that in Fig. 6.3c the heavy- and light-hole effective mass of 0.35 and 0.043 times the free electron mass were used ($\vec{B}\|\langle 110\rangle$). This picture can explain the range of the E/B-field ratios of Ge:Be lasers but not the limited range found for Ge:Ga lasers.

Next a mechanism that is based on magneto-phonon, magneto-impurity, and resonance effects is described. It explains emission peaks as depending on the E/B-field ratio for lasers with different dopant species. In particular, it will become clear why Ge:Ga lasers lack the peak at $E/B = 0.96$ kV/(cmT). It is a result of the hole, especially light-hole, dynamics in momentum space and their interaction with resonance energies due to energy transitions in acceptor dopants and with optical phonons.

6.2.2 Magneto-Phonon and Magneto-Impurity Effects

As described above, holes oscillate at a constant angular frequency of $\omega_{CR} = eB/m^*$ between $W_{min} = 0$ and $W_{max} = 2m^*\upsilon_D^2$. The largest probability to observe a hole within a certain constant energy interval is found at W_{min} and W_{max}. The latter is only distinctive for the light and heavy hole, so this energy and the corresponding E/B-field ratio are considered.

The formation of the population inversion can be divided into two processes: first, the pumping of the upper laser level by converting heavy holes into light holes by optical phonon emission, and second, the loss of light holes from the upper laser level due to non-radiative relaxation, such as by impurity interaction.

Heavy-Hole Magneto-Phonon Interaction
Several resonances that coincide with emission peak intensities are observed in experiments [23,38]. They are first-order and higher order resonances in

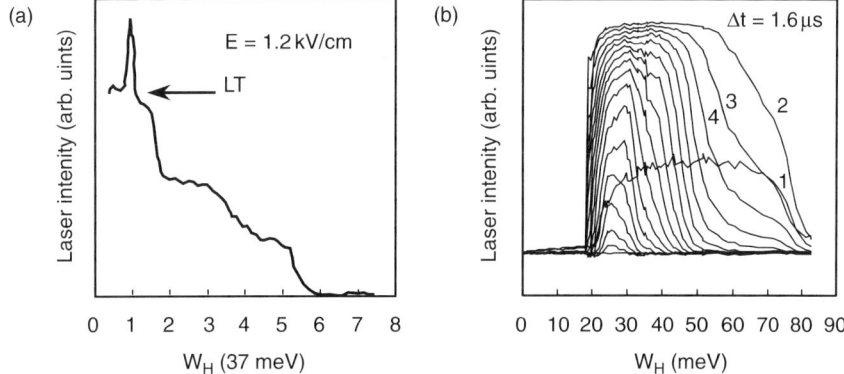

Fig. 6.4 (A) Emission of a small 5 mm³ Ge:Be laser (*L*14, Table 6.1) at $E = 1.2$ kV/cm close to the laser threshold (LT) indicated by the arrow as a function of the magnetic field, plotted as a function of the parameter W_H in units of the optical phonon energy (37 meV). (B) Time dependent laser emission of a Ge:Be laser (*L*2, Table 6.1) at time intervals of 1.6 μs after excitation as a function of the parameter W_H. The first four intervals after excitation are numbered from 1 to 4 as a guide to the eye.

which the maximum heavy-hole energy W_H coincides with one, two, three, or more optical phonons:

$$W_H = nW_{OP} \qquad (n = 1, 2, 3, \ldots). \tag{6.9}$$

The resulting E/B-field ratio is found as $E/B = 0.96\sqrt{n}$ kV/(cmT), which is approximately 1.0, 1.4, 1.7, and 2.0 kV/(cmT) for $n = 1, 2, 3$, and 4. However, the probability for higher order resonances and multiple optical phonon emission should decrease, so the strongest emission should always occur at the first resonance at 1 kV/(cmT).

This is true for Ge:Be and Ge:Zn lasers [21]. In Fig. 6.4*a* a step structure is clearly visible for a small Ge:Be laser that corresponds to the different resonances. The maximum emission appears at 1 kV/(cmT). The laser crystals discussed in this chapter are listed in Table 6.1.

Heavy-Hole Magneto-Impurity Interaction

Ge:Be laser emission measured at different time during the pulse rises equally over the range from $W_H = 30$ to 70 meV immediately after electrical excitation (see Fig. 6.4*b*). During time the laser heats up and the emission for parameters with lower gain disappears. First, the peak at $2W_{OP}$ weakens, and then the peak at W_{OP} decays, so a peak around 25 meV survives. The ionization energy of beryllium acceptors in germanium is 25 meV. Therefore this peak can be related to magneto-impurity interaction of heavy holes with beryllium acceptors. Note that a larger W_H value corresponds to a larger E/B-field ratio and higher electric field at a constant magnetic induction. This leads to a higher input power, faster heating, and faster reduction of the gain due to an increased acoustical phonon interaction of light holes.

TABLE 6.1 Examples of germanium laser crystals grown in E. E. Haller's group, Berkeley.

	A	N_A ($10^{14}\,\text{cm}^{-3}$)	L (mm)	b (mm)	d (mm)	GC
L0	Be	4.2	5.1	5.1	7.6	vac
L1	Be	1.4	3.0	2.0	3.0	vac
L2	Be	1.4	18	3.0	3.0	vac
L3	Be	0.3	12	2.0	2.0	vac
L4	Cu	15	7.5	3.0	2.0	diff
L5	Al	1.2	3.0	2.0	4.0	vac
L6	Be	1.4	10	2.4	0.75	vac
L7	Be	1.4	20	4.0	4.0	vac
L8	Be	1.4	3.0	3.0	3.0	vac
L9	Be	0.8	25	3.0	3.1	D_2
L10	Tl	0.35	20	2.9	3.7	H_2
L11	Ga	0.45	30	3.4	4.2	H_2
L12	Zn	2.0	3.0	3.0	5.3	vac
L13	Be(P)	7.0	2.0	3.0	6.0	vac
L14	Be(P)	7.0	0.9	0.9	5.8	vac
L15	Be	1.4	1.0	2.0	3.0	vac

Note: Acceptor species A, acceptor concentration N_A, length L, width b, intercontact distance d, and growth condition GC (in vacuum, in D_2 or H_2 gas, by diffusion). For L13 and L14 phosphorus was used to compensate the shallow hydrogenic acceptors.

Current-Voltage Characteristics. Figure 6.5 displays the current density j and the laser intensity as a function of the electric field E for differently doped germanium crystals ($B = 0.9\,\text{T}$). The shape of the j–E-curve depends on the dopant species and can be modeled using a semiclassical description. The heavy-hole trajectories, mainly originating in the center of the velocity space, can reach a maximum kinetic energy W_H. Impact ionization is not possible if this energy is below the ionization energy of the acceptor. The ionization energies of aluminum, beryllium, and copper are 11 meV, 25 meV, and 43 meV, respectively. If W_H is sufficiently large to initiate ionization and subsequent optical phonon scattering, the number of free holes increases and a rapid increase of the current density is observed.

The ionization energies coincide with W_H at $E = 0.47\,\text{kV/cm}, 0.71\,\text{kV/cm}$, and $0.94\,\text{kV/cm}$ (see Fig. 6.5a), respectively. A sufficient current density for laser action is also observed at lower electric fields for aluminum and beryllium-doped germanium. I attribute this to resonances due to hole transitions from the acceptor ground state to bound excited states absorbing the kinetic energy W_H. This excitation followed by ionization can increase the number of free holes.

The carrier interaction process can be illustrated in a model similar to a Monte Carlo method. A carrier transport Monte Carlo simulation is based on repetitive cycles of short, random length, free motion of the carrier in external fields that terminates due to an instantaneous scattering event leading to

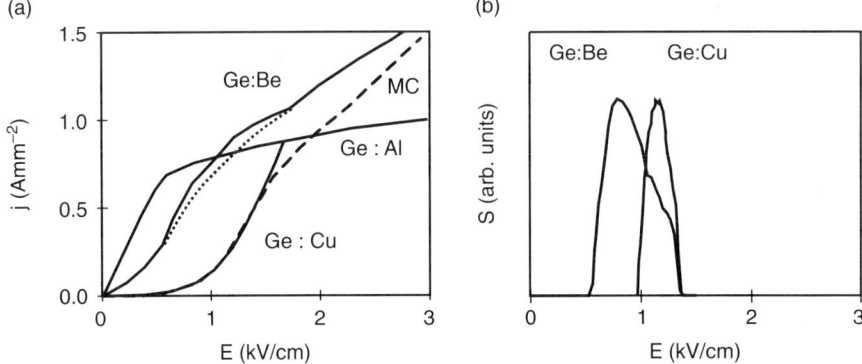

Fig. 6.5 Reprinted from Infrared Phys. Technol. 40(3), E. Brüdermann et al., Novel design concepts of widely tunable germanium terahertz lasers, 141–151, Copyright 1999, with permission from Elsevier Science. (A) Current density j as a function of the applied electric field E for a Ge:Be, Ge:Cu, and Ge:Al laser (solid lines, corresponding to laser $L1$, $L4$, and $L5$ in Table 6.1) and a Monte Carlo (MC) simulation including acoustical and optical phonon interaction at 10 K but without impurity scattering (dashed line). The dotted line in the j-E-curve of the Ge:Be laser illustrates the current excess during stimulated emission. (B) Laser emission of the Ge:Be ($L1$) and Ge:Cu ($L4$) laser corresponding to the j-E-curves shown in (A).

new phase-space coordinates. For a certain E/B-field ratio heavy holes can reach energies that can ionize beryllium acceptors. Such holes can be transferred into the light-hole band due to inelastic impurity scattering or they can elastically scatter into a different point in momentum space. This initial point for the next trajectory can have new coordinates relative to P_D outside the disc or spindle of accumulated holes (for the definition of the point P_D, see Eq. 6.5). Then the hole can quickly accelerate, emit an optical phonon, and populate the light-hole band.

The ionization and scattering events lead to a measurable current flow and current density. The lowest resonance energy of 6.6 meV (Al), 20 meV (Be), and 39 meV (Cu) corresponds to the G-line, the absorption line of an impurity from the ground to the first excited state (see Section 6.2.8). Equating this energy to the parameter W_H disregarding nonlinear splitting effects due to the magnetic and electric field gives electric fields of $E = 0.37$ kV/cm, 0.64 kV/cm, and 0.89 kV/cm, respectively (see Fig. 6.5a).

The incomplete ionization of copper-doped germanium (Ge:Cu) lasers is clearly reflected in the current density. At $E < 0.9$ kV/cm the number of free holes and the current density are negligible for the Ge:Cu laser (see Fig. 6.5a; $L4$ in Table 6.1). The number of free holes is too small to allow laser emission. Laser emission was only measured at $E > 0.9$ kV/cm when the energy W_H is large enough to ionize copper acceptors. At $E = 1.5$ kV/cm the current density is comparable to the value of a 10 times lower doped Ge:Be crystal ($L1$ in Table 6.1). Obviously sufficient holes are necessary to form a hole inversion at the optimal condition $E/B = 0.96\,\text{kV}/(\text{cmT})$ (see Fig. 6.5a). However, for higher copper acceptor concentrations a higher temperature for diffusion is necessary that will introduce additional problems during fabrication, such as activation of shallow hydrogenic acceptor contaminants (see Section 6.3.3).

Light-Hole Magneto-Impurity Interaction

Shallow hydrogenic acceptor-doped germanium lasers, such as Ge:Ga and Ge:Al lasers, typically emit laser radiation at $E/B \gg 1\,\text{kV}/(\text{cmT})$. This is not the consequence of insufficient free holes during laser operation but a result of a magneto-impurity interaction of the shallow hydrogenic impurities. This result should not be confused with absorption of generated photons from the Landau level transitions by the populated states of the impurities (see Section 6.2.8). Here I mean the loss mechanism in which light holes lose their kinetic energy to initiate an impurity level transition. Besides this inelastic scattering process the light hole can scatter elastically with new coordinates relative to P_D similar to the heavy-hole magneto-impurity interaction described above. However, the latter process involving heavy holes is beneficial, whereas the former is not.

The impurity transitions of boron, aluminum, and gallium cover an energy range from 6.2 to 11.3 meV. The loss mechanism is then active in the range from $E/B = 1.1$ to $1.5\,\text{kV}/(\text{cmT})$ by comparing this energy range to the light hole parameter W_L. The transitions to the first four excited levels, especially those corresponding to the C- and D-lines, seem to be most efficient. This range unfortunately coincides with the optimum parameters for pumping heavy holes into the light-hole band by optical phonon scattering or by heavy-hole magneto-impurity interaction.

Figure 6.6a shows an example of magneto-impurity interaction of light holes in a Ge:Tl laser. Thallium acceptors have an ionization energy of 13 meV. Increasing the repetition rate of the laser increases the temperature and the acoustical phonon interaction of light holes, which will result in a gain reduction. The above-described loss mechanism becomes clearly visible at high repetition rates. A strong effect of the D-line magneto-impurity interaction appears that separates the emission into two peaks. At high temperature close to the laser threshold a depression of the remaining peak is observed due to the thallium G-line magneto-impurity interaction.

The emission around $1.4\,\text{kV}/(\text{cmT})$, related to the second resonance of heavy holes with optical phonons, and the intensity suppression at and below $1\,\text{kV}/(\text{cmT})$ are a result of a light-hole magneto-impurity interaction in germanium lasers doped with shallow hydrogenic acceptors. These effects are reduced for low acceptor concentrations. Therefore Ge:Ga laser emission can be found at a E/B-field ratio close to but still above $1\,\text{kV}/(\text{cmT})$ for the IVB-type lasers and well below $1\,\text{kV}/(\text{cmT})$ for the LHCR-type laser.

Optogalvanic Effect

Stimulated transitions of accumulated light holes (see Fig. 6.3) into lower laser states in the heavy-hole band increase the number of free, heavy holes that

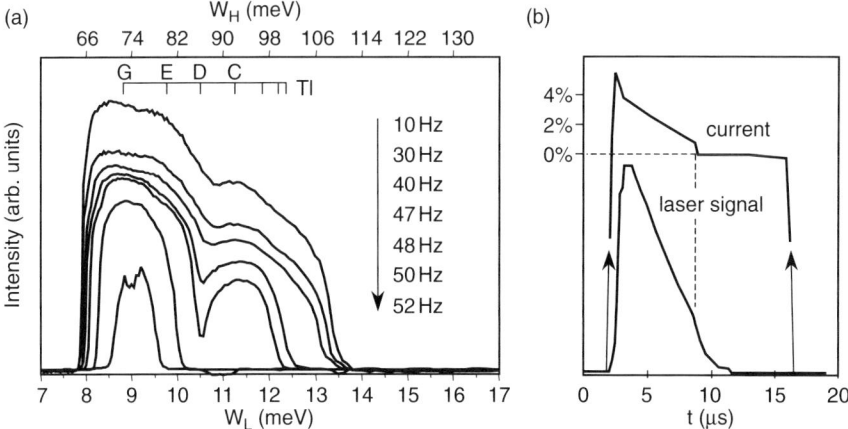

Fig. 6.6 (A) Laser emission of a Ge:Tl laser ($L10$, Table 6.1) at different repetition rates as a function of the magnetic field for pulses of 5 µs length. The emission is plotted as a function of the parameters W_H and W_L. Impurity transition energies are indicated for the thallium acceptor. (B) Current enhancement due to stimulated emission in the Ge:Tl laser. The detector signal shows an exponential decay when laser emission terminates which depends on the detector electronic circuit and the cavity ring down.

scatter by optical phonon emission. Hence a current enhancement is measured if the pulse generator is a constant voltage source during the applied voltage pulse. This enhancement can reach up to 8% for Ge:Be lasers [39] (see Fig. 6.5a). A maximum of 4% has been found for gallium, aluminum, and thallium-doped germanium lasers (see Fig. 6.6b). The reduction of light hole–impurity interaction in Ge:Be lasers in comparison to lasers doped with shallow hydrogenic acceptors leads to a more efficient accumulation. Subsequently the depletion by stimulated emission of the trapped and accumulated light holes and the current enhancement are larger.

The same reasoning that is used for the Ge:Be laser also holds for zinc-doped germanium (Ge:Zn) lasers. The ionization energy of zinc acceptors is 33 meV. The onset of laser emission is shifted to a higher E/B-field ratio with respect to Ge:Be lasers, but laser emission is still observed for a E/B-field ratio less than 1 kV/(cmT) [21]. Figure 6.7 shows a measurement of a Ge:Zn laser. The electrical excitation pulse was modified to achieve a slow field decay. A double-peak structure appears. This structure is a result of the changing electric field strength sweeping through different E/B-field ratios at a constant magnetic field. If the double-peak data are plotted against the parameter W_H, two peaks at the corresponding resonances of heavy holes with one and two optical phonons are obtained (see Fig. 6.7b).

Detector Characteristics
Figure 6.8 displays two examples of laser emission as a function of the applied fields and dopant species. Note that the frequency dependent sensitivity of

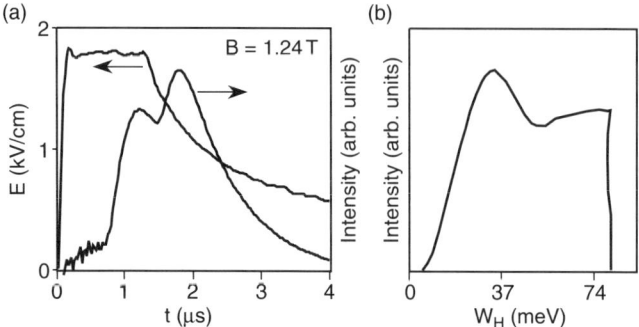

Fig. 6.7 (A) Applied electric field E and laser emission intensity of a Ge:Zn laser ($L12$, Table 6.1) at a magnetic induction of $B = 1.24$ T. (B) The double pulse structure from (A) as a function of the parameter W_H.

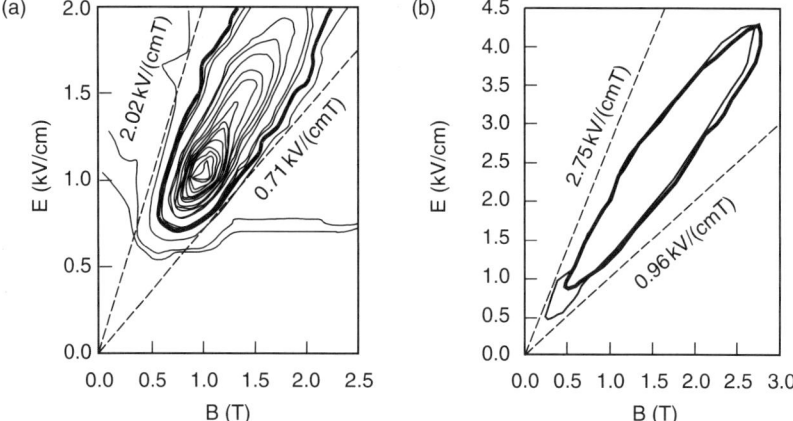

Fig. 6.8 (A) Isointensity contour lines in the E-B-plane for a Ge:Be laser ($L0$, Table 6.1). The thick line defines the laser threshold which separates the spontaneous emission from strong laser emission. The dashed lines enclose the theoretical area where laser action is possible derived from a semi-classical description including magneto-impurity interaction of light and heavy holes with beryllium. (B) Contour lines of the laser threshold in the E-B-plane for a Ge:Ga laser measured with different photoconductors, unstressed Ge:Ga detector (thick line) and stressed Ge:Ga detector (thin line). The dashed lines enclose the theoretical area where laser action is possible derived from a semi-classical description without magneto-impurity interaction (Ge:Ga laser sample and detectors from N. Hiromoto's group, CRL, Tokyo).

the detectors is folded into the measured data during the measurement (see Fig. 6.8b).

Using different detectors can give insight into the laser mechanism (see Fig. 6.9). A NbN superconducting hot electron bolometer (HEB) has a wider frequency response [40] in comparison to a Ge:Ga photoconductor, which is only

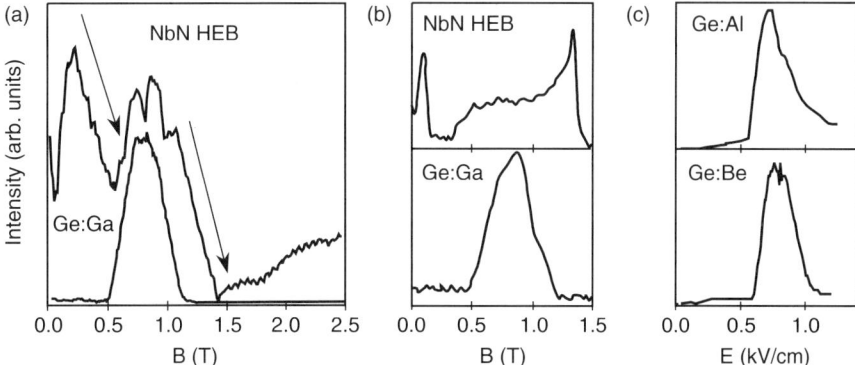

Fig. 6.9 Comparison of different detectors using a Ge:Be laser source ($L0$, Table 6.1): (A) NbN hot electron bolometer (HEB) detector and a Ge:Al photoconductor with the laser operated at $E = 0.8\,\text{kV/cm}$. (B) Another NbN HEB detector and Ge:Al photoconductor with the laser operated at $E = 0.9\,\text{kV/cm}$. (C) Two germanium photoconductors with different dopant species, aluminum and beryllium, with the laser operated at $B = 0.7\,\text{T}$.

very sensitive in the 70 to $120\,\text{cm}^{-1}$ frequency range. Very long wavelength radiation which appears at low magnetic fields can be observed by using a hot electron bolometer (see Fig. 6.9a). This radiation can originate from Bremsstrahlung of hot carriers. The emission around $2.75\,\text{kV/(cmT)}$ falls off if the magnetic field is increased due to light-hole accumulation.

It again rises if a population inversion forms that leads to far-infrared emission. The Ge:Ga photoconductor only detects this far-infrared laser emission. A decrease of the signal at $1\,\text{kV/(cmT)}$ is measured if the magnetic field is further increased. This is a result of the accumulation of heavy holes and, subsequently, the reduced conversion of heavy into light holes.

The emission rises for higher magnetic fields because of a population inversion formation between different Landau levels of light holes. The latter leads to low-frequency emission at the cyclotron resonance frequency of light holes. This frequency is proportional to B and has a value of $20B\,\text{cm}^{-1}/\text{T}$ ($\vec{B}\|\langle 110\rangle$).

The frequency sensitivity can be shifted using different dopants in photoconductors. Gallium or aluminum dopants in germanium are ionized with radiation in excess of 11 meV. Beryllium acceptors in germanium require 25 meV photons for ionization. However, the detected laser emission is similar using a Ge:Al or a Ge:Be photoconductor detector (see Fig. 6.9c).

No high-frequency emission is expected above $140\,\text{cm}^{-1}$ or 17 meV, which was determined by frequency-resolved measurements using a germanium bolometer with a wide and more flat frequency response [25]. The similar result can be explained by considering two-photon excitation in the Ge:Be photoconductor, which is possible due to the strong coherent laser emission. These two photon excitations also allow the measurement of laser emission at very long wavelength, for example, 200 to 220 μm, with unstressed Ge:Ga photoconductors.

6.2.3 Scattering Mechanisms

Three major scattering mechanisms are relevant to the hot-hole laser: optical phonon scattering, acoustical phonon scattering, and Coulomb scattering. They depend on the carrier mass and therefore on the different effective mass of the holes. Further, neutral scattering, which is proportional to the concentration of neutral dopants, is important in highly doped Ge:Cu lasers. Polar scattering is important in III–V materials like GaAs.

Optical phonon emission and the excitation of an optical lattice wave requires a threshold energy, the optical phonon energy (37 meV in germanium). In lightly doped germanium at a low lattice temperature, the scattering probability of carriers is relatively low if their kinetic energy W is smaller than the optical phonon energy in germanium, namely $W < W_{OP}$. If the kinetic energy is above the threshold energy ($W > W_{OP}$), then the scattering cross section increases by several orders of magnitude because optical phonon emission becomes possible. This strong interaction of holes with optical phonons in p-type germanium is important for the formation of a population inversion as described above.

Essential for the laser mechanism is the coexistence of light and heavy holes with different effective mass m^*_L and m^*_H, respectively. For the condition $W_H > W_{OP} > W_L$ a majority of heavy holes performs streaming motion, frequently emitting optical phonons with a time interval roughly given by

$$T^H_{OP} = \frac{\sqrt{2m^*_H W_{OP}}}{eE}. \tag{6.10}$$

The laser mechanism crucially depends on orthogonal electric and magnetic fields. An electric field component E_z parallel to the magnetic field leads to an acceleration of the holes in the direction of the magnetic field. Light holes will therefore quickly reach the optical phonon energy within

$$\tau_{L\|z} = \frac{\sqrt{2m^*_L W_{OP}}}{eE_z} \approx 1\,\text{ps}\frac{\text{kV/cm}}{E_z} \tag{6.11}$$

and will emit optical phonons. Their lifetime is significantly reduced. However, this effect can be employed to modulate the gain by intentionally applying an additional, modulated electric field parallel to the magnetic field. This gain modulation was used to develop actively mode-locked germanium lasers. More details on mode-locked germanium lasers can be found in Section 6.4.4.

Light holes should not emit optical phonons and their average scattering time is given by

$$\tau_L = \left(\frac{1}{\tau_I} + \frac{1}{\tau_A} + \frac{1}{\tau_T} + \frac{1}{\tau_{L\|z}}\right)^{-1}. \tag{6.12}$$

$\tau_{L\|z}^{-1}$ is zero for perfectly crossed electric and magnetic fields (see Eq. 6.11). τ_I and τ_A are the average impurity and acoustical phonon scattering times, and $\tau^{-1}{}_T$ is an averaged probability for light holes to be transferred to the heavy-hole band through tunneling processes, for example, by mixing of light- and heavy-hole wavefunctions. Light holes can absorb optical phonons, which could reduce the light hole's lifetime. However, at low temperature and energies of the accumulated light holes, the equilibrium population of optical phonons is extremely small. Therefore it can be neglected.

If a heavy hole emits an optical phonon, it can be scattered into the light-hole band. The probability for this process can be estimated as

$$P_{H \to L} = \frac{m_L^{*3/2}}{m_H^{*3/2} + m_L^{*3/2}}, \tag{6.13}$$

which is approximately equal to $(m_L^*/m_H^*)^{3/2}$, the ratio of the density of states of the light- and heavy-hole bands without external applied fields. The ratio of the light- and heavy-hole distribution functions f_L and f_H can be approximated by

$$\frac{f_L}{f_H} \approx \frac{\tau_L}{T_{OP}^H}. \tag{6.14}$$

T_{OP}^H is 2–4 ps for typical electric fields E of 1–2 kV/cm. τ_L can reach 100 ps which leads to a population inversion $f_L/f_H > 1$.

In the following discussion, the textbook formulas for low-energy scattering of acoustical phonon emission and impurity scattering, such as in the form of Brooks-Herring and Conwell-Weisskopf, are used. However, these approximations can fail even if some corrections are made by adding higher order terms in the electron temperature. Simplifying the scattering probabilities by removing constant parameters and higher order terms leads to

$$\tau_I^{-1} \propto \frac{N_A}{\sqrt{m^* W^3}} \propto \frac{N_A}{m^{*2} v^3}, \tag{6.15}$$

$$\tau_A^{-1} \propto T\sqrt{m^{*3} W} \propto Tm^{*2} v, \tag{6.16}$$

with the carrier energy W, the concentration of ionized dopants N_A, and the temperature T. The last terms in the equations follow from using the kinetic energy $W = m^* v^2/2$. Changing the semiconductor host material will require one to consider the different parameters like the deformation potential and elastic constant for acoustical phonon emission and the relative dielectric constant relevant for impurity scattering.

6.2.4 Optical Gain

The absorption coefficient α can be written as the sum of direct intersubband transitions α_{LH}, indirect (Drude-type) absorption α_D by phonons and impurities, and lattice absorption α_{LT}. The latter is typically small [41]:

$$\alpha = \alpha_{LH} + \alpha_D + \alpha_{LT} + \frac{\ln(R_1 R_2)}{2L}. \qquad (6.17)$$

The last term in Eq. (6.17) describes the optical losses and is given by the resonator optical length L and the reflectivity R_1 and R_2 of the resonator mirrors. One mirror is ideally a 100% reflector ($R_1 = 1$). The other mirror controls the output.

However, high-duty cycle emission from Ge:Be lasers without external resonators has been observed [22,23]. This emission is a result of internal reflections under small angles from the germanium surfaces. The normal incidence reflectivity of the germanium surfaces is only 36% due to the high refractive index of germanium ($n_{Ge} = 4$), typically insufficient to support a laser mode. Germanium resonators will be discussed in Section 6.5.1.

The sum α has to be negative to provide amplification of radiation (see Eq. 6.17). The component that can only turn negative is α_{LH}; the amplification is due to a population inversion between the light- and the heavy-hole bands. The stimulated transition rate is proportional to $\langle f_L(\vec{p})\rangle(1 - \langle f_H(\vec{p})\rangle)$, namely to the product of the probability $\langle f_L \rangle$ that the upper state is occupied and the probability $(1 - \langle f_H \rangle)$ that the lower state is empty. The angle brackets $\langle \ldots \rangle$ denote averaging over directions of momenta.

In a two-level population inverted system a negative density value ($N_1 - N_2$) exists. For the germanium laser N_1 and N_2 are replaced by the heavy- and light-hole band distribution functions

$$N_1 - N_2 \rightarrow \frac{\rho(\vec{k})d\vec{k}}{V}(\langle f_H \rangle(1 - \langle f_L \rangle) - \langle f_L \rangle(1 - \langle f_H \rangle))$$
$$= \frac{\rho(\vec{k})d\vec{k}}{V}(\langle f_H \rangle - \langle f_L \rangle) \qquad (6.18)$$

with the density of states $\rho(\vec{k})$.

The rate of a transition involving a photon is always proportional to an integral over the crystal volume, which includes the product of the initial state wavefunction and the complex conjugate of the final state wavefunction. Such an integral is only significantly nonzero if the difference of the k-vectors of the initial and final state is small. Therefore, transitions are vertical and vertical arrows indicate laser transitions in the band diagram (see Fig. 6.1). The photon energy is then given as

$$\hbar\omega = p^2\left(\frac{1}{2m_L^*} - \frac{1}{2m_H^*}\right) \qquad (6.19)$$

with the same initial and final k-vector or p-momentum.

Hence α_{LH} in the isotropic approximation is given as a frequency dependent amplification coefficient

$$\alpha_{LH} = \frac{4\pi^3 e^2 \hbar}{c\sqrt{\varepsilon_r \varepsilon_0}} p(\langle f_H \rangle - \langle f_L \rangle). \tag{6.20}$$

The inversion $\langle f_L \rangle / \langle f_H \rangle > 1$ (see Eq. 6.14) leads to a negative α_{LH} and to amplification. Indirect transitions are essential in the low-frequency domain. The coefficient may be found in the Drude-type formula:

$$\alpha_D = \frac{4\pi e^2 N_A}{m_H^* c \sqrt{\varepsilon_r \varepsilon_0}} \left(\frac{T_{OP}^H}{2}\right)^{-1} \omega^{-2}, \tag{6.21}$$

with the velocity of light c, the hole concentration N_A, and the relative dielectric constant ε_r which connects to the refractive index of germanium ($n_{Ge} = \sqrt{\varepsilon_r}$). The characteristic time in Eq. (6.21) is approximated by $T_{OP}^H/2$.

The temperature dependence of τ_A (see Eq. 6.16) leads to a termination of laser action at high temperature, typically 20 to 40 K, because the ratio τ_L/T_{OP}^H (see Eqs. 6.12 and 6.14) decreases and therefore α turns positive. The terms $\langle f_H \rangle$ and $\langle f_L \rangle$ are proportional to the acceptor concentration N_A, meaning that a higher gain for a higher acceptor concentration is expected (see Eq. 6.20). The present minimum acceptor concentration for germanium lasers is approximately 5×10^{12} cm^{-3}.

Indirect transitions described by α_D are also proportional to N_A (see Eq. 6.21), which typically limits the concentration to a few times 10^{15} cm^{-3}. A high acceptor concentration leads to a stronger reduction of the low-frequency emission due to the frequency dependence of the term α_D. The low-frequency emission is typically connected to low magnetic and electric fields [25]. The observed area at low fields obtained by measuring the laser threshold in the E-B-plane (see Fig. 6.8) is therefore smaller for more highly doped lasers. Increasing the acceptor concentration N_A will also reduce τ_L/T_{OP}^H due to the increase of Coulomb impurity scattering expressed in the reduction of τ_I (see Eq. 6.15).

Both scattering rates, τ_I and τ_A, depend on the effective mass (see Eqs. 6.15 and 6.16). In addition the velocity of accumulated light holes is limited below υ_{OP}, whereas heavy holes in streaming motion can reach velocities larger than υ_{OP}. Therefore impurity scattering is significant for light holes (see Eq. 6.15). Heavy holes are predominantly affected by acoustical phonon scattering in addition to optical phonon scattering (see Eq. 6.16). Figure 6.10 shows a calculation of scattering rates for heavy and light holes including different scattering effects.

The mass and velocity dependence are especially important if light- and heavy-hole wavefunctions mix. The resulting wavefunction can be described by a hole with an intermediate, mixed effective mass m_M^* larger than the light hole's effective mass. This hole will have a higher kinetic energy $W_M = 2m_M^* \upsilon_D^2$ due to the applied fields and a shorter lifetime. The mixing of wavefunctions for light and heavy holes will be discussed in Section 6.2.6.

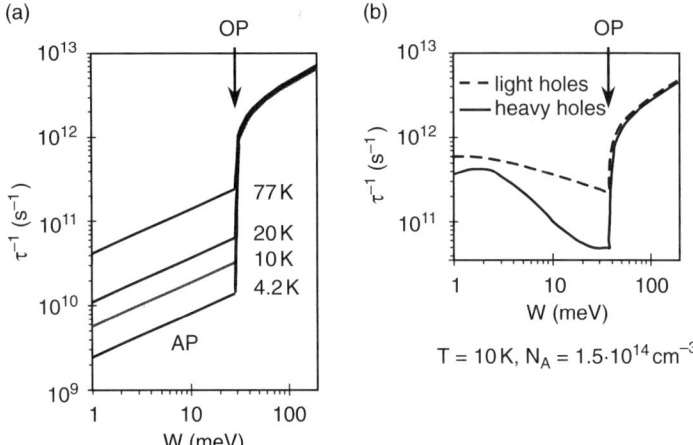

Fig. 6.10 (A) Scattering rate τ^{-1} of heavy holes with energy W which includes the acoustical (AP) and optical phonon scattering at different temperatures 4.2, 10, 20 and 77 K with the threshold of the optical phonon energy (OP) at 37 meV. (B) Scattering rate τ^{-1} of heavy and light holes which includes acoustical and optical phonon scattering at 10 K and impurity scattering with an acceptor concentration N_A of 1.5×10^{14} cm^{-1}.

Measured Gain Coefficients and Amplification Cross Sections

Various authors have measured the optical gain, in Ge:Ga lasers and give values of the net gain or the amplification cross section, the ratio of the net gain, and the acceptor concentration. The values can differ depending on the applied external fields and experimental conditions. Komiyama et al. [16] measured the laser pulse rise time to determine the small signal gain. They obtained rise times t_R of 18 and 5 ns for an acceptor concentration of 4.5×10^{13} and 1.7×10^{14} cm^{-3}, respectively. The small signal gain g can be calculated as

$$g = \frac{1}{t_R} \frac{n_{Ge}}{c}. \tag{6.22}$$

The amplification cross section is 1.6×10^{-16} cm^2 in both cases.

For LHCR lasers in the Faraday configuration a value of $(8 \pm 3) \times 10^{-16}$ cm^2 has been calculated [42]. Using a germanium laser to probe another germanium crystal in the Faraday and Voigt configuration resulted in amplification cross sections of 2 and 5×10^{-16} cm^2, respectively [43]. The gain coefficients were as high as 0.12 cm^{-1}. A maximum amplification cross section of 2×10^{-16} cm^2 has been obtained for mode-locked germanium lasers in the Voigt configuration [44].

For a germanium amplifier in the Faraday configuration values of $(4 \pm 1) \times 10^{-16}$ cm^2 have been found [45]. Keilmann et al. [46] used a gas laser to probe the germanium laser at different frequencies and found a short-lived (0.2 μs), maximum gain coefficient in the Faraday configuration of $0.1\,\text{cm}^{-1}$ corresponding to an amplification cross section of $1.4 \times 10^{-15}\,\text{cm}^2$.

The optical gain in Ge:Be laser material is expected to exceed the values of Ge:Ga material because of the typically 10 times higher duty cycle of Ge:Be lasers. However, we have to be very careful comparing measurements. The orientation of the crystallographic axes with respect to the applied fields is very important. Nonuniform electric fields, which are discussed next, will also influence the gain in the sample.

The field strength and the orientation of the total electric field, the sum of the applied and Hall electric field, depend on the laser shape. For example, a crystal of $50 \times 5 \times 5\,\text{mm}^3$ volume can have a very low electric field uniformity if the crystal is mounted in the Faraday configuration with the $5 \times 5\,\text{mm}^2$ surfaces perpendicular to the magnetic field. However, the uniformity can be very high in the Voigt configuration for the same crystal with a $50 \times 5\,\text{mm}^2$ surface perpendicular to the magnetic field. This could be one possibility to explain the higher amplification cross section in the Voigt configuration in some experiments.

Note that a Ge:Be laser crystal in a cubic shape where all germanium surfaces belong to the same crystallographic family, for example, (110), has the laser threshold at the same electric and magnetic fields in Voigt and in Faraday configuration (see Section 6.4.3). Detailed studies of the optical gain are still missing in which a variety of different germanium crystals are tested and one parameter is only changed at a time.

6.2.5 Hall Effect and Device Geometry

As seen in Section 6.2.2, optimal laser amplification relies on a special *E/B*-field ratio, which should be kept constant throughout the whole active laser volume. This is especially important if a resonant model is considered where the heavy-hole energy precisely coincides with the optical phonon energy ($W_H = W_{OP}$).

A homogeneous magnetic field can be generated by well-designed superconducting or permanent magnets. However, the electric field is a complicated function of the crystal geometry, the electrical contact geometry, and the Hall field component induced by the magnetic field. At a constant magnetic field, variation of the applied electric field changes the mixing of hole wavefunctions. As a result different Landau levels dominate the transition frequencies, and a range of frequencies is obtained [25].

In addition the Hall field component is altered if one of the parameters, E or B, is changed. The total electric field in the crystal is the vector sum of the applied and the Hall electric field. Therefore the total electric field changes both orientation and strength.

Stoklitskiy [47] calculated the lifetime of holes in light-hole Landau levels for a magnetic field parallel to the crystallographic axis $\langle 110 \rangle$ using different electric field orientations in the (110) plane. The longest lifetime for a light hole in a Landau level is achieved with the electric field pointing parallel to another $\langle 110 \rangle$ direction. For different field orientations in the (110) plane the light hole's lifetime can reduce by several orders of magnitude. Therefore a constant electric field orientation is necessary throughout the active laser volume. The decreasing light hole's lifetime is mainly due to the mixing of the hole wavefunctions, which strongly depends on the effective mass, the applied field directions, and the crystal symmetry (see Section 6.2.6).

Stoklitskiy's studies also showed that for the LHCR laser with a magnetic field parallel to a $\langle 110 \rangle$ direction, an optimal population inversion is achieved if the total electric field and the $\langle 001 \rangle$ direction enclose an angle of 20° in the (110) plane. Then the heavy holes flow along the direction with the highest effective mass.

Two rules for optimal laser performance can be derived. Rule (1): the magnetic induction B and the electric field E have to be *simultaneously* uniform and properly orientated with respect to the crystallographic axes throughout the laser volume to allow optimal amplification of the desired frequencies. Rule (2): the E/B-field ratio has to be constant and preferably equal to 1 kV/(cmT) throughout the laser volume to allow an optimal formation of the hole population inversion.

An experimental attempt to improve the electric field uniformity was made by Bespalov and Renk [48]. They used a rhombic cross section that was inclined by an estimated Hall angle to improve the uniformity of the electric field distribution due to parallel equipotential lines. They found that the cyclotron resonance line width of a LHCR germanium laser was narrower than that of a laser with a rectangular cross section, which they related to the enhanced electric field uniformity.

R. Strijbos et al. have calculated the electric field distribution for germanium lasers with different rectangular cross section [49] and with space charge effects due to current saturation [50]. Very nonuniform electric fields were found but only a small change in the gain coefficient was determined by Monte Carlo simulations.

The Hall angle was assumed to be in the range from 30° to 60°. However, these simulations can only serve as a guide because R. Strijbos et al. used a two band model disregarding the complex Landau level structure and the mixing of light- and heavy-hole wavefunctions. This mixing alters the lifetime of holes by many orders of magnitude, which will dramatically effect the gain coefficient.

They also included a fitting parameter to model the current saturation. Current saturation is reached in Ge:Ga lasers during laser operation because of the necessary larger E/B-field ratio in comparison to Ge:Be lasers. The latter typically operate on the steepest slope of the j-E-curve at $E/B = 1$ kV/(cmT).

In my view, the gain at a certain frequency will be very different even if the electric field only varies by a small amount, such as by 1% (see Section 6.2.6 for details). This is a consequence of the strong dependence of the light- and heavy-hole wavefunction mixing on the electric field strength and orientation with respect to the crystallographic axes. Mixing effects can only be taken into account if a full quantum mechanical model is solved which is not easy to implement in a Monte Carlo simulation.

Calculation of Potentials and Electric Fields

I have calculated the potential Φ and the electric field distribution for germanium lasers with rectangular cross sections and with other, more complex geometry by solving the Laplace equation with a finite difference technique [52]

$$\Delta\Phi = 0. \tag{6.23}$$

The space charge was assumed to be zero. If two ohmic contacts are involved in the laser geometry, then they were fixed at potentials Φ_1 and Φ_0. The other metal-free crystal surfaces have to comply with tangential current flow and the boundary conditions set by the Hall effect.

A classical model derived by Komiyama [51] determines the Hall angle θ_H as

$$\tan\theta_H = \frac{W_H}{2W_{OP}}\arccos\left(1 - \frac{2W_{OP}}{W_H}\right) - \sqrt{\frac{W_H}{W_{OP}} - 1}. \tag{6.24}$$

The Hall angle θ_H is $\arctan(\pi/2) \approx 60°$ in the limit $W_H = W_{OP}$, which is the boundary condition used here.

The current condition is introduced by the scalar product

$$\vec{I}\cdot\vec{n} = 0 \tag{6.9}$$

with the vector $\vec{n}$ defined normal to the crystal surface. The Hall effect leads to a relation for the electric field components parallel to the x- and y-axis for a perpendicular magnetic field parallel to the z-axis. The ratio $E_x = \partial\Phi/\partial x$ and $E_y = \partial\Phi/\partial y$ is a function F of the tangents of the Hall angle θ_H:

$$\frac{E_y}{E_x} = F(\tan(\theta_H)). \tag{6.26}$$

The finite mesh used in these calculations allows one to evaluate the electric field strength in the crystal volume by counting its occurrence in regular intervals.

Indeed, the resulting electric field strength and the orientation for the commonly used rectangular parallelepiped crystal geometry with a near-square cross section is very nonuniform (see Fig. 6.11). This geometry typically displays a uniformity ($\Delta E_T/E_T$) of less than 20%, more often less than 10% for an electric field deviation of 1% (see Fig. 6.11c). In fact one of our best Ge: Be lasers with a duty cycle of 2.5% [23] has a d/L-ratio of 0.5, and only 6% of

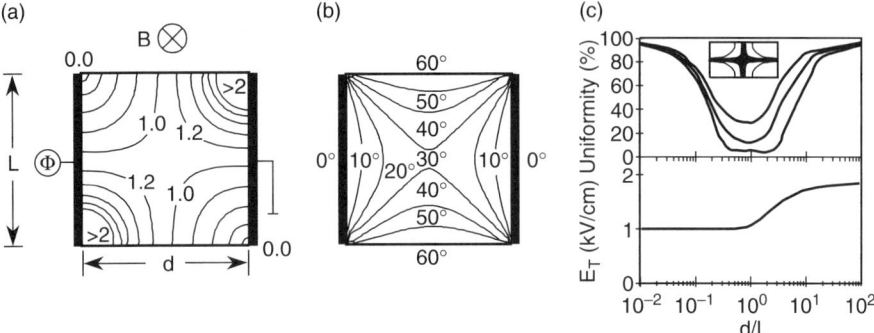

Fig. 6.11 (A) Electric field strength and (B) orientation in a laser crystal with a square cross section for a potential Φ perpendicular to the magnetic induction B. (C) Calculation of the electric field uniformity as a percentage of the crystal volume and total electric field E_T in the crystal center as a function of the cross section dimension ratio d/L. The applied field is $E = 1\,kV/cm$. The set of three lines corresponds to an electric field uniformity $\Delta E_T/E_T$ of 1%, 10%, and 40%. Inset: 1% (filled area) and 10% (contoured area) uniformity of a high duty cycle (2.5%) laser.

the volume has an electric field uniformity of better than 1%. Therefore the highest optical gain for a specific frequency only exists in a small fraction of the crystal volume.

The reader may ask: Why were those cross sections used if they show such a low electric field uniformity? This is partly due to various experimental setups in the past. To achieve high and uniform magnetic fields over sufficient length for the first germanium lasers with crystal lengths of 30 to 80 mm, it was necessary to use superconducting coils with very small center bores. To achieve a large d/L ratio would have meant polishing the narrow cross section of tall crystals, which is difficult to do. Therefore a near-square cross section was used. The radiation was coupled out in the Faraday configuration.

Nowadays, with permanent magnets of high magnetic remanence, it is possible to study crystals with large d/L ratio using the Voigt configuration. However, the problem of polishing small surfaces remains for high-duty cycle lasers, which require a small volume so that the germanium surfaces become small again for a high d/L ratio.

Laser Structures
Of special interest is a laser structure with a large d/L-ratio. Besides a high electric field uniformity, a higher total electric field can be generated inside the crystal for the same applied voltage due to the Hall effect (see Fig. 6.11c). A large d/L-ratio means a pin-like structure with small ohmic contacts provided that a standard laser is used with opposing contacts for a constant crystal volume defined by the desired power consumption. Such a structure is not very useful because the electrical contact surfaces, which are typically connected to heat sinks, will be very small and the whole laser is difficult to polish.

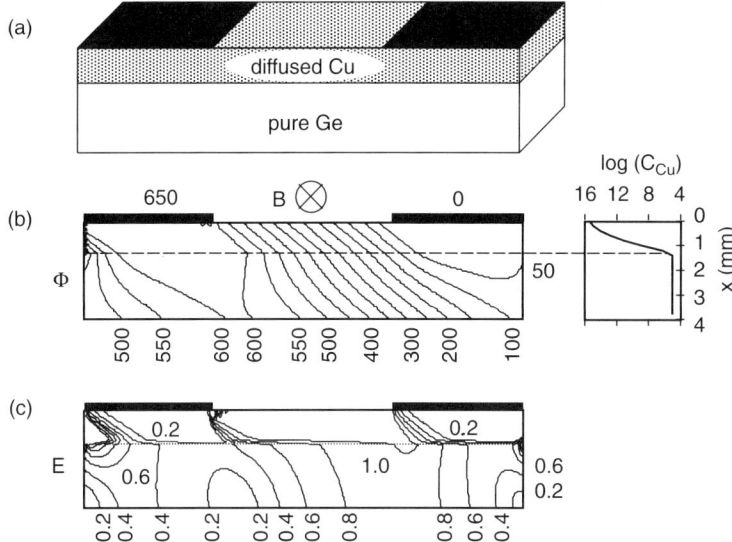

Fig. 6.12 (A) Design of a potential continuous wave laser with coplanar electrical contacts (black filled areas). A thin active layer is formed by diffusion, here copper diffusion. (B) Calculation of the potential Φ lines (on the left) for a calculated copper diffusion profile (on the right, copper concentration C_{Cu}) perpendicular to a magnetic induction B. Between the ohmic contacts is a uniform region. The potentials on the two contacts are 0 and 650 V. (C) Electric field distribution in 0.2 kV/cm intervals for a contact spacing of 10 mm.

A laser structure which could combine the benefits of a large d/L-ratio and a good heat sink is shown in Fig. 6.12a. This laser structure consists of a pure germanium crystal in which a thin layer of copper acceptors is diffused. The undoped germanium could serve as a heat sink. The calculation of the potential and the electric field distribution shows a very uniform field in the doped layer between the contacts.

A potential difference of $U = \Phi_1 - \Phi_0 = 650$ V was used in the calculation. The thin doped layer naturally leads to a large d/L ratio so that the electric field is very uniform between the ohmic contacts. The copper doping profile (see Fig. 6.12b) was calculated by using an infinite copper source on the crystal surface.

A diffusion profile can also be designed by using a finite source, for example, by implantation of beryllium. The passivation of beryllium acceptors at the surface will lead to a reduced concentration there. A doping profile could be achieved in which at some depth below the surface a suitable concentration of beryllium acceptors exists.

A suitable active layer ($N_A > 10^{13}$ cm^{-3}) with a thickness of 20 μm is obtained by diffusing for three days at 800°. The maximum concentration of this layer is located 8 μm below the surface. The initial implant was chosen to give a peak acceptor concentration of 5×10^{17} cm^{-3} at 100 nm.

The successful operation of complex laser structures relies on the use of germanium laser material doped with nonhydrogenic acceptors. Laser photons are not absorbed in these impurities. Crystal volumes without electric fields can be considered as pure germanium substrates.

This new concept in germanium laser engineering may lead to a variety of new laser designs. It is possible to integrate optical resonators on the same crystal. The inherent self-absorption in shallow hydrogenic acceptor-doped germanium lasers requires that the laser crystal is active in the whole crystal volume, meaning that electric contacts typically cover two opposite crystal surfaces *completely*.

Optical resonators, typically structured metal layers, have therefore been suspended from the laser crystal by pure silicon, germanium, or sapphire substrates to avoid electrical short circuits and a field disturbance. An integration of these resonators without interfaces, which are between the traditionally used substrates, and the active laser medium reduces the possibility of undesired reflections (for details on optical resonators, see Section 6.5.1).

In addition the cooling can be improved by coupling the heat sink without interfaces to the active laser medium. Here the heat sink is the inactive germanium. Thermal effects will be discussed in Section 6.2.10.

6.2.6 Quantum Mechanical Model

Theoretical study of germanium, as one of the group IV semiconducting elements, has been intense since the 1950s. Several authors tackled the problems of a nonparabolic and nonisotropic band structure and the influence of magnetic and electric fields [53,54]. The description of these effects will be limited here, but detailed work on quantum effects in germanium lasers will be referenced.

Band Structure and Crystallographic Effects

The germanium band structure is nonparabolic and not isotropic. Energy surfaces of the light- and heavy-hole bands can be described by the equation

$$W_{+,-}(\vec{k}) = \frac{\hbar^2}{2m_{e^-}}(Ak^2 \pm g(k_x, k_y, k_z)), \qquad (6.27)$$

with $g(k_x, k_y, k_z) = \sqrt{B^2 k^4 + C^2(k_x^2 k_y^2 + k_y^2 k_z^2 + k_z^2 k_x^2)}$. The positive and the negative root describe the light and heavy holes, respectively. The constants A, B, and C are −13.38, −8.48, and 13.14 for germanium. The hole velocity is given by

$$\vec{v}(\vec{k}) = \hbar^{-1} \nabla_k W(\vec{k}) \qquad (6.28)$$

or

$$v_i = \frac{\hbar k_i}{2m_{e^-}} \left(2A \pm \frac{2B^2 k^2 + C^2 (\Sigma_{j=x,y,z, j \neq i} k_j^2)}{g(k_x, k_y, k_z)} \right). \tag{6.29}$$

The equation of motion can then be described by

$$\frac{d\vec{k}}{dt} = \frac{e}{\hbar^2} \left(\nabla_k \left(W(\vec{k}) + \frac{\hbar}{B^2} \vec{k}(\vec{B} \times \vec{E}) \right) \right) \times \vec{B}, \tag{6.30}$$

which is, in general, very complicated. It can only be solved numerically. However, the description is still not complete because the magnetic field leads to quantum effects.

Landau Levels and Stark Effect

The magnetic induction B parallel to the z-axis introduces a quantization in the x-y-plane. The energy levels in a spherical and isotropic model are found to be

$$W_{L,H}(N) = \hbar \frac{eB}{m^*_{L,H}} \left(N + \frac{1}{2} \right) + \frac{p_z^2}{2m^*_{L,H}} - \frac{m^*_{L,H}}{2} \left(\frac{E}{B} \right)^2 + p_x \frac{E}{B} \tag{6.31}$$

with p_z the momentum parallel to the magnetic field. The characteristic energy of the Landau levels can be calculated as

$$W_{CR} = \hbar \omega_{CR} = \hbar \frac{eB}{m^*} = 0.1152 \frac{\text{meV}}{\text{T}} \left(\frac{m_{e^-}}{m^*} \right) B, \tag{6.32}$$

which can be related to the thermal energy

$$k_B T = 0.08617 \frac{\text{meV}}{\text{K}} T. \tag{6.33}$$

This relation gives a criterion for the occurrence of quantum effects.

The constant part $p_x E/B$ is neglected because for photon emission only energy differences count. The largest accumulation area is in the x-y-plane ($p_z = 0$), so the second term in (see Eq. 6.31) will also be neglected. The problem is further simplified. Equation (6.31) is divided by $\hbar \omega_{e^-}$ ($\omega_{e^-} = eB/m_{e^-}$), which results in a dimensionless equation

$$\frac{W_{L,H}(N)}{\hbar \omega_{e^-}} = \frac{m_{e^-}}{m^*_{L,H}} \left(N + \frac{1}{2} \right) - \frac{m^*_{L,H}}{m_{e^-}} \frac{m^2_{e^-}}{2 \hbar eB} \left(\frac{E}{B} \right)^2. \tag{6.34}$$

The last term can be evaluated as $245(m^*_{L,H}/m_{e^-})(E^2/B^3)$ for B in units of Tesla and E in units of kV/cm. Figure 6.13a illustrates the Landau level dependence on this term. The mass factor $(m^*_{L,H}/m_{e^-})$ is 0.35 for heavy and 0.043 for light holes ($\vec{B} \| \langle 110 \rangle$, $\vec{E} \| \langle 001 \rangle$). Therefore the Landau levels are spaced by 1/0.043 and 1/0.35, and the level slopes are shallow for light and steep for heavy holes, respectively.

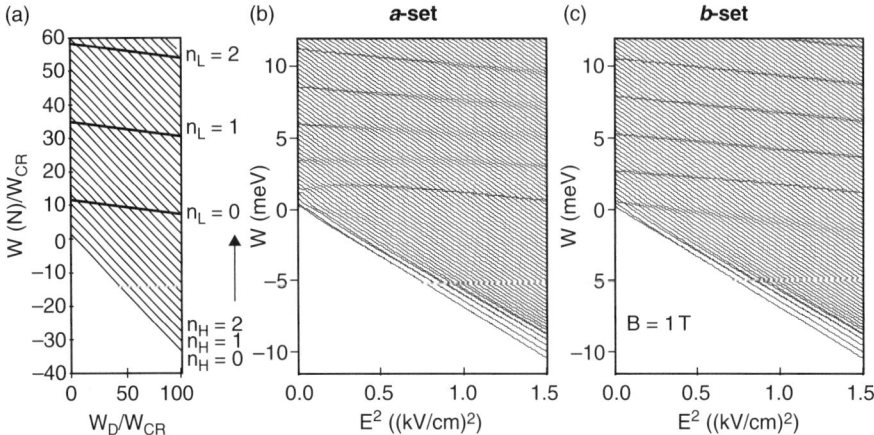

Fig. 6.13 (A) Illustration of the Landau energy level structure $W(N)$ as a function of W_D/W_{CR}, the ratio of the electric field dependent drift energy and the cyclotron resonance energy of free holes, under the assumption of orthogonal wavefunctions and an isotropic parabolic band structure for a magnetic induction parallel to a $\langle 110 \rangle$ crystallographic axis. The light and heavy hole Landau levels are indexed by n_L and n_H. (B) Calculation of the Landau energy level structure with spin according to the *a*-set using a six by six Hamiltonian in germanium at $B = 1\,\text{T}(\vec{B} \| \langle 110 \rangle)$ and as a function of the total electric field $E(\vec{E} \| \langle 001 \rangle)$. (C) Landau energy level structure of the *b*-set, parameters the same as in (B).

However, a correct calculation has to consider the full nonparabolic and nonisotropic Hamiltonian. Such calculations were performed by a number of authors using several degrees of complexity in the Hamiltonian: namely four by four Hamiltonian of light and heavy holes [55], six by six Hamiltonian including the split-off band [56], eight by eight Hamiltonian including the conduction band and warping terms [57], and additional terms that describe uniaxial stress [58].

The eight by eight Hamiltonian, four bands with two spin states each, transform into an eight by eight matrix using harmonic oscillator functions to describe the Landau levels. This matrix can easily be solved at $E = 0$. Different states indexed by n couple due to nonzero matrix elements of the form

$$\langle n-1|eEx|n \rangle = \langle n|eEx|n-1 \rangle = eEL_M\sqrt{\frac{n}{2}} \qquad (6.35)$$

if an electric field parallel to the *x*-axis is introduced with the magnetic length

$$L_M = \sqrt{\frac{\hbar}{eB}} = 26\,\text{nm}\sqrt{\frac{T}{B}}. \qquad (6.36)$$

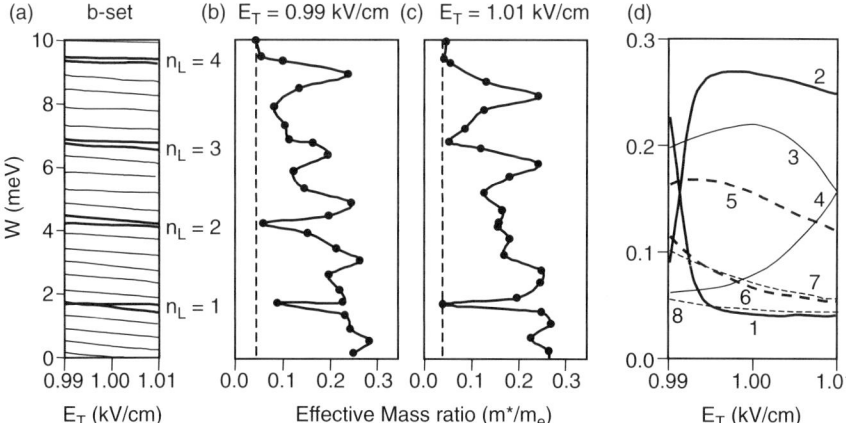

Fig. 6.14 (A) Small section of the Landau level b-set structure showing anticrossing phenomena. Electric field variations due to the Hall effect can have dramatic effects even for a field deviation of 1%. The approximate position of unmixed light hole Landau levels n_L are indicated. Here, the lowest level $n_L = 0$ is not shown. (B) Effective mass determined from the slope of the Landau levels at $E_T = 0.99$ kV/cm and (C) at $E_T = 1.01$ kV/cm. The lines are guides to the eye. (D) Effective mass determined from the slope of the Landau levels for the eight thick lines from (A) numbered sequentially from bottom to top. Adjacent levels (1,2), (3,4), (5,6), and (7,8) are drawn with the same line style.

The diagonal matrix elements only introduce a common shift of all levels, which can be neglected.

As a consequence an infinite-dimensional, symmetric band matrix has to be diagonalized, which consists in the most complete model of eight by eight matrices on the diagonal for each Landau level. Additional terms due to the electric field and, if introduced, due to warping couple the adjacent eight by eight blocks. In practice, the matrix is truncated at a dimension where the light-hole Landau level structure below the optical phonon energy converges with the desired accuracy.

Obviously these calculations are very useful to determine the laser emission frequencies and are quickly obtained at high magnetic fields when only a few levels below the optical phonon energy exist. If the electric and magnetic fields are applied along crystallographic directions of high symmetry, such as along ⟨110⟩ or ⟨100⟩, the problem is further simplified because the eight by eight matrix separates into two independent four by four matrices describing different spin states. They are called the a- and b-set.

In the Figs. 6.13b and 6.13c a calculation of the a- and b-set are given using a six by six Hamiltonian neglecting the conduction band contribution. Detailed information on the calculation of germanium and silicon laser Landau levels can be found in the references [55,56,57,58,59].

TABLE 6.2 Effective mass factor of heavy (m_H^*/m_{e^-}) and light (m_L^*/m_{e^-}) holes in germanium for magnetic fields parallel to different crystallographic axes.

	m_H^*/m_{e^-}	m_L^*/m_{e^-}
⟨100⟩	0.284	0.0438
⟨110⟩	0.352	0.0430
⟨111⟩	0.376	0.0426

The electric field variation due to the Hall effect (see Section 6.2.5) can have a dramatic effect. A small electric field variation of 1% can lead to a significant change of the hole's lifetime.

Figure 6.14a shows a small section of the Landau level structure for different electric fields. The degree of mixing of the light- and heavy-hole wavefunctions is determined by extracting the effective mass from the slope of the corresponding levels (see Eq. 6.34). A light-hole Landau level that supports an emission frequency can transform into a heavy-hole Landau level due to a change in the electric field strength of 1% (see Fig. 6.14b–d), whereas another heavy-hole level transforms into a light-hole level simultaneously. The two levels have an intermediate mass in the intermediate range. The higher effective mass will lead to stronger scattering and will reduce the lifetime of the hole. Both levels will not support an emission frequency. Repulsive shifts in the energy of both levels can also be observed, an anticrossing behavior, so a frequency shift results. Figure 6.14 definitely illustrates the intimate relationship between quantum effects, electric field distribution, device geometry, and laser performance.

Crystallographic orientations play an important role. For instance, the effective mass of the holes depends on the orientation of the magnetic field with respect to the crystallographic axes (see Table 6.2). The reduction of the light hole's lifetime in a Landau level, as described by Stoklitskiy [47] (see Section 6.2.5), is connected to the mixing of hole wavefunctions. This mixing also depends on the orientation of the applied fields, the crystal symmetry, and the different effective mass of light and heavy holes.

A multilevel laser process which illustrates the complexity of the germanium laser mechanism combining the information on Landau levels and scattering is shown in Fig. 6.15. Two sets of light-hole Landau levels and only one set of the heavy-hole Landau levels are drawn. The light-hole Landau level sets are offset from each other by approximately $10\,\text{cm}^{-1}/\text{T}$. Transitions between these sets involve a spin-flip [25,60]. The electric field disturbs or lifts the selection rules which can result in harmonics of the cyclotron resonance frequency [61]. Scattering and tunneling, such as in mixed states, are an integral part of the laser process.

6.2.7 Uniaxial Stress

Uniaxial stress is an additional parameter that can be altered within the experiment. One important effect is the lifting of the band degeneracy at the Γ point.

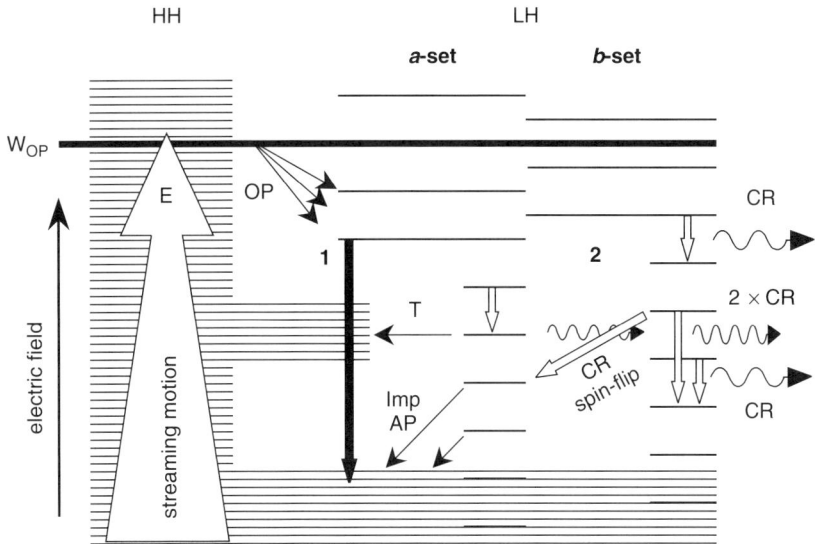

Fig. 6.15 Laser emission in a multilevel laser process including heavy hole (HH) and light hole (LH) Landau levels. Heavy holes are streaming due to the electric field E. Optical phonon scattering (OP) redistributes a few percent of heavy holes into light hole Landau levels below the optical phonon energy W_{OP}. Two light hole Landau level sets with different spins (a-set and b-set) for constant electric field and magnetic induction are shown. Laser emission can occur via light to heavy hole transitions (process 1, thick filled arrow) or by transitions between light hole Landau levels (process 2, thick open arrows). The latter can occur as cyclotron resonance (CR), harmonics of cyclotron resonance ($2 \times$ CR), spin-flip, and combined cyclotron resonance and spin-flip transitions. Light hole levels can be emptied by tunneling (T) processes or mixing between heavy and light hole levels, by impurity (Imp) scattering and acoustical phonon emission (AP).

In the eight by eight Hamiltonian as studied by Murdin et al. [58] additional terms due to stress were included. A detailed study of stress effects on the Landau level structure is also given in the articles of Suzuki and Peter [62]. In early studies the germanium lasers were stressed up to 1.3 kbar but more typically up to 0.5 kbar. It was observed that the absorption features due to the gallium dopant change because uniaxial stress decreases the acceptor ionization energy and the acceptor-related transition frequencies [63,64,65,66,67].

A remarkable experiment was recently performed by N. Hiromoto et al. [68]. They applied uniaxial stress up to 5.5 kbar to a small Ge:Ga laser. The reduction of impurity scattering due to the uniaxial stress was clearly demonstrated because the laser did not operate at zero stress but at stress values above 1.3 kbar with an optimum performance at 4 kbar.

Chamberlin et al. [69] observed a reduction of the emission for Ge:Be lasers while increasing the uniaxial stress from 0 to 0.6 kbar. A small 6 mm^3 Ge:Be laser which operated with a duty cycle close to 2% at zero stress showed no laser emission at 2 kbar [70], which can partly be explained by the reduction of the beryllium acceptor ionization energy. A Ge:Be laser under high uniaxial stress behaves as an unstressed Ge:Ga laser and the loss mechanism due to magneto-impurity effects becomes important.

A second mechanism that can reduce or terminate the laser emission is related to the transition energies of the impurity, which can coincide with the frequency range of the generated laser photons. This self-absorption mechanism is described next.

6.2.8 Impurities and Self-absorption Processes

Impurity atoms within a host semiconductor material give rise to localized electronic states within the band gap. The effective mass approximation can be used for shallow level donors and acceptors associated with substitutional impurities whose valency differs by one. An atom that lacks one valence electron with respect to the host lattice can easily attract an electron from the valence band, leaving a mobile hole and acquiring a negative charge. This mobile hole moves in the long-range attractive coulombic potential analogous to the hydrogen atom, except that in the immediate vicinity of the origin the potential departs from a simple coulombic form. The energy levels of the excited states indexed by n neglecting core effects for a spherical band at $\vec{k}=0$ follow as

$$W_{Imp}(n) = -\frac{R^*}{n^2} \qquad R^* = \frac{m^* e^4}{8(\varepsilon_r \varepsilon_0)^2 h^2} = \frac{1}{\varepsilon_r^2} \frac{m^*}{m_{e^-}} R, \qquad (6.37)$$

where ε_0 and h are the dielectric and Planck's constants.

The effective Rydberg energy R^* only deviates from $R = 13.6\,\text{eV}$ of the hydrogen atom by a factor due to the density of states effective mass of the valence band m^*/m_{e^-} and the relative dielectric constant ε_r of the host material. The effective Rydberg energy is determined as $18.6\,\text{meV}$ using $\varepsilon_r \approx 16$ and $m^* = m_H^*$, the effective mass of heavy holes, for an approximation. A more accurate treatment gives a value of $11.2\,\text{meV}$ using screened potentials and a proper weighting to obtain the hole's effective mass [71].

The effective Bohr radius r_B^* follows as

$$r_B^* = \frac{4\pi\varepsilon_r \varepsilon_0 \hbar^2}{m^* e^2} = \varepsilon_r \frac{m_{e^-}}{m^*} r_B. \qquad (6.38)$$

The hydrogen Bohr radius $r_B = 0.0529\,\text{nm}$ is only modified by the effective mass and the relative dielectric constant of the host material, which gives $r_B^* = 2.4\,\text{nm}$, again using $m^* = m_H^*$ and $\varepsilon_r \approx 16$.

This value can be related to the average acceptor distance for typical acceptor concentrations of 10^{13} to $10^{15}\,\text{cm}^{-3}$. The probability that an acceptor lies within a shell defined by the radius r and thickness dr will be $P_A(r)dr$. The term P_A can be calculated as

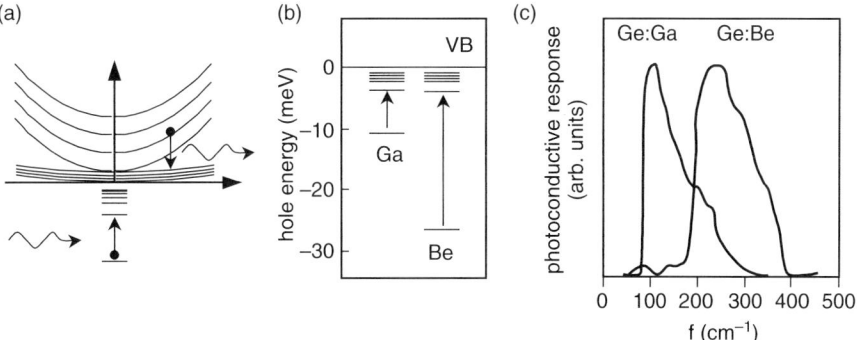

Fig. 6.16 (A) Emission generated in the valence bands can be reabsorbed at impurity sites by transitions from the impurity ground state to excited states. (B) Ground and excited states of gallium (Ga) and beryllium (Be) acceptors below the valence band (VB) of germanium. (C) Photoconductive response of Ge:Ga and Ge:Be detectors.

$$P_A(r) = N_A 4\pi r^2 \exp\left(-N_A \frac{4\pi r^3}{3}\right), \tag{6.39}$$

using the average acceptor density N_A.

The boundary condition $P_A(r \to 0) \to 0$ leads to an average acceptor distance of

$$\langle r \rangle = \int_0^\infty r P_A(r) dr = \sqrt[3]{\frac{3}{4\pi N_A}} \Gamma\left(\frac{4}{3}\right) \approx 0.55 \,\mathrm{m} N_A^{-1/3}. \tag{6.40}$$

The last approximation results from the evaluation of the Γ-function with N_A in units of m^{-3}.

An average acceptor distance of 255 to 55 nm is obtained for the typical acceptor concentrations ranging from 10^{13} to 10^{15} cm^{-3}, whereas the germanium atoms are spaced on average by 0.57 nm. For completeness, the equation for the Debye length L_D is given:

$$L_D = \sqrt{\frac{\varepsilon_r \varepsilon_0 k_B T}{e^2 N_A}} \approx 28 \,\mathrm{nm} \sqrt{\frac{T}{N_A} \frac{10^{20} \,\mathrm{m}^3}{\mathrm{K}}}. \tag{6.41}$$

The estimation of the energy levels using the effective mass approximation is quite good for excited states for which core effects are negligible. Therefore the level structure is nearly independent of the chemical nature of these hydrogenic acceptors (see Fig. 6.16).

The excited states for multiple, nonhydrogenic acceptors, such as beryllium and copper, can be approximated by exchanging the electronic charge e by Ze

TABLE 6.3 Transition energies and ionization energy W_i of hydrogenic acceptors in germanium [72,73].

	B (meV)	Al (meV)	Ga (meV)	In (meV)	Tl (meV)
W_i	10.82	11.15	11.32	11.99	13.43
I_1	10.534	10.864	11.018		
I_2		10.757	10.958		
I_3		10.669			
I_4		10.595			
I_5	10.198	10.533	10.674		
I_6	10.139	10.474	10.625		
I_7	10.048	10.382	10.536		
I_8	9.989	10.320			
A_1	9.863	10.198	10.360	11.033	
A_2	9.785	10.130	10.287	10.955	12.43
A_3	9.655	9.995	10.152	10.828	12.26
A_4	9.568	9.927	10.091	10.746	
B	9.320	9.654	9.814	10.506	11.92
C	8.686	9.025	9.185	9.864	11.32
D	7.936	8.272	8.437	9.113	10.57
E	7.57		8.02	8.42	9.83
G	6.215	6.565	6.720		8.87

TABLE 6.4 Transition energies and ionization energy W_i of nonhydrogenic acceptors and acceptor hydrogen complexes A(Be,H), A(Zn,H) in germanium [72,73,74].

	Be (meV)	Zn (meV)	Cu (meV)	A_1(Be,H) (meV)	A_2(Be,H) (meV)	A(Zn,H) (meV)
W_i	24.81	32.98	43.21	11.29	10.79	12.53
I	24.22					
A	23.73	31.84				
A'			42.27			
A''			42.07			
B	23.35	31.48	41.76			
C	22.7	30.86	41.12	9.2	8.6	
D	21.92	30.10	40.37	8.4	7.9	
E						
G	19.9	28.27	38.67			

Note: For the (Be,H) acceptor complex two sets of lines, A_1 and A_2, appear because of a trigonal symmetry C_{3V}, which splits the ground state into two levels.

in the Eqs (6.37) and (6.38) where Z counts the positive charges. However, the ground state energy is then significantly underestimated. Tables 6.3 and 6.4 give an overview of energy transitions and ionization energies of different dopants.

Doping of germanium with boron, aluminum, gallium, indium, or thallium leads to optically active transitions between these energy states in the energy range from 6 to 13 meV. However, this energy range coincides with the possible emission frequencies of the germanium laser due to the population inversion between the hole bands. Characteristic gaps in the emission spectrum from 60 to 80 cm^{-1} are the result of self-absorption, especially at the frequencies which correspond to the C and D-type optical transitions. Several papers have dealt in detail with the various absorption features with respect to the electric and magnetic fields [20,63,64,75,76].

The laser emission not only varies due to the electric and magnetic field dependent Landau level transitions but also varies due to the absorption in impurity transitions. In addition the energy levels of the impurities split due to magnetic and electric field effects.

Obviously this laser is a very complicated system due to the vast amount of possible transitions and the difficulty of measuring the emission spectra with high resolution for a multitude of parameters, such as for different electric and magnetic fields and temperature. Further the explanation above of self-absorption might suggest a quasi-static system in which some acceptors are not and others are ionized.

However, the system is dynamic and consists of recombination processes, ionization, carrier transport from Landau level to Landau level by instantaneous scattering events and radiative transitions, transitions from Landau states into impurity states, and so on. I prefer to think of a dynamic system in which the number of net holes on acceptor sites, for example, increases for nonhydrogenic acceptors in comparison to hydrogenic acceptors because the probability of a hole to *stick* to a nonhydrogenic acceptor is higher due to the larger ground state energy. A certain percentage of the total number of holes can stick to acceptor sites, which might be considered as bound, justifying a quasi-static model argument.

In germanium lasers with nonhydrogenic acceptors self-absorption processes are prevented because the ionization and the optically active transition energies of the dopants do not overlap with the energy range of the emission spectrum. These lasers allow an easier analysis of the transitions within the Landau levels of the valence band because the acceptor states do not interfere.

6.2.9 Monte Carlo Simulations

A Monte Carlo simulation is a useful tool to analyze the transport of carriers, here holes, in semiconductors [77,78]. The occupation probability of a state at the position $\vec{r}$, the wavevector $\vec{k}$, and the time t is given by the distribution function $f_{DF}(\vec{r}, \vec{k}, t)$ of light and heavy holes within the germanium band structure. This function multiplied by the transition probability gives the gain, one of the most important parameters of a laser. The distribution function can be obtained by solving the Boltzmann transport equation. The total differential (df_{DF}/dt) is then equal to

$$\nabla_r f_{DF} \frac{\partial \vec{r}}{\partial t} + \nabla_k f_{DF} \frac{\partial \vec{k}}{\partial t} + \frac{\partial f_{DF}}{\partial t} = \left(\frac{\partial f_{DF}}{\partial t}\right)_c. \tag{6.42}$$

In bulk germanium there is negligible r dependence so the first term can be neglected. The partial derivative of $\vec{k}$ follows from Eq. (6.30), which includes the external forces. If the distribution function does not change locally in time, namely in steady state, then the third term, the partial differential $(\partial f_{DF}/\partial t)$ is zero. The total differential of the distribution function has to be equal to the changes of the distribution function caused by collisions $(\partial f_{DF}/\partial t)_c$. This last term includes all the scattering processes.

Using a Monte Carlo method to solve this function is more or less a semiclassical approach with a classical, although random length, carrier motion according to the external fields possibly including the crystal band structure. Scattering events terminate this motion instantaneously. However, a full implementation of a quantum mechanical model with the Landau level structure depending on the electric fields for a real device size is difficult. Typically the scattering events are modeled using low field-scattering functions, a fixed temperature (e.g., to define acoustical phonon scattering), fixed electric and magnetic fields, and, if photon interaction is considered, a single frequency possibly with a certain narrow spread.

A germanium laser device can generate multiple frequencies unless a single mode resonator is used. The temperature increases during the applied electric field, which itself is not uniform across the sample. The large number of parameters and multiple possible transitions clearly show that Monte Carlo simulations can only give qualitative information on how to optimize the laser. Convergence of such simulations especially of those complicated hole motions due to the band structure require considerable CPU time which limits the range of samples taken in the parameter space.

Nevertheless, several studies have been performed to determine the optimal laser parameters [79,80]. R. Strijbos et al. used Monte Carlo simulations to obtain some information on the gain over a germanium laser cross section with different electric fields [49]. They also studied the effect of an additional ultra high frequency modulation for mode-locking [81]. Drijksta et al. [82] used Monte Carlo simulations to study silicon hot-hole lasers.

6.2.10 Thermal Effects

The thermal cycle can be separated into two time intervals: the crystal is heated during the electric pulse for a time t, and the crystal is cooled after the pulse has been switched off until another excitation occurs after a time τ. The repetition rate is defined as

$$f = (\tau + t)^{-1}. \tag{6.43}$$

The electric field E, given by the voltage U across the ohmic contacts separated by the distance d, leads to a current I through the contact area A. The current density

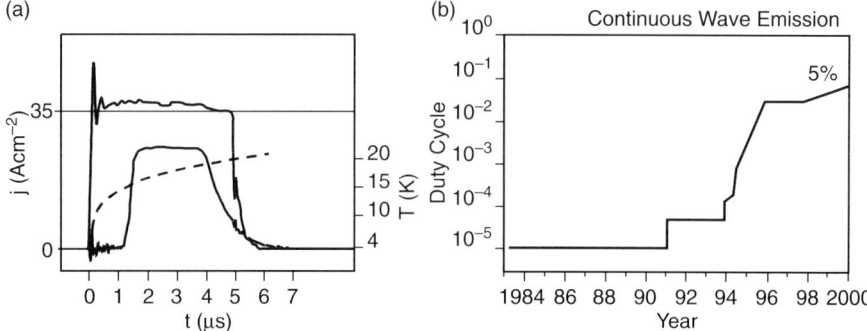

Fig. 6.17 (A) Laser emission pulse (lower solid line), current density j (upper solid line), and calculated laser temperature T (dashed line) of a Ge:Tl laser ($L10$, Table 6.1). (B) Development of the laser duty cycle since the invention of germanium lasers. Note the logarithmic scale. Major improvements were recently made by introducing new laser materials in 1995. Continuous excitation experiments have begun.

$$j = \frac{I}{A} = eN_A \upsilon \qquad (6.44)$$

is determined by the acceptor doping concentration N_A, the hole charge e, and the carrier velocity υ. The electrical input power P for a crystal volume V is

$$P = IU = jAEd = jEV = eN_A\upsilon EV. \qquad (6.45)$$

The strongest emission for Ge:Be lasers is found for the field ratio $E/B \approx 1$ kV/(cmT) [21]. Optical phonon scattering is the dominant scattering process, mainly affecting heavy holes. They are the main constituent of the current. The same carrier velocity υ is obtained for the same E and B fields. The power density P/V is proportional to the acceptor concentration N_A.

The crystal temperature can be modeled by an adiabatic heating process [83], using the nonlinear specific heat capacity function c_V of germanium [84]:

$$Pt_M = V \int_{T_B}^{T_M} c_V dT. \qquad (6.46)$$

Laser action will terminate after the time t_M at the temperature T_M. The latter is between 20 and 40 K [23] using experimental data U, I, t_M, V, and the bath temperature T_B. This temperature increases with the doping concentration, which suggests a higher optical gain but the pulse length t_M decreases because the limiting temperature is reached earlier. The maximum laser pulse length t_M of 32 µs was found for a low acceptor concentrations of 2×10^{13} cm^{-3} [23]. Figure 6.17a displays the current density and the laser pulse of a Ge:Tl laser. The temperature curve is modeled by using Eq. (6.46).

Three modifications can reduced the total power consumption. A smaller crystal volume V leads to a smaller ohmic contact area A and/or to a smaller

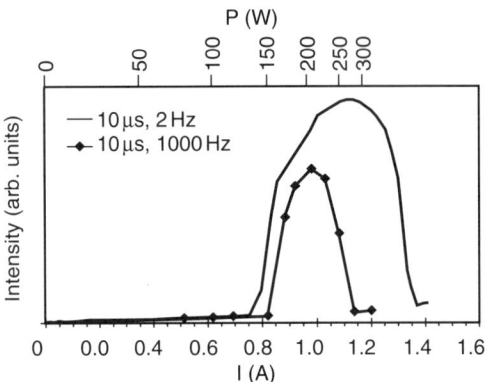

Fig. 6.18 Laser emission of a small 6 mm³ Ge:Be laser ($L15$, Table 6.1) as a function of the current I at low and high repetition rate, 2 and 1000 Hz, for an excitation pulse of 10 μs length. The threshold current can be as low as 800 mA for an input power P of 150 W during the pulse.

intercontact distance d. The current I through the device and the applied voltage U will be reduced, but the electric field $E = U/d$ and the current density $j = I/A$ remain unchanged. A lower acceptor concentration N_A reduces the current density and the current. A higher conversion efficiency will also reduce the power consumption. The last option involves a range of possibilities: an optimized laser geometry for a uniform electric field, constant magnetic and electric fields, and optimal orientation with respect to the crystallographic axes.

The duty cycle is the product of the pulse length t and the repetition rate f. The average power consumption which leads to the average heating of the crystal lattice is then given by $\overline{P} = Ptf$. Figure 6.18 shows the laser emission of a small Ge:Be laser with 6 mm³ volume and a doping concentration of 1.5×10^{14} cm⁻³. The small contact area A of 2 mm² allowed a very low driving current of 800 mA. The limiting duty cycle was found at 1.5%. The average input power $\overline{P}$ at a duty cycle of 1% was 3 W. Larger lasers with a higher surface quality can operate at an average input power in excess of 20 W.

For continuous excitation small Ge:Be lasers with volumes as small as 0.5 mm³ were fabricated. The power consumption P was reduced to 4 W [23]. Such a heat load can be cooled below 10 K using modern closed cycle refrigerators. Laser emission was found in a pulsed mode. Only weak, but nonthermal emission under continuous excitation was recently detected [22], which suggests that the coupling to the heat sinks and the surface preparation of the small surfaces were not optimal.

Heat Sinks

The heat capacity does not play a role at continuous excitation but the heat conductivity does. If a heat sink is used, it is beneficial to use pure copper with

TABLE 6.5 Heat conductivity of different materials at different temperatures T.

T (K)	99.999% Cu W/(cm K)	99.95% Cu W/(cm K)	Ge W/(cm K)	In W/(cm K)	He W/(cm K)
4	70	3.2	8	8	2.6×10^{-4}
10	135	8	18	4	1.7×10^{-4}
20	90	13	15	2	2.6×10^{-4}
30	40	14	10.5	1.2	3.4×10^{-4}

Note: The values for germanium are given for an acceptor concentration of $N_A \approx 10^{14}\,\mathrm{cm}^{-3}$.

a copper content of 99.999% because the heat conductivity is larger than the maximum heat conductivity of the laser at 10 K (see Table 6.5).

Obviously materials that have a higher or the same conductivity than germanium should be used, such as high purity germanium or copper. Pure copper heat sinks have indeed enhanced the duty cycle [22]. However, the design of an optimal heat sink requires the inclusion of the nonlinear functions of the heat conductivity and the geometry of the heat sink and the laser. Some studies of thermal problems in germanium lasers consider different crystal shapes and resonators [85].

Figure 6.17b summarizes the development of the laser duty cycle since the first experimental demonstration in 1984. The discovery of high-performance Ge:Be lasers in 1995 has led to a dramatic improvement of the duty cycle.

6.3 LASER MATERIAL FABRICATION

Germanium laser fabrication begins with pure polycrystalline germanium bars, which are purchased from commercial suppliers. These bars are zone refined to obtain the best possible purity. The total electrically active impurity concentration in the refined bar should be less than $10^{12}\,\mathrm{cm}^{-3}$. Zone refining of the polycrystalline germanium bar is performed, for example, in a graphite boat with a thin carbon soot coating inside a quartz tube. A zone of the bar is melted with a single turn radio frequency coil, which is moved along the bar. Most impurities prefer to stay in the liquid zone and are moved to the far or *dirty* end of the bar.

During zone refining, an inert gas such as argon or nitrogen is continuously flown through the tube to prevent oxidation and to flush away impurities. The refining process is repeated several times, always starting from the same end. The final pure, zone refined bars are cut with a diamond saw into pieces suitable for crystal growth.

6.3.1 Growth of Germanium Laser Material

Single crystal, germanium laser material is grown using the Czochralski technique. The crystal puller consists of a graphite susceptor in which the high purity charge is melted. The susceptor is heated by a radio frequency field. A small piece of highly doped germanium with beryllium or zinc doping around $10^{18}\,cm^{-3}$ and a predetermined weight to obtain the desired acceptor concentration is added. The acceptable range of concentrations for the Ge:Be and Ge:Zn lasers lies between 5×10^{12} and $10^{15}\,cm^{-3}$. The germanium charge and the small doping piece are melted at around 936°C in a vacuum or an ultra-pure hydrogen atmosphere. The hydrogen exchange gas allows a better thermal control during growth.

A vacuum atmosphere is preferred, in general, because in a hydrogen atmosphere additional beryllium or zinc-hydrogen complexes form. Their concentration is about 1% of the final beryllium acceptor concentration, and their ionization energy lies in the range of the shallow hydrogenic acceptors around 10 meV (see Table 6.4). These complexes should be avoided because as long as hydrogen complexes are present self-absorption can occur, which will reduce the laser performance. However, at a certain annealing temperature the hydrogen can diffuse out.

Special precautions for beryllium doping must be taken regarding any traces of oxygen (O_2 and H_2O) because of the extremely strong binding of beryllium and oxygen. Oxidized beryllium does not form a double acceptor in germanium.

6.3.2 Characterization

Before laser preparation, the crystals are characterized by variable temperature Hall effect (VTHE) measurements and photothermal ionization spectroscopy (PTIS) to determine the concentration and the type of the different impurities present in the crystals. The intentional beryllium concentration of Ge:Be crystals is typically $10^{14}\,cm^{-3}$, and the unintentional shallow hydrogenic acceptors, mostly boron and aluminum, have concentrations less than 1%, namely less than $10^{12}\,cm^{-3}$. The donor concentration, mostly phosphorus, is also less than $10^{12}\,cm^{-3}$.

A small piece, of germanium laser material is cut for a VTHE measurement, such as of size $4 \times 4\,mm^2$ cross section and 1 mm thickness. On the four corners of one side of the $4 \times 4\,mm^2$ germanium piece small ohmic contacts are formed, which allows the measurement of the Hall effect by the van der Pauw method [86].

Figure 6.19 shows an Arrhenius plot and a PTIS spectrum of typical Ge:Be laser material. Photoconductivity peaks correspond to optical transitions between the acceptor states. They are indicated for the residual shallow hydrogenic impurities and beryllium. The photoconductivity peaks of the deeper acceptor, beryllium, are more pronounced at a higher temperature.

6.3.3 Doping by Diffusion

Ge:Cu lasers are fabricated by diffusing copper into pure or ultra-pure germanium. Copper has a large diffusion coefficient that can range from $5 \times 10^{-8}\,cm^2\,s^{-1}$ to $4 \times 10^{-5}\,cm^2\,s^{-1}$ for germanium free of dislocations and for highly dislocated germanium, respectively [87]. This allows uniform doping in a relatively short time. A copper atom diffuses by the dissociative mechanism:

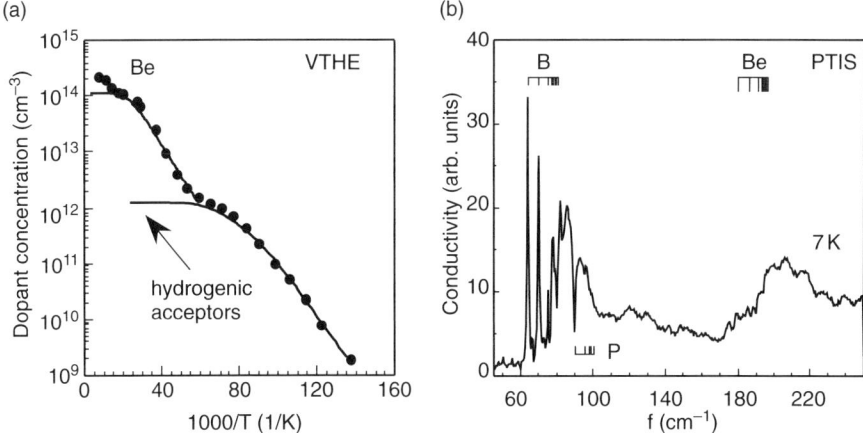

Fig. 6.19 (A) Variable temperature Hall effect (VTHE) measurement for a typical Ge:Be laser. The concentration ratio of the unintentional hydrogenic acceptors and the double acceptor is 1%. The shallow hydrogenic acceptor and donor concentrations are similar. (B) Photothermal ionization spectroscopy (PTIS) at 7 K. The unintentional hydrogenic acceptors and donors are identified as boron and phosphorus. The beryllium acceptor transitions appear above 180 cm^{-1}.

$$Cu_i + \text{vacancy} \leftrightarrow Cu_s. \tag{6.47}$$

The solubility of interstitial copper Cu_i is much lower than that of substitutional copper Cu_s, but it is much more mobile. The interstitial copper diffuses through the crystal until it reacts with a vacancy, thereby entering a substitutional site.

Typically germanium wafers of 4 to 6 mm thickness are cut with a low shallow hydrogenic acceptor concentration of 10^{11} cm^{-3} or less. These wafers are lapped in a 600 mesh SiC powder/water slurry, polish etched in a 4:1 HNO_3:HF mixture, and crystallographically oriented.

A 100 nm layer of copper is radio frequency sputtered onto the wafers as a diffusion source. The thickness is chosen to be in excess of the necessary acceptor concentration, which can be considered as an infinite source during the diffusion process. The wafers are cut with a diamond saw to produce a sample geometry with dimensions slightly larger than the final device.

Each laser crystal is cleaned and annealed separately in an ampoule sealed under vacuum. Copper is diffused for 40 hours at a fixed temperature ranging from 600 to 700°C. In this temperature range, the solubility of substitutional copper acceptors is known to vary from 0.4 to 4×10^{15} cm^{-3}. The ampoules are rapidly quenched in ethylene glycol to reach the desired concentrations of substitutional copper in the laser crystals.

The rapid quenching freezes the copper atoms on their equilibrium sites at the chosen annealing temperature, so the concentration of electrically active copper will be close to the concentration of the substitutional copper at the respective temperature.

The substitutional copper concentration measured by VTHE varies by less than 20% for each annealing temperature. The relatively high temperature during diffusion can result in an increase of the electrically active initial shallow hydrogenic acceptor concentration, such as from 2×10^{11} to $10^{12}\,\text{cm}^{-3}$.

Finally the laser crystal surfaces are lapped sequentially in 600 and 1900 mesh SiC powder/water slurries and are polish etched in a $4:1$ $HNO_3:HF$ mixture.

6.3.4 Ohmic Contacts for Germanium and Silicon

Ohmic contacts are an important ingredient of a laser device. For hot-hole lasers it is especially important that such contacts are stable at cryogenic temperature. The typical contact for germanium lasers was made by indium or aluminum evaporation onto the germanium surface with subsequent annealing at 300°C. Wires were then soldered with indium to those contacts. However, this contact can lead to a voltage drop at the contact. As a consequence discharges can form on the germanium surface due to the applied high voltage pulses. That is especially significant for silicon hot-hole devices, which need higher electric fields in comparison to germanium lasers. Current contacts are based on investigations of low noise ohmic contacts with a negligible voltage drop used for far-infrared photodetectors and bolometers [88].

The multistep process of ohmic contact formation consists of a brief HF etch to remove the oxide surface layer, followed by boron ion (B^+) implantation, metallization, and annealing. Silicon samples are additionally cooled to liquid nitrogen temperature during implantation.

Recipe for Germanium Hot-Hole Lasers
Ohmic contacts are formed by B^+-implantation with the sample at room temperature using two doses of 1×10^{14} and $2 \times 10^{14}\,\text{cm}^{-2}$ at 33 and 50 keV, respectively. Then 20 nm of palladium and 400 nm of gold are radio frequency sputtered onto the implanted surfaces. Annealing for one hour at 300°C in a nitrogen ambient atmosphere is performed to remove implantation damage and fully activate the boron acceptors in the implanted layer [86]. The annealing temperature needs to be less than the temperature where gold and germanium form an eutectic.

Recipe for Silicon Hot-Hole Lasers
Ohmic contacts are formed by B^+-implantation with the sample at 77 K using a dose of $3 \times 10^{15}\,\text{cm}^{-2}$ at 33 keV. This is followed by annealing at 500°C for one hour [89]. Finally the implanted surfaces are metallized by sputtering 20 nm chromium, 20 nm palladium, and 200 nm gold.

6.3.5 Laser Devices with Opposing Contacts

Traditional germanium lasers are made from rectangular parallelepipeds with ohmic contacts completely covering two opposing surfaces. The resulting

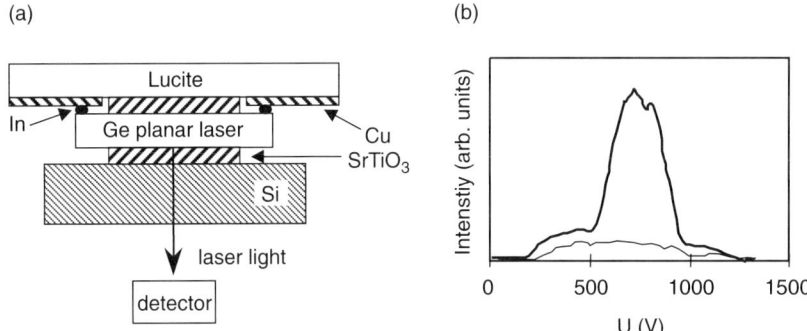

Fig. 6.20 (A) Germanium (planar) laser using $SrTiO_3$ reststrahlen mirror. The coplanar ohmic contacts are connected via indium (In) to copper foils glued to Lucite. Laser emission is detected through the silicon heat sink by a Ge:Al photoconductor. (B) Ge:Be planar laser emission as a function of the applied voltage with (upper line) and without (lower line) $SrTiO_3$ mirrors. Data courtesy of D. R. Chamberlin.

electric field depends on the application of the magnetic field as described in Section 6.2.5.

One case is especially interesting. If the contact distance d is very small and the crystal length L perpendicular to the magnetic induction is large, a very small d/L ratio is obtained: the so-called short Hall sample geometry, which leads to a negligible Hall voltage. The applied field is then equal to the total electric field in orientation and strength along the crystallographic direction determined during processing of the crystal and the ohmic contacts. This geometry is a good model to study the influence of electric and magnetic fields with respect to the crystallographic axes.

6.3.6 Laser Devices with Coplanar Contacts

A new contact geometry using coplanar contacts has recently been studied [90]. Large d/L-ratios, naturally obtained for this structure, improve the electric field uniformity. This geometry can also reduce the necessary applied electric fields due to the Hall effect [90] (see Section 6.2.5).

Figure 6.20 shows an example of a germanium laser with coplanar contacts: a planar laser. Four surfaces are lapped to suppress internal reflection modes. The laser emission appears due to two mirrors which form a Fabry-Perot resonator. The $SrTiO_3$ crystals of 0.5 mm thickness have a high reflectivity in the far infrared because of their reststrahlen bands [91]. The transmission of the crystals is approximately 0.2% at 4 K over the frequency range of interest. A $SrTiO_3$ crystal was previously used in combination with a mesh output coupler [92].

6.3.7 Laser Devices with Multiple Contacts

Multiple contacts have been fabricated to allow mode-locking of germanium lasers. Two additional contacts are used to apply a ultra-high-frequency electric field parallel to the magnetic field. The frequency is matched to the roundtrip frequency of the resonator. Muravjov et al. [93] studied multiple contacts in detail.

6.4 TECHNOLOGY

Germanium lasers require an electric field, a magnetic field, and a cooling system. The laser emission is analyzed in the time and frequency domain. These topics will now be addressed in detail.

6.4.1 Electric Field Generation

Germanium lasers come in different sizes and with different excitation power requirements. Until recently laser samples have been rather large with dimensions of typically 20 to 60 mm length and 10 to 50 mm^2 cross section, where the ohmic contact distance is 1 to 5 mm. Due to the typical acceptor concentration of 10^{14} cm^{-3} current densities of up to 100 Acm^{-2} and electric field strengths of up to 4 kV/cm need to be supplied.

I use a very simple pulse generator based on a capacitor discharge through a high voltage switch (made by Behlke Electronics GmbH, Frankfurt, Germany) with a fast rise time of 20 ns. A power supply charges a capacitor of 10 μF through a resistor to limit the charge current to between 10 and 100 mA. A TTL pulse controls the high voltage pulse structure of the fast high voltage switch. The discharge passes through the crystal and generates an electric field. This field can be pulsed with variable pulse lengths (0.2 μs–∞) and repetition rates (1 Hz–100 kHz).

A series resistor protects the high-voltage switch from accidental short circuits that can occur during discharges across germanium laser surfaces or from other possible faults. The resistor also determines the minimum discharge time. An RC-circuit in parallel to the switch limits the fly-back voltages especially during the switch-off phase.

Small crystals have a resistance of 10 to 300 Ω, which gives, neglecting the series resistor, a $1/e$ voltage drop between 100 and 3000 μs that is long in comparison to the current maximum laser emission pulse length of 32 μs. Smaller crystals have lower power requirements, which results in square-like waveforms for excitation. Impedance matching of the crystal and the pulse generator is also more easily realized.

6.4.2 Magnetic Field Generation

The laser requires magnetic fields up to several Tesla. Typically, superconducting magnets are used. They can provide a high field uniformity, such as 0.1% over a volume of 1 cm^3, and a high magnetic field strength. I used, for example, a small superconducting coil to generate a variable, magnetic induction up to 3 T [21]. Fine wire of 0.125-mm diameter allowed the winding of a magnet, which generates a magnetic field with a rating of 0.5 T/A. The magnet has an inner bore of 10 mm and an outer diameter of 30 mm.

However, the need for liquid helium is an inconvenience. There have been germanium laser experiments using electromagnets that come at the cost of large size and limited magnetic field strengths. The improvement of permanent magnetic materials based on NdFeB alloys has opened a new way to small, compact, and high field applications. Kim et al. [94] have demonstrated a germanium laser using permanent magnets with soft iron concentrators. They achieved a magnetic field of 0.35 T across the laser with a field uniformity of 3%. The emission is coupled out perpendicular to the magnetic field resulting in the Voigt configuration.

I have picked up the idea and used a simpler setup that relies on the holding force of the magnets, making a separate holder for the magnet system obsolete [23,24]. The magnets are placed at opposite surfaces of the laser crystals separated by the crystal thickness. Magnetic inductions up to 0.7 T across a 3-mm distance and close to 0.9 T across a distance of 2 mm can be reached by using two small magnets each of $20 \times 10 \times 10\,\text{mm}^3$ volume. The remanent magnetization of 1.0 to 1.1 T of these off-the-shelf NdFeB magnets is perpendicular to a $20 \times 10\,\text{mm}^2$ cross section. The cost of these two magnets is approximately \$25. The development of magnet materials has already reached remanent magnetizations of close to 1.5 T.

"Magic Rings"

Obviously current driven magnets allow an easy field variation by changing the superconducting current. Halbach [95] has suggested "magic rings" made from permanent magnets to build tunable magnetic fields. Leupold [96] demonstrated permanent magnet fields up to 4 T that are tunable by rotation of one magic ring within another magic ring that is fixed in position.

The magnetic induction in the center of one magic ring can be calculated as

$$B = B_R \eta \ln\left(\frac{r_o}{r_i}\right), \quad \eta = \frac{\sin(2\pi/M)}{2\pi/M}, \qquad (6.48)$$

with the remanent magnetic induction B_R, the outer (r_o) and inner (r_i) radius of the ring, and an integer number M counting the segments from which the ring is constructed. The factor η will be one for $M \to \infty$, but practical realization will have 8 or 16 segments with η equal to 0.90 or 0.97, respectively.

Figure 6.21a shows a possible realization of a tunable magic ring system. Rotation of one ring leaving the other fixed in position changes the center magnetic induction. Two configurations were calculated by using finite elements and a remanent magnetization of 1.1 T. In the first configuration (see Fig. 6.21b), the rings had the same position to each other (0°) as shown in Fig. 6.21a. In the second configuration (see Figs. 6.21c), the inner ring was rotated by 90°. Different angles in between 0° and 90° allow one to tune the field from 0.2 to 1.5 T.

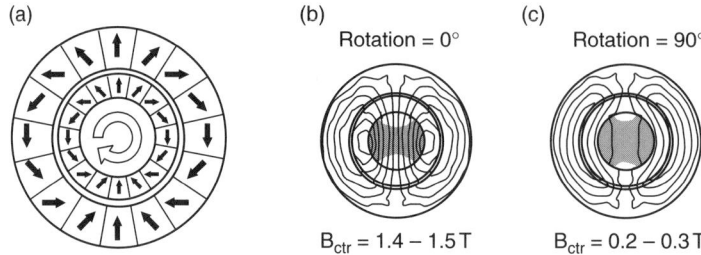

Fig. 6.21 (A) Magic ring system in which two magic rings can be rotated against each other consisting of permanent magnet cylinders with a predefined magnetization indicated by filled arrows. (B) Finite element calculation of the magnetic field lines of the unrotated structure in (A) for an outer diameter of 22 mm, an inner diameter of 8 mm, and a remanent magnetization of 1.1 T. The center field B_{ctr} (screened area) is 1.4–1.5 T. (C) Same structure as in (B) but one ring is rotated by 90°. The center field B_{ctr} (screened area) reduces to 0.2–0.3 T.

Ge:Be lasers require a magnetic induction of up to 1.7 T to cover the frequency range up to 140 cm^{-1}, which can be realized with commercially available NdFeB material ($B_R = 1.37$ T) and a radius ratio r_o/r_i of 3.5. The resulting laser system is very compact.

One advantage of permanent magnets is the possibility of using closed cycle refrigerators, especially at higher temperature, since the germanium laser can in principle operate up to 40 K [23]. There have also been reports of germanium laser emission up to 80 K [97]. At these high temperatures no superconducting wires for coil winding are available because common materials like NbTi and Nb$_3$Sn have Curie temperatures of 8.3 and 17.2 K for a magnetic induction of $B = 2$ T.

6.4.3 Cooling Systems

The classic setup of the germanium laser is immersion in liquid helium (LHe) at 4.2 K. The lasers were mounted in a holder which was centered in a superconducting coil. The whole system, magnet and laser, was immersed. Laser emission could easily be detected by a cooled Ge:Ga detector, which can be immersed in the same dewar.

Immersion in liquid helium has one major problem. If a tunable mechanical resonator is used [37,98,99,100], careful construction of such a system is necessary to provide proper ventilation to extract air remnants and clean surfaces for smooth mechanical operation. Vorobjev et al. [101] mounted a sample in vacuum where one side was attached to a liquid helium cooled surface.

In 1999 the first operation of a germanium laser in a closed-cycle refrigerator was realized [39], as suggested five years earlier [102]. A small laser cube is mounted on a copper heat sink in between two small permanent magnets with a volume of $10 \times 10 \times 6$ mm^3 each. This setup is cooled on the cold head of a closed cycle refrigerator (see Fig. 6.22).

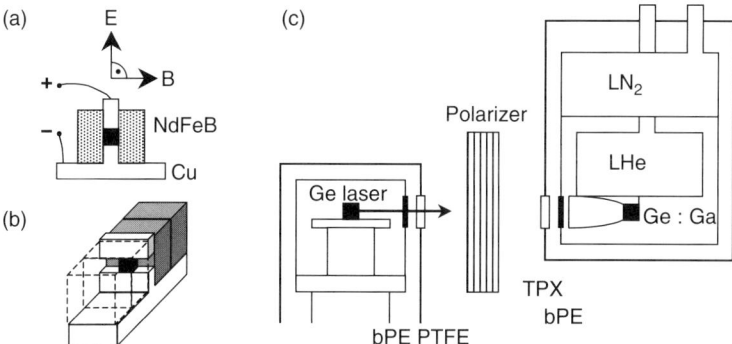

Fig. 6.22 (A) A Ge:Be laser cube of 3 mm length (filled square) mounted between two NdFeB magnets of $10 \times 10 \times 6\,mm^3$ volume on a copper heat sink drawn to scale in crossed electric E and magnetic B fields. (B) By adding additional magnets the field strength is increased. (C) Laser set-up as shown in (A) and (B) mounted in a closed cycle refrigerator (on the left). Additional filters and windows are used: black polyethylene (bPE), Teflon (PTFE), and TPX. The radiation is detected by a Ge:Ga photoconductor in a separate liquid helium (LHe) dewar precooled by liquid nitrogen (LN2). The polarization is measured by inserting a polarizer made from a wire grid.

Figure 6.23 shows a few measurements. The polarization of the radiation emitted in the Voigt configuration is linear and parallel to the magnetic field direction (see Fig. 6.23b). In the Faraday configuration the polarization is elliptical. However, the polarization can vary in a complicated manner for Ge:Ga lasers [103,104].

The closed cycle refrigerator had a minimum cold head temperature of 14 K (see Fig. 6.23c). A closed cycle refrigerator (SRDK-408D made by Sumitomo Heavy Industries, Ltd., Japan) which we recently tested has a cooling power of 1 W at 4.2 K and can dissipate a heat load of 42 W at 20 K. Newer models have a 50% improved performance and can cool loads of 1.5 W to 4.2 K.

A technological advantage of the closed-cycle refrigerator, besides the short distance to the laser source, is the possibility to include mechanical and optical elements close but without contact to the laser sample or the cold head. The full range of commercially available, room-temperature, and vacuum-fitted systems can be used, such as micrometers, feed-throughs, and O-ring sealed movable axes.

Laser cooling also requires one to study the heat conductivity and the heat capacity of the materials that contact the laser and define the heat sink geometry. Typically copper heat sinks are attached to the ohmic contacts of the lasers.

In the past, thin indium sheets were inserted between the contacts and the heat sinks to increase the contact area and to fill in voids in the Ge/Cu joint that may occur due to surface roughness. However, the heat conductivity of indium decreases above 8 K (see Table 6.5). Therefore current lasers are operated without indium sheets because the germanium crystal can be pressed in the copper heat sink, which results in efficient cooling.

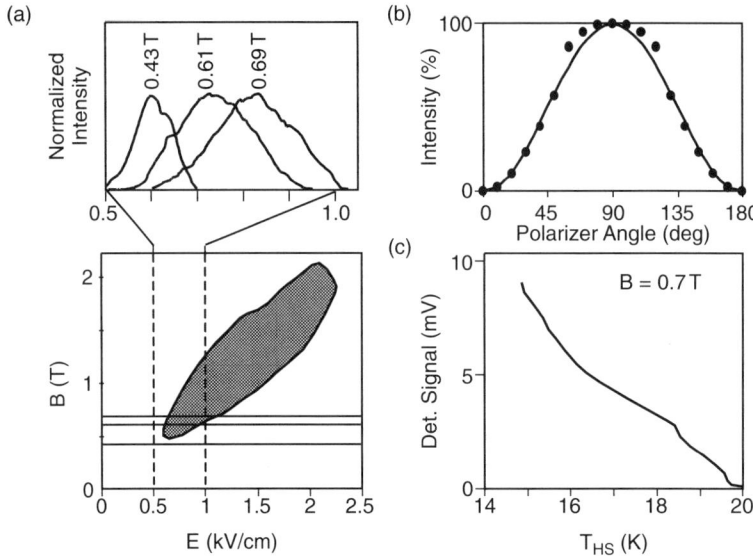

Fig. 6.23 (A) Laser emission in the Voigt configuration of a Ge:Be laser cube (L8, Table 6.1) for different magnetic fields and number of magnets. For comparison the area in the E-B-plane where laser emission in the Faraday configuration was found is shown using a superconducting magnet. Horizontal lines correspond to the magnetic fields in the Voigt configuration. The electric and magnetic fields at laser threshold are the same for both configurations. In the Voigt configuration a detector amplifier was used which resulted in the detection of the long wavelength emission at $B = 0.43$ T. (B) Linear polarization parallel to the magnetic field measured in the Voigt configuration. (C) Decreasing laser emission intensity with increasing cold head, heat sink temperature T_{HS} at a magnetic induction of $B = 0.7$ T.

6.4.4 Mode-Locking

The emission spectrum of germanium lasers without a high quality external resonator is typically 10 to 30 cm^{-1} wide. With heterodyne mixing spectroscopy the existence of several hundred equidistant longitudinal modes within such a spectrum has been shown [99].

Further, a narrow line at the roundtrip frequency of the laser cavity was measured [99]. The width of such a line is close to the Fourier transform-limited value due to the laser pulse length t, namely FWHM $\approx \pi/t$. The measured line width ranged from 0.9 to 6 MHz, depending on the laser pulse length from 3.7 to 0.5 µs, respectively. This means that the coherence length of the laser emission in each mode is defined by the laser pulse duration.

Because of the large number of modes, the wide amplification bandwidth, and the temporal coherence the germanium laser is expected to be an ideal device for the generation of short pulses on a picosecond time scale. Early experimental evidence of a fast modulation was given by studying the emission pulse detected by a Schottky diode. Intense subnanosecond spikes on the

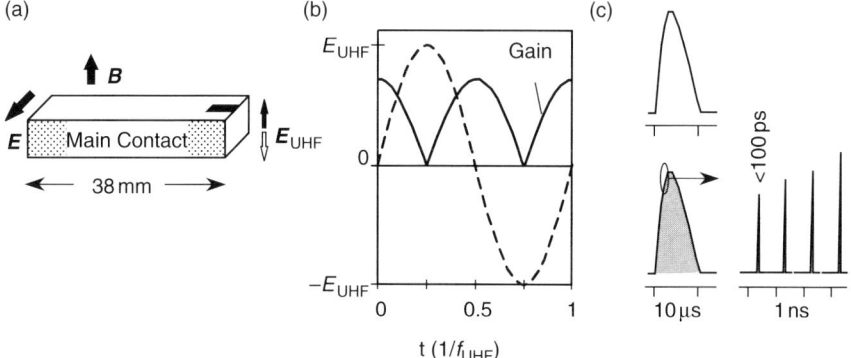

Fig. 6.24 (A) Mode-locking by using a ultra high frequency field E_{UHF} parallel to the magnetic field B coupled into the laser by two additional, opposing small contacts (black, filled area). The electric field E is applied to the two main opposing contacts. (B) The gain is maximized if the field strength E_{UHF} is zero and modulated at $2 \times f_{UHF}$. (C) A 10 μs long pulse is strongly modulated by mode-locking which results in narrow pulses with the roundtrip frequency with time intervals of 1 ns for a 38 mm long laser crystal. Pulse lengths of 63 ps have been obtained.

macropulse were observed [92,99]. Short (several tens of picosecond long) pulses can be obtained by active and passive mode-locking.

Active Mode-Locking

R. Strijbos et al. [81] suggested mode-locking of the germanium laser by actively modulating the gain. Figure 6.24 shows the principle. An ultra-high-frequency f_{UHF} is applied to two additional contacts in a short section, such as in the first 5 mm, of a 50-mm-long laser crystal parallel to the applied magnetic induction B. The gain can locally be modulated using these contacts.

Modulating at half the cavity roundtrip frequency f_{RT} leads to maximum gain suppression at maximum ultra-high-frequency field strength. Maximum gain is obtained if the ultra-high-frequency field strength is zero. The roundtrip frequency is easily determined as $f_{RT} = c/(2n_{Ge}L)$ with the speed of light c, the refractive index of germanium n_{Ge}, and the crystal length L. The roundtrip frequency is 1.3 to 0.76 GHz for a 30- to 50-mm-long crystal, which leads to an ultra high frequency $f_{UHF} = f_{RT}/2$ of 650 to 380 MHz. Mode-locked germanium lasers were demonstrated in 1997 [105].

Cross-saturation experiments [106] showed that the germanium laser has a homogeneously broadened light- to heavy-hole transition of at least 70 cm^{-1} due to the magnetically induced hole circulation in momentum space. This homogeneous broadening also explains why the integrated emission power of the germanium laser is the same if it is operated with a nonselective or single-mode resonator. Therefore a single-mode resonator can increase the spectral intensity by many orders of magnitude [99].

The achievable pulse length τ with active mode-locking at the modulation frequency in a homogeneous medium with a bandwidth Δf can be estimated by

$$\tau \approx \frac{0.5}{\sqrt{f_{RT}\Delta f}}. \qquad (6.49)$$

A typical value of $f_{RT} = 1\,\text{GHz} = 0.033\,\text{cm}^{-1}$ and $\Delta f = 2.1\,\text{THz} = 70\,\text{cm}^{-1}$ leads to pulse lengths of $\tau \approx 10\,\text{ps}$.

Experimental investigations have shown that these pulse lengths are indeed within reach of current lasers [105]. Actively mode-locked germanium lasers can emit pulse lengths down to 63 ps, which is the current limit of the real-time pulse acquisition in use [107]. Note that this value was obtained for Ge:Ga lasers that show self-absorption in the frequency range from 60 to 80 cm^{-1}, limiting the amplification bandwidth. With beryllium, zinc, or copper-doped germanium lasers a wider gain bandwidth and shorter pulse lengths could be expected.

Passive Mode-Locking

Muravjov et al. [108] have observed passive mode-locking by using a laser resonator in which two types of modes, longitudinal modes in a Fabry-Perot resonator and total internal reflection modes due to the four polished germanium surfaces, can oscillate simultaneously. Very short spikes with less than 100 ps pulse length were observed.

Amplification of Short Pulses

Germanium laser material can be used as an amplifier if all self oscillating modes are suppressed, e.g., by a special surface preparation. Such an amplifier can conserve the fast pulse structure of a second, seeding germanium laser proving the fast response of the amplifying medium [45].

6.5 LASER EMISSION

We have analyzed the emission characteristics of the germanium laser by using a grating spectrometer with a low resolution of 0.5 to 1 cm^{-1}, a Fourier-transform spectrometer with a resolution of better than 0.1 cm^{-1}, and a heterodyne spectrometer with a resolution of better than 3×10^{-5} cm^{-1} (1 MHz). But first, germanium laser resonators are discussed (see Fig. 6.25).

6.5.1 Germanium Laser Resonators

A variety of optical resonators have been used with germanium lasers.

Resonators with Total Internal Reflection Modes

The simplest resonator to fabricate uses the polished surfaces of the germanium laser crystal (see Fig. 6.25a–d). The germanium sample is cut as a rectangular parallelepiped and the surfaces are polished. The latter are typically parallel within 30 arcsec [12] to 3 arcmin [68]. The high refractive index of germanium (3.925 at 100 μm and 1.5 K [109]) enables total internal reflection (TIR) modes.

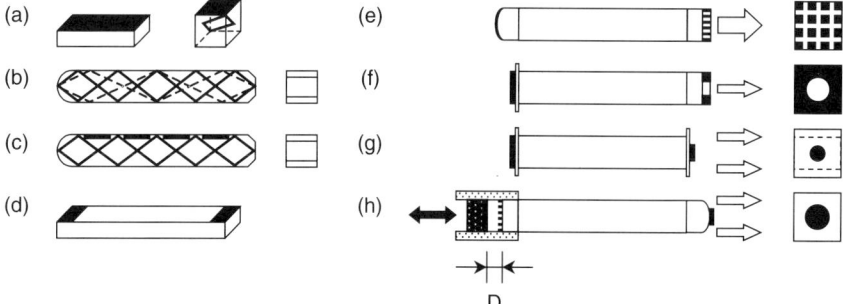

Fig. 6.25 Germanium laser resonators: (A) Resonator using total internal reflections (TIR). A TIR mode is shown in a cubic crystal (thick solid line). Opposing ohmic contacts are indicated (filled areas). (B) Resonator using TIR modes with a different number of reflections (thick dashed and solid line). The ohmic contacts are perpendicular to the plane. The structure of the output coupler cross section is shown. (C) Same resonator as in (B) which only allows a TIR mode with five reflections on one side due to four rough sections in between polished sections (indicated by filled areas). (D) TIR modes also appear in lasers with coplanar ohmic contacts. Contacts are indicated as filled areas. Here, all six sides can contribute to different modes. (E) Semiconfocal resonator consisting of a curved mirror and a metal mesh on silicon substrates. The mesh structure typically consists of 29 μm gold squares separated by 1 μm gaps for a 99% reflection around 100 cm^{-1}. The cross section of the emitting end is shown on the right. (F) Fabry-Perot resonator consisting of a metal mirror insulated by a thin PTFE (or PE) foil and a metal hole output coupler on a silicon substrate. (G) Fabry-Perot resonator consisting of two metal mirrors insulated by a thin PTFE (or PE) foil. One mirror is smaller than the cross section of the laser which results in spatial ring modes. (H) Tunable resonator consisting of a second Fabry-Perot resonator which can be coupled to the Fabry-Perot resonator of the laser crystal to obtain narrow line emission. The intersection at the gap D consists of a lamella structure covering 50% of the laser cross section with a grating constant of 0.2 mm. The radiation which is coupled out into the tunable cavity (length D) is reflected on the mechanically movable back mirror and interferes with the immediately reflected radiation at the lamella structure.

The limit angle is given as $\sin(\alpha_{Ge}) = 1/n_{Ge} \approx 14.8°$. These TIR modes are contained within the crystal if the condition

$$\tan(\beta_N) = N\frac{w}{l} \qquad (6.50)$$

is fulfilled with the length l and the width w of the crystal. An integer number N is counting the reflections on the side of length l. The angle β_N describes the angle between the beam propagation and the main axis parallel to the side of the crystal with length l. The limit angle leads to the condition

$$0.26\frac{l}{w} \approx \tan(\alpha_{Ge})\frac{l}{w} < N < \tan(90° - \alpha_{Ge})\frac{l}{w} \approx 3.80\frac{l}{w}. \qquad (6.51)$$

Obviously a large number of closed modes, such as with a different number of reflections on each surface, are possible for a given device geometry.

A cubic sample can ideally support modes with one reflection on each side incident under 45° (see Fig. 6.25a). Keeping two opposite sides fixed at one reflection, up to three reflections on the remaining surfaces are possible. Very high numbers of reflections can be excluded because with each reflection at the surface some radiation is lost due to non-ideal (e.g., rough) surfaces.

Longer lasers with a l/w ratio of 10 can support modes with 3 to 37 reflections. The frequency spacing of TIR modes can be calculated by the equation

$$\Delta f = \frac{c}{2n_{Ge}L} = \frac{c}{2n_{Ge}l}\cos(\beta_N), \qquad (6.52)$$

with the beam path length L inside the cavity given by $L^2 = l^2 + (Nw)^2$.

With heterodyne mixing spectroscopy frequencies corresponding to $N = 7$ to 14 were observed for a sample with $l = 51.4$ mm and $w = 4.77$ mm [99] (see Fig. 6.25b). The strongest line with a width of about 2 MHz was found for $N = 10$ at 545 MHz, which corresponds to the calculated value for this angle ($\beta_{10} = 42.8°$).

A second sample was fabricated with five total reflection mirrors on one of the longer surfaces of the germanium sample divided by rough parts [99] (see Fig. 6.25c). Only one line was found at 671.7 MHz with a line width of 1.3 MHz for a 2.5 µs laser pulse. This frequency corresponds to the predetermined value $\beta_5 = 25.0°$ for five mirrors. The calculated value is 671.6 MHz leading to an angle uncertainty below 0.1%.

These measurements directly confirm the hypothesis, that stimulated emission on total internal reflection modes is only possible for discrete angles β_N of the beam propagation. The angle β_N has to satisfy the condition of total internal reflection in Eq. (6.51). These discrete values of the propagation angle are caused by the fact that the total internal reflection modes with the highest Q-factor are those modes that avoid reflection close to the edges of the crystal.

It is worth noting that the exact value of the beam propagation angle β in such a cavity was used for an intra-cavity frequency selection by a Bragg mirror on the larger germanium surface of an active sample [17,110]. A single emission line was the result.

Fabry-Perot Resonator

The second resonator type consists of a rectangular parallelepiped where two opposite germanium surfaces are polished flat and parallel to form a Fabry-Perot (FP) resonator (see Fig. 6.25e–h). The remaining germanium surfaces are roughed (e.g., by sand blasting or by lapping with large grains of SiC) to prevent reflections. Weak laser emission in such samples is only observed for very long samples of about 50 mm length. To improve the 36% reflection of the Ge/LHe or Ge/vacuum interface additional elements can be attached.

The simplest method uses copper disks that are pressed against the polished germanium surfaces using these surfaces as reference planes for a resonator alignment (see Fig. 6.25g). One copper disk covers the full surface of the laser, and the other only covers the center of the polished germanium surface.

Typically the ohmic contacts fully cover two opposing germanium surfaces. Therefore it is necessary to electrically insulate the metal disks, so a thin (e.g., 20-μm) Teflon foil is sandwiched in between the germanium crystal and the mirror. A less absorbing polyethylene (PE) foil with a smaller thickness of 4 μm can also be used. Note that the plastic foil is inside the resonator and will act as an additional selective Fabry-Perot element. A 20 μm foil with a refractive index of about 1.5 will introduce an interference minimum every 60 μm influencing the emission spectra of the laser that can emit at wavelengths from 70 to 350 μm.

Semiconfocal Resonator

Silicon has little absorption in the far infrared if the impurity concentration is low. The absorption is further reduced if the carriers freeze out on their impurity sites. That can be realized by cooling silicon to the operation temperature of the germanium laser. The impurity levels in silicon are at much higher energies than the energy range of the germanium laser emission. Silicon can therefore be used as a substrate for metal mirrors formed by evaporation or sputtering. The opposite, pure silicon side is pressed against the Fabry-Perot resonator of the germanium laser crystal serving, simultaneously, as an electrical insulator.

Plane parallel substrates will introduce additional Fabry-Perot resonances because of the refractive index discontinuity (n_{Ge} = 3.925, n_{Si} = 3.384 at 100 cm^{-1} and 1.5 K [109]), although small, at the Si/Ge interface. The emission spectrum of a laser in Fig. 2 of Ref. [99] shows this sinusoidal Fabry-Perot modulation by a silicon substrate (see Fig. 6.25f).

The mode spacing is determined by the optical length. The optical length is here a sum of products, each consisting of the refractive index n and the geometric length L of the materials, for example,

$$\Delta f = \frac{c}{2(n_{Ge}L_{Ge} + n_{Si}L_{Si} + n_{PE}L_{PE})}. \tag{6.53}$$

The advantage of silicon substrates is the possibility of custom polishing the surface which can be metallized. By using a curved surface, a semiconfocal resonator (SCR) can be built that allows one to place the laser beam waist at the outcoupling surface for a small mode divergence (see Fig. 6.25e).

Germanium substrates would be a better choice to minimize the refractive index change at the laser/mirror interface, but germanium has impurities unintentionally introduced during growth, typically boron. Of course, the impurity concentration can be low, such as 10^{10} cm^{-3}, but the ionization energies (boron: 10.8 meV) lie in the range of the germanium laser emission. This absorption will limit the laser performance. However, this effect could be used to build a saturable absorber for mode-locking.

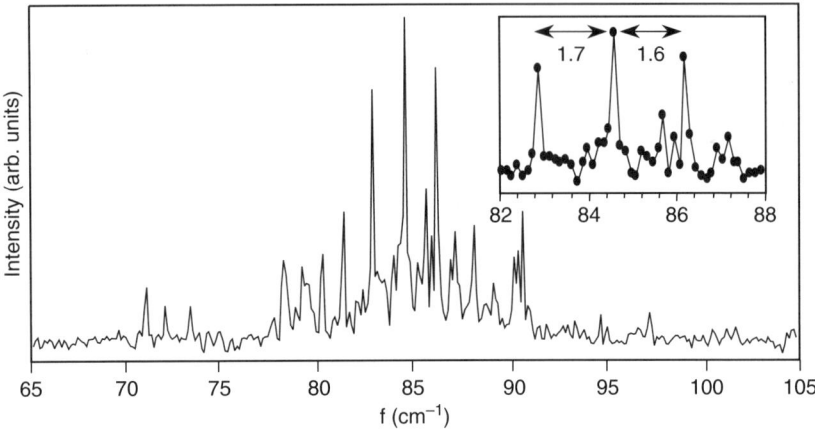

Fig. 6.26 Laser emission of a Ge:Be laser ($L7$, Table 6.1) at $B = 1\,\text{T}$, $E = 1\,\text{kV/cm}$, 12 µs pulse length, and 5–300 Hz. The inset shows the central mode structure with a frequency resolution of $0.13\,\text{cm}^{-1}$ in detail.

The measured free space beam profile of a Ge:Ga laser with an external resonator using a mesh output coupler had a divergence of 4.5° at FWHM [94]. A mode radius in the resonator of 1.3 mm was previously calculated from the transverse mode structure [92].

Single-Mode Tunable Resonator

Applications typically require a tunable laser line with a narrow line width. Single-line operation has been demonstrated for Ge:Ga lasers [37,98,99,100,111,112]. The single-mode emission is realized by coupling an additional Fabry-Perot resonator to the existing Fabry-Perot resonator of the polished end faces of the germanium crystal (see Fig. 6.25h). Laser emission is only possible if both resonance conditions are fulfilled. As a result one longitudinal mode of the Fabry-Perot germanium resonator is singled out.

Brewster Angle Resonators

Resonators that use a Brewster angle geometry to remove the influence of the germanium as a cavity and to only control the emission by the external mechanical resonator are currently under investigation. Two forms of Brewster angle resonators are conceivable. In the first geometry the laser medium is traversed on the center axis, which requires an off-axis resonator construction because of the refractive index change at the surfaces. The second geometry has an on-axis resonator but can lead to a complicated beam path inside the laser medium. A Brewster angle geometry is only sensible in the Voigt configuration because the transitions have to be linearly polarized. In the Faraday configuration the radiation has an elliptical polarization.

6.5.2 Spectra and Mode Structure

The laser emission depends on the applied magnetic and electric field. The magnetic field broadly tunes the laser gain curve, which can result in laser emission from 30 to 140 cm^{-1} [25]. Figure 6.26 gives an example of the laser emission at $B = 1$ T. Here the mode structure is a result of total internal reflection modes. The spectrum, however, consists of very narrow, unresolved resonator modes.

In a Fabry-Perot resonator longitudinal modes exist with a spacing defined by the optical length, the product of the refractive index and the geometric length of the resonator (see Fig. 6.25g). Heterodyne spectroscopy reveales very narrow line widths of this resonator (see Fig. 6.27).

Figure 6.28 shows a spectrum of a Ge:Zn laser obtained with a grating spectrometer. The frequency resolution is 0.5 to 1 cm^{-1}. The emission shows dominant peaks that can be attributed to different transitions within the Landau levels. The laser emission frequency depends on the magnetic field due to the energy dependence of the light-hole Landau levels, which are the upper laser states. They are approximately spaced with the cyclotron resonance energy $heB/m_L^* \approx 20B$ cm^{-1}T^{-1}.

The a- and b-set of the light-hole Landau levels are offset by 10 cm^{-1}T^{-1} to each other. Considering the possibility of harmonics of cyclotron resonance transitions with and without spin-flip a peak structure with a 10 cm^{-1}T^{-1} spacing should appear. Regular peaks with a spacing of 11 cm^{-1} can be seen for the Ge:Zn laser at a magnetic induction of $B = 1.15$ T (see Fig. 6.28a).

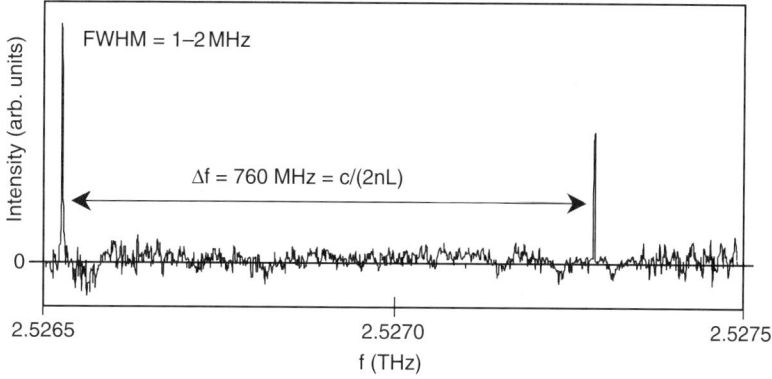

Fig. 6.27 Ge:Ga laser emission measured with heterodyne spectroscopy. The line width is pulse length limited to FWHM = 1–2 MHz. The longitudinal mode spacing Δf results from the optical length nL of the resonator. The frequency resolution is 3×10^{-5} cm^{-1} (Ge:Ga laser from V. N. Shastin's group, IPM, Nishni Novgorod). Reprinted from Infrared Phys. Technol. 36(1), E. Bründermann et al., Mode fine structure of the FIR p-Ge intervalenceband laser measured by heterodyne mixing spectroscopy with an optically pumped ring gas laser, 59–69, Copyright 1995, with permission from Elsevier Science.

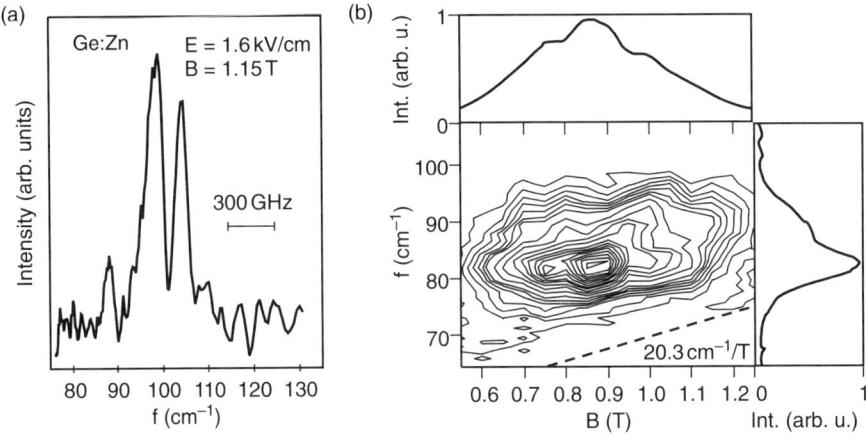

Fig. 6.28 (A) Laser emission of a Ge:Zn laser ($L12$, Table 6.1). (B) Reprinted from Infrared Phys. Technol. 40(3), E. Brüdermann et al., Novel design concepts of widely tunable germanium terahertz lasers, 141–151, Copyright 1999, with permission from Elsevier Science. Isointensity contour plot and total intensities of the emission of a Ge:Be laser ($L0$, see Table 6.1) as a function of the magnetic induction B and emission frequency f at an applied electric field of $E = 0.83\,\text{kV/cm}$. A general shift of the emission with the cyclotron resonance spacing of the light hole Landau levels of $20\,\text{cm}^{-1}/\text{T}$ is observed.

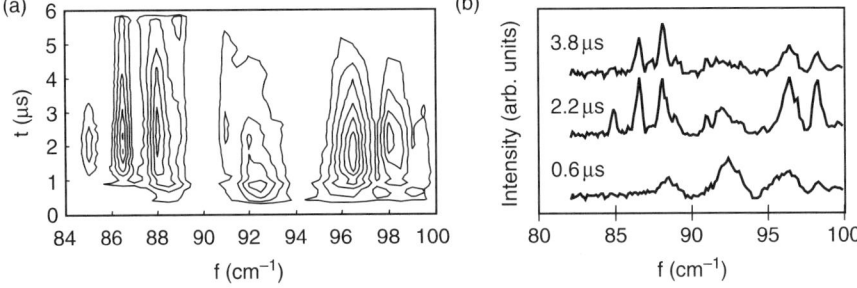

Fig. 6.29 (A) Isointensity contour lines for a Ge:Be laser ($L7$, Table 6.1) at $B = 1\,\text{T}$ and $E = 1\,\text{kV/cm}$ as a function of the emission frequency f and time t after excitation. (B) Intensity lines as a function of the emission frequency at different times after laser excitation at 0.6, 2.2, and 3.8 µs.

The spectra for a Ge:Be laser using different magnetic fields at a constant applied electric field show the frequency and magnetic dependence of the light-hole Landau levels with $20\,\text{cm}^{-1}\text{T}^{-1}$ (see Fig. 6.28b).

Time-resolved spectroscopy indicates the complex dynamics of different laser modes (see Fig. 6.29). A few broad peaks appear immediately after excitation of the laser. In time the peaks become narrower which is a result of mode competition in the homogeneously broadened laser medium. The temperature rises during the pulse so that different upper and lower laser levels can contribute to emission lines. For example, the peak at 85 cm^{-1} only exists between 1.1 and 3.2 μs, whereas the peak at 88 cm^{-1} exists between 0.5 and 6 μs reaching a maximum intensity at a later time.

6.5.3 Output Power

Power measurements in the far infrared are difficult. If a germanium photodetector is placed close to the laser crystal the detector is easily saturated even without using an external laser resonator or a detector amplifier. This observation is true for crystals with volumes as small as 5 mm^3 and an electrical input power of approximately 200 W, which results in a detected output power of approximately 0.1 W. However, the laser emits radiation through all four germanium surfaces.

External mirrors can lower the beam spread which further enhances the collected power. Very small lasers with a volume of 0.5 mm^3 using 4 W electrical input power lead to detector voltage drops of a few mV that correspond to a detected power of a few tens of μW [23] or a conversion efficiency in excess of 10^{-6}. However, it is difficult to provide the same high-quality surfaces for such small crystals in comparison to larger crystals with a surface cross section in excess of 5 mm^2.

Andronov et al. [11] claimed an output power of approximately 10 W for a crystal volume of 1 cm^3 that corresponds to a conversion efficiency in excess of 10^{-4}. LHCR lasers have shown emission powers of 300 to 500 mW for crystal volumes of approximately 1.4 cm^3 using an external resonator [42,57].

Using a Golay detector, Keilmann et al. [106] deduced a detected output power of 5 W for a 0.4-μs-long laser pulse. The signal was measured at the room-temperature end of a light pipe that was connected to the laser sample immersed in liquid helium in a storage vessel. An external resonator was attached. The crystal with a volume of 1.1 cm^3 required an electrical pump power of 100 kW. The lower limit of the conversion efficiency is 5×10^{-5}.

Lower limits of the emission power were given at 4.2 K and 15 K for a Ge:Ga laser [24]. The detected power after passage through several windows and an air gap was at least 20 μW at 15 K and 1.3 mW at 4 K for a sample volume of 0.43 cm^3 using 12.4 kW input power. No external resonator was used. The conversion efficiency is in excess of 10^{-7} at 4.2 K.

Using Schottky diode detectors at the end of a room-temperature light pipe, we measured voltage drops up to 1 V. This voltage drop corresponds to a laser power of 25 mW coupled into the antenna of the mixer block. The electrical input power of approximately 100 kW for a 1 cm^3 crystal results in a conversion efficiency of 2.5×10^{-7}.

The variety of values indicates that shaping the beam profile and efficient radiation collection are necessary to achieve conversion efficiencies comparable to those where the detector is close to the source. Theoretical studies estimate a maximum conversion efficiency of 5×10^{-3} or 0.5% [80,113].

6.6 FUTURE TRENDS

Beryllium-doped germanium laser material with an acceptor concentration of $N_A = 10^{14}$ cm^{-3} allows maximum emission at the optimal field ratio $E/B = 0.96$ kV/(cmT). In addition lasers of this material can be operated at a lower E/B-field ratio, and subsequently lower electric field values E and lower current densities j in comparison to other dopant species in germanium. That is especially important for achieving continuous wave emission which requires a low input power $P = jEV$ because most of this power has to be dissipated in the heat sink. Here are a few guidelines on what to look for in new far-infrared hot-hole lasers:

- *Light* light holes and *heavy* heavy holes (i.e., unmixed hole wavefunctions)
- High ratio of m_H^*/m_L^* for a well-defined population inversion at a low E/B-field ratio
- Doping with acceptors that have their lowest absorption frequency, the G-line, above the laser frequencies
- Acceptor concentrations of 10^{13} to 10^{15} cm^{-3} for low impurity scattering
- Low compensation by donors, typically $\leq 1\%$ and preferably none
- Magnetic induction B perpendicular to a crystal plane with a mirror symmetry (without this requirement a Sasaki-Shibuya-type effect and a parallel Hall effect occurs [114])
- Electric field E parallel to the *heaviest* hole effective mass direction [57] (Landau level mixing might change this requirement)
- Long Hall sample geometry and strong Hall field to reduce the required applied voltage [90]

6.6.1 Continuous-Wave Germanium Lasers

Continuous-wave operation of germanium lasers will require a small crystal volume (see Section 6.2.10) and a high conversion efficiency, such as that due to a uniform electric field (see Section 6.2.5).

Pulsed laser emission was already detected from a small 0.5 mm^3 Ge:Be laser which only uses 4W electrical power when continuously excited [23]. Even nonthermal emission from a Ge:Be laser under continuous excitation was found [22]. A power consumption of 4W can easily be cooled by current commercially available closed cycle refrigerators to temperatures below 10K.

However, the small samples are difficult to polish, and further work to improve the surface quality is necessary. Lasers with coplanar contacts simplify this problem. The opposing large, wafer-like surfaces are easily polished and can be used to attach external, ideally dielectric mirrors like SrTiO$_3$ crystals as shown in Fig. 6.20. The small surfaces can be rough because they are not part of the laser cavity.

The coplanar contact design inspired a new approach to reach continuous-wave operation by designing an active laser volume with a uniform field

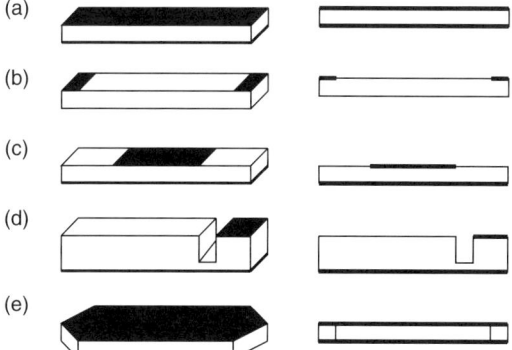

Fig. 6.30 The filled areas on the left and the thick lines on the right indicate ohmic contacts. (A) Traditional laser with opposing contacts. (B) Laser with coplanar contacts. (C) Laser with a contact structure which can be seen as inverted from (B) [39]. (D) Laser with a small active part (cubic part) and a connected larger non-active germanium piece. (E) Laser with four 45° corners removed to allow a rectangular internal reflection mode.

distribution in a larger germanium substrate. The inactive, electric field-free germanium laser material can be used as a heat sink (see Fig. 6.12).

Figure 6.30 shows an overview of lasers with different geometry that we have tested. The sample shown in Fig. 6.30d was made from a solid crystal in which a slot of 80% depth was cut. The laser, the cubic part, is three times smaller in volume compared to the attached inactive germanium piece. This laser showed a three times longer pulse length relative to the pulse length obtained with the cubic laser after it was completely separated from the inactive piece. The longer pulse length indicates an efficient cooling of the cubic volume because of the connection to the inactive germanium heat sink via a thermal interface-free section. A laser with the dimensions $20 \times 3\,\text{mm}^2$ and a contact distance of 0.5 mm is shown in Fig. 6.30e. Rectangular internal reflection modes on four sides can occur which are inclined by 45° to the main axis.

These few examples indicate the wide variety of possible new laser structures. New geometric forms of ohmic contacts entirely rely on the use of germanium laser material doped with nonhydrogenic acceptors, such as beryllium, zinc, and copper, because undoped material or unexcited material without an electric field will have negligible absorption.

6.6.2 Picosecond Pulsed Germanium Lasers

Impressive progress has been made with mode-locked Ge:Ga lasers. Active mode-locking, passive mode-locking, and amplification of short pulses has been obtained. Trains of short pulses of 63 ps at the current limit of real time pulse acquisition have been observed [107]. Germanium lasers doped with nonhydrogenic acceptors promise to generate very short pulses due to an increased amplification bandwidth.

An important question remains. The pulses of a few tens of picoseconds emitted by the germanium laser are separated by the roundtrip time, the inverse of the roundtrip frequency, which would range for 30- to 50-mm-long laser cavities from 0.9 to 1.3 ns. For such a pulse train to be used in applications, the processes under investigation have to be relaxed before the next pulse peak arrives. A dramatic improvement could be made if a cavity dumping technique is employed to isolate a single pulse.

6.6.3 Hot-Hole Silicon, Diamond, and III-V Lasers

The hot-hole laser mechanism relies on the presence of two valence bands, the light- and heavy-hole band, which are typically degenerated at the center of the Brillioun zone. These exist in all semiconductors. A few potential hot-hole laser materials are discussed here [15].

Hot-Hole Silicon Lasers

Several theoretical papers [56,80,82,115] report calculations of silicon hot-hole laser parameters. The advantageous predictions for silicon in comparison to germanium lasers involve a wider frequency tuning range up to 10 THz [56,80] and optical gain at temperatures as high as 77 K [56].

Starikov and Shiktorov [80] performed Monte Carlo simulations of hot holes in silicon. They included the warped silicon valence bands but neglected the Landau level structure. Their predictions involve emission over a wide energy range from 2 to 20 meV (0.5–5 THz). Gavrilenko [115] predicted that a heavy-hole current flow especially in the vicinity of the heaviest hole mass direction along the $\langle 110 \rangle$ axis, which is preferable for a population inversion.

A silicon crystal was cut parallel to $\langle 100 \rangle$ directions and the magnetic induction and the electric field were applied along the $\langle 100 \rangle$ and $\langle 010 \rangle$ directions, respectively [116]. Due to the Hall field component the heavy-hole current is expected to flow along the $\langle 011 \rangle$ direction. The experiments with such silicon crystals [116] revealed optical gain at 3 and 5 T between 0.6 and 1.0 kV/(cmT), which lies well within the predicted range [80]. However, a detailed experimental study of such lasers is still missing and may offer interesting effects.

It is of interest to note that bulk silicon material with shallow impurities can also serve as a laser material if it is optically pumped. A CO_2 laser can excite the electrons in a Si:P crystal from the $1s(A)$ ground state into the conduction band. This is followed by a fast relaxation via an optical phonon and a cascade of acoustical phonon-assisted relaxation processes. Electrons that relax into the $2p_0$ state of the phosphorus impurity have a long lifetime and a population inversion can appear between the $2p_0$ state and the $1s$ states [117]. Laser emission pulses of 0.3 µs length were detected at a CO_2 pump intensity in excess of 7 kWcm^{-2} for CO_2 laser pulse lengths longer than 0.3 µs. Filters were used to determine the wavelength of 59 µm, which results from the neutral donor intracenter transition $2p_0 \to 1s(E)$ [118].

Hot-Hole Diamond Lasers

The high Debye temperature of diamond (1936 K) would suggest that a population inversion could be realized even at room temperature because the hole population is not hot and optical phonons are not excited thermally. However, it seems that significant advances in the material purity are required before hot-hole diamond lasers can be realized. Diamond is typically doped with boron, an acceptor in diamond.

Hot-Hole III-V Lasers

The ratio of the heavy- and light-hole effective mass is high in InSb, but purity and low compensation is difficult to achieve. GaSb and InAs might be very interesting materials to study because some parameters, such as the light- and heavy-hole effective mass, are very similar to germanium. These materials are now frequently employed in molecular beam epitaxy and purity might be sufficient. GaAs is another interesting material for a hot-hole laser that could allow the integration of fast terahertz Schottky diodes on the same chip.

6.6.4 Lasers without Magnetic Field under Uniaxial Stress

Uniaxial stress along a crystallographic axis splits the valence bands and lifts the degeneracy between the light- and heavy-hole band at the Γ point in the center of the Brillioun zone. A hot-hole laser was demonstrated based on intra-impurity transitions between the split acceptor levels ranging from 10 to 37 meV [66,119]. This laser mechanism does not need a magnetic field. Frequency tunability is achieved by variation of the uniaxial stress.

However, the power conversion efficiency seems to be small because a very high surface quality of these germanium lasers is necessary. Parallelism of 20 arcsecs and better are needed which is difficult to achieve for small crystals with a typical volume of 5 mm^3. This leads to a very small yield during laser fabrication. High uniaxial stress of more that 3 kbar up to 10 kbar also requires a very careful design of the stressing rig to avoid breakage. However, continuous wave emission of these lasers with an output power in excess of one μW seems possible [120].

6.6.5 Lasers in Parallel Electric and Magnetic Fields

NEMAG masers [66] using an inverted heavy-hole distribution have recently been studied in the Voigt configuration [121]. The millimeter emission is strongly enhanced by using external resonators, which are easily implemented in the Voigt configuration. A wider range of electric and magnetic fields is available in this configuration. The duty cycle can be very high. The maser power of 5 mW did not decrease even at a duty cycle of 10^{-3}, which was the limit duty cycle of the pulse generator in the experimental setup. Continuous-wave emission at 4 K seems to be possible. Using silicon instead of germanium as the host material promises to extend the emission to shorter wavelengths. In silicon the band structure is strongly warped in comparison to germanium.

6.7 SUMMARY

The recent improvement of the germanium laser material combined with technological advances in the laser operation, such as using permanent magnets in a closed-cycle refrigerator and mode-locking, have made this laser an interesting candidate for far-infrared and time-resolved spectroscopy.

The emission is widely tunable from 30 to 140 cm^{-1} with narrow line widths below 1 MHz (3×10^{-5} cm^{-1}). Linear polarization can be achieved in the Voigt configuration. The pulse length can be several tens of picoseconds to several tens of microseconds. The pulsed output power can reach 10 W. An even higher peak power can be expected with mode-locking and cavity dumping.

Under continuous-wave conditions an output power of several tens of μW to tens of mW can be expected. Nonthermal emission has already been found for small continuously excited Ge:Be crystals with a volume of 0.5 mm^3. These small crystals can emit pulsed laser emission consuming 4 W of electrical power during the pulse. The theoretical conversion efficiency of 0.5%, if achieved under experimental conditions, can lead to 5 mW generated output power for an electrical pump power of 1 W. Modern closed-cycle refrigerators still reach a temperature of less than 4.2 K at a heat load of 1 W.

The very nonuniform electric field distribution of current lasers largely depends on the chosen crystal dimensions and the geometry of the ohmic contacts. It shows that there are multiple possibilities to dramatically enhance the laser conversion efficiency.

Germanium doped with beryllium seems to be the laser material of choice. It allows one to construct quite arbitrary shapes because the generated photons cannot be absorbed by impurity related transitions. Unexcited germanium material within the laser sample can act as transparent material for the far infrared, or it can act as a heat sink. A laser consisting of a thin active, doped layer introduced by diffusion in a larger pure germanium substrate seems to be an ideal candidate for a continuous-wave germanium laser.

In addition the germanium laser is an interesting physical system to study hot-hole phenomena. Except for a promising result in silicon, hot-hole lasers in other semiconductor materials have not been realized.

Future applications of germanium lasers in far-infrared spectroscopy of chemical and biological species [122] will certainly show the uniqueness and the value of this tunable semiconductor laser source in the far infrared.

ACKNOWLEDGMENTS

The author thanks G. N. Gol'tsman and A. D. Semenov for the NbN hot electron bolometers and L. Schloss for the SrTiO$_3$ substrates.

All germanium laser crystals discussed in this chapter were grown in E. E. Haller's group, LBNL, Berkeley, except for two Ge:Ga lasers. These two lasers came from V. N. Shastin's group, IPM, Nishni Novgorod, and N. Hiromoto's group, CRL, Tokyo.

The author gratefully acknowledges the intensive collaboration, assistance, and comments during the past five years of current and former members of the Berkeley team, especially, D. R. Chamberlin, O. D. Dubon, J. W. Beeman, H. Bracht, W. L. Hansen, and E. E. Haller. The support by the Alexander von Humboldt Foundation through a Feodor Lynen Fellowship from 1997 to 1999 in Berkeley is also gratefully acknowledged.

Numerous discussions with V. N. Shastin, A. A. Andronov, M. S. Kagan, R. C. Strijbos, J. N. Hovenier, W. T. Wenckebach, S. Komiyama, C. Kremser, F. Keilmann, C. R. Pidgeon, and R. E. Peale, among others, led to new ideas.

In addition the author has enjoyed fruitful collaborations with A. V. Muravjov, H.-W. Hübers, G. W. Schwaab, S. G. Pavlov, W. Heiss, J. J. Koning, L. A. Reichertz, G. C. Sirmain, M. Fujiwara, I. Hosako, and N. Hiromoto.

The author also thanks M. Havenith who made writing this chapter possible, and a very special thanks go to M. F. Kimmitt for numerous and fruitful discussions over the last 10 years and for careful reading of the manuscript.

REFERENCES

1. W. Shockley, "Hot electrons in germanium and Ohm's law," *Bell System Tech. J.* **30**, 990 (1951).
2. H. Krömer, "Proposed negative-mass microwave amplifier," *Phys. Rev.* **109**, 1856 (1958).
3. T. Kurosawa and H. Maeda, "Monte Carlo calculation of hot electron phenomena. I. Streaming in the absence of a magnetic field," *J. Phys. Soc. Jpn.* **31**, 668 (1971).
4. H. Maeda and T. Kurosawa, "Hot electron population inversion in crossed electric and magnetic fields," *J. Phys. Soc. Jpn.* **33**, 562 (1972).
5. A. A. Andronov, V. A. Kozlov, L. S. Mazov, and V. N. Shastin, "Amplification of far IR radiation in Ge on 'hot' hole population inversion," *Pis'ma Zh. Eksp. Teor. Fiz.* **30**, 585–589 (1979) [*Sov. Phys.-JETP Lett.* **30**, 551 (1979)].
6. L. E. Vorob'ev, F. I. Osokin, V. I. Stafeev, and V. N. Tulupenko, "Discovery of hot hole population inversion in Ge," *Pis'ma Zh. Eksp. Teor. Fiz.* **34**, 125 (1981) [*Sov. Phys.-JETP Lett.* **34**, 118 (1981)].
7. S. Komiyama and R. Spies, "Hot-carrier population inversion in p-Ge," *Phys. Rev.* **B23**, 6839 (1981).
8. S. Komiyama, "Far-infrared emission from population-inverted hot-carrier system in p-Ge," *Phys. Rev. Lett.* **48**, 271 (1982).
9. S. Komiyama, "Streaming motion and population inversion of hot carriers in crossed electric and magnetic fields," *Adv. Phys.* **31**, 255 (1982).
10. L. E. Vorob'ev, F. I. Osokin, V. I. Stafeev, and V. N. Tulupenko, "Discovery of long-wavelength IR radiation generation-by hot holes in Ge in crossed electric and magnetic fields," *Pis'ma Zh. Eksp. Teor. Fiz.* **35**, 360 (1982) [*Sov. Phys.-JETP Lett.* **35**, 440 (1982)].
11. A. A. Andronov, I. V. Zverev, V. A. Kozlov, Yu. N. Nozdrin, S. A. Pavlov, and V. N. Shastin, "Stimulated emission in the long-wavelength IR region from hot holes in

Ge in crossed electric and magnetic fields," *Pis'ma Zh. Eksp. Teor. Fiz.* **40**, 69 (1984) [*Sov. Phys.-JETP Lett.* **40**, 804 (1984)].

12. S. Komiyama, N. Iizuka, and Y. Akasaka, "Evidence for induced far-infrared emission from p-Ge in crossed electric and magnetic fields," *Appl. Phys. Lett.* **47**, 958 (1985).

13. Yu. L. Ivanov, "Rise of thermoluminescence in transverse magnetic field," *Pis'ma Zh. Eksp. Teor. Fiz.* **34**, 539 (1981). [*Sov. Phys.-JETP Lett.* **40**, 515 (1981)]. Y. L. Ivanov, "Generation of cyclotron resonance radiation by light holes in germanium," *Opt. Quantum Electron.* **23**, S253 (1991).

14. A. A. Andronov, A. M. Belyantsev, V. I. Gavrilenko, E. P. Dodin, E. F. Krasil'nik, V. V. Nikonorov, S. A. Pavlov, and M. M. Shvarts, "Germanium hot-hole cyclotron-resonance maser with negative effective hole masses," *Pis'ma Zh. Eksp. Teor. Fiz.* **90**, 367 (1986) [*Sov. Phys.-JETP* **63**, 211 (1986)].

15. A. A. Andronov, "Population inversion and far-infrared emission of hot electrons in semiconductors," *Infrared Millim. Waves* **16**, 149 (1986).

16. S. Komiyama and S. Kuroda, "Far-infrared laser oscillation in p-Ge," *Solid State Commun.* **59**, 167 (1986).

17. V. A. Kozlov, "Tunable far infrared semiconductor lasers," *Phys. Scripta* **T19A**, 215 (1987).

18. Special issue on "Far-infrared semiconductor lasers," *Opt. Quantum Electron.* **23**, S111 (1991).

19. Special issue: *Semicond. Sci. Technol.* **7**, B604–B654 (1992).

20. W. Heiss, K. Unterrainer, E. Gornik, W. Hansen, and E. Haller, "Influence of impurity absorption on germanium hot-hole laser spectra," *Semicond. Sci. Technol.* **9**, B638 (1994).

21. E. Bründermann, A. M. Linhart, L. Reichertz, H. P. Röser, O. D. Dubon, W. L. Hansen, G. Sirmain, and E. E. Haller, "Double acceptor-doped Ge: A new medium for inter-valence-band lasers," *Appl. Phys. Lett.* **68**, 3075 (1996). US Pat. 6,011,810 (Jan. 4, 2000), E. E. Haller and E. Bründermann, "Doping of germanium and silicon crystals with non-hydrogenic acceptors for far infrared lasers," (to The Regents of the University of California, Oakland).

22. E. Bründermann, D. R. Chamberlin, and E. E. Haller, "High duty cycle and continuous terahertz emission from germanium," *Appl. Phys. Lett.* **76**, 2991 (2000).

23. E. Bründermann, D. R. Chamberlin, and E. E. Haller, "Thermal effects in widely tunable germanium terahertz lasers," *Appl. Phys. Lett.* **73**, 2757 (1998).

24. E. Bründermann and H. P. Röser, "First operation of a far-infrared p-germanium laser in a standard closed-cycle machine at 15 K," *Infrared Phys. Technol.* **38**, 201 (1997).

25. L. A. Reichertz, O. D. Dubon, G. Sirmain, E. Bründermann, W. L. Hansen, D. R. Chamberlin, A. M. Linhart, H. P. Röser, and E. E. Haller, "Stimulated far-infrared emission from combined cyclotron resonances in germanium," *Phys. Rev.* B**56**, 12069 (1997).

26. R. Titz, M. Birk, D. Hausamann, R. Nitsche, F. Schreier, J. Urban, H. Küllmann, and H. P. Röser, "Observation of stratospheric OH at 2.5 THz with an airborne heterodyne system," *Infrared Phys. Technol.* **36**, 883 (1995).

27. R. T. Boreiko and A. L. Betz, "Far-infrared spectroscopy of C II and high-J CO emission from warm molecular gas in NGC 3576," *Astrophys. J. Suppl.* **111**, 409 (1997).

28. D. Grischkowsky, S. Keiding, M. van Exter, and C. Fattinger, "Far-infrared time-domain spectroscopy with terahertz beams of dielectrics and semiconductors," *J. Opt. Soc. Am.* B**7**, 2006 (1990).
29. M. C. Nuss, P. M. Mankiewich, M. L. O'Malley, E. H. Westerwick, and P. B. Littlewood, "Dynamic conductivity and 'coherence peak' in $YBa_2Cu_3O_7$ superconductors," *Phys. Rev. Lett.* **66**, 3305 (1991).
30. J. E. Pedersen and S. Keiding, "THz time-domain spectroscopy of nonpolar liquids," *IEEE J. Quantum Electron.* **28**, 2518 (1992).
31. H. Harde and D. Grischkowsky, "Coherent transients excited by subpicosecond pulses of terahertz radiation," *J. Opt. Soc. Am.* B**8**, 1642 (1991).
32. R. A. Cheville and D. Grischkowsky, "Far-infrared terahertz time-domain spectroscopy of flames," *Opt. Lett.* **20**, 1646 (1995).
33. A. Roitberg, R. B. Gerber, R. Elber, and M. A. Ratner, "Anharmonic wave functions of proteins: quantum self-consistent field calculations of BPTI," *Science* **268**, 1319 (1995).
34. L. Young, V. V. Prabhu, E. W. Prohofsky, and G. S. Edwards, "Prediction of modes with dominant base roll and propeller twist in b-DNA poly(dA)-poly(dT)," *Phys. Rev.* A**41**, 7020 (1990).
35. H. P. Röser, "Heterodyne spectroscopy for submillimeter and far-infrared wavelengths from 100μm to 500μm," *Infrared Phys.* **32**, 385 (1991).
36. V. N. Shastin, S. G. Pavlov, A. V. Muravjov, E. E. Orlova, R. Kh. Zhukavin, and B. N. Zvonkov, "Far-infrared hole absorption in $In_xGa_{1-x}AsGaAs$ MQW heterostructures with δ-doped barriers," *Phys. Status Solidi* B**204**, 174 (1997). Ya. V. Aleshkin, A. A. Andronov, A. V. Antonov, N. A. Bekin, V. I. Gavrilenko, A. V. Muravev, S. G. Pavlov, D. G. Revin, V. N. Shastin, I. G. Malkina, E. A. Uskova, and B. N. Zvonkov, "Far infrared emission and absorption (amplification) under real space transfer and population inversion in shallow multi-quantum wells," *Phys. Status Solidi* B**204**, 563 (1997).
37. S. Komiyama, H. Morita, and I. Hosako, "Continuous wavelength tuning of intervalence-band laser oscillation in p-type germanium over range of 80–120μm," *Jpn. J. Appl. Phys.* **32**, 4987 (1993).
38. E. Bründermann, A. M. Linhart, L. A. Reichertz, H. P. Röser, O. D. Dubon, W. L. Hansen, G. C. Sirmain, and E. E. Haller, (Edited by: M. Scheffler, R. Zimmermann), "A semiclassical description of the optimum p-type germanium laser emission visible under high repetition rate excitation," *Proc. 23rd Int. Conf. Phys. Semicond.*, Vol. 4, Berlin, Germany, 21–26 July 1996, World Scientific, Singapore, 1996, pp. 3179–3182.
39. E. Bründermann, D. R. Chamberlin, and E. E. Haller, "Novel design concepts of widely tunable germanium terahertz lasers," *Infrared Phys. Technol.* **40**, 141 (1999).
40. A. V. Bespalov, G. N. Gol'tsman, A. D. Semenov, and K. F. Renk, "Determination of the far-infrared emission characteristic of a cyclotron p-germanium laser by use of a superconducting Nb detector," *Solid State Commun.* **80**, 503 (1991).
41. R. Brazis and F. Keilmann, "Lattice absorption of Ge in the far infrared," *Solid State Commun.* **70**, 1109 (1989).
42. K. Unterrainer, C. Kremser, E. Gornik, C. R. Pidgeon, Yu. L. Ivanov, and E. E. Haller, "Tunable cyclotron-resonance laser in germanium," *Phys. Rev. Lett.* **64**, 2277 (1990).

43. L. E. Vorob'ev, S. N. Danilov, Yu. V. Kochegarov, D. A. Firsov, and V. N. Tulupenko, "Amplification of radiation in the far infrared range by hot holes in germanium in crossed electric and magnetic fields," *Fiz. Tekh. Poluprovodn.* **31**, 1482 (1997) [*Semiconductors* **31**, 1280 (1997)].
44. J. N. Hovenier, T. O. Klaassen, W. T. Wenckebach, A. V. Muravjov, S. G. Pavlov, and V. N. Shastin, "Gain of the mode locked p-Ge laser in the low field region," *Appl. Phys. Lett.* **72**, 1140 (1992).
45. A. V. Muravjov, S. H. Withers, S. G. Pavlov, V. N. Shastin, and R. E. Peale, "Broad band p-Ge optical amplifier of terahertz radiation," *J. Appl. Phys.* **86**, 3512 (1999).
46. F. Keilmann and H. Zuckermann, "Transient gain of the germanium hot hole laser," *Opt. Commun.* **109**, 296 (1994).
47. S. A. Stoklitskiy, "Quantum states interaction in hot carriers accumulation and stimulated emission processes in p-Ge," *Semicond. Sci. Technol.* **7**, B610 (1992).
48. A. Bespalov and K. Renk, "Electric field dependence of the cyclotron p-Ge laser line," *Semicond. Sci. Technol.* **9**, B645 (1994).
49. R. Strijbos, J. Dijkstra, J. Lok, S. Schets, and W. Wenckebach, "Effects of sample shape in p-Ge 'hot-hole' lasers," *Semicond. Sci. Technol.* **9**, B648 (1994).
50. R. C. Strijbos, A. V. Muravjov, and W. T. Wenckebach, "Effects of space charge on the electric field distribution in p-Ge hot hole lasers," *Proc. Hot Carriers in Semiconductors*, Chicago, IL, 31 July–4 August 1995, Plenum Press, New York, 1996, pp. 627–629.
51. S. Komiyama, T. Masumi, and K. Kajita, "Streaming motion and population inversion of hot electrons in silver halides at crossed electric and magnetic fields," *Phys. Rev.* B**20**, 5192 (1979).
52. G. H. Shortley and R. Weller, "The numerical solution of Laplace's equation," *J. Appl. Phys.* **9**, 338 (1938).
53. J. M. Luttinger, "Quantum Theory of Cyclotron Resonance in Semiconductors: General Theory," *Phys. Rev.* **102**, 1030 (1956).
54. J. C. Hensel and M. Peter, "Stark Effect for Cyclotron Resonance in Degenerate Bands," *Phys. Rev.* **114**, 411 (1959).
55. S. Kuroda and S. Komiyama, "Quantum-mechanical calculations of population inversion and amplification of far-IR laser oscillation in p-Ge," *Semicond. Sci. Technol.* **7**, B618 (1992). S. Kuroda and S. Komiyama, "Far-infrared laser oscillation due to cyclotron emission in p-type germanium," *Int. J. Infrared Millim. Waves* **12**, 783 (1991).
56. A. V. Muravjov, R. C. Strijbos, W. T. Wenckebach, and V. N. Shastin, "Population inversion of Landau levels in the valence band of silicon in crossed electric and magnetic fields," *Phys. Status Solidi* B**205**, 575 (1998).
57. P. Pfeffer, W. Zawadzki, K. Unterrainer, C. Kremser, C. Wurzer, E. Gornik, B. Murdin, and C. Pidgeon, "p-type Ge cyclotron-resonance laser: Theory and experiment," *Phys. Rev.* B**47**, 4522 (1993).
58. B. N. Murdin, C. R. Pidgeon, C. Kremser, K. Unterrainer, E. Gornik, P. Pfeffer, and W. Zawadzki, "High intensity p-Ge tunable cyclotron resonance laser," *J. Mod. Opt.* **39**, 561 (1992).
59. A. V. Bespalov, A. Schilz, W. Prettl, W. W. Fischer, and K. F. Renk, "Nonlinear magnetic field dependence of the emission frequency and bandwidth of a far-infrared p-type Ge cyclotron laser," *Phys. Rev.* B**47**, 6312 (1993).

60. I. Hosako, S. Kuroda, and S. Komiyama, "Cyclotron resonance laser in p-Ge in the Voigt configuration," *Conf. Digest 20th Int. Conf. Infrared Millimeter Waves*, 11–14 December 1995, Orlando, FL, 1995 pp. 325–326.

61. A. V. Murav'jov and V. N. Shastin, "Landau quantization and hot hole stimulated FIR emission in crossed electric and magnetic fields," *Opt. Quantum Electron.* **23**, S313 (1991).

62. K. Suzuki and J. C. Hensel, "Quantum resonances in the valence bands of germanium. I. Theoretical considerations," *Phys. Rev.* B**9**, 4184 (1974). J. C. Hensel and K. Suzuki, "Quantum resonances in the valence bands in germanium. II. Cyclotron resonances in uniaxially stressed crystals," *Phys. Rev.* B**9**, 4219 (1974).

63. S. Demihovsky, A. Muravjov, S. Pavlov, and V. Shastin, "Stimulated emission using the transitions of shallow acceptor states in germanium," *Semicond. Sci. Technol.* **7**, B622 (1992).

64. V. N. Shastin, E. E. Orlova, A. V. Muravjov, S. G. Pavlov, and R. H. Zhukavin, "Influence of shallow acceptor states on the operation of the FIR hot hole p-Ge laser," *Int. J. Infrared Millim. Waves* **17**, 359 (1996).

65. S. Komiyama and S. Kuroda, "Remarkable effects of uniaxial stress on the far infrared laser emission in p-type Ge," *Phys. Rev.* B**38**, 1274 (1988). S. Kuroda and S. Komiyama, "Far-infrared laser oscillation in p-type Ge: remarkably improved operation under uniaxial stress," *Infrared Phys.* **29**, 361 (1989).

66. V. I. Gavrilenko, N. G. Kalugin, Z. F. Krasil'nik, V. V. Nikonorov, A. V. Galyagin, and P. N. Tsereteli, "Inverted distributions of hot holes in uniaxially stressed germanium," *Semicond. Sci. Technol.* **7**, B649 (1992).

67. Yu. A. Mityagin, V. N. Murzin, O. N. Stepanov, and S. A. Stoklitsky, "Anisotropy and uniaxial stress effects in submillimetre stimulated emission spectra of hot holes in germanium in strong E perpendicular to H fields," *Semicond. Sci. Technol.* **7**, B641 (1992).

68. N. Hiromoto, I. Hosako, and M. Fujiwara, "Far-infrared laser oscillation from a very small p-Ge crystal under uniaxial stress," *Appl. Phys. Lett.* **74**, 3432 (1999).

69. D. R. Chamberlin, O. D. Dubon, E. Bründermann, E. E. Haller, L. A. Reichertz, G. Sirmain, A. M. Linhart, and H. P. Röser, "Multivalent acceptor-doped germanium lasers: a solid-state tunable source from 75 to 300 µm," in *Infrared Applications of Semiconductors II*, Symp, *Mater. Res. Soc.* 1998, Warrendale, PA, (Infrared Applications of Semiconductors II Symposium, Boston, 1–4 Dec. 1997), 1998, pp. 177–182.

70. E. Bründermann and I. Hosako, experiments performed at communications Research Laboratory, Tokyo (March 2000).

71. A. Baldereschi and N. O. Lipari, "Spherical model of shallow acceptor states in semiconductors," *Phys. Rev.* B**8**, 2697 (1973). A. Baldereschi and N. O. Lipari, "Binding energy of shallow acceptors in group IV elements and III-V compounds," *J. Luminescence* **12–13**, 489 (1976).

72. E. E. Haller and W. L. Hansen, "High resolution Fourier transform spectroscopy of shallow acceptors in ultra-pure germanium," *Solid State Commun.* **15**, 687 (1974).

73. R. E. McMurray Jr., N. M. Haegel, J. M. Kahn, and E. E. Haller, "Beryllium-hydrogen and zinc-hydrogen shallow acceptor complexes in germanium," *Solid State Commun.* **61**, 27 (1987).

74. P. Fisher and H. Y. Fan, "Absorption spectra and Zeeman effect of copper and zinc impurities in germanium," *Phys. Rev. Lett.* **5**, 195 (1960).
75. C. Kremser, W. Heiss, K. Unterrainer, E. Gornik, E. E. Haller, and W. L. Hansen, "Stimulated emission from p-Ge due to transitions between light-hole Landau levels and excited states of shallow impurities," *Appl. Phys. Lett.* **60**, 1785 (1992).
76. P. D. Coleman and C. Moe, "Laser mechanisms and processes in the p-Ge far IR laser," *Int. J. Infrared Millim. Waves* **14**, 903 (1993). P. D. Coleman and J. J. Wierer, "Establishment of a dynamic model for the p-Ge far IR laser," *Int. J. Infrared Millim. Waves* **16**, 3 (1995). P. D. Coleman and D. W. Cronin, "Experimental-theoretical Landau energy level data and its use in making p-Ge laser transition assignments," *Int. J. Infrared Millim. Waves* **17**, 973 (1996).
77. P. J. Price, "Monte Carlo calculation of electron transport in solids," in R. K. Willardson and A. C. Beer, eds., *Semiconductors and Semimetals*, Vol. 14, Academic Press, New York, 1979, pp. 249–308.
78. C. Jacoboni and L. Reggiani, "The Monte Carlo method for the solution of charge transport in semiconductors with applications to covalent materials," *Rev. Mod. Phys.* **55**, 645 (1979).
79. V. N. Shastin, "Hot hole inter-sub-band transition p-Ge FIR laser," *Opt. Quantum Electron.* **23**, S111 (1991).
80. E. V. Starikov and P. N. Shiktorov, "Numerical simulation of far infrared emission under population inversion of hole sub-bands," *Opt. Quantum Electron.* **23**, S177 (1991).
81. R. C. Strijbos, J. G. S. Lok, and W. T. Wenckebach, "A Monte Carlo simulation of mode-locked hot-hole laser operation," *J. Phys.: Condens. Matter* **6**, 7461 (1994).
82. J. E. Dijkstra and W. Th. Wenckebach, "Theoretical investigation of a current in the direction of B in p-type silicon in $E \perp B$ fields," *J. Phys.: Condens. Matter* **9**, 10373 (1997).
83. V. I. Gavrilenko, A. L. Korotkov, Z. F. Krasil'nik, V. V. Nikonorov, and S. A. Pavlov, "Far IR luminescence of hot holes in Ge: Diagnostics of intersubband population inversion and effects of uniaxial stress," *Sol. State Electron.* **31**, 755 (1988).
84. P. Flubacher, A. J. Leadbetter, and J. A. Morrison, "The heat capacity of pure silicon and germanium and properties of their vibrational frequency spectra," *Philos. Mag.* **4**, 273 (1959).
85. P. D. Coleman and D. W. Cronin, "Problems of realizing a long pulse length, high duty cycle p-Ge Landau level FIR laser," *Int. J. Infrared Millim. Waves* **18**, 1241 (1997).
86. K. S. Jones and E. E. Haller, "Ion implantation of boron in germanium," *J. Appl. Phys.* **61**, 2469 (1987).
87. H. Bracht, N. Stolwijk, and H. Mehrer, "Diffusion and solubility of copper, silver, and gold in germanium," *Phys. Rev.* B**18**, 14465 (1991).
88. E. E. Haller, "Physics and design of advanced IR bolometers and photoconductors," *Infrared Phys.* **25**, 257 (1985).
89. B. A. Young and K. M. Yu, "Ion-implanted charge collection contacts for high purity silicon detectors operated at 20 mK," *Rev. Sci. Instrum.* **66**, 2625 (1995).
90. D. R. Chamberlin, E. Bründermann, and E. E. Haller, "Planar contact geometry for far-infrared germanium lasers," *Appl. Phys. Lett.* **74**, 3761 (1999).
91. W. G. Spitzer, R. C. Miller, D. A. Kleinman, and L. E. Howarth, "Far infrared dielectric dispersion in $BaTiO_3$, $SrTiO_3$, and TiO_2," *Phys. Rev.* **126**, 1710 (1962).

92. A. V. Bespalov, "Temporal and mode structure of the interband p-germanium laser emission," *Appl. Phys. Lett.* **66**, 2703 (1995).
93. A. V. Muravjov, S. H. Withers, R. C. Strijbos, S. G. Pavlov, V. N. Shastin, and R. E. Peale, "Actively mode-locked p-Ge laser in Faraday configuration," *Appl. Phys. Lett.* **75**, 2882 (1999).
94. K. Park, R. E. Peale, H. Weidner, and J. J. Kim, "Submillimeter p-Ge laser using a Voigt-configured permanent magnet," *IEEE J. Quantum Electron.* **32**, 1203 (1996). US Pat. 5,784,397 (Jul. 21, 1998), J. J. Kim, R. E. Peale, and K. Park, "Bulk semiconductor lasers at submillimeter/far infrared wavelengths using a regular permanent magnet."
95. K. Halbach, "Design of permanent multipole magnets with oriented rare earth cobalt material," *Nucl. Instr. and Meth.* **169**, 1 (1980). K. Halbach, "Specialty magnets," *Am. Inst. Phys. Conf. Proc.* **2**, no. 153, 1277 (1987). K. Halbach, "Use of permanent magnets in accelerator technology: Present and future," *Mat. Res. Soc. Symp. Proc.* **96**, 259 (1987).
96. H. A. Leupold, E. Potenziani II, and A. S. Tilak, "Adjustable multi-tesla permanent magnet field sources," *IEEE Trans. on Magnetics* **29**, 2902 (1993). H. A. Leupold, E. Potenziani II, J. P. Clarke, and D. J. Basarab, "Use of permanent magnets in accelerator technology: Present and future," *Mat. Res. Soc. Symp. Proc.* **96**, 279 (1987).
97. L. E. Vorob'ev, S. N. Danilov, Yu. V. Kochegarov, D. A. Firsov, and V. N. Tulupenko, "Characteristics of a far-infrared germanium hot-hole laser in the Voigt and Faraday field configurations," *Fiz. Tekh. Poluprovodn.* **31**, 1474 (1997) [*Semiconductors* **31**, 1273 (1997)].
98. A. V. Murav'ev, I. M. Nefedov, S. G. Pavlov, and V. N. Shastin, "Tunable narrowband laser that operates on interband transitions of hot holes in germanium," *Kvantovaya Elektron.* **20**, 142 (1993) [*Quantum. Electron.* **23**, 119–124 (1993)].
99. E. Bründermann, H. P. Röser, A. V. Muravjov, S. G. Pavlov, and V. N. Shastin, "Mode fine structure of the FIR p-Ge intervalenceband laser measured by heterodyne mixing spectroscopy with an optically pumped ring gas laser," *Infrared Phys. Technol.* **36**, 59 (1995).
100. A. V. Muravjov, S. H. Withers, H. Weidner, R. C. Strijbos, S. G. Pavlov, V. N. Shastin, and R. E. Peale, "Single axial-mode selection in a far-infrared p-Ge laser," *Appl. Phys. Lett.* **76**, 1996 (2000).
101. L. E. Vorobjev, S. N. Danilov, and V. I. Stafeev, "Generation of far-infrared radiation by hot holes in germanium and silicon in E perpendicular to H fields," *Opt. Quantum Electron.* **23**, S221 (1991).
102. E. Bründermann, Ph.D. thesis, "Mode fine structure and mode dynamics of the pulsed and tunable p-Germanium single crystal intervalence band laser," University Bonn and Max-Planck-Institute for Radioastronomy, Germany, 23 March 1994 (in German), pp. 1–97.
103. I. Hosako and S. Komiyama, "p-type Ge far-infrared laser oscillation in Voigt configuration," *Semicond. Sci. Technol.* **7**, B645 (1992).
104. S. Komiyama, S. Kuroda, and T. Yamamoto, "Polarization of the far-infrared laser oscillation in p-Ge in Faraday configuration," *J. Appl. Phys.* **62**, 3552 (1987).
105. J. N. Hovenier, A. V. Muravjov, S. G. Pavlov, V. N. Shastin, R. C. Strijbos, and W. T. Wenckebach, "Active mode locking of a p-Ge hot hole laser," *Appl. Phys. Lett.* **71**, 443 (1997).

106. F. Keilmann, V. N. Shastin, and R. Till, "Pulse buildup of the germanium far-infrared laser," *Appl. Phys. Lett.* **58**, 2205 (1991).
107. J. N. Hovenier, TU Delft, personal communication, November 16, 1999.
108. A. V. Muravjov, R. C. Strijbos, C. J. Fredricksen, H. Weidner, W. Trimble, S. H. Withers, S. G. Pavlov, V. N. Shastin, and R. E. Peale, "Evidence for self-mode-locking in p-Ge laser emission," *Appl. Phys. Lett.* **73**, 3037 (1998).
109. E. V. Loewenstein, D. R. Smith, and R. L. Morgan, "Optical constants for far infrared materials. I. Crystalline solids," *Appl. Opt.* **12**, 398 (1973). J. Leotin, S. Barre, C. Laverny, M. Goiran, and J. R. Birch, "The far-infrared optical constants, quantum efficiency and internal quantum yield of detector-quality gallium-doped germanium," *Infrared Phys.* **28**, 165 (1988).
110. A. A. Andronov, V. A. Kozlov, S. A. Pavlov, and S. G. Pavlov, "Bragg selection in hot hole FIR lasers," *Opt. Quantum Electron.* **23**, S205 (1991).
111. K. Unterrainer, M. Helm, E. Gornik, E. E. Haller, and J. Leotin, "Mode structure of the p-germanium far-infrared laser with and without external mirrors: Single line operation," *Appl. Phys. Lett.* **52**, 564 (1988).
112. S. Komiyama and S. Kuroda, "Far-infrared laser oscillation in p-Ge using external reflectors," *Jpn. J. Appl. Phys.* **26**, L71 (1987).
113. L. E. Vorobjev, S. N. Danilov, and V. I. Stafeev, "Distribution function, population inversion and FIR gain of hot holes in germanium in crossed electric and magnetic fields," *Opt. Quantum Electron.* **23**, S195 (1991).
114. R. C. Strijbos, S. I. Schets, and W. T. Wenckebach, "Appearance of a large 'Hall' current component parallel to B in p-Ge in strong E perpendicular to B fields," *Proc. Hot Carriers in Semiconductors*, Chicago, IL, 31 July–4 August 1995, Plenum Press, New York, 1996, pp. 469–471.
115. V. I. Gavrilenko, E. P. Dodin, Z. F. Krasil'nik, and M. D. Chernobrovtseva, *Fiz. tekh. Poluprov*, **21**, 484 (1987) [*Sov. Phys.-Semicond.* **21**, 299 (1987)].
116. E. Brüdermann, E. E. Haller, and A. V. Muravjov, "Terahertz emission of population-inverted hot-holes in single-crystalline silicon," *Appl. Phys. Lett.* **73**, 723 (1998).
117. H.-W. Hübers, K. Auen, S. G. Pavlov, E. E. Orlova, R. K. Zhukavin, and V. N. Shastin, "Population inversion and far-infrared emission from optically pumped silicon," *Appl. Phys. Lett.* **74**, 2655 (1999).
118. S. G. Pavlov, R. Kh. Zhukavin, E. E. Orlova, V. N. Shastin, A. V. Kirsanov, H.-W. Hübers, K. Auen, and H. Riemann, "Stimulated emission from donor transitions in silicon," *Phys. Rev. Lett.* **84**, 5220 (2000).
119. I. V. Altukhov, E. G. Chirkova, M. S. Kagan, K. A. Korolev, V. P. Sinis, and I. N. Yassievich, "Intracenter population inversion of strain-split acceptor levels in Ge," in: M. Scheffler and R. Zimmermann eds., *Proc. 23rd Int. Conf. Phys. Semicond.*, Vol. 4, Berlin, Germany, 21–26 July 1996, World Scientific, Singapore, 1996, pp. 2677–2680.
120. Yu. P. Gousev, I. V. Altukhov, K. A. Korolev, V. P. Sinis, M. S. Kagan, E. E. Haller, M. A. Odnoblyudov, I. N. Yassievich, and K.-A. Chao, "Widely tunable continuous-wave THz laser," *Appl. Phys. Lett.* **75**, 757 (1999).
121. I. Hosako and N. Hiromoto, "p-type Germanium sub-terahertz maser oscillation in the Voigt configuration," *Proc. 7th Int. Conf. Terahertz Electronics*, IEEE, 25–26 November 1999, Nara, Japan, 1999, pp. 193–194.
122. "Http://homepage.ruhr-uni-bochum.de/Erik.Bruendermann/"

CHAPTER 7

Continuous THz Generation with Optical Heterodyning

J. C. PEARSON, K. A. MCINTOSH, and S. VERGHESE

7.1 INTRODUCTION

7.1.1 Scientific Interest in THz Waves

For decades, chemists and physicists have worked in the submillimeter-wave or terahertz (THz) region between 30 and 1000 µm because of its richness of spectral features in molecules and condensed matter. The lack of convenient sources, however, has limited this scientifically important region to a resourceful but small community of scientists who allocate significant time and resources to instrumentation development. A convenient instrument would have a source and detector pair that is small, coherent, widely tunable, and operates close to room temperature.

Advances in laser technology are facilitating a variety of new possibilities for research in physics, and chemistry that involve THz spectroscopy. The generation of coherent light waves in the terahertz or far-infrared frequency region has been investigated by many researchers with varying degrees of success and presently forms a frontier in optical science [1].

In general, THz frequencies are suitable for the study of low-energy light-matter interactions, such as phonons in solids, rotational, torsional and rovibrational transitions in molecules, vibration-rotation-tunneling transitions in weakly bound clusters, electronic fine structure in atoms, thermal imaging of cold objects and plasma dynamics. In remote sensing applications, sources such as dense interstellar molecular clouds contain significant amounts of cold dust and are optically thick (opaque). As a result these sources absorb short-wavelength radiation and reradiate most of it in the THz region, making THz observations of interstellar material in our own galaxy and external galaxies

Long-Wavelength Infrared Semiconductor Lasers, Edited by Hong K. Choi
ISBN 0-471-39200-6 Copyright © 2004 John Wiley & Sons, Inc.

essential for the study of the origin and evolution of the cold universe. Another important remote sensing application is THz studies of both the terrestrial and planetary atmospheres, which yields critical insights into the nature of the atmospheric chemistry and dynamics.

7.1.2 Source Problem

In order to realize the applications described above, various types of THz sensors have been proposed and developed for both high and low frequency-resolution. In the area of high-resolution sensors, progress in cryogenic and noncryogenic heterodyne mixer technology has opened up the possibility of complete coverage of the THz spectral region with low-noise heterodyne receivers for remote sensing should suitable local oscillators be developed. For conventional spectroscopy using direct detection, a wide range of well-developed detectors operating at ambient and cryogenic temperatures are available. The technical challenge has always been in the area of sources suitable for use as local oscillators or as coherent sources for spectrometers. Molecular gas lasers are available at moderate power and a few discrete frequencies throughout the THz region, and electron beam tubes such as backward-wave oscillators have been demonstrated to 1.5 THz with milliwatt output powers and 10% bandwidth. P-type germanium lasers also provide milliwatt output power and reasonable tuning but currently are limited to pulsed operation and imprecise frequency control. Free-electron lasers can cover the entire THz spectral region with plenty of power, but these devices are massive, power hungry, and limited to pulsed operation with little opportunity for precise frequency control. On the other hand, harmonic up-conversion from phase-locked microwave source provides excellent frequency control and continuous-wave operation. However, to date this approach has suffered from poor conversion efficiency, with bandwidths limited by the high-Q varactor diode circuits used and the resulting low output power and bandwidths limited to about 10%. Optical down conversion or photomixing quickly solves the bandwidth limitation of conventional multiplier circuits, but all spectrometer or local oscillator systems have rather stringent requirements on source frequency control, frequency accuracy and spectral purity. As a result a photomixing source must address the spectral requirements as well as generate sufficient power for the application.

Direct detection spectroscopy has the least stringent power requirements with a few nW being sufficient for detection of strong spectral features. Cryogenic heterodyne receivers require 100 nW to a few μW of power, while room-temperature mixers typically require several mW of power. As sources, photomixers have been used for local oscillators with cryogenic THz heterodyne detectors [2] and for high-resolution gas spectroscopy when used in conjunction with liquid-helium-cooled bolometers [3,4]. This chapter discusses a system suitable for either spectroscopy or a broad-band heterodyne local oscillator.

7.1.3 THz Generation with Photomixers

Optical heterodyne conversion with a photoconductive switch (photomixing) is analogous to the operation of a transistor amplifier. In a transistor, a small RF signal applied to the gate modulates the conductance of a switch under a relatively large dc bias. The modulated output power is drawn from the source providing the dc bias. In photomixing, the beating of the two near optical lasers modulates the photoconductance and the output THz power is generated from current drawn from the battery that provides a dc bias between the photoconductor electrodes.

Photomixing is fundamentally different than difference-frequency generation using a $\chi^{(2)}$ process in a material such as $LiNbO_3$. In this type of difference-frequency generation the output THz power is generated from the optical photons, and only one THz photon can be created in each pair of optical photons. The fact that two hot photons are required to create one cold photon results in an efficiency penalty that makes photomixing more efficient between 1 and 3 THz. At frequencies above several THz, however, the $\chi^{(2)}$-mixing process becomes more efficient than photomixing, since photomixers suffer from parasitic impedances as $\chi^{(2)}$-mixing process become more efficient. This trade-off has been discussed in more detail by Brown et al. [5].

Photomixers are compact, solid-state sources that combine a pair of laser frequencies to generate a difference frequency by photoconductive mixing. If a THz difference frequency is desired, the mixing needs to be done in a very fast low capacitance structure fabricated out of a material like low-temperature-grown (LTG) GaAs [6,7,8]. The intrinsic bandwidth limitation of a photomixers is only limited by the speed of the material and the antenna structure produced with the device. As a result it is possible, with careful selection of lasers, to have a difference frequency that is tunable over several THz. A number of photomixer theory and design details are covered in subsequent sections of this chapter.

7.2 REQUIREMENTS FOR PHOTOMIXING SYSTEMS

The rotational transitions of molecules in the THz region of the spectrum typically have natural line widths of a few Hz. As a result Doppler or pressure broadening dominates the observed line profile unless a sub-Doppler technique is used. Typical pressure broadening for a 1 Debye permanent dipole is on the order of 10 MHz per Torr and molecules with larger or smaller permanent dipole moments tend to broaden faster and slower, respectively. Doppler broadening is about 3 MHz full-width at half-maximum (FWHM) near 1 THz and ambient temperature for a light molecule like water. The center of low-pressure molecular absorption or emission features can easily be determined to 10% of the FWHM and to better than 1% should the line profile and instrument baseline effects be carefully accounted for. As a result direct absorption or emission spectroscopy in the THz region requires a frequency accuracy of at least one part in 10^7, with one part in 10^8 being desirable. If sub-Doppler techniques are to be employed, then the linewidth is on the order of 50 kHz. In the sub-Doppler case an accuracy of at least one part in 10^9 is necessary, with one part in 10^{10} being desirable.

The molecular line widths place a requirement on the spectral purity of the source. In general, the spectral purity or effective spectral width of the source needs to be about one order of magnitude better than the spectral width of the observed feature. If the source purity is known, the source contribution to the observation can be deconvolved, but in all cases when the source is nearly as broad or broader than the observed feature the resolution of the system is degraded. As a result a FWHM line width of less than 1 MHz is necessary at 1 THz for conventional microwave style spectroscopy or heterodyne astronomy, and a FWHM of less than 10 kHz is necessary for sub-Doppler applications. The last requirement is source stability. All remote sensing and spectroscopic instruments should be able to sit and integrate for hours and achieve the radiometric noise expectation that the observed signal-to-noise ratio increases as the square root of integration time.

The spectral purity and stability of a signal generated by photomixing is almost entirely determined by the laser system supplying the photons to the photomixer. In the case of GaAs-based photomixers the lasers need to be in the 870- to 700-nm-wavelength range for optimal radiation coupling to the photomixer. Fortunately, nearly all the commercially available near-IR lasers in the frequency range suitable for GaAs-based photomixing are widely tunable. Diode-laser-based systems are particularly promising in this type of application, since they combine low-power consumption and relatively long lifetime in an inexpensive and compact package. Such systems have already been applied to laboratory spectroscopy by several authors [7,9]. However, the frequency accuracy, spectral purity or stability of the THz-wave output obtained is generally insufficient for high-resolution laboratory spectroscopy or heterodyne remote sensing applications. In the subsequent section a variety of techniques for the generation and precision control of a precision THz difference frequency are presented. The techniques presented are not limited to the near-IR wavelengths useful to GaAs photomixing and can be adapted to other wavelengths.

7.2.1 Laser Selection

The critical step in production of any local oscillator system is careful selection of the fundamental oscillator to meet the spectral purity requirements and the bandwidth requirements, and to facilitate proper control of the system. Because photomixers have large intrinsic bandwidths, the laser must have significant tuning bandwidth. There are really only three options presently available in the near infrared region appropriate for GaAs photomixing: argon-ion pumped Ti-sapphire lasers, pumped dye lasers, or solid-state diode lasers. Due to the significant cost and complexity of the pumped systems, it is often highly desirable to use diode lasers if possible. The rest of this discussion will focus on diode lasers, but the frequency calibration techniques could be applied to pumped laser systems as well.

Diode lasers, in general, have very high gain, which makes extended high-Q optical resonators unnecessary. Unfortunately, this causes a problem with diode lasers because the laser cavity empties itself of photons very rapidly and is refilled with gain on another spontaneous emission event. The laser literature typically discusses this as a coherence time. For most diode lasers the coherence times are on the order of a microseconds. The problem is actually subtler than the coherence model, because diode lasers have been shown [10] to emit at more than one closely spaced frequencies simultaneously. Due to this behavior there is no linear electrical feedback circuit that can lock and narrow the intrinsic line width of the laser. The only solution is to employ optical feedback, which seeds the laser with photons from a previous coherence time and effectively extends the resonator cavity. If the level of feedback is sufficient, then the emission of other closely related frequencies is suppressed. This typically occurs in the high feedback regime where a reflective element of reflectance >10% recycles some of the laser power back into the laser cavity. The spontaneous emission is only partially suppressed when the optical feedback is at a lower level. In general, external cavity lasers operate in the high feedback regime while distributed Bragg reflector (DBR) or distributed feedback (DFB) lasers operate in the low feedback regime of <1% optical feedback. Progress in the development of DFB lasers has made them comparable to the external cavity lasers at some wavelengths. A number of newer solid-state lasers like fiber-ring lasers, which have a circulating pool of photons, have very long coherence times.

The spectral purity of the free-running laser selected dictates the control strategy needed in laser system design. In general, it is highly recommended to start with the best possible spectral purity, which is easily available to simplify the system design though high-quality systems can be built imperfect diode lasers.

To first order, the apparent frequency motion of a free-running GaAs-based diode laser is about what is expected due to temperature fluctuations in the small active area at the 30 GHz per degree C modulation sensitivity. The fluctuations and resulting jitter are generally bulk thermal in nature (about 1 mK in magnitude) and can be controlled with standard servo techniques having loop bandwidths on the order of a few kHz. Control of the intrinsic linewidth or spectral purity is far more difficult and must generally have the optical component previously discussed. Even with optical feedback, good external cavity lasers have a residual line width of about 100 kHz and the other kinds of low feedback lasers range from 0.5 to 10 MHz. Even the best ring lasers have line widths of a few kHz. As long as there is a dominant single emission feature, the laser can be narrowed by electronic locking. For electrical narrowing, the two different modulation mechanisms in the diode laser need to be kept in mind. The first is the thermal tuning of the cavity, which is close 30 GHz per degree C in GaAs-based lasers. Since the laser active areas are small, the bandwidth of this is between 100 kHz and about 1 MHz. The second effect is a change in the index of refraction due to injection of electrons into the laser cavity. This effect is very rapid with the bandwidth limited only by the capacitance of the laser device itself. The difficulty is that these two modulations have opposite phases. There is no null in the modulation, since the temperature effect is an order of magnitude larger and has response time due to the limited bandwidth, which causes a phase change allowing it to smoothly blend into the index of refraction effect with a 180 degree phase shift at a few MHz modulation frequency.

Practical electrical locking and narrowing of the diode laser line can be achieved if the available lock loop bandwidth is about ten times the line width of the signal to be narrowed and the signal has one tone. As a result it is feasible to narrow the intrinsic laser emission should the laser be spectrally pure enough to meet this general guideline. Clearly, more complicated control schemes can be used to avoid the 180-degree phase shift, but these require far more elaborate electronics and a detailed understanding of the actual laser to be used. Before developing such a scheme, it is strongly advised that the line profile be studied on very short (less than 1 microsecond) time scales to ensure that the approach is feasible.

7.2.2 Frequency and Spectral Control

Construction of a THz synthesizer is possible using photomixers and diode lasers. However, there are always a number of practical problems that arise with the actual implementation of such a system. In this section the design of such a system is discussed, and in subsequent sections the practical problems and pitfalls are addressed. The system example provided was previously described in [11], but the discussion provided here focuses more on the practical implementation and the general approach, since the implementation chosen would not be the first choice in hindsight. The system constructed employed DBR lasers near 850 nm. These were chosen largely for their relatively high power and low cost. However, they posed a number of challenges that had to be overcome in the system design.

Once the lasers have been selected and characterized, a method of determining and controlling the absolute difference frequency must be selected. Traditional microwave techniques rely on a down conversion chain to a reference oscillator at 5 or 10 MHz. There are two problems with this technique in a photomixing setup. First, if a suitable harmonic generation chain with the necessary bandwidth could be build, there would be no need for the photomixer system. Second, the power available from the photomixer is generally low enough so that very little would be available for mixing in the down conversion lock. As a result it is very desirable to use the optical domain for the frequency calibration and control. There are really only two ways to calibrate difference frequencies in the near infrared. The first is the use of a frequency standard like the Cs D-line at 351 THz and measure the difference with a transfer oscillator of some sort. The second is to use transfer oscillator without direct absolute reference. In either case the transfer oscillator must be carefully calibrated. For photomixing only the difference is important, so the use of a calibrated transfer oscillator is the most direct route, especially if the transfer oscillator can be calibrated to the necessary precision. Fabry-Perot etalons with known or actively measured free spectral ranges are a simple way to build high accuracy and stable transfer oscillators.

Fabry-Perot etalons have discrete frequency comb determined by their length with one mode every free spectral range. If complete coverage is required, then a method of filling in the gaps between the etalon orders is necessary. In many remote-sensing applications, a second local oscillator can be applied to the intermediate frequency of the mixer making full coverage possible without filling the gaps. For applications where the source is used with a direct detector, a third laser with a tunable offset lock to one of the etalon modes or a way to measure and adjust the etalon free spectral range is needed. The offset lock can easily be implemented with commercial microwave equipment, while the tuning of the free spectral range is another exercise in control theory and requires a way to actively measure to the necessary precision the free spectral range of the etalon.

The Three-Laser System

A practical THz synthesizer capable of full frequency coverage and precise frequency calibration requires a three-laser system. The light source of the difference frequency system consists of three diode lasers, as is depicted in Fig. 7.1a. In this case DBR lasers were outfitted with a collimating lens, a length-tunable external cavity assembly and the necessary optics to circularize and fiber couple the optical radiation. Alignment is maintained by a compact rigid rail structure. All the fiber optical signal processing components are implemented in polarization-maintaining (PM) single-mode fiber. The optical components and their layout are shown in Fig. 7.1. Commercial components were used whenever available, but some home-made components were necessary for the fiber to free space transitions.

Fiber-optic components offer flexibility, compactness, insensitivity to vibration, ease of optical alignment, and eye protection, but the entire system can be equally well implemented in free space as long as temperature dependent alignment problems are carefully taken care of. The optical fiber serves as a spatial filter facilitating combination of two different lasers with nearly perfect spatial and mode overlap. The spatial overlap is critical in the Fabry-Perot etalon alignment, in achieving equal amplification in the optical amplifier and in efficient photomixer operation. The drawback of the fiber-based optical system is relatively poor optical insertion into the PM fiber from free space. The best coupling ever achieved was slightly more than 50%, but thermal drifts make the losses significantly more with time. As a result the typical output

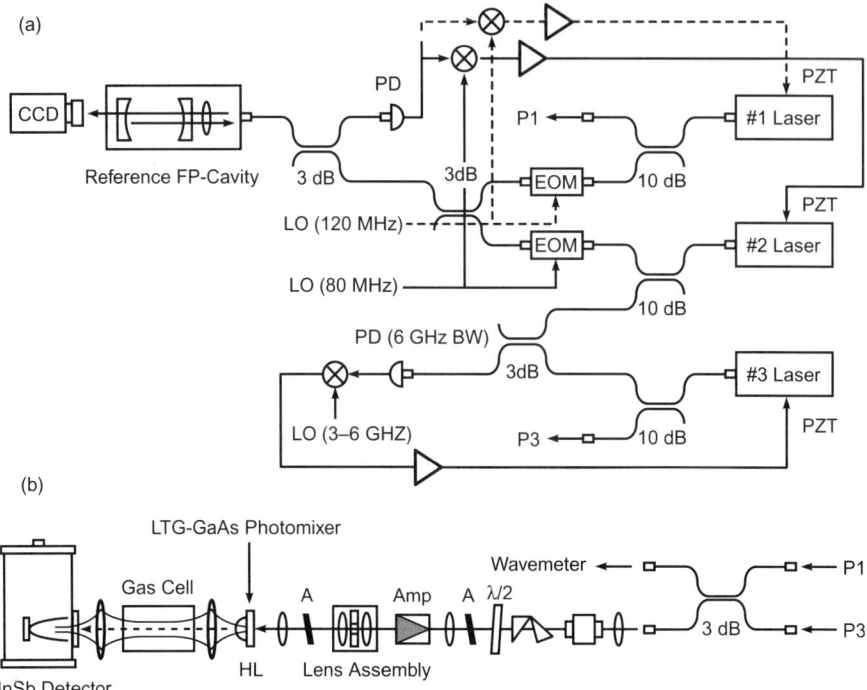

Fig. 7.1 Schematic diagram of (*a*) the three DBR laser system that synthesizes a precise difference frequency, and (*b*) the tapered-amplifier (TA) system and the setup for spectroscopy.

power of the present laser system was insufficient to optimally pump the photomixer used to generate THz-waves despite the 150 mW DBR output power levels. As a result a semiconductor optical tapered amplifier (TA) was employed as the final optical element before the photomixer after it was shown that the spectral properties of the seed lasers were preserved in the amplification process [12].

Frequency control is achieved by locking two of the lasers 1 and 2 to two different longitudinal modes of an ultra-low-expansion (ULE) Fabry-Perot (FP) etalon. This is done with crossed polarizations to allow easy verification of proper laser cavity alignment. The difference frequency between the two cavity-locked lasers is discretely tunable in steps of the free spectral range (FSR) of the cavity. The third laser 3 is phase locked to one of the cavity-locked lasers 2 with a tunable (3–6 GHz) microwave synthesizer serving as an offset frequency. The difference frequency between the 1 and 3 is determined by the sum of integral multiples of the 3 GHz FSR of the reference cavity and the microwave offset frequency. The important practical consideration is to make sure that the FSR is large enough to determine the order number with some kind of wave meter. The accuracy of the difference frequency is determined by the accuracy of the FSR measurement and any dc offset in the electronics of the lock loops. The microwave offset frequency may be locked to a high accuracy (one part in 10^{10} or better) reference and counted directly by a counter locked to the same reference measuring in real time any electrical offset in that lock loop. The ULE etalon material has a thermal expansion coefficient at room temperature of $\alpha = -2 \times 10^{10}/C$, which is comparable to the stability of a temperature-stabilized quartz reference oscillators in conventional microwave sources. Operation of the etalon in a vacuum or humidity- and density-controlled environment is necessary, since the index of refraction in air changes the FSR order number by about seven. If a nitrogen-filled sealed box is used to house the etalon, care should be taken to stabilize the box temperature, since the thermal expansion and contraction of the box will change the internal gas density and FSR by an easily measurable amount.

Laser Assembly and Control

The details of the external cavity added to the DBR diode lasers are presented in Fig. 7.2. The assembly consists of an SDL5722 852-nm, 150-mW DBR diode laser, an $f = 4.5$ mm collimating lens, an external cavity comprised of a 4% partial reflector mounted on a piezoelectric transducer (PZT), an anamorphic prism pair, a 60-dB optical isolator, an $f = 8$ mm optic, and an adjustment flexure for fiber coupling. The actual external cavity is formed by an roof-top mirror glued onto a 20% beam-splitter cube. The edge of the roof-top mirror is oriented along the short axis of the elliptical beam from the DBR laser, which is more sensitive to the tilt angle than is the long axis. The round-trip mirror configuration offers relatively robust optical feedback. The reflector and the DBR laser constitute an external cavity with a length of approximately 50 mm. The anamorphic prism pair circularizes the 3:1 aspect ratio beam generated by the DBR, which achieves better spatial overlap with the PM fiber and improves the fiber coupling efficiency. These components are assembled in a 250-mm-long rigid rail to maintain alignment. A homemade aluminum

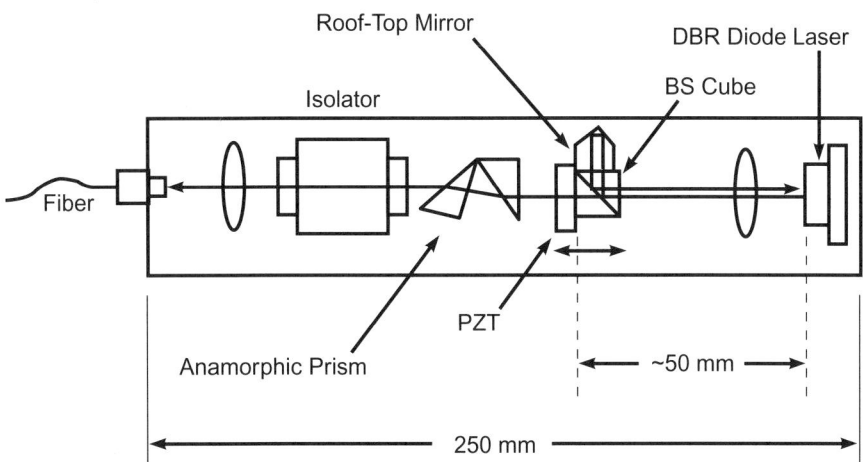

Fig. 7.2 Schematic diagram of the external-cavity DBR laser assembly.

structure was used, but commercial units designed for this general purpose exist and could easily be adapted.

The typical optical power coupled into the PM fiber was approximately 30% of the original power emitted from the DBR laser. The transmission of the free-space optics in the laser assembly was measured to be 70% and the fiber-coupling efficiency was typically 50% when fully optimized. The theoretical maximum fiber-coupling efficiency calculated from the measured beam size and the fiber core size is 60% of the DBR power. The best coupling ever achieved by this optical system was 55%. Other power losses include the use of a passive 3-dB directional coupler to combine the output laser beams and small losses at each fiber connector. As a result the typical power available at the TA is near 30mW, with the best ever achieved for 55mW. The difference in the typical and maximum power is due to slight thermal deformations in the rail assembly and the slow buildup of dust and dirt in the optics and exposed fiber connectors over time.

The DBR diode laser coupled to the external cavity oscillates at the cavity mode, which is the closest to the gain maximum of the DBR. The FSR of the external cavity was also about 3 GHz, corresponding to the separation between the partial reflector and the DBR active area. The DBR laser frequency is continuously tunable, since the external cavity's FSR can be tuned by the PZT, at the rate of 160MHz/V. The continuous PZT tuning range is limited by mode hops to different transverse external cavity modes favored by the gain profile of the DBR. This effect for a DBR laser operating at a fixed temperature and current is shown in Fig. 7.3 by the triangles. In order to avoid these mode hops,

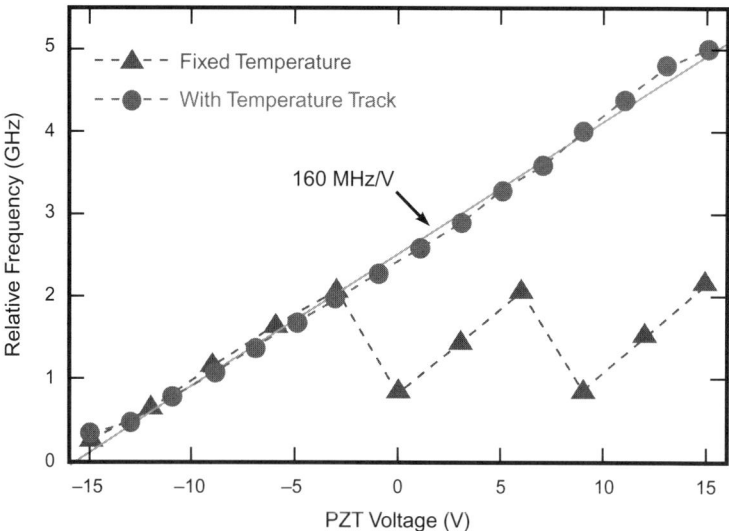

Fig. 7.3 The frequency of the external-cavity laser plotted against the PZT control voltage. The triangles represent the tuning curve during which the temperature of the DBR laser is fixed, and the circles represent that with temperature tracking.

the laser's temperature is adjusted to maximize its gain at the external cavity mode frequency, completely suppressing the mode hops and facilitating continuous tuning over the PZT tuning limit of 5 GHz, which corresponds to ±15 V, as shown in Fig. 7.3 by the circles. Coarse frequency tuning spanning approximately 700 GHz is available to DBR lasers by changing the laser temperature over the range of 5 to 30°C (or 27 GHz/K) with the internal thermoelectric cooler. Although the injection current also affects the laser frequency by approximately 600 MHz/mA, the current was normally set to a constant value unless fine or fast frequency tuning (e.g., a phase lock) was necessary.

Stable diode-laser operation requires that both the injection current and the temperature be precisely controlled. To satisfy these criteria, a low-cost high-performance control circuit has been designed. The current controller was adapted from Libbrecht and Hall's work [13] so as to suit the anode-ground configuration of the DBR lasers. The laser current is controlled by comparing a reference voltage to a voltage drop produced by the laser current through a series resister. The circuit components of this current source were carefully chosen to maximize the dc and thermal stability. The voltage reference was an Analog Devices 584L with a temperature coefficient of 3 ppm/°C, while the sense resistor was a 33.3-Ω Vishey resistor with a temperature coefficient of 0.5 ppm/°C. The sensing circuit was designed with filters to eliminate response above 10 kHz and low-noise, low-offset voltage op-amps (Analog Devices 784B) to drive a FET (IRFD-120) providing the 0 to 300 mA. An average current noise of $7 nA/Hz^{1/2}$ over the range of 5 to 2000 Hz at 150 mA DC current has been measured. The noise declines to immeasurable levels at 10 kHz. A bridge feedback circuit using the DBR laser's built-in thermistor and thermo-electric cooler is used to stabilize the temperature. The thermal time constant of a DBR laser package mounted in the aluminum rail used was measured to be 24 s, but the temperature control circuit achieves a loop bandwidth of 12 Hz. The temperature can be set in 16-bit steps from 3 to 30°C by a serial-controlled D/A converter. As with the current controller, careful selection of low-noise circuit components was essential, and a temperature stability of better than 1 mK has been achieved. The observed long-term unlocked frequency stability (~1 h) of the DBR is around 300 MHz. The current supply, the temperature controller, and the PZT controller, including the frequency stabilization circuit, are fabricated on a 4 by 6 inch circuit board and interfaced with a computer through a serial interface for monitor and control purposes. This single board design has the advantages of compactness, high performance, and low cost.

Despite the much lower level of optical feedback than that of conventional external cavity diode lasers, a significant narrowing of the laser linewidth was realized. The linewidth measured by the delay-line self-heterodyne technique was narrower than 500 kHz FWHM, which is near the spectral resolution limit imposed by the optical delay length of 200 m. This linewidth is a significant improvement over the observed (several to several tens of MHz) linewidth of a free-running DBR laser. According to a previous study on the frequency stability of diode lasers at various feedback levels [14], our case falls into the weak feedback regime where stable single-mode narrow-line oscillation should be observed. If the feedback level is higher than this regime, but still lower than the DBR internal laser-chip facet's feedback, the oscillation becomes unstable and the coherence collapses, resulting in a drastic line broadening. This effect has been observed at a feedback level of 20% with the same optical configuration. If much more intense feedback is applied or if the laser chip facet is AR-coated, stable narrow-line oscillation in strong feedback regime could again be obtained with somewhat improved linewidths, but such intense feedback would impose a steep cost on the available output power. It should be noted that the optical feedback in this system is electrically equivalent to a very fast feedback circuit. The reasons for using optical feedback instead of fast electrical feedback will be addressed in the next section.

Frequency Stabilization

The P1 and P2 lasers were cross-polarized locked to the different longitudinal modes of the ULE etalon with the Pound-Drever-Hall method [15,16]. FM sidebands were generated on the two cavity-locked lasers with electro-optic phase modulators (EOM) at 80 and 120 MHz. The modulation index was chosen to maximize the phase shift seen in the reflection from the cavity. The coupling to the fundamental longitudinal cavity mode was verified by monitoring the light transmitted through the cavity with a polarizing grid and CCD camera. Since pressure and humidity changes in air can cause FSR fluctuations through refractive index changes, the cavity was installed in a sealed box, which was evacuated and then back filled with dry nitrogen so that the pressure inside the box was roughly the ambient pressure. This box was placed on a temperature controlled chill plate along with the laser assemblies.

The phase of the beam reflected from the cavity was compared to the modulation frequency in a frequency multiplier (Analog Devices 834Q). When the laser frequency is within the modulation frequency of the cavity resonance, the output of the frequency multiplier produces a dc frequency dispersion, which crosses zero at the cavity resonance. Figure 7.4 shows the dispersion curve at

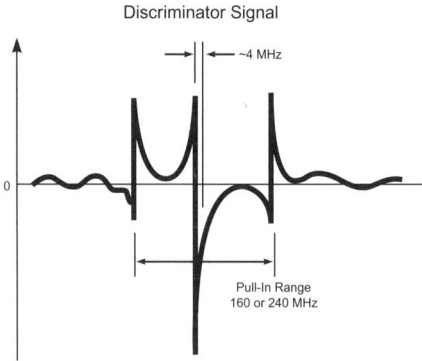

Fig. 7.4 The dispersion signal of the cavity fringe used for frequency stabilization.

the mixer output obtained by sweeping the laser frequency through the cavity resonance. The central linear portion of the dispersion curve centered at the cavity is used to generate an error signal voltage, which is fed back to the PZT of the external laser cavity with a simple servo electronic circuit-consisting of an integrator and an amplifier. The linear part of the frequency-discriminator curve shown in Fig. 7.4 is about 4 MHz wide, in accordance with the FSR of the cavity of 3 GHz and its finesse of 750. As can be seen in Fig. 7.4, the pull-in range of this locking technique is limited by the distance to the upper and lower sidebands or twice the modulation frequency 160 and 240 MHz, respectively. The loop bandwidth of the PZT control circuit was limited to 3 kHz to avoid acoustic resonances in the support structure of the partial reflector and the PZT. Only 200 Hz of bandwidth was required for stabilization of the laser, but the additional bandwidth was needed to facilitate sweeping the offset laser.

The offset-locked laser is locked to a beat signal between lasers 2 and 3. The beat signal is detected by a 6 GHz bandwidth photodetector and compared to tunable frequency generated by a microwave synthesizer in a double-balanced mixer. In order to implement a phase lock with the same electronics as the cavity lock, the signal was split with one arm sent through a delay line, and the phase shift between the two signals was used to generate the error signal for the PZT of laser 3. The delay line has an effective reference frequency of 71.9 MHz; however, concerns about the accuracy and stability of this effective lock frequency led to the use of a counter. The offset frequency is measured precisely by a microwave counter locked to a high precision reference, making any drifts or offsets in the phase lock scheme irrelevant for the system frequency calibration. The offset frequency can be continuously tuned over 5 GHz by stepping the synthesizer frequency and tracking the PZT control voltage. The maximum sweep rate of approximately 100 MHz/s is limited by the feedback loop's bandwidth.

The overall system performance was assessed by observing a beat note between lasers 1 and 3 with a 25-GHz bandwidth photodetector and a spectrum analyzer. Figure 7.5 shows a 12 GHz beat spectrum for a one-second integration time and a spectral resolution of 100 kHz. The FWHM spectral power bandwidth is approximately 800 kHz. This result confirms the self-heterodyne linewidth measurements of 500 kHz. The short-term linewidth of each laser is determined entirely by the optical feedback from the external cavity because the 3-kHz bandwidth of the lock loop circuit is much less than the laser linewidth. Linewidths of 50 kHz have been achieved by weak optical feedback from FP-cavities [3]. The additional narrowing is due to 180-degree phase shift over a narrow (~60 MHz) cavity width compared to the same phase shift over 1.5 GHz in the scheme presented here. Fast electronic feedback to the laser current can also be used to narrow the laser's linewidth [17], but DBR lasers have some characteristics that make this impractical.

Stabilization and narrowing of 850-nm DBR lasers with electrical feedback was attempted in the first design iteration of this laser system. The DBR lasers have two modulation mechanisms previously discussed. The Ohmic heating of the active region by the injected current has a magnitude of 27 GHz per degree

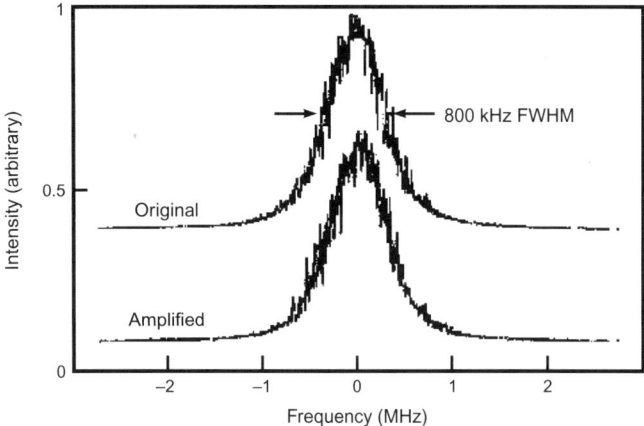

Fig. 7.5 The spectrum of the beat note signal between lasers 1 and 3 before (*upper*) and after (*lower*) amplification with the optical tapered amplifier.

Celsius and has a 3 dB knee near 600 kHz. The change in index of refraction due to injection of electrons is 580 MHz/mA with the temperature control loop in operation. This effect was observed to have no phase shift in the 200 MHz bandwidth of our network analyzer and has a magnitude of only 10% of the heating effect, but it has a 180-degree phase difference. Due to the differing magnitudes of the two modulation effects, there is no null in the modulation response, but there is a 180-degree phase shift in the modulation response between dc and 1.2 MHz. Unfortunately, this is considerably slower than the observed free-running laser linewidth of 3 to 10 MHz FWHM, depending on where in relation to the ideal current the laser is operating. The DBR lasers could be stabilized with current feedback applied with less than 10 kHz of lock loop bandwidth, but no linear lock loop up to the 1.2 MHz 180-phase shift point could be developed to narrow the laser linewidth. It should be possible to use more sophisticated control approaches to narrow the DBRs, but each DBR laser is slightly different, requiring individually tuned feedback loops. As a result optical feedback was adopted as the preferred method for narrowing a DBR laser. A combination of optical and electrical feedback might ultimately prove to be the best method for narrowing a DBR laser, but selection of a more spectrally pure laser would probably be a much easier approach.

One problem experienced in the system design was the relatively large temperature changes in our laboratory and the change of temperature equilibrium in the external cavity and the laser itself. The PZT voltage generally compensates these drifts, but the drifts were on many occasions large enough to exceed the PZT error signal limit. This, coupled with the DBR lasers propensity to start up in one of two modes separated by approximately 30 GHz, led to actively stabilizing the temperature of the laser mounting structure. Once this was done, all-day-long cavity-locks of the lasers external cavity were routinely achieved without adjusting the temperature of the laser itself. When the active temperature stabilization the lasers turn on and equilibrated, the lasers typically turned on to within a few MHz of where they were last used. The long-term frequency stability of the system when locked to the ULE etalon is as good as our ability to determine the center of the 500 kHz laser linewidth, and it is has been repeatable to that accuracy for months.

Dual-Frequency Tapered Amplifier Operation

Since the typical output power of the fiber-coupled laser system is on the order of 30 mW, it is not sufficient to efficiently generate THz radiation with the photomixing technique employed. As a result the two frequencies used in the photomixing were amplified in a tapered amplifier. Dual-frequency TA operation provides over 20 dB of small signal gain on each frequency and provides up to 500 mW of output power while preserving the spatial overlap of the launching fiber, which is essential for efficient photomixing.

The TA system is shown in lower half of Fig. 7.1b. The amplifier is a single traveling-wave 850 nm semiconductor tapered optical amplifier, which is the active component of the SDL8630 single-mode laser system. The output from the fiber was collimated by an f = 8 mm collimating lens, passed through a 60-dB optical isolator, and sent to a half-wave plate for fine adjustment of the polarization. The circular beam was transformed into the required elliptical shape by an anamorphic prism pair to match the amplifier's $1 \times 3\,\mu m$ input facets spatial mode. After appropriate attenuation the beam is injected into the optical amplifier chip through an f = 8 mm focusing lens. The output beam from the amplifier is spatially filtered and collimated to a 3-mm-diameter Gaussian beam by the telescope provided with the SDL8630 laser system. A detailed description of the dual-frequency TA system has been given by Matsuura et al. [12].

The amplifier was operated under highly saturated conditions at an input laser power of 10 mW. As a result the output power is relatively insensitive to both the input power and frequency. Figure 7.6 shows the output power of the amplifier as a function of the frequency difference between the two injected signals. Due to the highly saturated operation, the output power was constant within 5% over the entire 1.3 THz range of the difference frequencies. The injection seeding bandwidth of these amplifiers is the same as their 20 to 25 nm bandwidth [18], making two frequency amplification feasible to offsets of 10 to 15 THz with appropriate seed lasers.

The other desirable TA property for photomixing is that the gain on each of the two injected frequencies is very nearly equal. The output power ratio between the two frequency components, as measured with a scanning FP spectrometer, is plotted in Fig. 7.5. As expected, the P3/P1 ratio was close to unity over a wide range of difference frequencies from 10 GHz to 1.3 THz. The amplifier used was a commercial external-cavity laser, so one of the amplifier chip facets was antireflection coated but not the other. The small variations in

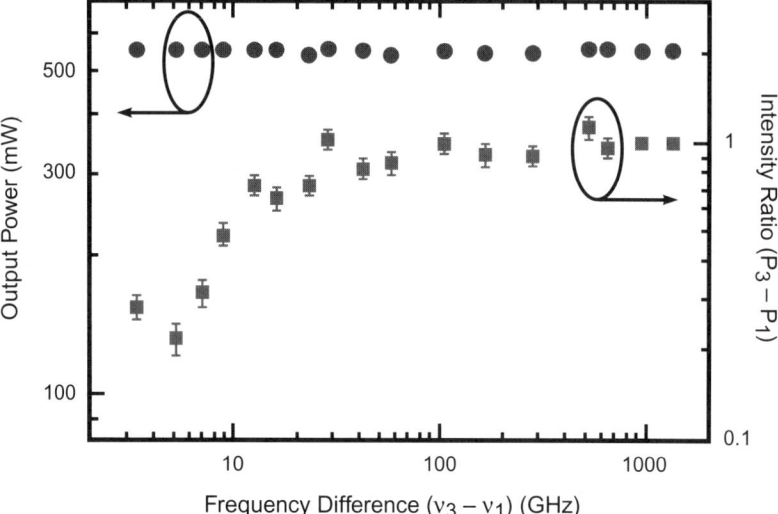

Fig. 7.6 The total output power of the dual-frequency TA (*circles*) and the intensity ratio between the two frequency components, P3/P1, (*squares*) as a function of the difference frequency.

the output power and the P3/P1 ratio shown in Fig. 7.5 are caused by standing waves in the chip mode itself, and have the expected 15 GHz period. This small frequency dependence should be reduced to negligible levels by AR-coating both of the amplifier chip facets.

Unbalanced amplification between the two frequencies occurred only at difference frequencies lower than 10 GHz. This behavior arises from four-wave mixing, which is an interaction between the two frequency components through the refractive index change induced by the carrier density modulation at the difference frequency [12]. Therefore the lower frequency limit of dual-frequency amplification is determined by the carrier lifetime of the amplifier. As shown in Fig. 7.6, the spectrum of beat signal between the two frequency components was identical to that of the seed laser as long as the difference frequency was greater than 10 GHz.

Two Laser Systems

In applications with a second local oscillator, this system would work equivalently well with only two lasers and no offset laser. This would not allow full coverage with the source, but the IF bandwidth of the mixer would facilitate the full coverage. In the two-laser case calibration of the etalon FSR would require precise knowledge of the second local oscillator frequency as well as the ability to change cavity modes and tune the second local oscillator the desired frequencies for calibration as described in the frequency calibration section.

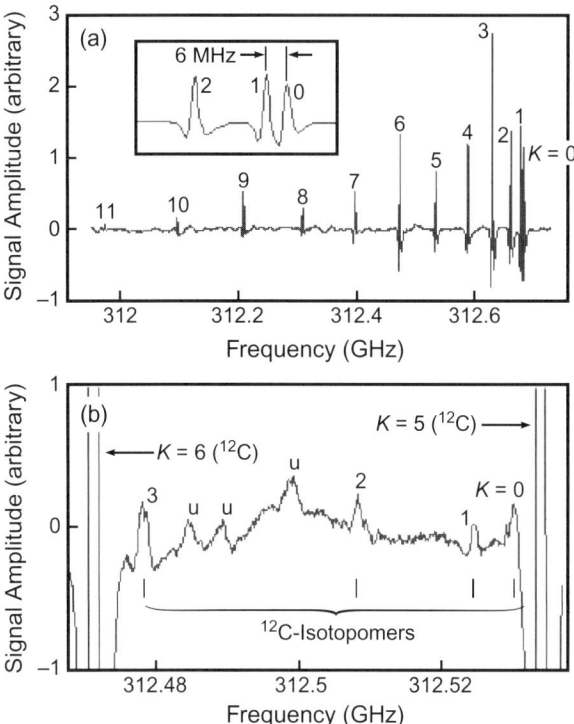

Fig. 7.7 The second-derivative absorption spectrum of CH3CN JK = 17K to 16K rotational transitions near 312 GHz. (a) The spectrum for 12CH312C14N. The inset is expanded view of the K = 0–2 lines. (b) The K = 0–3 lines of the 13C-isotopomer. Features denoted by "u" are unidentified lines.

7.2.3 THz-Wave Verification

Verification of the spectral purity of the THz waves generated by difference frequency mixing can be conveniently demonstrated through rotational spectroscopy of the simple molecules like acetonitrile (CH_3CN) and carbon monoxide (CO). These molecules have well-known transition frequencies and line profiles. Spectroscopic measurement of known features provides an excellent diagnostic of both frequency and spectral purity. This is an essential verification if a difference frequency source is going to be useful as a local oscillator or as a coherent source for a spectrometer. It is also significantly easier than generating a beat signal of known spectral purity in the THz and performing a direct comparison.

Spectroscopic Verification

An experimental setup for spectroscopy is shown in Fig. 7.1*b*. The difference frequency output from the TA is appropriately attenuated and focused onto the photomixer. The THz-wave output beam is collimated with a combination of the silicon hyper-hemispherical lens and an external Teflon lens, passed through an 8-cm-long 1-inch-diameter gas cell fitted with polyethylene windows. The beam transmitted through the cell is focused with a Teflon lens into a 4.2-K InSb hot-electron bolometer or a 1.8-K composite Si bolometer. The tone-burst modulation [19] was used to obtain the absorption spectra of various molecules. The advantage of the tone burst modulation compared to traditional FM modulation is that sensitive detection with slow detectors such as silicon composite bolometers typically used for THz detection can be achieved. The injection current of the cavity-locked laser 1 was modulated with a 2-MHz tone, above the cavity-lock loop bandwidth, at a 10-kHz burst rate. A lock-in amplifier, detecting at 10 kHz, was used to demodulate the detector signal, generating a second-derivative molecular absorption features. The modulated laser is the fixed laser, which proved to be better for long-term lock stability than modulating the swept laser.

Figure 7.8a presents the absorption spectrum of the CH_3CN $J_K = 17_K$ to 16_K rotational transitions near 312 GHz. The spectrum taken with a sweep rate of 2 MHz/s is plotted as a function of the microwave offset frequency. The data was recorded at 7 sample/s with a lock-in amplifier time constant of 0.3 s. The spectrum shows the well-known K-structure of a symmetric top, with the K components from $K = 0$–11 assigned in the spectrum. As seen in the inset of Fig. 7.8a, the $K = 0, 1$ lines, which are separated by 6 MHz, are clearly resolved.

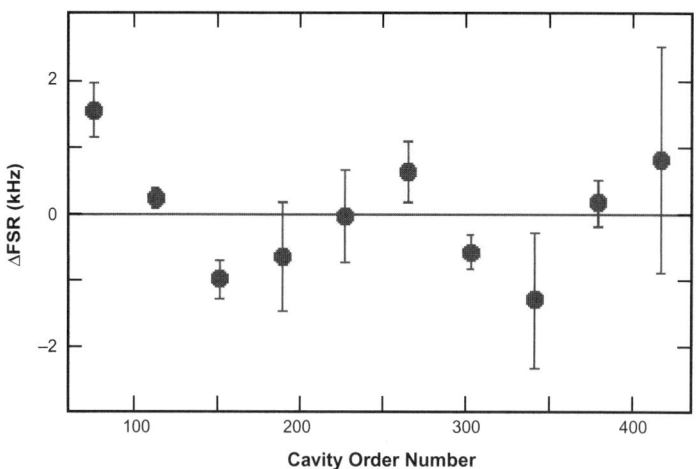

Fig. 7.8 The result of the FSR measurements by CO lines for Jlower = 1–10. (a) The predicted frequency for each CO line subtracted from the microwave offset frequency is plotted against the cavity order difference. The solid line represents the result of linear fit to the data. (b) The deviation of the FSR values from their average are shown as a function of the cavity order difference.

The gas pressure was 60 mTorr, and the observed line widths are consistent with a convolution of pressure-broadened linewidth of a very large dipole moment and the instrument response. The minimum detectable absorption of this system is estimated to be 10^{-5}, and is detector-noise limited. As shown in Fig. 7.8b, the spectrum lines of the ^{13}C-isotopomer of acetonitrile, which has a natural abundance of roughly 1%, could be also detected with a signal-to-noise ratio of 20, consistent with the detection limit. Some of the features were clearly identified as the $K = 0$–3 lines from line position and intensity ratio. The remaining features, denoted by u in the figure, are from the excited vibrational state and have not been assigned. These results indicate that the spectral purity, frequency control, and the output power of this system are sufficient for the laboratory spectroscopic study of molecules at THz frequencies, as well as many local oscillator applications.

The 800-kHz difference frequency width is insufficient for observing very narrow features. Either better lasers or a more elaborate control strategy is needed. Either of these improvements can be implemented in a straightforward way as was described previously.

Spectroscopic Frequency Calibration

For further spectroscopic measurements such as in the search for unknown molecular lines and for use in astronomical observations, absolute frequency calibration of the difference frequency is necessary. Since the accuracy of the difference frequency is defined by the reference Fabry-Perot cavity, the frequency calibration precision is determined by the accuracy of the FSR of the known cavity. Once the value of the FSR is obtained, the difference frequency can be determined to within an accuracy of 10^{-10}, if temperature fluctuations of the cavity are kept below 1°C.

The FSR of the Fabry-Perot cavity was measured by detecting the beat signal between the two cavity-locked lasers 1 and 2 with a microwave counter. In this measurement, the FSR was determined to be 2996.71 ± 0.05 MHz. Well-known molecular lines in the THz region, such as the rotational transitions of carbon monoxide (CO), are suitable for more accurate calibration. The frequencies of these THz molecular transitions correspond to 300 times the FSR and can be easily measured to with an accuracy of 10^{-7}. A number of measurements and the careful use of statistics allows for rapid calibration to an accuracy of 10^{-8}.

To verify this approach, the pure rotational transitions of CO were measured using the same configuration as the acetonitrile measurements. A composite silicon bolometer was used in these measurements, because of its improved responsivity at frequencies above 1 THz. According to the theory for diatomic molecules like $^{12}C^{16}O$, the rotational transition should appear at frequencies of $v = (W_{J+1} - W_J)/h$, where $W_J/h = BJ(J + 1) - DJ^2(J + 1)^2 + HJ^3(J + 1)^3$, and where h is Planck's constant, and J is an integer [20]. For $^{12}C^{16}O$, $B = 57635.9660$ MHz, $D = 0.1835053$ MHz, $H = 1.731 \times 10^{-7}$ MHz. Absorption measurements for CO lines with $J_{\text{lower}} = 1\text{--}10$ over the range of 230 GHz to 1267 GHz were carried out by measuring the microwave offset frequency ν_{offset} and counting the number of cavity orders between the two cavitylocked lasers. The line position was determined by fitting a parabola to the center of the second derivative line profile. In Fig. 7.9a, the line position predicted by the 3 GHz cavity FSR measurement is subtracted from the observed frequency, and $\nu - \nu_{\text{offset}}$ is plotted against the cavity order difference.

The linear coefficient of this plot represents the correction to the 3 GHz FSR measurement of the cavity. The small nonzero intercept is the dc offset in the lock loops, which was calculated to be about 4 kHz. From this data set, the average FSR value for all CO line measurements was determined to be 2,996,757.48 ± 0.10 kHz. The solid line in the figure corresponds to the average FSR. Figure 7.9b shows the deviation of the FSR values from their average as a function of the cavity order difference. If the scatter of the data around the average value is real, the frequency dependence of the FSR over a 1.3 THz span is constant to within 1 kHz. In this measurement the frequency accuracy is limited by our ability to determine the center of the measured line, which in turn depends on the instrumental resolution and the signal-to-noise ratio of the observed spectrum. Increasing the source power and/or improving the detector sensitivity could improve the calibration accuracy. The final measured frequencies were obtained by using the calibrated FSR value of the cavity. The small offset was statistically insignificant and was ignored. It should be noted that the FSR could be a function of cavity order due to dispersion of the refractive index of the cavity coatings. However, none of this is observed, and the effect is expected to be smaller than the current precision.

Direct Calibration of the Free Spectral Range

The FSR of the etalon can be measured with a double modulation technique. In this case the cavity signal at the sum or difference of the two modulations is observed. When the larger frequency modulation is tuned to the FSR of the cavity, a null is observed. This null is independent of any offsets in the lock loops and can be measured to a part in 10^8 or better even without a very high finesse etalon. Additionally the frequency can be compared to an absolute reference like the Cs D-line on a higher cavity order. Any drift in the cavity FSR will cause a shift in the observed offset. The impact on the FSR is the total frequency shift divided by the number of FSRs to reach the 351 THz frequency. This should allow for drifts on the order of one part in 10^9 to be easily tracked.

7.3 DESIGN TRADE-OFFS FOR PHOTOMIXERS

Once the laser stabilization scheme is optimized, one needs to optimize the photomixer on which the laser excitation is focused. The basic design

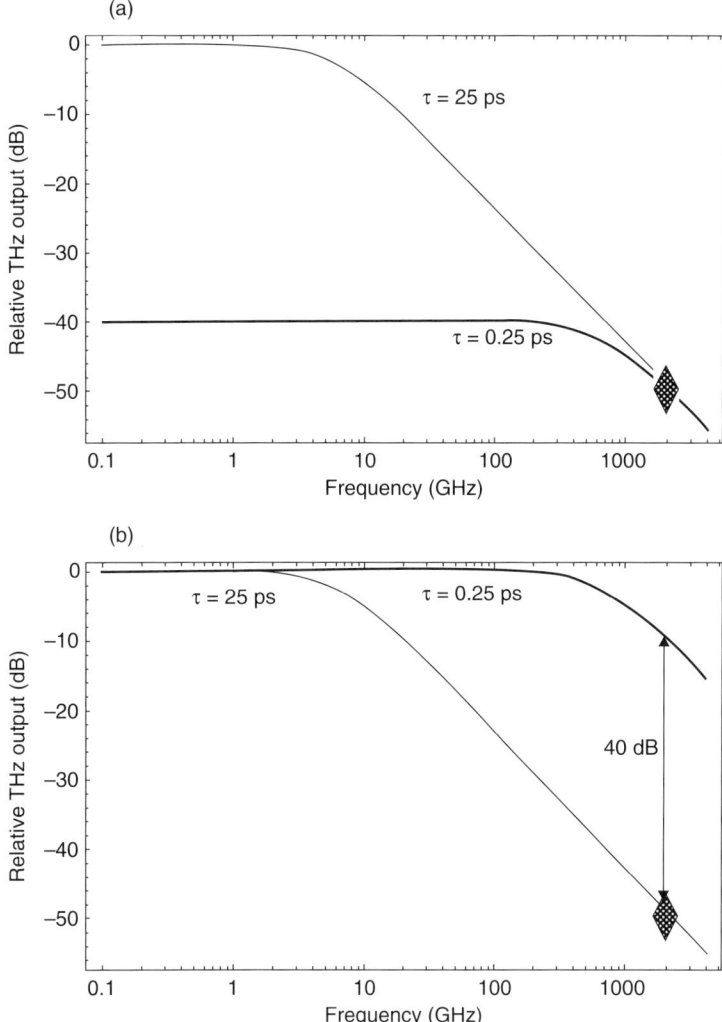

Fig. 7.9 Calculated THz output power comparisons. (a) Two photomixers using photoconductors with different values of carrier lifetime τ_e and illuminated by the same laser power. (b) Two photomixers with different τ_e operating at the same DC photocurrent but with different laser illumination powers. These photomixers have similar amounts of ohmic heating due to DC photocurrent but have very different THz output powers.

trade-offs for lumped-element photomixers are constrained by the following: optical spot size, thermal limits from optical and ohmic heating, capacitance, and the photoconductor's RF resistance.

7.3.1 Basic Operation

A photomixer comprises a thin photoconducting film of low-temperature-grown (LTG) GaAs, submicron electrodes, and a radiating antenna also used to provide a dc bias. In the photoconducting film the incident light modulates the conductivity through the generation of photocarriers. The photomixing process occurs during illumination of an electrode region with two single-mode lasers with average powers P_1 and P_2 and frequencies ν_1 and ν_2, respectively. The instantaneous optical power incident on the photomixer contains mixing terms (beats) between the two optical frequencies:

$$P_i = P_1 + P_2 + 2\sqrt{P_1 P_2}(\cos 2\pi(\nu_1 - \nu_2)t + \cos 2\pi(\nu_1 + \nu_2)t), \quad (7.1)$$

where $f = \nu_1 - \nu_2$ is the difference frequency. The $\nu_1 + \nu_2$ term occurring at the sum frequency does not couple to the THz antenna and can be neglected.

The dc photocurrent generated by the optical power is

$$I_o = \eta_e \frac{e}{h\nu} P_o, \quad (7.2)$$

where $e/h\nu$ is a constant of 0.69 $(V)^{-1}$ for an optical wavelength near 860 nm (the band gap of GaAs) and η_e is the external quantum efficiency. Using the two equations above, the amount of THz power transmitted out of the radiating antenna is then

$$P_w = \frac{0.5 I_o^2}{1 + w^2 \tau_e^2} \eta_{ant} \operatorname{Re}(Z_{ant}), \quad (7.3)$$

where $\omega = 2\pi f$, τ_e is the photocarrier lifetime, η_{ant} is an antenna efficiency that accounts for rf absorption, and $\operatorname{Re}(Z_{ant})$ is the real part of the antenna impedance and includes all the parasitic reactances associated with the electrodes. Note that the effective THz current is only a fraction of the dc photocurrent

$$I_w = \frac{I_o}{\sqrt{1 + w^2 \tau_e^2}}. \quad (7.4)$$

A detailed analysis of the effect of photocarrier lifetime on photomixer performance is available in [21,22]. Here we present a simplified argument that explains the importance of a short carrier lifetime. From Eq. (7.2) it is clear that the dc photocurrent is proportional to the external quantum efficiency, which in a simplified one-dimensional case can be written as

$$\eta_e = \eta_{opt} \frac{\mu E \tau_e}{L(1+(\mu E)/(v_{sat}))}, \qquad (7.5)$$

where η_{opt} is the absorbed fraction of the optical power, μ is the effective carrier mobility, E is the electric field, L is the distance between electrodes, and v_{sat} is the saturated velocity of the carrier. The second factor in (7.5) is the photoconductive gain. In a more accurate model, the strength of E decays with depth into the sample and only a fraction of the carriers travel at v_{sat}. This has led some [23,22] to design photomixers with film thickness nearly 0.2 µm, considerably thinner than the absorption depth (~1 µm) so that all of the carriers move at v_{sat}. In this case η_e and therefore I_o are proportional to τ_e and inspection of (7.3) suggests that the photomixer output power is independent of τ_e. Figure 7.9a depicts this relationship in a plot of THz output power versus frequency for various values of τ_e.

The preceding argument does not take into account the dissipated power limit imposed on photomixers by their failure mechanism [24,22,23]. Measurements at different labs suggest that photomixers fail thermally because of the combination of optical heating and ohmic dissipation. Ohmic dissipation is particularly severe since the power deposited per unit volume is proportional to $j \cdot E$ (where j is the current density) and both j and E are highly concentrated near the edge of a given electrode. The main point is that while the THz power can be independent of τ_e, the dc ohmic power deposited near the electrodes is proportional to τ_e. Therefore a given electrode geometry with a given voltage bias has a maximum sustainable value of I_o imposed by thermal limits. Reduction of τ_e allows the optical power and dc bias to be increased, thereby increasing the THz output power just below the maximum sustainable value of I_o. Figure 7.9b shows the maximum THz power available given the constraint on total I_o. Note that if one is willing to increase the diode laser power, a gain of 40 dB is achievable by reducing τ_e by a factor 0.01 (e.g., 0.25 ps instead of 25 ps). The optimum lifetime τ_e for a given operating frequency f is approximately the value of τ_e that satisfies the equation $2\pi f \tau_e \approx 1$. This represents a near optimal trade-off between reduced dc power dissipation and efficient generation of THz photocurrent.

7.3.2 Role of Photocarrier Lifetime

The value of τ_e for a particular wafer is determined by the growth conditions in the molecular beam epitaxy (MBE) apparatus. Figure 7.10 shows the photocarrier lifetime—measured by optical pump-probe reflectometry—as a function of substrate temperature during the MBE process [25]. The lifetime measured with the optical pump-probe technique was consistent with the

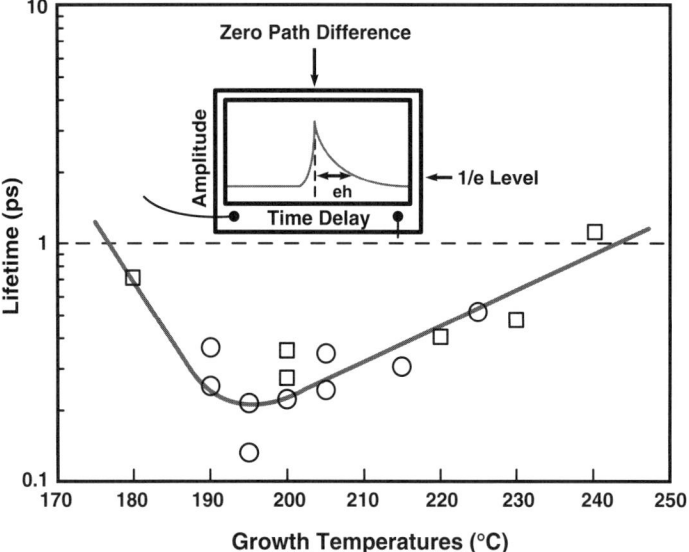

Fig. 7.10 Optical pump-probe measurements of the photocarrier lifetime for different LTG-GaAs wafers. Also shown are typical values of LTG-GaAs electronic properties compared to conventional epitaxial GaAs.

value inferred at low-voltage bias from autocorrelating two nearly 100-fs laser pulses on a LTG-GaAs photoconductor [26].

According to the argument in the section above, a photomixer designed to operate at a particular frequency has a corresponding optimum carrier lifetime. Since τ_e is a material property, this implies that a given LTG-GaAs wafer is best suited for one operating frequency. Experimental results, however, suggest that LTG-GaAs with short τ_e produces high output power at high frequencies and can still produce high output power at lower frequencies. A pos-

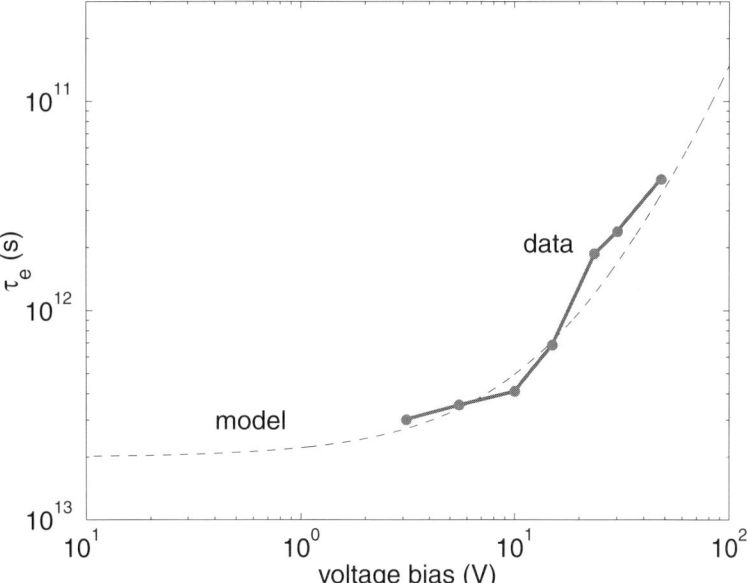

Fig. 7.11 Measured values of photocarrier lifetime inferred from the decay time resulting from correlating two fs laser pulses on a LTG-GaAs photoconductor. The increase in lifetime with electric field is attributed to a barrier-lowering model in which the deep traps become less efficient at trapping carriers under high electrical bias.

sible explanation of this effect is based on the measurements of Zamdmer [21,27] who showed how the photocarrier lifetime is lengthened at high values of dc electric field. These measurements were performed in the time domain by measuring the decay time of the photoconductive response of a LTG-GaAs switch to a mode-locked laser pulse. Figure 7.11 shows measured lifetime as a function of dc electric field. This effect was simulated assuming a barrier-lowering model in a material with many electron traps. Figure 7.12 shows how this effect manifests itself in the frequency domain in photomixer output power. In a simple model the THz power should scale with the dc photocurrent as $P_w \sim I_o^2$. The measured THz power at frequencies above 326 GHz, however, no longer increases with applied dc voltage and dc photocurrent. This suggests that $2\pi f \tau_e > 1$ for $f > 326$ GHz in this particular LTG-GaAs sample when biased at 30 V.

The preceding data suggest one possible strategy, which is to grow LTG-GaAs wafers with a short lifetime and then optimize the dc voltage bias for photomixers with antennas optimized for different frequencies. For example, use a larger voltage bias to operate a photomixer with an antenna designed for lower frequencies.

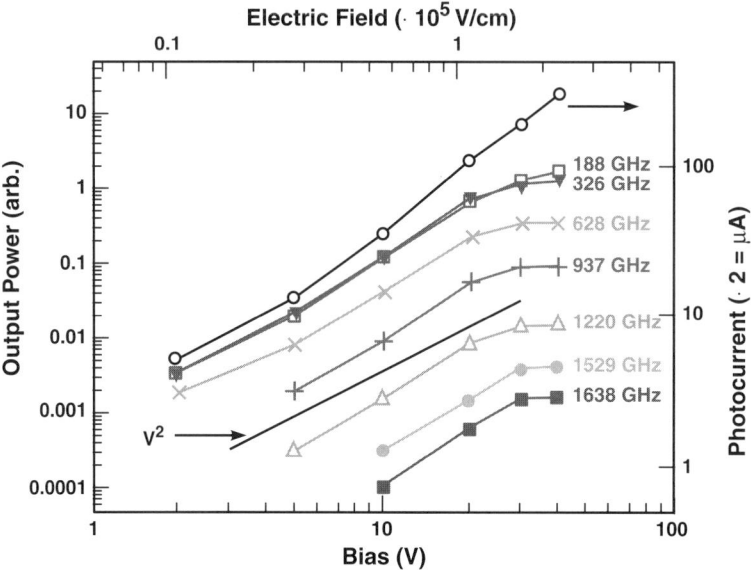

Fig. 7.12 Measured THz output power at various frequencies as a function of DC voltage bias. Note how the output saturates above 30 V for frequencies above 300 GHz.

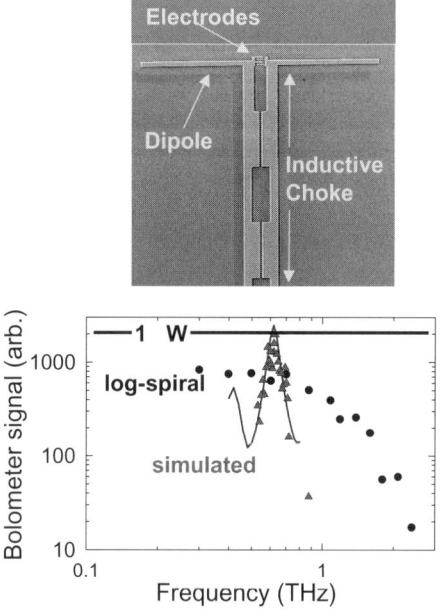

Fig. 7.13 Full-wave dipole antenna. Scanning electron micrograph and measured versus calculated performance. Measured data from a photomixer with a log-spiral antenna are included for comparison.

7.4 ANTENNA DESIGN

The section above described how to optimize the properties of the photoconductor. This section describes the basic antenna design problem of extracting THz power from the photoconductor using metallic electrodes. While illuminated, the effective RF resistance of the photoconductor is estimated to be >10 kΩ for typical operating conditions. To good approximation, Eq. (7.3) describes the output power of the photomixer where the effect of the electrodes is modeled as part of the complex impedance Z_{ant}. Ideally the antenna would present almost a 10 kΩ load impedance to the photoconductor to optimally transfer power out of the antenna. In practice, a broadband antenna—such as a log-spiral or log-periodic—on GaAs typically has a real impedance of around 70 Ω as well as capacitive reactance arising from the fringing fields between the interdigitated electrodes that collect the photocurrent. Resonant antennas are useful if the application of interest allows one to trade octaves of tuning bandwidth for increased power over a narrow band (typically ~20% fractional bandwidth). In principle, a resonant antenna could present a >10 kΩ impedance to the photoconductor. In practice, metallic loss at THz frequencies preclude the use of tuning circuits with the necessary Q-factor to achieve such a high antenna impedance.

The work by Duffy presents a detailed design procedure for photomixer antennas as well as the computational techniques used to verify them [28]. Here we describe some of the critical trade-offs used to design planar antennas for photomixers. We also discuss distributed designs that have more broadband performance.

7.4.1 Resonant Designs

Resonant single-element antennas offer some advantage over broadband antennas in power output. They have some disadvantages, however, like low H-plane gain and interference of the photoconductor electrodes with the antenna drive point. Some results on single-element antennas are shown in Fig. 7.13 and are detailed in [29,28,2].

Twin-element photomixers allow more design flexibility, and they have been demonstrated to have superior performance between 1 and 3 THz. Twin dipoles and slots have been analyzed for photomixers by Duffy et al. in [28]. Here we only discuss the twin dipoles. The twin elements allow more modular control of the various design aspects than single elements. For example, the electrodes are situated between the two antennas and away from their near-field distributions. Also the capacitance is tuned out with two lengths of line that function as single-section impedance transformers. RF chokes are used to block THz leakage down the bias lines. The twin dipole elements are driven in phase and are separated by roughly a half-wavelength. This improves the antenna gain of the H-plane pattern since the twin dipoles now act as a two-element phased array. Figure 7.14 shows the equivalent circuit and a SEM of a typical device.

378 CONTINUOUS THz GENERATION WITH OPTICAL HETERODYNING

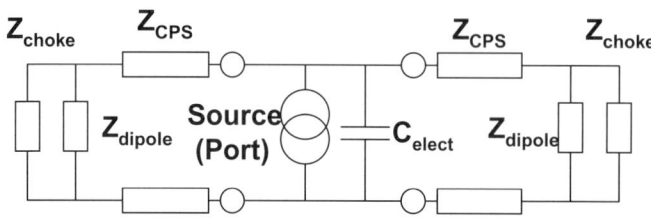

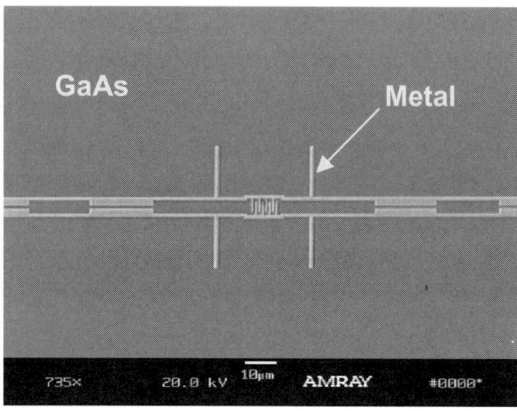

Fig. 7.14 Twin half-wave dipole antennas and their equivalent circuit. This design outperforms single-element designs at frequencies above 1 THz.

The photoconductor and its interdigitated electrodes are modeled as a current source with a capacitance C_{elect}. Between the photoconductor and each antenna element is approximately a quarter-wavelength of high-impedance transmission line with impedance Z_{cps}. The drive-point impedance of the dipoles and the input impedance of the chokes appear in parallel connection as Z_{dipole} and Z_{cps}. As discussed by Duffy, twin slot antennas can be analyzed in a similar fashion, but with the equivalent circuit elements in series connection.

Figure 7.15 shows the measured and calculated performance of several twin-dipole photomixers designed for various operating frequencies [28,22]. A broadband spiral antenna is shown for comparison with its −12 dB/octave rolloff caused by the combination of parasitic capacitance and photocarrier lifetime. The key indication that the twin-dipole photomixers were properly designed is that the line indicating a −6 dB/octave rolloff passes along the maximum outputs of the various antennas. This indicates that the twin-dipole photomixers were designed with the parasitic capacitance optimally tuned out at each frequency, leaving only the rolloff pole from carrier lifetime. We speculate that the 2.7 THz design was less optimal because the THz loss in the

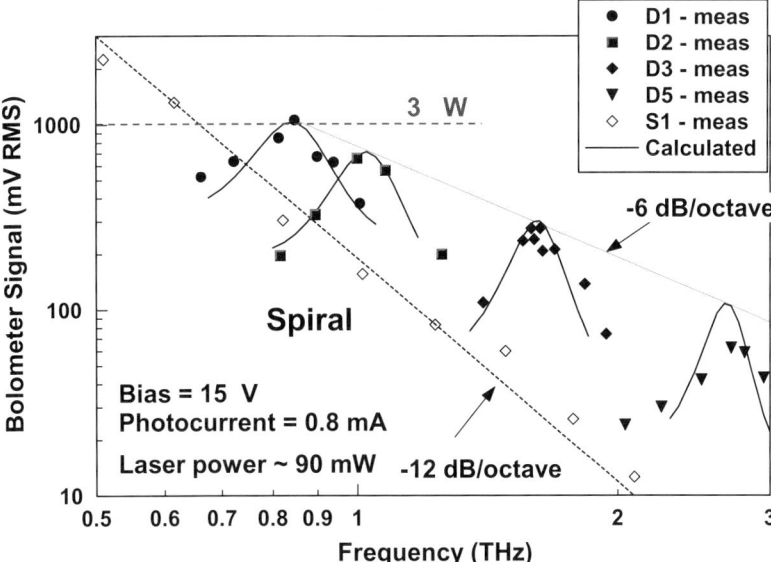

Fig. 7.15 Simulations and measurements for four different twin-dipole photomixers. The 6 dB/octave rolloff shows that the pole resulting from capacitive loading by the electrodes has been efficiently cancelled at the resonance frequency of each antenna. A log-spiral antenna that rollsoff at 12 dB/octave is shown for comparison.

Ti:Au metallization limited the quality factor Q needed to resonate out the electrode capacitance. The results shown here represent state-of-the-art performance for lumped-element photomixers.

7.4.2 Broadband Distributed Designs

An alternative approach to photomixer design is the distributed photomixer. These devices can potentially improve the bandwidth and output power of photomixers operating above 2 THz. The relevant analogy is the success of traveling-wave transistor amplifiers and nonlinear transmission lines in enhancing the high-frequency performance of transistors and Schottky diodes, respectively. Two approaches look encouraging. A surface-illuminated approach has been developed by Matsuura et al. that has demonstrated record output power above 1 THz [30]. This approach has the advantage of relatively simple fabrication. Velocity matching of the optical group velocity to the THz phase velocity is achieved by controlling the angle of incidence of the optical waves. A second approach has been described by Duerr et al. that is similar to the design of high-power photodiodes used for analog optical communication links [31].

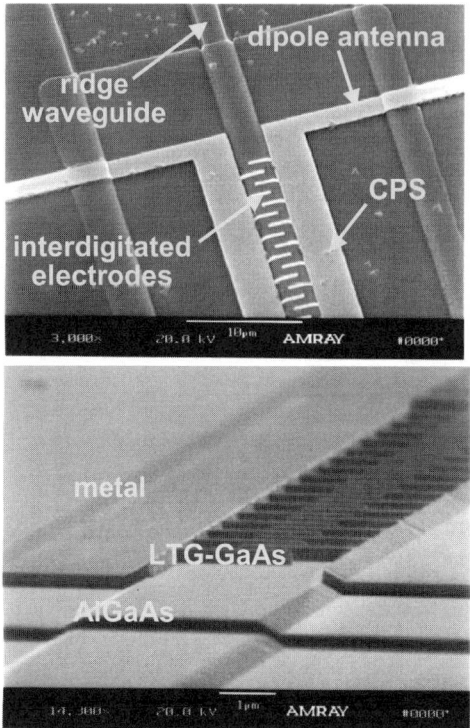

Fig. 7.16 Scanning electron micrographs of a distributed photomixer. Top view shows the AlGaAs ridge waveguide, the electrodes and coplanar strips (CPS), and the full-wave dipole antenna. Side view shows the LTG-GaAs film on top of the AlGaAs waveguide and the metal electrodes.

Figure 7.16 shows two SEMs of a distributed photomixer that is terminated by a full-wave dipole antenna [31]. Here an AlGaAs ridge waveguide couples the optical beat signal to a thin layer of LTG-GaAs photoconductor that is patterned on its surface. Interdigitated electrodes collect the photocurrent onto a coplanar stripline (CPS) and add just enough capacitance per unit length to velocity match the THz and optical waves. Figure 7.17 shows initial results from a velocity-matched distributed photomixer. Such a structure coupled to a broadband bow-tie antenna is consistent with the model and is limited only by the photocarrier lifetime and RF loss in the CPS. The structure coupled to a dipole antenna has its output spectrum band-limited by the antenna impedance. In theory, this device should outperform lumped-element photomixers at frequencies above 2 THz, since it can handle higher optical pump power and has wider bandwidth. In practice, engineering details such as efficient thermal management, optical facet coatings, and RF antenna design need to be optimized before this device realizes its potential.

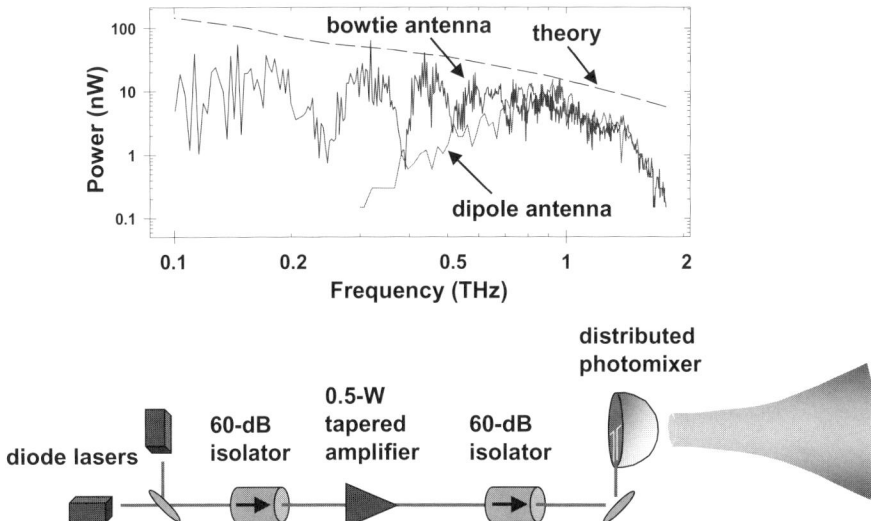

Fig. 7.17 Measured data for two distributed photomixers with different antennas. Simulations of the expected performance are also shown. A bolometer was used to detect the output power.

7.5 APPLICATIONS

7.5.1 Local Oscillators

Recent advances in superconducting heterodyne detectors promise significant scientific payoff in submillimeter-wave astronomy and the study of planetary atmospheres. Hot-electron bolometers (HEB) and superconductor-insulator-superconductor (SIS) mixers can achieve almost quantum-limited sensitivity at frequencies above 600 GHz. Such systems presently use local oscillators (LOs) consisting of one of the following: frequency-multiplied diode oscillators, vacuum tubes, or molecular lasers.

The photomixer is an interesting alternative LO technology. It can be assembled with no moving parts, is relatively low power, has a wide tuning range, and is based on commercial lasers and a custom photoconductor. For array applications, multiple photomixer LOs can be remotely located from the lasers, thereby easing signal distribution to remote antennas. The main challenges in making a viable LO out of photomixers are generating a coherent tone with minimal frequency jitter and amplitude noise, and generating sufficient power to overcome diplexer and other losses incurred when coupling a LO to a cryogenic heterodyne detector.

An experiment in collaboration with Harvard Smithsonian Center for Astrophysics demonstrated the use of a photomixer as a local oscillator at 630 GHz [2]. The heterodyne detector was a Nb SIS mixer mounted in a waveguide mixing block. The LO was a photomixer with a single-element full-wave dipole. Approximately 0.2 µW of RF power was coupled to the SIS mixer. The resulting double-sideband noise temperature was 331 K—in good agreement with the 323 K noise temperature obtained when a multiplied Gunn oscillator coupling in 0.25 µW was substituted for the photomixer.

The twin-element photomixers described above have higher output power and operate at higher frequencies. These device should be adequate for laboratory experiments that integrate HEB mixers with photomixer LOs. The photomixer technology should be a reliable LO technology as it becomes more robust against thermal failure and as the LO power consumption of the HEB mixers drops.

7.5.2 Transceiver for Spectroscopy

As was described in detail earlier, photomixers are attractive sources for high-resolution spectroscopy of gases. Gas spectrometers based on photomixers and bolometers exhibit high sensitivity, as well as frequency resolution that is unmatched by systems such as Fourier-transform spectrometers that do not use coherent sources. A practical disadvantage for fielded systems, however, is the requirement of a helium-cooled bolometer. The photomixer transceiver addresses this limitation [32]. At present, however, its detection sensitivity is not as good as the systems using helium-cooled bolometers.

The transceiver uses a pair of photomixers that are pumped by the same pair of diode lasers. One photomixer is a coherent transmitter of THz radiation. Its input consists of a dc bias voltage and the optical beat signal produced by the combined output of the two lasers. Its output is a coherent THz beam. The THz beam passes through a gas cell and is detected by the second photomixer. The second photomixer is a coherent homodyne receiver of THz radiation. Its input consists of a THz waveform and the optical beat signal. Its output is a dc current that is proportional to the amplitude of the incident THz electric-field strength.

Two photomixers with twin slot antennas with a center frequency of 1.4 THz were used to perform a demonstration of high-resolution spectroscopy on water vapor. Figure 7.18 shows successive measurements of the transmission at 1.411 THz through the gas cell with the water vapor pressure in a 50-cm-long gas cell varying between 0.1 and 1 Torr (from top to bottom). Note the narrowing of the line that occurs as the pressure is reduced as well as the ability of the transceiver to resolve it. This particular set of photomixers had sufficient THz power to measure the water-vapor spectrum from 1.15 to 1.5 THz (26% fractional bandwidth).

Recently, an on-chip photomixer transceiver was demonstrated, where the two photomixers were coupled by a coplanar waveguide rather than through free space [27]. The on-chip transceiver is especially useful for studying the frequency dependence of the photoconductor and interdigitated electrodes since propagation along a well-designed transmission line is mostly linear-

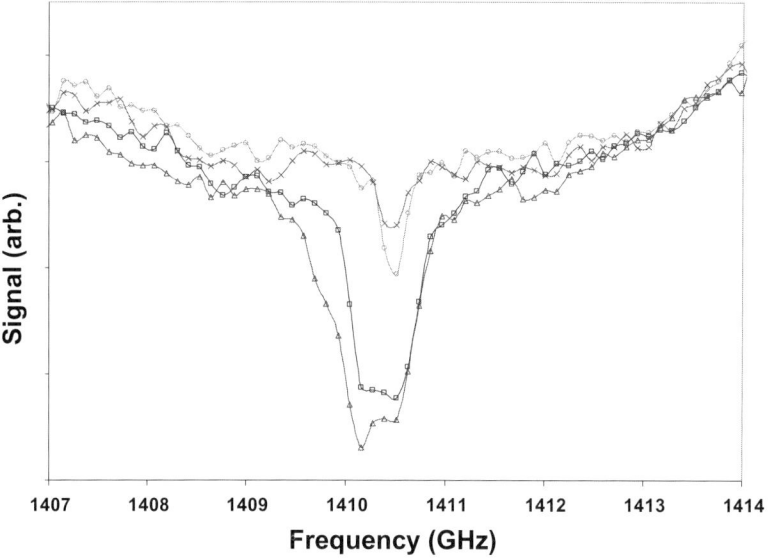

Fig. 7.18 Measurements of the 1.4 THz absorption line in water vapor. From the bottom trace to the top, the pressure in the gas cell was reduced from 1 Torr to 0.1 Torr. Further resolution in frequency can be obtained by stabilizing the diode lasers.

phase and can be de-embedded more accurately than with quasioptical systems. Analysis of the data from the on-chip transceiver confirmed the predictions of the model described by Zamdmer et al. for the electric-field dependence of the photocarrier lifetime [21].

7.6 SUMMARY

Photomixers are a frequency agile technology for generating continuous-wave THz radiation. Potential applications of photomixers include local oscillators and gas spectroscopy. The constraints of the particular application will determine the requirements on the pump laser system for output power, frequency accuracy, and tuning range. An understanding of photocarrier lifetime and of antenna design is necessary to fabricate high-performance photomixers. Newer structures that use a distributed design may ultimatedly outperform conventional structures for frequencies above 2 THz.

ACKNOWLEDGMENTS

This work was sponsored by the Department of the Air Force under United States Air Force Contract No. F19628-95-C-0002. Opinions, interpretations, conclusions, and recommendations are those of the author and are not necessarily endorsed by the United States Air Force. A part of this research was carried out at the Jet Propulsion Laboratory, California Institute of Technology, under a contract with the National Aeronautics and Space Administration.

REFERENCES

1. R. Datla, E. Grossman, and M. K. Hobish, eds. *Metrology Issues in Terahertz Physics and Technology*, **NIST-5701**, Gaithersburg, MD 103 (1995).
2. S. Verghese, K. A. McIntosh, S. D. Calawa, C.-Y. E. Tong, R. Kimberk, and R. Blundell, "A photomixer local oscillator for a 630-GHz heterodyne receiver," *IEEE Microw. Guided Wave Lett.* **9**, 245 (1999).
3. A. S. Pine, R. D. Suenram, E. R. Brown, and K. A. McIntosh, "A terahertz photomixing spectrometer: Application to SO_2 self broadening," *J. Mol. Spectrosc.* **175**, 37 (1996).
4. P. Chen, G. A. Blake, M. C. Gaidis, E. R. Brown, K. A. McIntosh, S. Y. Chou, M. I. Nathan, and F. Williamson, "Spectroscopic applications and frequency control of submillimeter-wave photomixing with distributed-Bragg-reflector diode lasers in low-temperature-grown GaAs," *App. Phys. Lett.* **71**, 1601 (1997).
5. E. R. Brown, S. Verghese, and K. A. McIntosh, "Terahertz photomixing in low-temperature-grown GaAs," *Proc. SPIE Conf. Adv. Technol. MMW, Radio, and Terahertz Telescopes* **3357**, 132 (1998).
6. E. R. Brown, F. W. Smith, and K. A. McIntosh, "Coherent millimeter-wave generation by heterodyne conversion in low-temperature-grown GaAs photoconductors," *J. Appl. Phys.* **73**, 1480 (1993).
7. K. A. McIntosh, E. R. Brown, K. B. Nichols, O. B. McMahon, W. F. Dinatale, and T. M. Lyszczarz, "Terahertz photomixing with diode lasers in low-temperature-grown GaAs," *Appl. Phys. Lett.* **67**, 3844 (1995).
8. S. Verghese, K. A. McIntosh, and E. R. Brown, "Highly tunable fiber-coupled photomixers with coherent terahertz output power," *IEEE Trans. Microw. Theory Tech.*, **45**, 1301 (1997).
9. S. Matsuura, M. Tani, and K. Sakai, "Generation of coherent terahertz radiation by photomixing in dipole photoconductive antennas," *Appl. Phys. Lett.* **70**, 559 (1997).
10. Y. R. Shen, *Principles of Nonlinear Optics*, J Wiley, New York, 1984.
11. S. Matsuura, P. Chen, G. A. Blake, J. C. Pearson, and H. M. Pickett, "A tunable cavity-locked diode laser source for terahertz photomixing," *IEEE Trans. Microw. Theory Tech.* **48**, 380 (2000).
12. S. Matsuura, P. Chen, G. A. Blake, J. C. Pearson, and H. M. Pickett, "Simultaneous amplification of terahertz difference frequencies by an injection seeded semiconductor laser amplifier at 850 nm," *Int. J. Infrared Millim. Waves* **19**, 849 (1998).
13. K. G. Libbrecht and J. L. Hall, "A low-noise high-speed diode laser current controller," *Rev. Sci. Instr.* **64**, 2133 (1993).
14. R. W. Tkach and A. R. Chraplyvy, "Regimes of feedback effects in 1.5-μm distributed feedback lasers," *IEEE J. Lightwave Technol.* **4**, 1655 (1986).

15. R. V. Pound, "Electronic frequency stabilization of microwave oscillators," *Rev. Sci. Instrum.* **17**, 490 (1946).
16. R. W. P. Drever, J. L. Hall, F. V. Kowalski, J. Hough, G. M. Ford, "Laser phase and frequency stabilization using an optical resonator," *Appl. Phys.* B **31**, 97 (1983).
17. L. Hollberg, V. L. Velichansky, C. S. Weimer, and R. W. Fox, "High-accuracy spectroscopy with semiconductor lasers: Application to laser-frequency stabilization," in M. Ohtsu, ed., *Frequency Control of Semiconductor Lasers*, Wiley, New York, 1996, pp. 73–93.
18. D. Wandt, M. Laschek, F. V. Alvensleben, A. Tünnermann, and H. Welling, "Continuously tunable 0.5 W single-frequency diode laser source," *Opt. Commun.* **148**, 261 (1998).
19. H. M. Pickett, "Determination of collisional linewidths and shifts by a convolutional method," *Appl. Optics* **19**, 2745 (1980).
20. I. G. Nolt, J. V. Radostitz, G. Dilonardo, K. M. Evenson, D. A. Jennings, K. R. Leopold, M. D. Vanek, L. R. Zink, A. Hinz, K. V. Chance, "Accurate rotational constants of CO, HCl, and HF: spectral standards for the 0.3- to 6-THz (10- to 200-cm^{-1}) region," *J. Mol. Spec.* **125**, 274 (1987).
21. N. Zamdmer, Q. Hu, K. A. McIntosh, and S. Verghese, "Increase in response time of low-temperature-grown GaAs photoconductive switches at high voltage bias," *Appl. Phys. Lett.* **75**, 2313 (1999).
22. A. W. Jackson, "Low-temperature-grown GaAs photomixers designed for increased terahertz output power," Ph.D. dissertation, Department of Materials Science, University of California, Santa Barbara, 1999.
23. E. R. Brown, "A photoconductive model for superior GaAs THz photomixers," *Appl. Phys. Lett.* **75**, 769 (1999).
24. S. Verghese, K. A. McIntosh, and E. R. Brown, "Optical and terahertz power limits in low-temperature-grown GaAs photomixers," *Appl. Phys. Lett.* **71**, 2743 (1997).
25. K. A. McIntosh, K. B. Nichols, S. Verghese, and E. R. Brown, "Investigation of ultrashort photocarrier relaxation times in low-temperature-grown GaAs," *Appl. Phys. Lett.* **70**, 354 (1997).
26. S. Verghese, N. Zamdmer, E. R. Brown, A. Förster, Q. Hu, "An optical correlator using a low-temperature-grown GaAs photoconductor," *Appl. Phys. Lett.* **69**, 842 (1996).
27. N. Zamdmer, Q. Hu, K. A. McIntosh, S. Verghese, and A. Förster, "On-chip frequency-domain submillimeter-wave transceiver," *Appl. Phys. Lett.* **75**, 3877 (1999).
28. S. M. Duffy, S. Verghese, K. A. McIntosh, A. W. Jackson, A. C. Gossard, and S. Matsuura, "Accurate modeling of dual dipole and slot elements used with photomixers for coherent terahertz output power," *IEEE Trans. Microw. Theory Tech.* **49**, 1032 (2001).
29. K. A. McIntosh, E. R. Brown, K. B. Nichols, O. B. McMahon, W. F. Dinatale, and T. M. Lyszczarz, "Terahertz measurements of resonant planar antennas coupled to low-temperature-grown GaAs photomixers," *Appl. Phys. Lett.* **69**, 3632 (1996).
30. S. Matsuura, G. A. Blake, R. A. Wyss, J. C. Pearson, C. Kadow, A. W. Jackson, and A. C. Gossard, "A traveling-wave THz photomixer based on angle-tuned phase matching," *Appl. Phys. Lett.* **74**, 2872 (1999).

31. E. K. Duerr, K. A. McIntosh, and S. Verghese, "Distributed photomixers," in *Conference on Lasers and Electro-Optics*, OSA Technical Digest, Optical Society of America, Washington DC, 2000, p. 382.
32. S. Verghese, K. A. McIntosh, S. D. Calawa, W. F. DiNatale, E. K. Duerr, and K. A. Molvar, "Generation and detection of coherent terahertz waves using two photomixers," *Appl. Phys. Lett.* **73**, 3824 (1998).

INDEX

Absorption, 1, 2, 7, 11, 34, 49, 69–70, 81, 83, 95, 102–103, 113–117, 145, 147, 159, 162, 165, 177, 179–180, 185, 201, 206, 218, 224–225, 227, 235, 237, 251, 260, 283, 291–292, 298, 306, 311–312, 315, 320, 330, 333, 338–339, 353, 367–369, 372–373, 383
  coefficient, 95, 158, 162, 164, 178–179, 185, 187, 202, 235, 236, 249–251, 259, 298–299
  edge, 159, 162, 179, 201
  intersubband, 6, 8, 113–114, 116, 218
  free-carrier, 76, 83, 102, 113, 115, 165, 225, 227, 237
  strength, 1, 11–12
Acceptor, 10, 148–149, 158, 160, 201, 280, 284, 288–293, 295, 299–300, 305–306, 311–315, 317–322, 324, 338–339, 341
Acetonitrile, 367, 369
Acoustical phonon, 279, 284, 289, 292, 296–297, 311, 316, 340
Aging, 53, 168–170
Amplification
  of picosecond pulses, 330
  See also Gain, 298–299
  cross section, 300–303, 331
Annealing, 43, 45, 150, 170, 320–322
Antimonide lasers, 6, 70, 74, 82, 85, 115, 125
Applications, 1–3, 7, 10–11, 19, 70, 146, 281–283
  far infrared, 4, 323, 333, 342
  terahertz, 2, 281–282, 350–351
Auger
  surface analysis, 156
  recombination, 5, 7, 15, 35–36, 56, 69, 71–72, 79, 83, 88, 95–96, 110, 112–114, 125–126, 138–140, 145, 157, 170, 176
  coefficient, 32, 35, 73, 76, 81, 95, 110, 112–113
AuIn, 205

BaF_2, 8, 146, 179, 182–184, 193, 197, 200–205
Band-gap
  energy, 2–4, 6–10, 21–32, 35, 37–38, 49–50, 58, 84–85, 149–150, 154, 156–158, 164, 167, 201–202, 204, 225
  energy shift induced by strain, 22, 28
  wavelength of strained quantum well, 29
Band offset, 6, 28–29, 72–74, 76, 89–90, 93–96, 108, 110–112, 126, 167, 174, 241, 243, 245
Band structure, 7, 13, 26–27, 29, 70–72, 111–113, 162–163, 167, 178–179, 182, 222, 224, 282–284, 306, 315–316
Barrier height, 21, 30, 34, 36–38, 75, 181, 243
Beam quality, 71, 77, 100, 102, 115, 117–118, 121, 123
Bolometers, 282, 294–295, 322, 342, 352, 368–369, 381–382
Bragg reflector stop band, 262
Bragg wavelength, 56, 188
Brillouin zone, 174
Broadended waveguide, 14, 76–78, 102–103, 111, 114
Buried heterostructure (BH), 43, 175, 185

CaF_2, 197, 203–204, 215
Carbon monoxide, 367, 369
Cascading, 6–8, 219–220, 226
Cascade laser,
  interband, 6, 71, 80–81, 98–99, 102–106, 126, 135–138, 218–220

387

intersubband, 3–4, 6, 8–9, 13, 80, 103–104, 106, 135–137, 141, 218–219, 222–223, 271
See also Quantum cascade laser
Cavity length, 34, 50–53, 65, 77–78, 89, 119, 224, 235, 237, 251, 262
Chalcogenide, 145, 148, 150
See also Lead-chalcogenide laser
Characteristic temperature (T_0), 55, 76, 79, 244
Chemical beam epitaxy (CBE), see Metalorganic molecular beam epitaxy
Closed-cycle refrigerator (cooler), 11, 281, 318, 326–327, 338, 342
Compositional modulation, 22–23, 25, 43, 45, 47
Conduction-band discontinuity, 218, 220, 234, 237, 239–240, 242–243
Cladding, 8, 37, 52, 54–55, 76, 79, 83–85, 87, 89–91, 96, 102, 113–114, 119, 124–125, 154, 165, 167, 173, 227, 229, 234–236, 257, 259, 263, 269
Compound semiconductor
III–V, 2–5, 12
IV–VI, 2–5, 144
Confinement, 28, 34, 39, 71–72, 74–76, 82–83, 88–90, 94–95, 97–98, 114, 119, 125, 154, 157, 160, 165, 167–168, 173–178, 190, 204, 211, 221, 228, 234–236, 242–243, 246, 269
Coupled mode theory, 185, 251
Critical layer thickness, 22, 25, 31–34, 43, 63, 239
Current-voltage characteristic, 92, 240, 241, 290
Cyclotron resonance frequency, 295, 310

Detectors, 11, 95–96, 133, 150, 152, 174, 230, 238, 255, 260, 293–295, 313, 326, 328, 337, 350–352, 357, 368–370, 381
far infrared, 295, 337, 345, 348, 350–351
HgCdTe (mercury cadmium telluride), 255
Schottky diode, 328, 337
DFB-SH, 195
Diode lasers, 5–6, 11, 69, 74–77, 90, 93, 97, 103, 105, 108, 110–111, 114, 145, 160, 165, 175, 177, 198, 202, 204, 219–220, 226–227, 264, 266, 283, 354–357, 359–362, 373, 382–385
frequency locking, 355
Differential quantum efficiency, 6–7, 34, 52, 75–76, 79, 89, 91, 153, 253
Diffusion, 66, 89, 147–149, 170, 208, 290, 292, 305, 320–322, 342
Dislocation, 32, 34, 39, 43
Distributed Bragg reflector (DBR) laser, 146, 184
Distributed-feedback (DFB), 5, 20, 49, 53–58, 118, 142, 146, 174, 184, 210, 213, 218, 248, 251, 355, 384
laser, 5, 20, 49, 53, 118, 146, 174, 184, 218, 248, 251, 355, 384
complex-coupled, 252, 256
index coupled, 251, 256
DNA, 2, 282, 345
Donor, 10, 148–149, 158, 160, 167, 284, 312, 320–321, 338, 340
Doping, 76, 84, 148–149, 157–158, 160, 165, 198, 200–201, 203, 223–224, 228–229, 234–235, 237, 271, 284, 305, 315, 317–318, 320, 338
by diffusion, 305, 320
by ion implantation, 148–149, 348
Double-heterostructure (DH) laser, 7–8, 22, 74–75, 82–85, 87–89, 91, 97, 100, 120, 131–132, 135, 141, 154, 156–157, 161–162, 165–168, 171–172, 175, 177, 195, 197, 201

Effect,
Hall, 301–304, 309–311, 320–321, 323, 338, 340
optogalvanic, 292
Stark, 282, 307
thermal, 316, 320
Effective mass, 10, 12, 28–29, 35, 89, 178, 182, 234, 237, 280, 285–286, 288, 296, 299, 302, 309–310, 312–313, 338, 341
Efficiency,
differential quantum, 6–7, 34, 52, 75–76, 79, 89, 91, 153, 253
slope, 34, 55–56, 80, 100, 105, 137, 230, 233–234, 244, 251–254, 269–270
wall-plug, 105

INDEX **389**

Electric field, 10, 12, 255, 278–279, 285–287, 289–291, 293–294, 296–297, 299, 301–306, 308–311, 315–316, 318, 322–324
  applied, 218, 240–242, 285, 291, 293–294, 299, 301–302, 304, 309–311, 316, 318, 322–323, 329, 335–336, 338, 340
  calculation, 299, 303–305, 308–309, 340, 346
  generation, 324
  Hall, 301–304, 309–310, 323
Embossing, 8, 195
Emission wavelength
  and number of quantum wells, 21
  current-tuning rate, 276
  range of III-V semiconductor lasers, 4–5, 20
  temperature dependence, 65, 245–246, 249
Epitaxy, 12, 21, 23, 38, 60–67, 138, 155–156, 174, 197, 200–203, 209–218, 273–274, 341, 373
  liquid phase (LPE), 23, 62, 67, 71, 155, 210
  gas-source molecular beam(GSMBE), 62, 66
  hot wall (HWE), 155, 210, 212, 214
  metalorganic molecular beam, 21, 38, 60–61, 63, 66–67
  metalorganic vapor phase, see Metalorganic vapor phase epitaxy
  molecular beam, see Molecular beam epitaxy

Fabry-Perot
  cavity, 77, 80, 119, 121, 184, 236, 247, 251, 254, 267, 357–358, 369
  interference, 333
  interferometer, 249, 250
  laser, 21, 57, 77, 80, 117, 119, 121, 123–124, 165, 179, 184, 201, 204
  resonator, 323, 331, 333, 335
Faraday configuration, 301, 304, 328, 334, 349
Far-field, 121–123, 209, 262
Far-infrared (FIR), 2, 4, 279–283, 295, 322, 338, 351
  laser, 2, 4, 16–17, 279–283, 295–296, 351
  continuous wave, 13, 275, 345
Fourior Transform Infrared (FTIR), 149, 179–180, 182, 204, 230
Free-carrier absorption, 83, 102, 113, 115, 140, 227, 237

GaAs, 8, 11, 19–22, 60, 62, 64, 69, 71, 106–108, 207, 221–222, 225, 228–229, 233–235, 237, 240–241, 243–246, 256, 258, 263, 266, 268, 271, 296, 341, 353–356, 372, 374–375, 377, 380
  photomixers, 11, 13, 17, 353, 356–372, 375, 377–380, 384
  low-temperature-grown (LTG), 17, 353, 372, 374–376
  carrier lifetime, 11, 139, 372, 374
  quantum-cascade laser, 17, 273–276, 278
Gain, 11, 35, 56, 75, 77–79, 90, 95, 97, 100, 103, 107, 111, 113, 117–119, 121, 173, 175–176, 184, 188, 194, 220, 226, 231, 236–238, 241, 246–248, 254, 258, 260, 262–263, 265, 280–281, 283, 289, 292, 296, 298–302, 304, 315–317, 329–330, 335, 340, 355, 360–361, 365, 373, 377
  coefficient, 236, 298–299, 300–302
  modal, 35, 78, 111, 117, 130, 236–238, 254, 258
  modulation, 296
GaInAsSb or InGaAsSb, 5–6, 73–84, 88, 125
GaInSb, 6, 72, 75, 85, 96–107, 115, 117
Gas analysis, 208
GaSb, 4–7, 20, 69, 71–76, 80–83, 87, 89, 91, 94–97, 101, 104, 106–109, 112, 114–115, 117, 119–120, 174, 210, 341
Gas source molecular beam epitaxy (GSMBE), 62
Germanium,
  acceptors, 284, 289–293, 295, 305–306, 315, 320–322, 338–339
  band structure, 306, 308, 315–316, 341
  crystal growth, 319–320
  doping by diffusion, 305, 320
  donors, 321, 338
  dopants, 288, 295–297, 314–315

hydrogenic acceptor, 280, 290, 292–293, 306, 313–315, 320–322
impurities, 292, 298, 306, 312, 315, 319, 320, 333, 340
nonhydrogenic acceptor, 280, 306, 313–315, 339
ohmic contacts, 303–305, 316, 320, 322–323, 327, 331, 333, 339, 342
under uniaxial stress, 341, 347
Germanium laser, 10–11, 280–284, 290, 292–293, 296, 298–300, 302–304, 306, 310–311, 315–317, 319–320, 322–326, 328–331, 333, 338–342
tunable, 331, 334, 336, 342

Hakki-Paoli method, 243
Hall effect, 282, 301, 303–304, 309–310, 320–321, 323, 338
electric fields, 279, 290–291, 297, 299, 301–303, 306, 310, 316, 318
variable temperature, 320–321
Hamiltonian, 27, 29, 36, 114, 308–309, 311
Heat,
capacity, 317–318, 327, 348
conductivity, 233, 318–320, 327
equations, 317
sink, 87, 123, 125, 147, 151, 161–162, 168, 170, 192, 195, 197, 199, 202, 241, 244, 249–250, 254, 257, 259–261, 269, 304–306, 318–319, 323, 326–328, 338–339, 342
Heterodyne, 184, 281–282, 328, 330, 332, 335, 344–345, 349, 352–354, 361, 363, 381, 384
High-resolution x-ray diffraction (HRXRD), 179, 204
Hole velocity, 285, 306
Homostructure laser, 146–148, 153–154
Hot hole laser, 4, 10–11, 279–350
diamond, 340–341
germanium, see Germanium laser
III-V, 340
quantum mechanical model, 306–310
silicon, 340–342
semiclassical model, 281, 283–306
under uniaxial stress, 341

InAs, 6–7, 12–13, 63, 69, 74, 82–83, 89–90, 95, 97–98, 102, 104–105, 108, 131, 341

InAsSb, 6, 70, 72, 82–91, 94–96, 100, 102, 106–108, 112, 117–118, 120–121, 125–127, 131–134, 137–142
Infrared,
definition, 1–2
long wavelength,
mid (MIR), 1–6, 10, 19, 69–71, 74, 81, 84, 87, 90, 110, 113, 117, 119, 123, 145, 157, 165, 171–172, 184, 198, 200, 206, 225, 227, 235
near (NIR), 1, 6, 70, 80, 145
far (FIR), 1–3, 8, 74, 227, 279–281, 294, 322, 342, 351
InGaAs, 5, 8–9, 12, 19, 21–27, 29–32, 34–40, 42–53, 55
InGaAs/InAlAs, 8, 218, 220–222, 228–229, 234, 237, 239, 241
InGaAs/InGaAs quantum wells, 39–40, 42–43, 46–50
InGaAs/InGaAsP quantum wells, 43, 46–49, 53, 60, 64–65, 67
InGaAsP, 21–22, 24
InP, 4, 5, 8, 12–13, 19–22, 24, 28, 31–34, 38–39, 41, 45–47, 50, 52–55, 57, 71–72, 84, 86, 90–91, 102, 106–107, 115, 217–218, 220, 222, 228–229, 237, 239–240, 244, 249, 258, 263–264, 267–269
InSb, 6, 12, 81–82, 85, 87, 108, 130–131, 152, 212, 341, 368
Interband cascade lasers, 6, 13, 15, 103
Interband transition, 180, 218, 220, 349
Intersubband transition, 8–9, 80, 113–114, 136, 218–219, 225, 245, 258, 273, 275, 298
Integrated absorber, 100
Internal loss, 34–35, 49–50, 64, 69, 71, 76–77, 84, 89, 99–100, 111, 113–117, 125–126, 138, 140–141, 165, 237
Intervalence-band (IVB) laser, 280
IV-VI laser, 146, 184, 200, 205

Kramers-Kronig, 117, 179

Landau levels, 10, 280, 295, 301, 307–311, 315, 335–336
Lasers,
antimonide-based, 5–8, 69–126
diode, 3, 5–7, 11, 19, 60, 69, 71, 74–77,

INDEX **391**

    82, 84–85, 87, 89, 93, 97, 100, 102–103, 108, 110, 113–114, 121, 145, 152–153, 160, 165, 175, 177, 206, 219, 220, 226–227, 264, 266, 328, 337, 341, 352, 354–357, 359–362, 373, 381–385
  double-heterostructure, 7, 8, 22, 71, 82, 84–85, 117, 125, 146, 154, 171, 175, 275
  interband, 3–8, 10, 12–13, 71, 80–81, 85, 91, 95, 98–99, 103–105, 113–115, 118, 162, 218–220, 231, 263
  interband cascade, 6, 7, 71, 103, 105
  intersubband, 3, 4, 6, 8, 9, 13, 80, 103, 104, 106, 113, 135, 137, 172, 218–219, 223, 227, 231, 245, 271, 274, 278, 298
  intervalence-band, 34, 49, 64, 114, 349
  lead-chalcogenide, 210–211, 213–214
  mid-infrared (MIR), 1–6, 8, 10, 12–13, 19, 69–75, 81–82, 84–85, 87, 90–91, 102, 105, 108, 110, 113–114, 117–119, 121, 123, 145–146, 157, 165, 171, 174, 179, 183, 200, 218, 227, 231, 235, 246, 250, 266, 283
  near-infrared (NIR), 1–4, 6, 13, 69, 70, 79–80, 82, 85, 110–111, 115, 117, 119, 121, 145, 219, 227, 231, 250, 354
  quantum cascade, 3, 4, 6, 8–10, 13–17, 19, 60, 80, 99, 103, 143, 168, 172, 217–223, 226, 227, 229, 231, 233–235, 237, 239–246, 248–252, 254, 256, 261, 263–270, 272–278
  type-I, 4, 6, 12–13, 70–74, 83, 88–89, 91, 94–95, 98, 103, 105–107, 111–112, 114, 117, 125–126, 167, 171
  type-II, 4, 6, 13, 71–72, 74–75, 80, 91, 94–100, 102–103, 105, 108–112, 114–115, 117–118, 122, 167
  vertical-cavity surface-emitting (VCSEL), 80–81, 99, 119, 175, 183–184, 219
  "W", 97
Lattice constant, 5, 21–22, 25–27, 61, 69, 71, 85, 107, 146–147, 159, 164–165, 239
Lattice match, 5, 96, 155–158, 161, 175–176, 197, 200–201, 220–222, 239
Lattice mismatch, 22–25, 63, 157, 161, 164, 167–168, 174, 182

Lead-chalcogenide laser, see Lead-salt laser
Lead-salt laser, 2, 7–8, 145, 151, 165, 168, 171, 173–174, 206–207
Lifetime, 8–11, 75, 104, 110, 221–224, 266–267, 272, 287, 296, 373
Light-hole cyclotron resonance (LHCR) laser, 280, 300, 302, 337
Linewidth, 10–13, 57, 81, 115, 118–119, 121, 154–155, 165, 167, 176–177, 210, 220, 225, 231, 248–251, 261, 354–355, 361–365, 369, 385
Linewidth enhancement factor, 71, 115, 117–118, 126, 250
Liquid nitrogen, 120, 145, 150, 155, 168, 184, 259, 322, 327
Liquid phase epitaxy (LPE), 23, 71, 155, 167, 197, 200–201, 204–205
Lorenz force, 12
Loss,
  mirror, 35, 111, 224, 235, 238, 263, 266–267
  optical, 7, 34, 37, 224–225, 227–229, 233–235, 242–243, 252, 254–255, 263, 298
  waveguide, 224–225, 227–229, 233–237, 243, 248, 252, 254–257, 259–261, 263, 267–268
L-valley, 174, 178–179, 183

Magnet,
  magic rings, 325–326
  permanent, 11, 281, 301, 304, 325–326, 342, 349
  superconducting, 324–326
  tunable, 11, 281, 325–326, 349
Magnetic field, 10–17, 81, 85, 279–280, 282–283, 285, 289, 293, 295–297, 301–304, 307, 309–310, 315–316, 323–329, 335–336, 341
Magnetoluminescence, 162
Maximum operating temperature, 7–8, 10, 83, 85, 89–91, 96–97, 102, 105, 126, 149, 153–154, 156–157, 159, 161, 165, 167, 171–173, 175, 177, 184, 195, 200, 206, 239–241, 243, 245–246, 248
Mesa structure, 175, 188
Metalorganic chemical vapor deposition

(MOCVD), 71, 75, 83, 89–91, 93, 107–108, 127, 132–133, 268
Metalorganic molecular beam epitaxy (MOMBE), 12, 14, 21, 38–39, 43, 49, 54
Metalorganic vapor phase epitaxy (MOVPE), 23, 38–39, 52, 54, 102, see also Metalorganic chemical vapor deposition
Mirrors, DBR, 80–81
Mid-infrared (MIR) laser, 1–5, 8, 10, 12–13, 19, 66, 127, 145, 214–215, 225, 231, 246, 250, 266, 274, 276
Miscibility gap, 6, 22–25, 43, 45, 62, 71, 73, 78, 88–89, 108
Misfit dislocation, 29, 31–34, 43, 63
Mismatch strain, 25–28, 31, 34, 41, 45, 58–59, 215
Mixers,
  optical, see Photomixers,
  superconducting, 381
Mobility, 37, 157, 160, 165, 174, 279, 373
Modal charts, 152
Mode competition, 167–168, 175, 337
Molecular absorption, see THz, molecular absorption
Molecular beam epitaxy (MBE), 12, 21, 23, 38, 61–63, 66–67, 71–72, 75–76, 81, 83, 85, 89, 91, 95–96, 107–108, 155–159, 165–167, 173, 175–176, 179, 188–190, 195, 200–201, 203–206, 209–216, 218, 229, 239, 258, 267–268, 341, 373
Monochromator, 150, 152, 177
Monolithic laser array, 20, 52
Monomode, 185, 192, 195, 197, see also Single-mode
Monte Carlo simulation, 224, 279, 283, 290, 302–303, 315–316, 340
Multiple quantum wells (MQW), 39–48, 177–180
  lasers, 14, 20–22, 25, 29–30, 34, 36–43, 49–52, 54, 57–62, 65–67, 89–90, 121, 177, 179, 183, 200–202, 204–206, 212, 215, 345
  strain-compensation, 25, 57
  thermal stability, 43, 66

N-containing III-V materials, 59

Near-infrared (NIR) laser, 1–4, 6, 13, 127, 145, 231, 250, 354, 357
Negative differential resistance (NDR), 244–245
Noise, 8, 151, 154, 167–168, 175, 177, 322, 352, 354, 361, 369–370, 381–382, 384
Nomarski microscopy, 205
Nonradiative recombination, 3, 35, 39, 43, 110, 157, 176

Ohmic contact, 147–150, 229, 269, 303–305, 316–317, 320, 322–324, 327, 331, 333, 339, 342
Optical heterodyning, see also Photomixers
Optical (dipole) matrix element, 95, 97, 98, 223–224, 242, 273, 353, 369, 376, 377–380, 382, 384–385
Optical phonon, 10, 234, 271
  absorption, 224, 283, 291–292, 311
  emission, 224, 279–280, 282–284, 286–288, 291–293, 296–297, 299, 311, 317, 340–341
  scattering, 10, 221, 231, 244, 279–280, 282–284, 286–287, 290–293, 296–297, 299–300, 311, 317
Optical pumping, 6, 52, 70, 85, 88–89, 91, 93–94, 98–99, 108, 120, 125–126
Optical pumping injection cavity (OPIC), 99–101, 115–116, 126, 135
Overlap factor (Γ), 34, 224, 226, 235–237, 241, 243, 245

$Pb_{1-x}Ca_xTe$, 155
$Pb_{1-x}Eu_xSe$, 155, 157–158, 160–164, 168
$Pb_{1-x}Eu_xSe_yTe_{1-y}$, 155
$Pb_{1-x}Sn_xSe$, 204
$Pb_{1-x}Sn_xSe_{1-y}$, 201
$Pb_{1-x}Sn_xTe$, 208
$Pb_{1-x}Sr_xSe$, 201, 209
PbCdSSe, 149
PbEuSeTe, 7, 15, 156–158, 176–177, 211
PbS, 2, 4, 146, 157, 167
$PbS_{1-x}Se_x$, 209
PbSe, 2, 4, 8, 15, 146, 148–149, 157, 160–167, 172–177, 179–184, 188–189, 195, 197, 199–206
PbSnSe, 149, 153, 167, 170, 174, 176, 195, 198, 204

INDEX **393**

PbSnTe, 148–149, 167, 174–175, 177, 197
PbSrSe, 8, 164–166, 173, 175–177, 179–180, 182–184, 200–201, 205–206
PbSSe, 148–149, 153
PbTe, 4, 7, 15, 130, 146, 148, 156–158, 160, 167–168, 176–177, 179, 184, 197
$PbTe_{1-x}S_x$, 147
Phonon
  acoustical, see Acoustical phonon
  longitudinal optical (LO), see Optical phonon
  optical, see Optical phonon
Photoluminescence (PL), 23, 39, 73, 200
  intensity and barrier strain, 42, 46
  spectra and InGaAsP barrier compositions, 44–45
  spectra and number of quantum wells, 39
Photomixers, 11–12, 351–386
  GaAs, 11, 354, 372, 385
  laser selection, 354
  design trade-offs, 371
  theory, 372
  antenna design, 376
  resonant antennas, 377
  distributed, 379
  output power, 379
  local oscillator, 382
  transceiver, 382
Photothermal ionization spectroscopy (PTIS), 320
Plasmon emitter, 3–4, 12
Population inversion, 8–10, 178, 183, 217, 220–221, 223, 226, 244, 270–271, 278, 280, 283–284, 286–288, 295–299
Power,
  cooling, 150, 327
  electrical input, 76, 132, 317, 337
  output, 7–8, 11, 20, 34, 37–38, 49, 52–55, 75–77, 79–81, 83–91, 93–94, 98–100, 103–107, 113, 119–121, 150, 152–154, 170–175, 218, 230, 232, 240, 243, 245, 255, 260–261, 266, 269–271, 280, 337, 341–342, 352–354, 358, 360, 362, 365–367, 369, 371, 373–377, 379–385
  thermal, 254–255, 318,
Propagation constant, 188, 228

Quality control, 168, 170
Quantum cascade (QC) laser, 3–4, 6, 8, 10, 19, 60, 80, 99, 103, 136–137, 143, 168, 217–218, 221–222, 251, 273–278
  buried heterosturcure, 268–271
  chirped superlattice active region, 233
  continuous operation, 233, 249
  diagonal transition, 221, 223, 246–247
  distributed feedback (DFB), 251–263
  electrically tunable, 248
  GaAs-based, 234–246
  injection/relaxation region, 220–224
  superlattice active region, 233, 271
  three-quantum-well active region, 9, 222–223, 229, 246, 252, 264, 270–271
  vertical transition, 223, 229, 239, 255
Quantum engineering, 217
Quantum size effect, 28–30, 182
Quantum well, 3–4, 19, 38–39, 43, 50, 58–67, 71–72, 74, 170–180, 205–206, 211–212, 215, 218, 221, 223, 226, 241–242, 271–273, 283
  electron overflow, 30, 37
  nonuniform carrier distribution, 30, 37–38, 49, 58
  strained, 3, 5, 19, 25–49, 72, 75–85, 88–103
  subband filling against injection current, 37, 50, 58
  thickness undulation, 39, 43, 139, 431
  type-I, 4, 13, 70–74, 136, 181
  type-II, 4, 13–14, 17, 60, 71–72, 74–75, 212
Quantum-well lasers, type-I, 4, 88
Quantum-well lasers, type-II, 4, 94, 97

Recombination
  Auger, 5–7, 35–37, 56, 72, 79, 83–84, 88–89, 94–96, 110–114, 125–126, 145, 156, 171–172, 178, 207, 231
  nonradiative, 3, 35–36, 39, 43, 110, 157, 176
  radiative, 110, 178
  Shockley-Read, 79, 110
Reflection high-energy electron diffraction (RHEED), 32–33, 66, 156, 203
Reflectivity, 34, 76, 185, 203, 224, 266, 298, 323

Refractive index, 56, 114, 117, 121, 146–147, 149, 154, 174, 201, 204, 224, 228–229, 234–236, 249–250, 258–259, 261, 263, 265, 267, 298, 329, 330, 333–335, 370
Reliability, 13, 52, 67, 168
Resonator, 13, 83, 175, 184, 188, 194, 263, 266, 277, 281, 283, 298, 306, 316, 319, 323, 324, 326, 328–329, 330–335, 337, 341, 355, 385
  Bragg mirror, 332
  Brewster angle geometry, 334
  external, 281, 283, 298, 316, 328, 334, 337, 341, 355
  Fabry-Perot, 323, 330–335
  integrated, 188, 329
  modes, 282, 323, 328, 330–332, 335–336
  optical losses, 298
  semi-confocal, 331
  single mode, 281, 316
  total internal reflection, 330–332, 335
  tunable, 281–283, 326, 331, 334, 336, 342
  whispering disk, 263

Scanning tunneling microscopy, cross-sectional, 108, 138
Scattering,
  optical phonon, 10, 221, 231, 244, 279–280, 282–284, 286–287, 290–292, 296–297, 299–300, 309, 311, 317
  nonradiative, 35, 104, 221, 264
Scattering mechanism, 283, 296
Semiconductor laser, 1–2, 19, 20, 34, 60, 64–65, 67–69, 71, 75, 80, 82, 95, 99, 115, 119, 121, 123, 145, 171, 177, 184, 207, 210, 214, 217–218, 227, 263, 278–279, 342, 344, 351
  electrically pumped, 81–82, 171–172
  optically pumped, 71, 81, 98, 115–116, 124, 131, 135
Sensing, chemical, 131, 251
Separate confinement heterostructure (SCH), 39, 64, 128, 136
Separate-confinement buried heterostructure (SCBH), 15, 176, 211
Side-mode suppression ratio (SMSR), 56, 77, 80, 152, 258, 261

Silicon
  laser, 339–340
  ohmic contacts, 322
  growth of lead salts, 197, 202–205
Single-heterostructure (SH), 15, 95, 126, 149, 185–186, 195, 213, 365, 372
Single-mode lasing, 19, 53, 78–80, 95, 118–121, 153, 168, 176–178, 184, 218, 249–263, 355–365
Slope efficiency, 34, 55–56, 80, 100, 105, 107, 137, 223, 225–227, 230, 233–235, 239, 244, 251–254, 269–270
SnSe, 146
SnTe, 146
Solder, 123, 125, 143, 147, 229, 322
Stark effect, 282, 307, 346
Strain, 3, 5, 12, 19–22, 25–34, 36–51, 57, 72–78, 80–82, 88–94, 96, 107–108, 111–112, 114, 125, 161, 174, 182, 201, 204, 214–216, 235, 239–240
  compressive, 20, 26–27, 29, 36, 43, 50, 73–78, 80, 83, 88–90, 93, 108, 125, 145, 147, 182, 239
  tensile, 26–27, 43, 47, 73–74, 89, 91, 93, 108, 182, 204, 239
  compensated, 25, 40, 44, 75, 91, 108, 239–240, 274
Streaming motion, 279–280, 296, 299, 343, 346
Submillimeter waves, see THz
Superlattice, 9, 17, 91–93, 95–97, 102–104, 115–116, 120, 218, 233–234, 271

T_0, see Characteristic temperature
Temperature tuning, 57, 162, 202, 204, 261
Terahertz laser, 287, 291, 336, 344–345
  application, 340, 342
  continuous wave, 341–342, 345
Terahertz (THz), 2, 281–282, 287, 291, 336, 341, 350–351
  scientific interest in, 351
  photomixers, 17, 352, 383–386
  sources, 1–3, 282–283, 337, 342, 351
  difference-frequency generation, 353
  molecular absorption, 353
  molecular linewidths, 354
  frequency calibration, 1
  spectroscopy, 2, 281, 336, 351, 385

Thermal conductivity, 19–20, 123–124, 157, 174, 235, 269
Thermal cycling, 148, 168
Thermal effects, 17, 283, 306, 316, 344
Thermal management, 123, 380
  continuous wave, 70
Thermal resistance, 1, 124–125, 249
Threshold, 277–279, 289, 292, 294, 296, 299–301, 318, 328
  current, 5, 20–21, 34–38, 92, 150–153, 161–162, 165–166, 170, 173–178, 183, 195, 218, 243–247, 250–253, 257, 259–267, 269, 271
  current density, 6, 10, 35–36, 49–50, 73, 75–77, 82–85, 89–90, 95–96, 102–103, 111, 224, 226–227, 231, 233, 235–240, 243–246, 253–255, 264–265, 269, 271
  energy, 295–297
  gain, 35, 103, 117, 121, 194
Transfer matrix method, 185
Transmission electron microscopy (TEM), 23, 39, 108, 138
Transition
  band-to-band, see interband
  diagonal, 221, 223, 245–246
  interband, 3–4, 8, 10, 80, 95, 98, 103–106, 113–115, 118, 162, 180, 218–220, 231
  intersubband, 3–4, 8–9, 103–104, 106, 113–114, 218–220, 222, 224, 227, 245, 271, 298
  vertical, 212, 219, 221, 223, 229, 239, 245, 255
Tuning range, 153, 184, 192, 195, 197, 213, 248, 254, 281, 340, 360, 381, 383
Tunneling
  resonant, 218, 225, 230, 273, 276
  photon-assisted, 271
Two photon excitation, 295
Type-I, 4, 12–13, 70–74, 83, 88–89, 91, 95–96, 98, 103, 105–107, 110–114, 117, 167, 171, 181, 211
  interband cascade lasers, 13, 71, 103, 105, 126, 136
  quantum well lasers, 4, 12, 14, 127, 211

Type-II, 4, 6, 60, 71–72, 74–75, 80–81, 91, 94–103, 105–108, 110–118, 122, 167
  interband cascade lasers, 6, 13, 17, 71, 103, 105, 126, 136–138
  quantum well lasers, 4, 14, 60, 80, 94, 97, 127, 129, 130, 134–135, 137–138, 140–141, 212
  single heterojunction, 95, 126
  superlattice, 17, 91, 93, 95–97, 102–103, 115–116, 212

Uniaxial stress, 282, 308, 310–312, 341, 347–348
Unipolar laser, 8–9, 218

van de Waals bonding energy, 282
Variable temperature Hall effect (VTHE), 320–321
Vertical cavity surface-emitting laser (VCSEL), 15, 80, 130–131, 175, 212, 214, 219, 274
Voigt configuration, 300–301, 304, 325, 328, 334, 341–342, 349

"W" lasers, 97–102
Waveguide, 39, 43–44, 49–50, 52, 54, 76–79, 82, 89, 102–103, 111, 114, 119, 121, 125, 175–177, 185–191, 195, 217, 224–229, 233–237, 239, 243, 248, 252, 254, 256–261, 263, 267–268, 295, 382
  loss, 49–50, 76–77, 111, 114, 119, 121, 125, 224–225, 227–229, 233–237, 243, 248, 252, 254, 256–257, 259–261
  ridge, 20, 49–50, 77–79, 89, 119, 121, 175, 189, 210, 259–260, 268, 380
Wavelength tuning, 118, 121, 345

X-ray diffraction, 23, 43, 45, 66, 108, 204
  high-resolution (HRXRD), 108, 179, 204

Yield, 72, 77, 83–84, 87, 97–98, 102, 107, 111, 115, 118, 126, 149, 161, 167, 175, 179, 218, 226, 239, 249, 252, 267, 269, 341, 350, 352